AFTER THE DARK AGES: WHEN GALAXIES WERE YOUNG (THE UNIVERSE AT $2 < z < 5$)

AFTER THE DARK AGES: WHEN GALAXIES WERE YOUNG (THE UNIVERSE AT 2 < z < 5)

Ninth Astrophysics Conference

College Park, Maryland   October 1998

EDITORS
Stephen S. Holt
Eric P. Smith
NASA/Goddard Space Flight Center
Greenbelt, Maryland

AIP CONFERENCE
PROCEEDINGS 470

American Institute of Physics   Woodbury, New York

Editors:

Stephen S. Holt
NASA/Goddard Space Flight Center
Code 600
Greenbelt, MD 20771
U.S.A.

E-mail: steve.holt@gsfc.nasa.gov

Eric P. Smith
NASA/Goddard Space Flight Center
Code 681
Greenbelt, MD 20771
U.S.A.

E-mail: Eric.P.Smith.1@gsfc.nasa.gov

L.C. Catalog Card No. 99-61604
ISBN 1-56396-855-X
ISSN 0094-243X
DOE CONF- 981092

Printed in the United States of America

CONTENTS

4. Galaxy Formation Renaissance

5. Largest Structures

6. Galaxy Formation and Mergers

7. QSOs, AGN, and the CXRB

8. Star Formation History

9. Gamma Ray Bursts

10. Next Generation Capabilities

11. Conference Summary

Appendix A: Conference Programme

Appendix B: List of Attendees

Author Index

Subject Index

Preface

This is the ninth increment in the current series of annual October Astrophysics Conferences in Maryland. These conferences are organized by astrophysicists at the Goddard Space Flight Center and the University of Maryland. The topic for each conference is selected with the help of an International Advisory Committee, the current membership of which is:

Marek Abramowicz, Göteborg	*Sir Martin Rees*, Cambridge
Roger Blandford, Pasadena	*Vera Rubin*, Washington
Claude Canizares, MIT	*Joseph Silk*, Berkeley
Arnon Dar, Haifa	*David Spergel*, Princeton
Alan Dressler, Pasadena	*Rashid Sunyaev*, Moscow
Guenther Hasinger, Potsdam	*Alex Szalay*, Budapest
Steve Holt, Greenbelt	*Yasuo Tanaka*, Tokyo
Dick McCray, Boulder	*Scott Tremaine*, Princeton
Jim Peebles, Princeton	*Simon White*, Garching

The subject chosen for this conference is *"The Universe at $2 < z < 5$: After the Dark Ages, When Galaxies Were Young"* with its program developed by the Scientific Organizing Committee:

Guenther Hasinger	*Joe Silk*	*Virginia Trimble*
Steve Holt	*Eric Smith*	*Dan Weedman*
Dick McCray	*David Spergel*	*Bob Williams*

In the spirit of this series, where we attempt to identify "hot" topics with broad appeal to both observers and theoreticians, the subject of early galaxy formation clearly qualifies as an appropriate conference theme. A few measurements barely scraping $z = 5$ had been announced at the time that the conference was first being planned, but by the time that the conference took place the $z = 5$ threshold was thoroughly shattered.

The conference began with a welcome from *Dan Mote*, the new president of the Univerity of Maryland at College Park. We followed with introductory invited reviews by *Virginia Trimble* on the history of the search for the earliest galaxies, and by *Martin Rees* on our current understanding of the emergence from "the Dark Ages." The conference then proceeded through the next two days with a series of non-paralleled sessions, each devoted to a specific topic with two or three invited talks and an extensive discussion period. In earlier conferences we have allowed the session chairs to "promote" one or two short oral contributions from the poster papers. Last year, instead, we instituted a procedure to formalize the opportunity for each conference attendee to advertise his/her poster paper with one minute to present one viewgraph.

The banquet at the conclusion of the second day of these conferences generally features a particularly distinguished speaker. This year our speaker was *Vera Rubin*, whose subject was *"Astrophysics from Antarctica: a Visit to the South Pole"*. Vera mentioned to her dinner companions that she had previously awarded $100 for the first verified measurement of a galactic rotation curve in excess of 600 km/s, and the subject of this conference cried for a similar prize for the discovery of a galaxy with $z > 7$. The hat was passed for the endowment of such a prize ($125), which will be held by *Marvin Leventhal* until such time as the prize committee (*Vera Rubin, Bruce Partridge,* and *Carlos Frenk*) decides that a worthy recipient has been found. Without fear of copyright violation, Bruce suggested that the award be dubbed *"The Seven-Up Prize"*.

As usual, *John Trasco* and *Susan Lehr* made sure that all the logistics were handled flawlessly. Thanks also to *Eric Smith*, the co-editor of these Proceedings, for making sure that all the details that I usually watch out for myself at these conferences were not ignored. The fact is that I was not in attendance for the whole conference. I left for Boston before it was not quite half over in order to be present at the birth of my first grandchild, *Blake Ross Krantz*. The rest of the conference was wonderful, as evidenced by these Proceedings, but the part of that week that I'll remember best took place in Boston.

Steve Holt
January 1999

1. Introductory Overviews

Beyond the Bright Searchlight of Science: The Quest for the Edge of the World

Virginia Trimble

*Astronomy Department, University of Maryland, College Park, MD 20742
and Physics Department, University of California, Irvine, CA 92697*

Abstract. Human efforts to probe the extent of the cosmos clearly date back to pre-literate times, with significant progress occurring in early literate societies, the Renaissance to 18th century, the 19th and 20th centuries, but with some issues remaining for the millennium. The important issues are the size of the earth, the distance to the moon, the earth-sun distance, the distances to the stars, the size of the Milky Way, and the size of the universe as a whole, plus its age. It is perhaps significant that, with one exception, each question was definitively, answered in a later period than the one that first asked it.

INTRODUCTION AND AGES

It has been many generations since those raised in Eurocentric countries have risked anything more than our reputations by asking fundamental questions about the universe. The downside is that we can no longer expect to find definitive answers in the form of seeing the wheels that make the whole cosmic clock work. While estimates of the size of the universe have increased more or less monotonically with time, those of the age have cycled between wide limits. Many preliterate cultures imply in their mythologies a time scale that is 10's to 100's of human generations (and educated individuals can sometimes recite the names of their ancestors back to the supposed beginning).

Of the early literate communities, the writers of Genesis incorporated a total time scale somewhat less than 10,000 years. Byzantium (where this would be the year 7506) thought similarly (with assorted Mesopotamians probably the origin of both chronologies). In contrast, thinkers of the Indian subcontinent arrived at very long times scales, up to 10^{12} years, or perhaps infinite, and this wide range persisted through to the rise of modern science. Applications of basic physics, like conservation of energy and Newtonian gravitation, to astrophysical contexts led 19th century scholars to derive ages ranging from 10^7 yr (the Kelvin-Helmholtz time scale for the sun) up to 10^{12} years (the time needed to produce the observed

CP470, After the Dark Ages: When Galaxies were Young (the Universe at 2 < z < 5), edited by Stephen S. Holt and Eric P. Smith

distribution of binary star orbits and of clusters of stars and galaxies, starting with initial conditions that seemed relevant to James Jeans and others). As we approach the end of the 20th century (after yet a different zero point), very few astronomers would argue for an age much outside the range 10-20 Gyr, and this probably counts as progress.

THE SIZE OF THE EARTH

Many mythologies imply an earth that can be circumscribed by humans in years to generations and by gods in a day (consider Ra changing at dawn and dusk from his night boat to his day boat and back again, before sailing either above or below the earth, and Apollo and his chariot). Early maps invariably indicate that the center of the (flat) earth was somewhere quite close to where the map-maker lived.

In contrast, Eratosthenes of Cyrene (-276-195), in the one bit of Greek astronomy that most of us remember, set out to determine the size of a spherical world from the observation that, on 21 June, the sun cast no shadows at noon at Syene (Aswan, roughly), but 7° shadows at Alexandria. Clearly he assumed a distant sun, and we must assume a long-range collaboration, or measurements made over at least two years. With the distance between the two cities given as 5000 stadia, we arrive at a circumference of the earth of about 250,000 stadia. This is generally supposed to be quite accurate, based on the length of some particular representative stadium.

Early Chinese tales tell of a certain Pan-ku who shoved the heavens away from the earth by standing between them and growing 18,000 feet a day for 10 years. The product is about 12,400 miles (not necessarily the same as American statute miles!) while the earth's diameter is about 12,800 km. This is close enough to suggest that the writer had access to the results of some observation analogous to that of Eratosthenes.

FROM EARTH TO MOON

The distance to the moon is the only one of these scales for which the first people reported as having tried to make a measurement came close to the modern answer. The ratio of the size of the earth to the size of the moon's orbit is just the angular diameter of the earth as seen from the moon. Luckily you don't have to go there to look. The angular diameter of the earth's shadow as it crosses the moon during eclipse works just as well, leading to a ratio of about 1/60, which, again allowing for some uncertainty in the length of a stadium is probably pretty close. The idea is credited to Aristarchus of Samos (-310 to -230, since people lived backwards in those days). This is the one case I found where the people who first asked the question got a good answer.

ON TO THE SUN

Aristarchus and his contemporaries reached their level of incompetence in trying to get the solar distance from similar geometric methods. The idea is clever. If the sun is further away than the moon by a factor of 10 or 20 or 30, you will see differences in the lengths of times between the major moon phases (e.g. new to first quarter shorter than first quarter to full). Because the actual ratio is more like 400, the time difference would have been too small for the Greeks to record. But the moon's orbit is actually elliptical, so there is such a phase inequality of a day or so, leading the Greeks to put the sun at about 20 times the distance to the moon. Later, pre-telescopic geometrical attempts did not improve the situation. Thus Tycho and Kepler built their models of the solar system with quite good relative distance but absolute ones much too small.

An opportunity to pin down the absolute scale came in 1672, when Mars made a particularly close approach to the earth, permitting a determination of what is called geocentric parallax. One can do this two ways: with simultaneous measurements of the position of Mars in the sky made from distant points on earth, or from a single point on earth, but waiting for diurnal rotation to carry you over a significant fraction of the diameter. Cassini tried the simultaneous method from Paris with a collaborator in Cayenne, and, later-in the year, the earth rotation method with the assistance of Roemer (who later measured the speed of light from the timing of the eclipses of the moons of Jupiter).

The measurement is a very difficult one, and the close agreement of the three determinations on a parallax for Mars of about 20″ (solar parallax of 10-12″) must have had a contribution from chance and/or bandwagonism. In the first edition of his Principia, Newton actually used an even more erroneous value of solar parallax, 20″ (corresponding to 1 AU = 20,000 earth radii) and so obtained a mass for the sun that was too small by a factor about three.

An accurate value of the geocentric parallax of the sun of 8.4-8.8″ finally came from timing of the transits of Venus in 1769. The modern value is 8.794″, plus several more significant figures, and the main uncertainty in the mass of the sun comes from G, not the distance scale.

NEXT THE STARS

Implicit in many myths and explicit in the writings of Anaximander is a distance to the heavens comparable with the size of the earth. Other Greeks like Aristotle put the stars well outside the realm of the planets. Kepler arrived at a specific number by nesting the known planets in concentric Platonic solids (the ones whose faces are regular polygons) and putting the stars just outside. This leads to a sphere of the stars at about 20 Astronomical Units.

A sun-centered solar system firmly predicts stellar parallax (of the stars relative to your coordinate system, even if they are all at the same distance; otherwise of the stars relative to each other, which is easier to see). Not seeing parallax was one of the very strong arguments in favor of a Ptolemaic, earth-centered model, even with naked eye limits to parallax of a few degrees. Most Greek philosophers balked at the requirement that the stars be more than 100 times the distance to the sun.

By the time of Tycho, carefully-constructed instruments had reduced the upper limit on heliocentric parallax to about 25″, putting the stars at more than 3700 AU. Tycho himself constructed a curious cosmology in which most things orbited the sun, but the sun (and moon) in turn orbited the earth, leading to a prediction of no parallax.

A separate line of thought that implies very large stellar distances is the concept that "stars are suns." Since it is dark at night, despite the presence of thousands of stars in the sky, they must be very much more distant than the sun. Nicolas of Cusa (1401-1464, who became a cardinal and died of natural causes) is generally credited with the first comprehensible enunciation of this idea. He taught in addition that the sun, earth, and stars were all made of the same stuff (in contrast to the Aristotelian quintessence) and pointed out that, in an infinite universe, the center is everywhere and the circumference is nowhere.

The Englishman, Thomas Digges, published similar ideas in his <u>A Perfit Description of the Caelestiall Orbes</u> in 1576, but his drawing shows stars crowded up close together outside the orbit of Saturn. The caption says that (in modern spelling) "this orbe of stars fixed infinitely up extendeth itself in altitude spherically...", and perhaps he would have labeled the drawing "not to scale" if the concept had existed then. The best know propounder of "stars are suns" was, of course, Giordano Bruno, who was unlucky in his choice of time and place in which to propound, but who also seems to have been a rather poor politician and to have been condemned for his attitude toward his fellow churchmen as much as for his attitude toward the universe.

Invention of the telescope led to a new round of attempts to measure parallax. A couple of the unsuccessful ones were nevertheless astronomically useful. William Herschel, believing that all stars had the same intrinsic brightness, focused on close pairs of different apparent brightness, expecting them to be at different distances and so to reveal parallactic motion. What he actually discovered was orbital motion, thereby confirming the earlier claim by John Michell that the large number of close pairs could not be a chance occurrence. He announced the result in 1820. Bradley, in 1829, looked for parallax relative to an earth-centered coordinate system and so discovered the quaintly-named aberration of starlight (the tilt in arrival direction of starlight caused by the earth's orbital motion).

The winners, in 1836, were F. Bessel, Th. Henderson, and F. Struve, looking respectively at 61 Cygni (with a proper motion of 5.2″/yr), Alpha Centauri, and Vega (the latter chosen for brightness, 61 Cyg primarily for its large proper motion, implying a relatively small distance).

Parallax measurements and recognizable binary pairs both demonstrated that real stellar brightnesses cover a wide range, but some feeling that all are about like the sun appears to have carried over into Herschel's star gauging and to have contributed to his small scale for the size of the galaxy. Naked eye stars, or any magnitude limited sample, will be dominated by intrinsically bright ones, and the errors in distance made by assuming that a star is like the sun amount (roughly) to a factor of three for Procyon, 5 for Sirius, 10 for Arcturus, 30 for Canopus, and 100 for Rigel.

THE SIZE AND NATURE OF THE MILKY WAY

William Herschel's star gauging resulted in a somewhat flattened distribution of stars, with total diameter about 6 kpc, the sun very near the center, and one fairly deep gash along the central plane corresponding to the Cygnus rift. The thickness was about one-third the diameter. What all this means is that he was seeing to the edge of the galactic disk on the north-south directions, but not in the plane. And, as a result of neglecting interstellar absorption, he inevitably perceived the edge of the distribution to be at about the same distance in all directions, except toward the galactic center, where obscuration is the largest.

Many other astronomers arrived at rather similar conclusions for the size, shape, and sun-centeredness of the galaxy. Simon Newcomb, in an 1882 book, put regions of nebulae on either side of the Herschel "region of stars or galaxy." Eddington in 1912 put the sun slightly above the galactic plane (it still is), and, most remarkably, Cornelius Easton in 1900 sketched a galactic disk with spiral arms like those of the nebulae, with the center of the spiral galaxy correctly off toward Cygnus, but the sun still solemnly at the center of the coordinate system. Over the same century there was a low-key debate about whether the spiral nebulae might be other galaxies or "island universes." Majority opinion, including that of Lowell and others we still remember, said no; they might, rather, be new solar systems in formation along the lines indicated by the Kant-Laplace hypothesis. The main argument for the spirals being somehow an integral part of the Milky Way was that their positions in the sky avoided the galactic equator.

Harlow Shapley came on the scene near the end of World War I, using RR Lyrae stars in the globular clusters to determine their distances, and tying the scale to galactic plane Cepheids with distance measured from statistical parallax. To a certain extent, his neglect of interstellar absorption and his merging of variable stars of Population I and II into a single set compensated each other, though his "best bet" distance scale within the Milky Way was larger than the current one, putting the sun about 20 kpc (range 13 to 25) from the center of a spherical system with total radius 60 kpc or more. Thus, when Shapley and Heber Curtis faced off in the now well-known debate of 1920, Shapley was advocating a large galaxy, comprising in effect the entire stellar universe, with the sun far from its center. Curtis, in contrast, favored a much smaller galaxy, with the sun near the center, and many other similar stellar systems to be found outside. He noted that many nebulae seemed to have dust lanes in their central planes and suggested that a similar lane in the Milky Way could be the cause of our not seeing spirals in its plane. Jacobus Kapteyn, even as the debate was underway, was carrying out yet another very detailed count of stars as a function of magnitude, color, and position on the sky that would lead once again to a small, sun-centered system, the "Kapteyn universe." Its radius, down to 10% of the central stellar density was about 6 kpc and the half thickness about 2 kpc.

Since 1920, the distance of the sun from the center of the galaxy has undergone several oscillations of relatively small amplitude. Shapley said 20 kpc, Oort 6 and then 10 kpc in about 1926 and 1930. Baade moved us in to about 8.2 kpc in the early 1950s, and the IAU in a resolution (prompted by Oort) back out to 10 kpc in 1965. A later IAU resolution dropped this to 8.5 kpc, and many recent determinations, based on a wide range of stellar types and even the apparent proper motion of Sgr A* as seen from earth have found values of 7-8 kpc. Meanwhile, estimates of the total effective diameter have tended to reach back out to Shapley's 100 kpc and even beyond, with the recognition of outlying globular clusters, the dark halo, and so forth, in our own and other spiral galaxies.

Robert Trümpler is best known for his 1930 recognition of general interstellar absorption (arrived at by comparing distances to open clusters determined from their angular diameters with those determined from star brightnesses). The implications for the Milky Way took a little while to absorb, and, in the same year, he sketched a Milky Way with a spheroid 80 kpc across and centered 18 kpc from us (following Shapley), but, simultaneously, a small, thin Kapteyn universe centered near the sun and with its plane tilted slightly to the plane of the larger system, in the direction we associate with Gould's belt. By 1939, Plaskett could sketch an essentially modern galaxy, with thin plane of stars, still thinner plane of dust, a central bulge or spheroid and globular clusters concentrated toward the center but extending to large radii. He put the sun at Oort's 10 kpc.

The issue of the existence of external galaxies was resolved by Edwin Hubble's 1924-25 recognition of Cepheids in NGC 6822 and M31 soon after. Allan Sandage, however, remembers Milton Humason, toward the end of his life, saying that he, Humason, had marked some tentative Cepheids on a plate of M31 back in about 1921. Humason supposedly showed the plate to Shapley, who moistened a handkerchief and removed the identification markings from the back of the glass. Shapley could have saved himself a good deal of grief a few years later by having forgotten his handkerchief that morning, because, according to Cecilia Payne Gaposchkin, he regarded the Cepheids as "having destroyed his universe."

THE EXTRAGALACTIC DISTANCE SCALE

A number of authors had estimated distances to M31 (the Andromeda Galaxy) before the entire community agreed that this was a meaningful thing to do. For instances, Curtis, Lundmark, and Shapley (who thought the result was ridiculous) all placed it at 200-250 kpc on the basis of bright stars and novae, tied to distances within the Milky Way as then understood. The first attempt to double the distance to M31 came from Opik in 1922 (one of the very few of his important papers to appear in the Astrophysical Journal). He made the assumption that the mass to light ratio of the whole galaxy ought to be the same as that of the solar neighborhood (which had been estimated by Kapteyn and Jeans the same year), and using a published spectrogram of the Nebula taken by Pease on the Mt. Wilson 60″ telescope, came up with a distance of 450 kpc.

The first dozen velocities for "spiral nebulae" came from Vesto Melvin Slipher, working at Lowell Observatory. At least a dozen astronomers before Hubble had looked at them and attempted to derive a "solar motion" or a "solar motion with a K-term" (meaning one that showed constant expansion), or even a "solar motion with a distance-dependent K term", which is what Hubble reported in 1929. Lundmark even allowed for a term in the square of the distance (and, naturally found a better fit, as you nearly always do when adding a parameter). Hubble's work was the first to be taken seriously, in part perhaps because his manner was very convincing, but mostly because he seemed to have much more reliable distances (making use of his discovery of extra-galactic Cepheids) than the other analyzers.

Hubble's distances were consistent, but they were badly wrong. Thus his first paper and subsequent ones through 1936 reported values of what we now call H within 10% of 500 km/sec/Mpc. The period from 1929 to about 1965 was punctuated by large, downward steps in the generally-adopted value of H, each a result of re-interpretations of some critical observation or assumption. As early as 1931, Oort suggested 432 or 260 km/sec/Mpc. A careful recalibration of the Cepheid period-luminosity relation in 1941, by Mineur (who allowed for the interstellar absorption of about 1^m per kpc in V found by Trümpler in his study of star clusters) led to about 320. The German astronomer Behr, in 1951, "pre-discovered" what we now call the Scott effect. He pointed out that assuming a constant luminosity for the brightest galaxy in a cluster would inevitably lead to apparent distances smaller than the real ones by a factor that increases with distance. The point is that, far away, you will recognize only very rich clusters, which contain superluminous galaxies not in your local sample. His work implied H near 240 km/sec/Mpc. Since the effect Behr considered is completely separate from the Cepheid recalibration, a concordance of his work and Mineur's would have taken the Hubble parameter down to about 150 km/sec/Mpc. This did not happen, apparently because nobody brought the two ideas together, and very few people encountered even one of them (Mineur published in French in Annales d'Astrophysique, Behr in German in Astronomische Nachrichten.)

Thus at the Rome General Assembly of the International Astronomical Union, participants were duly astounded by the implications of Walter Baade's not seeing RR Lyrae stars in M31 and by A. David Thackeray's having seen them in the Magellanic Clouds. The distance scale doubled, H decreased, and the universe was suddenly twice its previous age. Humason, Mayall, and Sandage took us down to H = 180 in 1956, and a number of astronomers staked out numbers near 100 between 1958 and 1965. In the next decade, battle lines hardened between a "short" distance scale (H = 100 km/sec/Mpc, G. de Vaucouleurs, S. van den Bergh) and a "long" distance scale (H = 50, A. Sandage, G.A. Tammann). The number of astronomers estimating the Hubble constant has grown even faster than the size of the community as a whole over the past two decades, with 20 or more values published per year. The territory between 50 and about 90 filled in quickly. In recent years, the upper envelope has been moving downward and the lower one holding roughly steady. Thus the median H was 75 in 1992-93 and 60 for 23 papers published in 1998.

THE QUEST FOR LARGER REDSHIFTS

Slipher's first extragalactic spectrogram, in 1912, was that of M31. It revealed a heliocentric velocity of about -300 km/sec. This is made up partly of galactic rotation and partly of the Milky Way - M 31 orbit in the Local Group and is an authentic Doppler shift, not a cosmological redshift. Slipher achieved a "personal best" of +1800 km/sec for NGC 584, but found more distant galaxies beyond the reach of his 24″ telescope (he could, of course, have measured the redshift of 3C273!).

Milton Lassell Humason was the next major collector of redshifts. He quickly doubled Slipher's record, with +3779 km/sec for NGC 7619 and doubled it again to +7800 for NGC 7619 in the Coma cluster in 1929. An improved spectrograph on the 100″ telescope passed the 15,000 km/sec mark by 1934. Humason (unlike Hubble) lived to use the 200″ Hale telescope with some regularity. He retired in 1957, having reached 60,000 km/sec for galaxies in the Hydra cluster several years before. Some of the larger redshifts had been achieved using multi-night exposures, and Humason believed that one could go no further, even with the 200″, given the detectors of the time (photographic plates).

The discovery that strong radio sources were often associated with strong emission lines made a significant difference. Cygnus A, the first optical identification, had a redshift of only 0.06 (measured in 1954 by Baade and Minkowski), but 3C 295 took Minkowski to z = 0.46 shortly before he retired in 1960 (on his very last 200″ observing run, it is usually said).

The first quasar, 3C 273, also came out of the program to identify radio sources, as continued by Maarten Schmidt. The year 1963 saw z = 0.158 for it and z = 0.367 for 3C 48. The record for the largest redshift continued to be held by radio-loud quasars until the mid 1980s. Some steps were steep (to z = 2.01 in one fell swoop to 3C 9 in 1965 – the object that brought Lyman alpha into the visible part of the spectrum and led to tight limits on the amount of neutral hydrogen in intergalactic space). Then came some baby steps to 2.23 in 1967, and 2.36 in 1968 (Schmidt's last record). 1973 took us to 3.40 and 3.53, still for radio selected quasars, almost at once, then to 3.78 in 1982. The last time the most distant object was radio-loud was 1986, with a source at z = 3.80.

Radio-quiet quasi-stellars were bound to win out over radio loud ones, if only because there are so many more of them. The trick was to know where to look. And the answer came with objective prism or grism surveys. Meanwhile various other photoelectric detectors, culminating in CCDs, had begun to replace photographic emulsions as the spectroscopic materials of choice. Thus the lead in z-space was taken over by an optically-selected QSO at z = 4.014 in 1987, rising later the same year to 4.402 and to 4.733 in 1989. The current QSO record is z = 4.897, for PC 1247+3406, found in 1991, and the most distant radio-loud source is at z = 4.46.

Efforts to detect and recognize distant galaxies had, of course, gone on. Identification of 3C and other sources continued, largely under the leadership of Spinrad, and they reached z = 1.27 in 1982 and z = 1.82 for 3C 256 in 1985. Other groups, often using the Anglo-Australian Telescope, then got into the act, increasing the radio galaxy record to 2.3 in 1988 (actually a KPNO 4-m result) and z = 3.80 in 1990. This was followed by two medium-sized steps to 4.25 in 1994 and to 4.41 in 1997.

Ordinary galaxies are, of course, still commoner than either radio-emitting ones or active ones, and we expect this also to be the case at large redshift, though the galaxies or their precursors may not look like anything Hubble would have wanted to classify. The sky is, however, heavily smudged with faint, fuzzy things, and there will be a great many intrinsically faint, nearby galaxies for every bright, distant one. A few galaxies at redshifts near one had turned up in large surveys, and one quite recently beyond two. But, as usual, it was new ways of looking at things that made the difference. The first winner was to look for galaxies with strong Lyman alpha emission at the same places in the sky and with the same redshifts as Lyman-alpha absorbing clouds seen in the spectra of more distant QSOs. This was so obviously the right thing to do that the relative paucity of results led people to wonder whether an epoch of galaxy formation (when many spheroids made many of their stars in a short time) had ever actually happened (and the answer remains, "not entirely"). Searches for Lyman alpha emission through filters chosen to pick out a particular range of redshifts were similarly slow to yield results. But a handful of galaxies with strong Ly-alpha emission lines eventually turned up, with a current maximum redshift of 4.55 (for one not near a quasar).

Most recent and spectacularly successful is the technique called Lyman drop-out. The idea is that as 1216 A moves into a particular broad-band filter pass (U, B, V, etc.) the enormous forest of Lyman-alpha absorption lines eats most of the photons, and a galaxy becomes very faint in that band. Charles Steidel and his colleagues pioneered the method, looking first near distant QSOs, where they knew they would find something, then eventually at blank fields, including the Hubble Deep Field. Candidate galaxies must, of course, have their redshifts confirmed with spectra (required to show two lines or a line and a continuum break at a wavelength ratio recognizable as something you expect). But at the time the conference was being organized, the largest known z (apart from the cosmic background radiation!) belonged to a z = 5.34 galaxy found this way. It was 5.67 in October, with larger values definitely expected any day (at least in the windows where the lines can peek through the earth's OH and other atmospheric features - the next is at 6.7), and with several strong candidates for z = 6-7 in the HDF. Some are so faint that even

11

Keck and other 8-10m ground-based telescopes may not suffice to get redshifts, leaving us waiting for the NGST.

LOOKING AHEAD

In light of the rapid growth of maximum z and some theorists' predictions that we have already truly reached the era of first galaxy formation, the conference dinner speaker, Vera Rubin, suggested a modest prize, to be given to the first person to find an honest, spectroscopic redshift of 7.0 or larger, the prize to include an antarctic rock.

Of course, finding distant galaxies (etc.) is just the first step. Understanding is still more important, and we come back more or less to the conditions described by W.C.G. Whether in about 1920, from whose poem the title of this talk was taken:

Beyond the bright searchlights of science
Out of sight of the window of sense,
Old riddles still bid us defiance,
Old questions of Why and of Whence.

And there can be no better person to introduce you to those questions than the next speaker on the program, Sir Martin Rees.

Further details of the discovery of the cosmic distance scale and its changes can be found in PASP 108, 1073; in Space Science Reviews 79, 793, and in the 1997 proceedings of the STScI symposium on The Cosmic Distance Scale. It is reasonable to expect that fairly firm values of the cosmic distance scale and of the epoch at which the first galaxies, QSOs etc. formed will be established not too far into the 21st century.

The End of the 'Dark Age'

Martin J. Rees

Institute of Astronomy, Madingley Road, Cambridge CB3 0HA, UK

Abstract. At redshifts beyond 5 – perhaps even beyond 20 – stars formed within 'sub-galaxies' and created the first heavy elements; these same systems (together perhaps with 'miniquasars') generated the UV radiation that ionized the IGM, and maybe also the first significant magnetic fields. These uncertain processes set the backdrop for the phenomena at $z < 5$ that are the theme of the present conference.

INTRODUCTION

The Universe literally entered a dark age about half a million years after the big bang, when the primordial radiation cooled below 3000K and shifted into the infrared. Darkness persisted until the first non-linearities developed, and evolved into stars, galaxies or black holes that lit the Universe up again.

The combination of HST and Keck Telescope data is now elucidating the history of star formation, galaxies and clustering back to $z = 5$ – our knowledge of these eras is no longer restricted to 'pathological' objects such as extreme AGNs. The emergence and evolution of galaxies of all morphological types will be clarified by further Hubble Deep Fields, together with follow-up spectroscopy from the new generation of 10-metre telescopes. In addition, high-dispersion studies of absorption features in quasar spectra (the Lyman forest, etc), in conjunction with simulations, allow us to probe the clumping, temperature, and composition of diffuse gas on galactic (and smaller) scales, at least back to $z = 5$, rather as geophysicists can use ice cores to probe climatic history.

Within a few years, detailed sky maps of the microwave background (CMB) temperature (and perhaps its polarization as well) will offer direct diagnostics of the initial fluctuations from which the present-day large-scale structure developed. Most of the photons in this background have travelled uninterruptedly since the recombination epoch at $z = 1000$, when the fluctuations were still in the linear regime. We may also, in the next few years, discover the nature of the dark matter, and how it clusters gravitationally; computer simulations will incorporate gas dynamics and radiation in a sophisticated way.

CP470, After the Dark Ages: When Galaxies were Young (the Universe at 2 < z < 5),
edited by Stephen S. Holt and Eric P. Smith

But these advances may still leave us, several years from now, uncertain about the whole era from 10^6 to 10^9 years – the formation of the first stars, the first supernovae, the first heavy elements; and how and when the intergalactic medium was reionized. Even by the time Planck/Surveyor and NGST have been launched, we will probably still be unable to compute crucial things like the star formation efficiency, feedback from supernovae. etc – processes that current models for galactic evolution are forced to parametrise in a rather ad hoc way. And CMB fluctuations will still be undiscernible on the very small angular scales that correspond to sub-galactic structures, which, in any hierarchical ('bottom up') scenario would be the first non-linearities to develop.

COSMOGONIC PRELIMINARIES: MOLECULAR HYDROGEN AND UV FEEDBACK

Detailed studies of structure formation generally focus on some variant of the cold dark matter (CDM) cosmogony – with a specific choice for Ω, Ω_b and Λ. Even if its details prove incorrect, such a model offers a useful 'template' whose main features apply generically to any 'bottom up' model for structure formation. There is no minimum scale for gravitational aggregation of the CDM. However, the baryonic gas does not 'feel' the very smallest clumps, which have very small binding energies: pressure opposes condensation on scales below the baryonic Jeans mass.

The role of molecular cooling at early cosmic epochs has been considered by many authors, dating back to the 1960s; recent discussions are given in refs (1) and (2). This process allows clouds to contract if their temperature exceeds 500 K. The exact efficiency depends on the density, and therefore on the redshift when the first collapse occurs.

But even at high redshifts, H_2 cooling would be quenched if there were a UV background able to dissociate the molecules as fast as they form. Photons of $h\nu > 11.18$ eV can photodissociate H_2, as first calculated by Stecher and Williams (3). These photons can penetrate a high column density of HI and destroy molecules in virialised and collapsing clouds. (If the incident spectrum has a non-thermal component extending up to keV energies, as it might if there is a contribution from accreting compact objects, then there is a counterbalancing positive feedback because the hard photons penetrate HI, increasing the number of photoelectrons, and thereby enhancing molecule formation(4).)

Three 'cooling regimes' are relevant during successive phases of the cosmogonic process, each being associated with a characteristic temperature.

1. For a H-He plasma the only effective cooling at low temperatures ($< 10^3$ K) comes from molecular hydrogen. Even this process cuts off below a few hundred degrees; but above that temperature it allows contraction within the cosmic expansion timescale. The H_2 fraction is never high, and it is in any case not a very efficient coolant – indeed systems that collapse at $z < 10$ fail to form enough

molecules for effective cooling (eg Fig 1 of Ref 1) but molecular cooling almost certainly played a role in forming the very first objects that lit up the universe

2. When a UV background has built up that is strong enough, H_2 is prevented from forming, molecular cooling is suppressed,, and a H-He mixture behaves adiabatically unless T is as high as 8-10 thousand degrees, when excitation of Lyman alpha by the Maxwellian tail of the electrons provides efficient cooling whose rate rises steeply with temperature. Gas in this regime contracts almost isothermally, so that its Jeans mass decreases (allowing fragmentation) as the density rises.

3. The UV from early stars, and perhaps also from accretion onto compact objects, eventually becomes strong enough to ionize most of the diffuse gas. (This requires a much higher intensity than photodissociation of molecules). The HI fraction is then suppressed to a very low level, so there is no cooling by collisional excitation of Lyman lines; moreover the energy radiated whenever a recombination occurs is quickly cancelled by the energy input from a photoionization, so the only net cooling is via bremsstrahlung. The cooling is, in effect, then reduced by a factor of 100 (see, for instance, ref (5)). The minimum temperature (below which there is a net heating from the UV) depends on the UV spectrum, and on whether the He is doubly ionized: it is in the range 20-40 thousand degrees. When this third phase is reached, the thermal properties of the uncollapsed gas will resemble those of the structures responsible for the observed Lyman-forest lines in high-z quasars spectra – these are mainly filaments, draining into virialised systems. Such systems have velocity dispersions of 50 km/sec, and are destined to turn into galaxies of the kind whose descendents are still recognisable.

(These three regimes refer to a H He plasma. When heavy elements are present they can dominate the low-T cooling; ionization is still important in suppressing the most efficient channels for cooling.)

Only a small fraction of the UV that ionized the IGM can have been produced in systems where star formation was triggered by molecular cooling. Most must have formed in systems large enough to have been able to cool by atomic line effects.

The effect of photoionization is significant even for systems with virial velocities V up to 200 km sec (i.e. where T_v exceeds 10^6 K). Navarro and Steinmetz (6) show that the angular momentum ending up in a disc differs by 30 percent, depending on whether the gas is or is not photoionized. This difference comes about because discs are assembled from the inside out, the outer parts forming from cool-phase (but relatively diffuse) gas that falls in late.

THE EPOCH OF IONIZATION BREAKTHROUGH

The IGM remained predominantly neutral until a sufficient number of 'subgalaxies' had gone non-linear to provide the requisite O-B stars (or accreting black holes) that photoionized the IGM.

How many of these 'subgalaxies' formed, and how bright each one would be, depends on another big uncertainty: the IMF and formation efficiency for these

Population III stars. They form in an unmagnetised medium of pure H and He, bathed in background radiation that may be hotter than 50 K when the action starts (at redshift z the ambient temperature is of course $\sim 2.7(1 + z)$ K). Would these conditions favour a flatter or a steeper IMF than we observed today? This is completely unclear: the density may become so high that fragmentation proceeds to very low masses (despite the higher temperature and absence of coolants other than molecular hydrogen); on the other hand, massive stars may be more favoured than at the present epoch. Indeed, fragmentation could even be so completely inhibited that the first things to form are supermassive holes.

The gravitational aspects of clustering can be computed with convincingly high resolution. So also, now, can the dynamics of the baryonic (gaseous) component, including shocks and radiative cooling – indeed, impressive attempts are now being made to follow molecular cooling and radiative transfer in collapsing clouds. But the huge dynamic range of the star-formation process cannot be tracked computationally up to the densities at which individual stars condense out. Moreover, the first stars (or other compact objects) exert crucial feedback: the remaining gas is heated by ionizing radiation, and perhaps also by an injection of kinetic energy via winds and even supernova explosions – which is even harder to model, being sensitive to the IMF, and to further uncertain physics.

Three major uncertainties are:

(i) What is the IMF of the first stellar population? The high-mass stars are the ones that provide efficient (and relatively prompt) feedback. It plainly makes a big difference whether these are the dominant type of stars, or whether the initial IMF rises steeply towards low masses (or is bimodal), so that very many faint stars form before there is a significant feedback.

(ii) Quite apart from the uncertainty in the IMF, it is also unclear what fraction of the baryons that fall into a clump would actually be incorporated into stars before being re-ejected. The retained fraction would almost certainly depend on the virial velocity: gas more readily escapes from shallow potential wells.

(iii) The influence of the early stars depends on how much of their radiation escapes into the IGM. Much of the UV radiation could, for instance, be absorbed in the gas immediately surrounding the first stars, so that it exerts no feedback on the condensation of further clumps – the total number of massive stars or accreting holes needed to build up the UV background, and the concomitant contamination by heavy elements, would then be greater.

All these three uncertainties would, for a given fluctuation spectrum, affect the redshift at which molecules were destroyed, and at which full ionization occurred. Perhaps I'm being pessimistic, but I doubt that either observations or theoretical progress will have eliminated these uncertainties about the 'dark age' even by the time NGST flies.

How uncertain is the ionization epoch?

Even if we knew exactly what the initial fluctuations were, and when the first bound systems on each scale formed, the breakthrough redshift z_i is uncertain by a factor of 2, even if we postulate a 'standard' IMF. This can be easily seen as follows:

Ionization breakthrough requires at least 1 photon for each ionized baryon in the IGM (the extra photons are needed to balance recombinations, which are less important in underdense regions than in clumps and filaments). An OB star produces 10^4 10^5 ionizing photons for each constituent baryon, so the requisite UV could be supplied by 10^{-3} of the baryon turning into stars with a standard IMF. We can then contrast two cases:

(α) If the star formation were efficient, in the sense that all the baryons that 'went non-linear', and fell into a CDM clump larger than the Jeans mass, turned into stars, then only the rare 3-σ peaks in the initial fluctuation spectrum would contribute enough stars.

On the other hand:

(β) Star formation could plausibly be so inefficient that less than 1 percent of the baryons in these pregalactic systems condense into stars, the others being expelled by stellar winds, supernovae, etc., In this case, production of the necessary UV would have to await the collapse of more typical peaks (1.5-σ, for instance).

A 1.5-σ peak has an initial amplitude only half that of a 3-σ peak, and would therefore collapse at a value of $(1+z)$ that was lower by a factor of 2. For plausible values of the fluctuation amplitude this would change z_i from 15 (scenario α) to 7 (scenario β). There are of course other complications, stemming from the possibility that most UV photons may be reabsorbed locally; moreover in Scenario β the formation of sufficient OB stars might have to await the build-up of larger systems, with deeper potential wells, in which stars could form more efficiently.

If the IMF were biased towards low-mass stars, the situation resembles inefficient star formation in that a large fraction of the baryons (not just the rare 3-σ peaks) would have to collapse non-linearly before enough UV had been generated to ionize the IGM. By the time this happened, a substantial fraction of the baryons could have condensed into low mass stars. This population could be sufficient to contribute to the MACHO lensing events (see section VI)

What is the chance of detecting the ancient stellar systems that ionized the IGM at some redshift $z_i > 5$? The detectability of these early-forming systems, of subgalactic mass, depends which of the above extreme scenarios is nearer the truth. If β were correct, the individual high-z sources would have magnitudes of 31, and would be so common that there would be about one per square arc second all over the sky; on the other hand, option α would imply a lower surface density of brighter (and more readily detectable) sources for the first UV (7). There are already some constraints from the Hubble Deep Field, particularly on the number of 'miniquasars' (8). Objects down to 31st magnitude could be detected from the ground by looking at a field behind a cluster where there might be gravitational-lens magnification, but firm evidence is likely to await NGST.

Note that scenarios α and β would have interestingly different implications for the formation and dispersal of the first heavy elements. If β were correct, there would be a large number of locations whence heavy elements could spread into the surrounding medium; on the other hand, scenario α would lead to a smaller number of brighter and more widely-spaced sources.

The 'breakthrough' epoch

Quasar spectra tell us that the IGM is almost fully ionized back to $z = 5$, but we do not know when the diffuse IGM in effect became an HII region. The IGM would already be inhomogeneous at the time when the ionization occurred: most of the material would be concentrated in clumps and filaments filling a small fraction of the volume; most of the volume, on the other hand, would be pervaded by gas several times less dense than the mean. The traditional model of expanding HII region that overlap at a well defined epoch when 'breakthrough' occurs (dating back at least to the 1972 paper of Arons and Wingert (9)) is consequently rather unrealistic. The ionization of the underdense region can be indeed be viewed in this way. But HII regions in the 'voids' can overlap (in the sense that the IGM becomes ionized except for 'islands' of high density) before even half the material has been ionized. Thereafter, the overdense regions would be 'eroded away': Stromgren surfaces encroach into them; the neutral regions shrink and present a decreasing cross-section; the mean free path of ionizing photons (and consequently the value of J) goes up (10,11).

The thermal history of the IGM beyond $z = 5$ is relevant to the modelling and interpretation of the absorption spectra of quasars at lower redshifts. The recombination and cooling timescales for diffuse gas are comparable to the cosmological expansion timescale. Therefore the 'texture' and temperature of the filamentary structure responsible for the lines in the 'forest' yields fossil evidence of the thermal history at higher redshifts.

AGNs at high z?

By the epoch $z = 5$, some structures (albeit perhaps only exceptional ones) must have attained galactic scales. Massive black holes (manifested as quasars) accumulate in the deep potential wells of these larger systems. Quasars may dominate the UV background at $z < 3$: if their spectra follow a power-law, rather than the typical thermal spectrum of OB stars, then quasars are crucial for the second ionization of He. AGN formation may requires virialised systems with large masses and deep potential wells (cf refs (12,13)); if so, we would naturally expect the UV background at the highest redshifts to be contributed mainly by stars in 'subgalaxies'. However, this is merely an expectation; it could be, contrariwise, that black holes readily form even in the first 10^8 $M_\odot$ CDM condensations (this would be an extreme version of a 'flattened' IMF), Were this the case, the early UV production could be dominated by black holes. This would imply that the most promising high-z sources to seek at near-IR wavelengths would be miniquasars, rather than 'subgalaxies'. It would also, of course, weaken the connection between the ionizing background and the origin of the first heavy elements.

Distinguishing between objects with $z > z_i$ and $z < z_i$

The blanketing effect due to the Lyman alpha forest – known to be becoming denser towards higher redshifts, and likely therefore to be even thicker beyond $z = 5$ – would be severe, and would block out the blue wing of Lyman alpha emission from a high-z source. Such objects may still be best detected via their Lyman alpha emission even though the absorption cuts the equivalent width by half. But at redshifts larger than z_i – in other words, before ionization breakthrough – the Gunn-Peterson optical depth is so large that any Lyman alpha emission line is blanketed completely, because the damping wing due to IGM absorption spills over into the red wing (7,14). This means that any objects detectable beyond z_i would be distinguished by a discontinuity at the redshifted Lyman alpha frequency. The Lyman alpha line itself would not be detectable (even though this may be the easiest feature to detect in objects with $z < z_i$).

RADIO AND MICROWAVE PROBES OF THE IONIZATION EPOCH

CMB fluctuations as a probe of the ionization epoch

If the intergalactic medium were suddenly reionized at a redshift z, then the optical depth to electrons scattering would be $\sim 0.05\Omega_b\Omega^{-1/2}(1+z)^{3/2}$ (generalisation to a more realistic scenario of gradual reionization is straightforward). Even when this optical depth is far below unity, the ionized gas constitutes a 'fog' that attenuates the fluctuations imprinted at the recombination era; the fraction of photons that are scattered at $< z_i$ then manifest a different pattern of fluctuations, characteristically on larger angular scales. This optical depth is consequently one of the parameters that can in principle be determined from CMB anisotropy measurements. It is feasible to detect a value as small as 0.1 – polarization measurements may allow even greater precision, since the scattered component would imprint polarization on angular scales of a few degrees, which would be absent from the Sachs Wolfe fluctuations on that angular scale originating at t_{rec}.

21 cm emission, absorption and tomography

The 21 cm line of HI at redshift z would contribute to the background spectrum at a wavelength of $21(1 + z)$ cm. This contribution amounts to a brightness temperature of order $0.05\Omega_b(1 + z)^{1/2}$ – very small compared with the 2.7 K of the CMB; and even smaller compared to the non-thermal background, which swamps the CMB, even at high galactic latitudes, at the long wavelengths where high-z HI should show up. Nonetheless, inhomogeneities in the HI may be detectable because they would give rise not only to angular fluctuations but also to spectral structure. If the same strip of sky were scanned at two radio frequencies differing by (say) 1 MHz, the temperature fluctuations due to the CMB itself, to galactic thermal and synchrotron backgrounds, and to discrete sources would track each other closely. Contrariwise, there would be no correlation between the 21 cm contributions, because the two frequencies would be probing 'shells' in redshift space whose radial separation would exceed the correlation length. It may consequently be feasible to distinguish the 21 cm background, utilizing a radio telescope with large collecting area. The fact that line radiation allows 3-dimensional tomography of the high-z HI renders this a specially interesting technique.

For the 21 cm contribution to be observable, the spin temperature T_s must differ from that of the black-body cosmic background. The gas would be detected in absorption or in emission depending on whether T_s are lower or higher than T_{bb}. The hyperfine levels of HI are affected by the microwave background itself, by collisional processes, and by Lyman alpha (whose profile is itself controlled by the kinetic temperature). T_s will therefore be a weighted mean of the CMB and gas temperatures.

Before there has been any heat input due to the development of non-linear structures, the kinetic temperature would be lower than that of the radiation. However, the spin temperature would be coupled to the background radiation temperature, so the HI would show up neither in emission nor in absorption. If, however, Lyman alpha radiation penetrates the HI without heating it, it couples the spin and kinetic temperatures, and so can actually lower the spin temperature so that the 21 cm line becomes an absorption feature. On the other hand, when the kinetic temperature rises, the feature appears in emission. The kinetic temperature can rise due to the weak shocking and adiabatic compression that accompanies the emergence of the first (very small scale) non-linear structure (cf section II). When photoionization starts, there will also, around each HII domain, be a zone of predominantly neutral hydrogen that has been heated by hard UV or X-ray photons. This latter effect would be more important if the first UV sources emitted radiation with a power-law (rather than just exponential) component.

In principle, one might be able to detect incipient large-scale structure, even when still in the linear regime, because it leads to variations in the column density of HI, per unit redshift interval, along different lines of sight (15).

Because the signal is so weak, there is little prospect of detecting high-z 21 cm emission unless it displays structure on (comoving) scales of several Mpc (corresponding to angular scales of several arc minutes) According to CDM-type models, the gas is likely to have been already ionized, predominantly by numerous ionizing sources each of sub-galactic scale, before such large structures become conspicuous. On the other hand, if the primordial gas were heated by widely-spaced quasar-level sources, each of these would be surrounded by a shell that could feasibly be revealed by 21cm tomography using, for instance, the new Giant Meter Wave Telescope (GMRT). With luck, effects of this kind may be detectable. Otherwise, they will have to await next-generation instruments such as the Square-Kilometer Array.

VERY DISTANT SUPERNOVAE

If the reheating and ionization were due to OB stars, it is straightforward to calculate how many supernovae would have gone off, in each comoving volume, as a direct consequence of this output of UV and heavy elements: there would be one, or maybe several, per year in each square arc minute of sky (16). The uncertainty depends partly on the redshift and the cosmological model, but also on the uncertainties about the UV background, and about the actual production of heavy elements. (These may be overestimated because the heavy element distribution is 'patchy', and concentrated in the overdense regions that yield high-column density absorption features; on the other hand, they may be underestimated if the processed material remains concentrated in the sources. Moreover, if most of the UV is absorbed near its source, then more production is needed to generate the required intergalactic ionization.)

These high-z supernovae would be primarily of Type 2. The typical observed light curve has a flat maximum lasting 80 days. One would therefore (taking the time dilation into account) expect each supernova to be near its maximum for nearly a year. It is possible that the explosions proceed differently when the stellar envelope is essentially metal-free, yielding different light curves, so any estimates of detectability are tentative. However, taking a standard Type 2 light curve (which may of course be pessimistic), one calculates that these object should be approximately 27th magnitude in J and K bands even out beyond $z = 5$. The detection of such objects would be an easy task with the NGST. With existing facilities it is marginal. The best hope would be that observations of clusters of galaxies might serendipitously detect a magnified gravitationally-lensed image from far behind the cluster.

The first supernovae may be important for another reason: they may generate the first cosmic magnetic fields. Mass loss (via winds or supernovae permeated by magnetic flux) would disperse magnetic flux along with the heavy elements. The ubiquity of heavy elements in the Lyman alpha forest indicates that there has been widespread diffusion from the sites of these early supernovae, and the magnetic flux could have diffused in the same way. Stretched and sheared by bulk motions, this can be the 'seed' for the later amplification processes that generate the larger-scale fields pervading disc galaxies.

(Incidentally, it is now clear that the afterglows of gamma-ray bursts are 100 times brighter than supernovae – perhaps even more, if Fruchter (17) is correct in his recent claim that a strong burst has already been seen at a redshift of 5. The rate, however, is far below the supernova rate – even though the afterglow rate could exceed that of the bursts themselves if the gamma rays are more narrowly beamed than the slower-moving ejecta that cause the afterglow. Detection of afterglows from well beyond $z = 5$ would offer a marvellous opportunity to obtain a high-resolution spectrum of intervening absorption features.)

WHERE ARE THE OLDEST STARS?

The efficiency of early mixing is important for the interpretation of stars in our own galaxy that have ultra-low metallicity – lower than the mean metallicity that would have been generated in association with the UV background at $z > 5$. If the heavy elements were efficiently mixed, then these stars would themselves need to have formed before galaxies were assembled. To a first approximation they would cluster non-dissipatively; they would therefore be distributed in halos (including the halo of our own Galaxy) like the dark matter itself. More careful estimates slightly weaken this inference, This is because the subgalaxies would tend, during the subsequent mergers, to sink via dynamical friction towards the centres of the merged systems. There would nevertheless be a tendency for the most extreme metal-poor stars to have a more extended distribution in our Galactic Halo, and to have a bigger spread of motions. This is a project where NGST could be crucial, especially if it allowed detection of halo stars in other nearby galaxies.

The number of such stars depends on the IMF. If this were flatter, there would be fewer low-mass stars formed concurrently with those that produced the UV background. If, on the other hand, the IMF were initially steeper, there could in principle be a lot of very low mass (macho) objects produced at high redshift, many of which would end up in the halos of galaxies like our own.

SUMMARY

The gravitational aspects of clustering can all be modeled convincingly by computer simulations. So also, now, can the dynamics of the baryonic (gaseous) component – including shocks and radiative cooling. But the nature of the simulation changes as soon as the first stars (or other compact objects) form. The first stars exert crucial feedback – the remaining gas is heated by ionizing radiation, and perhaps also by an injection of kinetic energy via winds and even supernova explosions.

Perhaps only 5 percent of star formation occurred before $z = 5$ (the proportion could be higher if most of their light were reprocessed by dust). But at the conference on NGST held here in Maryland last year, Alan Dressler, in his concluding lecture, emphasised that these early stars were important, just as were the first 5 percent of humans. There is still a variety of models for cosmic structure, that seem consistent with the properties of our universe at the current epoch. Large-scale structure may be elucidated within the next decade, by ambitious surveys (2-degree field and Sloan) and studies of CMB anisotropies. But there will still be uncertainty about how the present structures emerged, and especially about the efficiency and modes of star formation in early structures on subgalactic scale.

The later talks in this conference will highlight the exciting progress and prospects in elucidating the formative stages of cosmic structure.

I am grateful to my collaborators, especially Zoltan Haiman, Martin Haehnelt. Avi Loeb, Jordi Miralda-Escude, Piero Madau, Priya Natarajan, and Max Tegmark, for discussion of the topics described here.

REFERENCES

1. Tegmark, M., Silk, J., Rees, M. J., Blanchard, A., Abel, T., & Palla, F. ApJ, 474, 1 (1997)
2. Haiman, Z., Rees, M. J., & Loeb, A. ApJ, 476, 458 (1997)
3. Stecher, T.P., & Williams, D.A.,Ap. J. (Letters) 149, L1 (1967)
4. Haiman, Z., Rees, M.J. & Loeb, A., Ap.J. 467, 522 (1996)
5. Efstathiou, G. , MNRAS, 256, 43p (1992)
6. Navarro, J. F., & Steinmetz, M. , ApJ, 478, 13 (1997)
7. Miralda-Escudé, J., & Rees, M. J., ApJ, 497, 21 (1998)
8. Haiman, Z., Loeb, A & Madau, P Ap. J in press
9. Arons, J., & Wingert, D. W. ApJ, 177, 1 (1972)
10. Gnedin, N. & Ostriker, J.P. Ap J 486, 581 (1998)
11. Miralda-Escudé, J., Haehnelt, M & Rees, M. J, Ap. J submitted
12. Haiman, Z & Loeb, A, Ap.J. 503, 505 (1998)
13. Haehnelt, M, Natarajan, P & Rees, M.J. MNRAS 300, 817 (1998)
14. Miralda-Escudé, J. Ap. J. 501, 15 (1998)
15. Madau, P, Meiksin, A & Rees, M.J. Astrophys. J. 475, 429 (1997)
16. Miralda-Escudé, J., & Rees, M. J., ApJ (letters) 478, L57 (1997)
17. Fruchter, A., Ap. J. submitted

2. The Earliest Structures

The Cosmic Microwave Background & The Epoch of Galaxy Formation

Siang Peng Oh & David N. Spergel

Department of Astrophysical Sciences
Princeton University
Princeton NJ 08544 USA

Abstract. Observations of the cosmic microwave background will be an important ingredient in our efforts to understand the epoch of galaxy formation. Microwave background observations are most sensitive to physical conditions near the surface of last scatter, $z \sim 1100$. MAP should enable cosmologists to infer many of the initial conditions for galaxy formation. The microwave background photons also illuminate the "dark age". Measurements of microwave background polarization can determine or at least constrain the redshift at which stars (or quasars) first reionize the universe. The observations are also sensitive to non-linear density fluctuations during this epoch. Microwave background measurements also probe the nearby universe. Low redshift objects produce additional fluctuations through gravitational lensing, the ISW effect and the Sunyaev-Zeldovich effect.

CMB & THE CLASSICAL AGE

During its early classic period, the universe was a simple place. It was composed primarily of photons, electrons, hydrogen and helium nuclei and dark matter. There were very small ($\sim 10^{-4}$) variations in its density and composition. Cosmologists [1–4,7] believe that they understand the basic physics that governs this classical age: linear gravity, radiative transfer and simple atomic physics. Theory has now advanced to the stage that cosmologists have the analytical tools need to compute the statistical properties (the multipole spectrum) of the microwave background to better than 1%.

Over the past few years, a number of experiments have detected microwave background fluctuations over a range of angular scales. The measured fluctuations appear to be consistent with the basic inflationary model and a flat universe; however, the current data is not yet definitive. Within two years, MAP[1] will begin its mission

---

[1] http://map.gsfc.nasa.gov

CP470, After the Dark Ages: When Galaxies were Young (the Universe at 2 < z < 5),
edited by Stephen S. Holt and Eric P. Smith
© 1999 The American Institute of Physics 1-56396-855-X/99/$15.00

of making a high angular resolution map of the microwave sky with minimal systematic errors. PLANCK will follow later in the coming decade with even higher resolution observations over a wider range of frequencies. This synergy between precise theoretical predictions and accurate experimental measurements promises to significantly advance our understanding of the early universe.

The observations will test the current standard model, the cold dark matter model. This model assumes a flat universe with most of the mass density of the universe is in the form of non-baryonic matter and with fluctuations seeded by inflation. In the next few years, we can anticipate a number of key tests of this model: measurements of the microwave background temperature and polarization fluctuations; measurements of the large scale structure power spectrum and perhaps, even the detection of the cold dark matter.

If the cosmic microwave background observations are consistent with the theoretical predictions, then we can use the data to accurately measure many of the fundamental cosmological parameters [6–10]: the geometry of the universe, Ω_{tot}, the ratio of matter density to radiation density, $\Omega_{tot}h^2$ the ratio of baryon density to radiation density, $\Omega_b h^2$ and the primordial power spectrum. Thus, MAP should enable cosmologists to infer many of the initial conditions for galaxy formation.

CMB & THE DARK AGES

The microwave background photons also illuminate the "dark age". Measurements of damping of small-scale anisotropies and microwave background polarization can determine or at least constrain the redshift at which stars (or quasars) first reionize the universe. Small scale fluctuations generated during inhomogeneous reionization can constrain the topology of the reionized IGM, while fluctuations in the free-free emission background probes the distribution of ionizing sources.

Suppression of Primordial Fluctuations

As photons free stream towards us after recombination, they Thompson scatter off free electrons in the reionized IGM. This has the net effect of smearing out small scale temperature fluctuations, since the temperature we observe in any given direction is now a weighted average of a patch of the $z=1000$ last scattering surface. The smearing occurs on an angular scale $\theta \approx \sqrt{\Omega_o/z_r}$ ($\sim 7°$), the angular size of the horizon at the most recent scattering event. The probability that the photon we see originated at the original surface of last scattering is $e^{-\tau}$, where typically the optical depth $\tau \sim 0.1$. The net result is that while the observed power on large scales is unaffected, the small scale power is uniformly suppressed by a factor $e^{-2\tau}$.

This damping of small scale power is likely to be observed by MAP and Planck, to a precision of $\Delta\tau \sim 0.1$ with temperature data alone and $\Delta\tau \sim 0.01$ with both temperature and polarization data [9]. Note that many different reionization histories (with different values of x_e and z_r) can produce the same optical depth, so this measurement alone cannot fully constrain the epoch of reionization.

Generation of Polarized Fluctuations

Early reionization increases the polarization amplitude on large scales [11]. The origin of the effect is as follows: prior to recombination, all multipole moments of the temperature distribution higher than the monopole are very small, due to the tight coupling between photons and electrons. After recombination, these higher order moments can grow due to photon free streaming. If there is early reionization with sufficient optical depth, the resultant non-zero quadrupole anisotropy is transformed into polarization, as Thompson scattering has a polarization dependent scattering cross section. The scale of the effect corresponds to the horizon size at recombination (degree scales). The polarization power spectrum peaks with an amplitude $\propto \tau$ and $l_{\text{peak}} \propto \sqrt{z_r}$. Not only will polarization measurements decrease the measurement errors on τ, they will also break the degeneracies between the τ and other parameters such as the ionization fraction x_e (which may be determined to an accuracy of 15%, a further constraint on reionization models) and the power spectrum normalization C_2 [9].

Generation of Small Scale Fluctuations

The onset of reionization also generates new small-scale anisotropies, due to the Doppler effect from CMB photons scattering off moving electrons. In an fully reionized IGM, the photons get nearly opposite Doppler shifts on different sides of a density peak, leading to a cancellation at first order [12]. However, at very small (arcminute) scales the second order Ostriker-Vishniac effect [13] generates new fluctuations, due to the coupling between large scale fluctuations in v_r and small scale fluctuations in δ. The signal depends primarily on the ionization fraction x_e and only secondarily on τ, so if detected, it could be used in concert with observations of small scale damping to distinguish between various reionization histories [14].

Another process which generates new anisotropies is inhomogeneous reionization. In the period when the IGM was only partially ionized, CMB photons scatter off patches of ionized gas where the velocity field is coherent, and this modulation of the ionization fraction prevents the cancellation of Doppler effects. This has been a subject of much recent interest [15–17]. If the ionized patches are randomly distributed, then the amplitude of the anisotropies is proportional to the typical size R of the ionized bubbles, with a white noise power spectrum on large scales and a peak corresponding to the angular scale of a typical patch ($l \sim 20000$ for $R = 1$ Mpc). If the ionizing sources are strongly clustered due to their high bias [18], then the coherence in the ionization field adds power on scales larger than the typical patch size of a single source [17]. As the new fluctuations depend on the unknown history and in particular topology of reionization, the predicted level of anisotropy is open to debate. All authors have assumed models of Stromgren spheres expanding into the neutral IGM, it is however, possible that the voids got ionized first. This would lead to a different ionization topology. Nonethless, even in extreme cases patchy reionization is likely to be only comparable to other foregrounds ($\Delta T/T \sim 10^{-6}$).

Free-Free Emission

Another probe of reionization comes from spectral distortions of the blackbody CMB spectrum. The COBE FIRAS constraint of the y-distortion $y \leq 1.5 \times 10^{-5}$ [19] constrains Compton scattering of CMB photons by hot electrons. Since $y \propto T_e n_e l$, the IGM at high redshift must not be very hot,$T_e \leq 10^5$K, or reionization must have taken place relatively recently,$z_r < 10$. Another spectral distortion comes from free-free emission of the warm IGM, $Y_{ff} \propto \int n_e^2 T^{-1/2}\, dl$. The distortion is characterized by a quadratic rise in temperature at long wavelengths, which is poorly constrained by the FIRAS measurements. One of the goals of the DIMES[2] satellite [20] is to measure this distortion and thus constrain the epoch of reionization z_r. The proposed 0.1 mK sensitivity will be able to detect distortions if $z_r \geq 25$.

However, due to the n_e^2 dependence of the signal, free-free emission from ionizing sources will swamp the signal from the ionized IGM [21], and cause a larger distortion. In addition, the discreteness of these sources create small scale anisotropies which peak at sub-arcsecond scales. Galactic foregrounds have much large coherence scales. By filtering out the long wavelength galactic foregrounds, we can potentially detect the fluctuating signal using the VLA observations at low frequencies ($\sim$ 5 GHz); a $10^9 M_\odot$ halo at z=10 creates $\Delta T/T \sim 10^{-5}$. We could further improve sensitivity by cross-correlate the signal with NGST observations of Hα emission, which share the same n_e^2 dependence. A measurement of the fluctuating free-free background will probe ionizing emissivity, gas clumping, and star formation in the early universe.

It is interesting to note that the kinetic Sunyaev-Zeldovich effect from these ionized halos is also potentially observable. The signal is typically $\Delta T/T \sim \tau v/c \sim 10^{-6}$ for a $10^9 M_\odot$ halo at z=10, and may be separated from the free-free signal by its different spectral dependence.

CMB & THE NEARBY UNIVERSE

Microwave background distortions are also affected by physical processes at low redshift. These processes include the Sunyaev-Zeldovich effect, the ISW effect and gravitational lensing. By detecting these effects, we can use microwave background observations as a tool for studying the low redshift universe.

---

[2] http://ceylon.gsfc.nasa.gov/DIMES

Sunyaev-Zeldovich Effect

When the microwave background photons scatter off of hot gas, they gain energy. This shift in photon energy reduces the number of low energy photons and increases the number of high energy photons. Thus, in the Rayleigh-Jeans regime, the SZ effect produces microwave background cold spots. While in the Wien limit, the SZ effect produces microwave background hot spots. The amplitude of the Sunyaev-Zeldovich depends upon the integrated gas pressure:

$$\frac{\delta T}{T} = \frac{2\sigma_T}{m_e c^2} \int p_e ds \tag{1}$$

where σ_T is the Thompson cross-section, m_e is the electron mass, c is the speed of light, p_e is the gas pressure and ds is the path length.

In the past few years, radio astronomers have detected Sunyaev-Zel'dovich distortions from dozens of clusters [22]. The SZ measurements have already been used to provide a distance-ladder independent measurement of the Hubble Constant and measurements of the baryon/dark matter in clusters [23] Since a cluster's SZ flux is independent of redshift, the SZ effect may be one of the best techniques for detecting high redshift clusters. The SZ counts are also potentially a powerful technique for distinguishing between different cosmological models [24].

While SZ observations from clusters will likely continue to be an important area of cosmology, SZ observations may also become an important probe of lower density regions. Unlike X-ray emissivity, which is proportional to n_e^2, the amplitude of the SZ effect scales as n_e. Thus, poor groups and filaments will make a significant SZ signal. Persi et al. [25] estimates that roughly half of the SZ signal comes from outside of rich clusters. We can potentially detect this signal by cross-correlating optical surveys with CMB maps [26].

ISW Effect

Variations in the gravitational potential along the line of sight are an additional source of microwave background fluctuations:

$$\frac{\delta T}{T} = 2 \int \dot{\Phi} d\eta \tag{2}$$

While CMB observations alone can not distinguish between fluctuations produced at the surface of last scatter and ISW fluctuations, it may be possible to detect these fluctuations through cross-correlating CMB maps with probes of large scale structure. Boughn et al. [27] have looked for cross-correlations between the HEAO1 X-ray maps and the COBE data. They have used the non-detection to place interesting limits on the cosmological constant. With MAP data and data from the new generation of X-ray satellites (ROSAT and ABRIXAS), we should be able to improve these measurements. This may lead to interesting constraints on Λ and other cosmological parameters.

Gravitational Lensing

Gravitational lensing can be viewed as a mathematical map from unlensed coordinates to lensed coordinates. This mapping does not produce new distortions in the microwave background; however, it does alter the appearance of the microwave sky. Temperature fluctuations behind high density regions are stretched, while temperature fluctuations behind low density regions are compressed.

Gravitational lensing produces a number of observable effects:

1. Lensing smoothes out the acoustic peaks in the microwave background spectrum [28] and by distorting the shape of the fluctuations increases the amplitude of fluctuations on small angular scales.

2. Large scale structure surveys will directly detect the superclusters responsible for gravitational lensing. Thus, we can detect the lensing effects by cross-correlating large scale structure surveys and microwave background maps. This cross-correlation should be detectable in the MAP and Sloan Digital Sky Survey (SDSS) data [29]. By measuring this effect, we can determine the bias factor of a galaxy population. We could also look for this cross-correlation effect in X-ray surveys.

3. Since the same superclusters distort the microwave background through lensing and through the Sunyaev-Zeldovich effect, the combination of these effects will produce non-Gaussian features in the microwave background [30]. The combination of the Sunyaev-Zelovich effect and the ISW effect may also lead to effects observable in MAP [31,32].

By detecting these effects, we can measure the amplitude of mass fluctuations, the galaxy and quasar bias factors, and detect the effects of a cosmological constant or quintessence.

ACKNOWLEDGEMENTS

DNS and SPO would like to thank the MAP project for support.

REFERENCES

1. Peebles, P.J.E. & Yu, J.T. 1970, Ap. J., 162, 815.
2. Wilson, M.L. & Silk, J. 1981, Ap. J., 243, 14.
3. Bond, J.R. & Efstathiou, G. 1984. Ap. J., 285, L45.
4. Bond, J.R. 1990. In *The Cosmic Microwave Background: 25 Years Later,* ed. N. Mandolesi and N. Vittorio, 45. Dordrecht: Kluwer.
5. Hu, W.. Zaldarriaga, M., White, M . & U. Seljak 1998, Phys. Rev. D, 57, 3290.
6. Spergel, D.N. 1994, Warner Prize Lecture, BAAS, 185.7301

7. Jungman, G., Kamionkowski, M., Kosowsky, A., & Spergel, D.N. 1996, Phys. Rev., D54, 1332.
8. Bond, J.R., Efstathiou, G., & Tegmark, M. 1997, Mon. Not. Royal Astron. Soc., 291, L33.
9. Zaldarriaga, M., Spergel, D.N., & Seljak, U. 1997, Ap. J., 488, 1
10. Wang, Y., Spergel, D.N. & Strauss, M.A. 1998, astro-ph/9802231, to appear in ApJ, 510.
11. Zaldarriaga, M., 1997, Phys. Rev. D, 55, 1822
12. Kaiser, N., 1984, Ap.J., 282, 374
13. Ostriker, J.P., & Vishniac, E., Ap.J., 306, L51
14. Jaffe, A., & Kamionkowski, M., 1998, Phys. Rev. D, 58, 043001
15. Aghanim, N., Desert, F.X., Puget, J.L., & Gispert, R., 1996, Astron. & Astrophys., 311, 1
16. Gruzinov, A., & Hu, W., 1998 Ap.J., 508, 435
17. Knox, L., Scoccimarro, R., & Dodelson, S., 1998, Phys. Rev. Letters, 81, 2004
18. Oh, S.P., 1998, these proceedings
19. Fixsen, D.J., Cheng, E.S., Gales, J.M., Mather, J.C., Shafer, R.A. & Wright, E.L. 1996, Ap. J., 473, 576.
20. Kogut, A., 1996, astro-ph/9607100
21. Oh, S.P., 1999, In preparation
22. Cooray, A.R., Carlstrom, J.E. Joy, M., Grego, L, Holzapfel, W.L. and Patel, S.K. 1998, astro-ph/9804149.
23. Grego, L., Carlstrom, J.E., Holder, G., Cooray, A., Reese, E., Holzapfel, W., Joy, M.K. & Patel, S.K. 1998, AAS 192.1707.
24. Aghanim, N., DeLuca, A., Bouchet, F.R., Gispart, R., & Puget, J.L. 1997 A&A, 325, 9.
25. Persi, F.M., Spergel, D.N., Cen, R. & Ostriker, J.P. 1995, Ap.J., 442, 1.
26. Refregier, A., Spergel, D.N., & Herbig, T. 1998, astro-ph/9810025, to appear in Ap.J.
27. Boughn, S.P., Crittenden, R.G., & Turok, N.G. 1998, New Astronomy, 3, 275.
28. Seljak, U. 1996, Ap.J., 460, 549.
29. Suginohara, M., Suginohara, T. & Spergel, D.N. 1998, Ap.J., 495, 511.
30. Goldberg, D. & Spergel, D.N.. 1998, astro-ph/9811252.
31. Spergel, D.N. & Goldberg, D.M. 1998, astro-ph/9811251.
32. Seljak, U. & Zaldarriga, M. 1998, astro-ph/9811123.

Empirical Constraints on the First Stars and Quasars

Z. Haiman[†] and A. Loeb[*]

[†]*Astrophysics Theory Group, Fermi National Accelerator Laboratory, Batavia, IL 60510*
[*]*Harvard-Smithsonian Center for Astrophysics, 60 Garden Street, Cambridge, MA 02138*

Abstract. Empirical studies of the first generation of stars and quasars in the Universe will likely become feasible over the next decade. The *Next Generation Space Telescope* will provide direct imaging and photometry of sub–galactic objects at $z \gtrsim 10$, while microwave anisotropy experiments, such as MAP or Planck, will set constraints on the ionization history of the intergalactic medium due to these sources. We describe the expected signals that will be detectable with these future instruments.

INTRODUCTION

The cosmic dark age [1] ended when the first gas clouds condensed out of the primordial fluctuations at redshifts $z = 10 - 20$ [2]. These condensations gave birth to the first star–clusters or mini–quasars in the Universe. Previous observations have not yet probed this epoch. Existing optical or infrared telescopes are capable of reaching out to redshifts $z \lesssim 6$, while current anisotropy experiments of the cosmic microwave background (CMB) probe only the recombination epoch at $z \sim 10^3$. The remaining observational gap between these redshift regimes is likely to be bridged over the coming decade. In particular, the planned launch of the infrared Next Generation Space Telescope (NGST) will enable direct imaging of sub–galactic objects at $z \gtrsim 10$, while microwave anisotropy satellites such as MAP or Planck will measure complementary signatures from the reionization of the intergalactic medium at these redshifts.

The detection [2,3] of Lyα emission from galaxies at redshifts up to $z = 5.7$ demonstrates that reionization due to the first generation of sources must have occurred at yet higher redshifts; otherwise, the damping wing of Lyα absorption by the neutral intergalactic medium would have eliminated the Lyα line in the observed spectrum of these sources [5,9]. In this review, we predict the redshift of reionization in popular cosmological models, and estimate the signals from this epoch that will become detectable with the NGST, MAP, or the Planck satellites.

CP470, After the Dark Ages: When Galaxies were Young (the Universe at 2 < z < 5),
edited by Stephen S. Holt and Eric P. Smith

PROPERTIES OF THE FIRST OBJECTS

The first objects left behind a variety of fossil evidence of their existence [7], including the enrichment of the intergalactic medium (IGM) with heavy elements, the reionization of the IGM, the distortion of the CMB spectrum by dust-processed radiation [8], and the production of stellar remnants. In the following, we explore the signatures of the early stars and mini-quasars using a simple semi-analytical model, based on the Press–Schechter formalism [12]. We calibrate the total amount of light that stars or mini–quasars produce based on data from redshifts $z \lesssim 5$. The efficiency of early star formation is calibrated based on the observed metallicity of the intergalactic medium [10,11], while the early quasars are constrained so as to match the quasar luminosity function [12] at redshifts $z \lesssim 5$, as well as data from the Hubble Deep Field (HDF) on faint point–sources. We focus on a particular cosmological model with a cosmological constant, $(\Omega_0, \Omega_\Lambda, \Omega_b, h, \sigma_{8h^{-1}}, n)$=(0.35, 0.65, 0.04, 0.65, 0.87, 0.96), named the "concordance model" by Ostriker & Steinhardt [3]. For a discussion of other cosmological models, as well as a more detailed description of our methods and results, we refer the reader to several papers [13–20]. More advanced 3–D numerical simulations have only now started to address the complicated physics associated with the fragmentation, chemistry, and radiative transfer of the primordial molecular clouds [21], as well as with the reionization of the IGM [5].

Following collapse, the gas in the first baryonic condensations is virialized by a strong shock [23]. The shock–heated gas can only continue to collapse and fragment if it cools on a timescale shorter than the Hubble time. In the metal–poor primordial gas, the only coolants that satisfy this requirement [24] are neutral atomic hydrogen (H) and molecular hydrogen (H_2). However, H_2 molecules are fragile, and are easily photo-dissociated throughout the universe by trace amounts of starlight [25,13] that are well below the level required for complete reionization of the Universe. Hence, most of the sources that ionized the Universe formed inside objects with virial temperatures $T_{vir} \gtrsim 10^4$K, or masses $\sim 10^8$ M$_\odot$, which cooled via atomic transitions. Depending on the details of their cooling and angular momentum transport, the gas in these objects fragmented into stars, or formed a central black hole exhibiting quasar activity. Although the first objects contained only a small fraction of the total mass of the universe, they could have had a dramatic effect on the subsequent evolution of the ionization and temperature of the rest of the gas [27]. Since nuclear fusion releases ~ 7 MeV per baryon, and accretion onto a black hole may release even more energy, and since the ionization of a hydrogen atom requires only 13.6 eV, it is sufficient to convert a small fraction of the baryonic mass into either stars or black holes in order to ionize the rest of the Universe.

The cooling gas clouds eventually fragment into stars [28]. Although the actual fragmentation process is likely to be rather complex, the average fraction f_{star} of the collapsed gas converted into stars can be calibrated empirically so as to reproduce the average metallicity observed in the Universe at $z \approx 3$. For the purpose of this calibration, we use the average C/H ratio, inferred from CIV absorption lines in

35

Lyα forest clouds [10,11]. The observed ratio is between 10^{-3} and 10^{-2} of the solar value [29]. If the carbon produced in the early star clusters is uniformly mixed with the rest of the baryons in the Universe, this implies $f_{star} \approx 2\text{--}20\%$ for a Scalo [30] stellar mass function. This number assumes inefficient hot bottom burning, i.e. maximal carbon yields [31], and includes a factor of ~ 3 due to the finite time required to produce carbon inside the stars (in a Press–Schechter star formation history, only a third of the total stellar carbon yield is produced and ejected by $z = 3$). Ultimately, 3-D simulations of the first generation of stars might be used to infer the expected star–formation efficiency in the first generation of gas clouds. Preliminary runs [32] imply that $\sim 1\%$ of the gas condenses into dense cores which could yield massive stars.

An even smaller fraction of the cooling gas might condense at the center of the potential well of each cloud and form a massive black hole, exhibiting mini–quasar activity. In the simplest scenario we postulate that the peak luminosity of each black hole is proportional to its mass, and there exists a universal quasar light–curve in Eddington units. This hypothesis is motivated by the fact that for a sufficiently high fueling rate, quasars are likely to shine at their maximum possible luminosity, which is some constant fraction of the Eddington limit, for a time which is dictated by their final mass and radiative efficiency. Allowing the final black hole mass M_{bh} to be a fixed fraction of the total halo mass M_{halo}, we find that there exists a universal light curve $[L(t) = L_{Edd} \exp(-t/t_0)$, with $t_0 \sim 10^6$ yr], for which the Press–Schechter theory provides an excellent fit to the observed evolution of the luminosity function (LF) of bright quasars between redshifts $2.6 < z < 4.5$. The required black hole to halo gas mass ratio is $M_{bh}/M_{gas} = 10^{-3.2}\Omega_0/\Omega_b = 5.5 \times 10^{-3}$, close to the typical value of $\sim 6 \times 10^{-3}$ found for the ratio of black hole mass to spheroid mass in a dozen nearby galaxies [22,16]. The existence of massive black holes in the centers of low–mass galaxies such as M32 or NGC 4486B [22,16] implies that the process of black hole formation does not discriminate against galaxies of this type. Since galaxies at $z \sim 10$ have a similar mass and velocity dispersion as these low–redshift examples, it is conceivable that low–luminosity quasars contributed significantly to the reionization of the Universe.

One does expect, however, that the ratio M_{bh}/M_{gas} would have a substantial intrinsic scatter. Observationally, the scatter around the average value of $\log(M_{bh}/L)$ is 0.3 [16], while the standard deviation in $\log M_{bh}/M_{gas}$ has been found to be $\sigma \sim 0.5$ [35]. Such an intrinsic scatter would flatten the predicted quasar LF at the high mass end, where the LF is a steep function of black hole mass. As an illustrative example, we show in Figure 1 the mass function of black holes in the Press–Schechter model of halos, with or without this scatter (solid versus short-dashed lines). In order to eliminate the flattening introduced by the scatter, we find that the average black hole to halo mass ratio must be reduced by $\sim 50\%$. The dot–dashed line in Figure 1 demonstrates that such a reduction would indeed compensate for the effect of an intrinsic scatter in the relevant mass range $(10^8 \, M_\odot \lesssim M_{bh} \lesssim 10^{10} \, M_\odot)$. Figure 1 also shows the effect of a more significant intrinsic scatter ($\sigma \sim 1$) on the black hole mass function (long-dashed lines). We

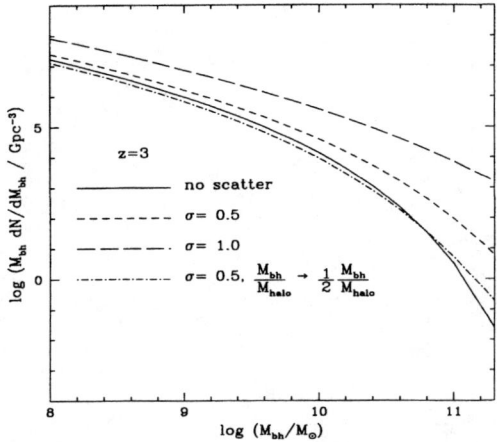

FIGURE 1. The comoving mass function of black holes at redshift $z = 3$ when a scatter is introduced to the logarithm of the ratio between the black hole and halo masses, $\log(M_{bh}/M_{halo})$.

find that the predicted black hole mass function in the presence of such a large scatter would be significantly different from any model with a constant value for M_{bh}/M_{gas}.

In reality, the relation between the black hole and halo masses may be more complicated than linear in reality. With the introduction of additional free parameters, a non–linear (mass and redshift dependent) relation between the black–hole and halo masses can also lead to acceptable fits [36] of the observed quasar LF. The nonlinearity in the relation must be related to the physics of the formation process of low–luminosity quasars, which was discussed in several papers [36,37,7]. If the black hole formation efficiency decreases in smaller halos, this would flatten the faint end of the LF, and therefore could not compensate the effect of a large intrinsic scatter of the type shown in Figure 1. Indeed, in order to fit the bright end of the LF in a model with a large intrinsic scatter, one must postulate that the black hole formation efficiency decreases in larger halos.

INFRARED NUMBER COUNTS

The Next Generation Space Telescope (*NGST*, [39]) will be able to detect the early population of star clusters and mini–quasars. *NGST* is scheduled for launch in 2007, and is expected to reach an imaging sensitivity of ~ 1 nJy (S/N=10 at spectral resolution $\lambda/\Delta\lambda = 3$) for extended sources after several hours of integration in the wavelength range of 1–3.5μm. Figure 2 shows the predicted number counts in the models described above, normalized to a $5' \times 5'$ field of view. This figure shows

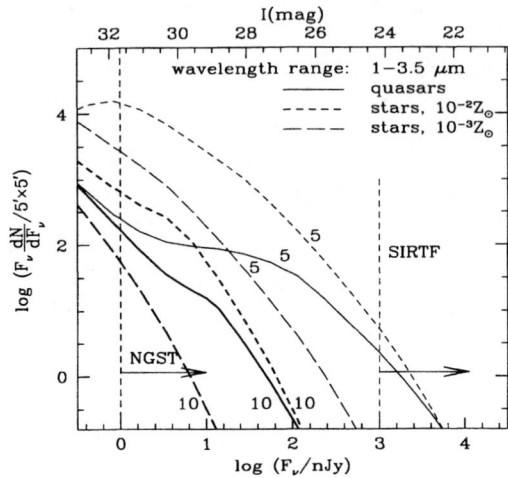

FIGURE 2. Infrared Number Counts. The solid curves refer to quasars, while the long/short dashed curves correspond to star clusters with low/high normalization for the star formation efficiency. The curves labeled "5" or "10" show the cumulative number of objects with redshifts above $z = 5$ or 10.

separately the number per logarithmic flux interval of all objects with redshifts $z > 5$ (thin lines), and $z > 10$ (thick lines). The number of detectable sources is high; *NGST* will be able to probe about ~ 100 quasars at $z > 10$, and ~ 200 quasars at $z > 5$ per field of view. The bright–end tail of the number counts approximately follows a power law, with $dN/dF_\nu \propto F_\nu^{-2.5}$. The dashed lines show the corresponding number counts of "star–clusters", assuming that each halo shines due to a starburst that converts a fraction of 2% (long–dashed) or 20% (short–dashed) of the gas into stars. These lines indicate that *NGST* would detect $\sim 40 - 300$ star–clusters at $z > 10$ per field of view, and $\sim 600 - 10^4$ clusters at $z > 5$. Unlike quasars, star clusters could in principle be resolved if they extend over a scale comparable to the virial radius of their dark matter halos [16]. The supernovae and γ-ray bursts in these star clusters might outshine their hosts and may also be directly observable [40,41].

CONSTRAINTS FROM THE HUBBLE DEEP FIELD

High resolution, deep imaging surveys can be used to set important constraints on semi–analytical models of the type described above. The properties of faint *extended* sources found in the Hubble Deep Field (HDF) [42] agree with detailed semi–analytic models of galaxy formation [43]. On the other hand, the HDF has revealed only a handful of faint *unresolved* sources, but none with the colors expected for high redshift quasars [44]. The simplest mini–quasar model described

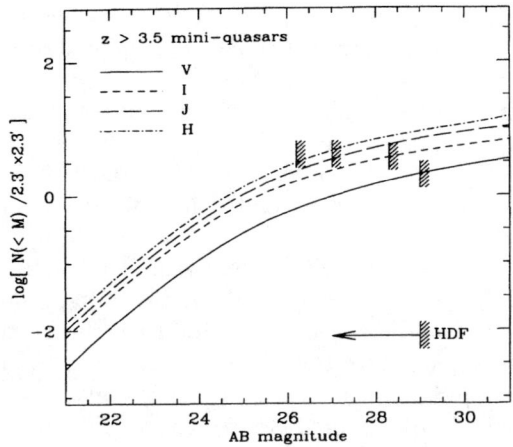

FIGURE 3. V, I, J, and H counts for mini–quasars in a model with a minimum halo circular velocity of $v_{\rm circ} = 75$ km s^{-1}. This model is consistent with HDF data in the V band at the 5% level. Sensitivities are shown in each band for the same signal–to–noise ratio and exposure time as the optical HDF.

above predicts the existence of ~ 10 B–band "dropouts" in the HDF, inconsistently with the lack of detection of such dropouts up to the $\sim 50\%$ completeness limit at $V \approx 29$ in the HDF. To reconcile the models with the data, a mechanism is needed for suppressing the formation of quasars in halos with circular velocities $v_{\rm circ} \lesssim 50-75$ km s^{-1}. This suppression naturally arises due to the photo-ionization heating of the intergalactic gas by the UV background after reionization [45,46]. Alternative effects could help reduce the quasar number counts, such as a change in the background cosmology, a shift in the "big blue bump" component of the quasar spectrum to higher energies due to the lower black hole masses in mini–quasars, or a nonlinear black hole to halo mass relation; however, these effects are too small to account for the lack of detections in the HDF [20].

The mini-quasars may not necessarily appear as point sources in the HDF if their extended host galaxies are actually resolved by *HST*. In fact, twelve candidate sources of activity in the nuclei of galaxies at high–redshifts ($z > 3.5$) have recently been identified [47] in the HDF. All of these point–like sources are embedded in extended host galaxies which are relatively bright ($V \sim 26 - 27$) and outshine their AGNs by typically 1 mag. As a consequence, these AGNs would have been missed by previous searches for isolated point–like sources [44]. In the models described above, the mini–quasars peak at a flux $\sim 1 - 3.5$ mag brighter in V than their host galaxy, which is assumed to undergo a starburst inside the same halo [18]. The twelve HDF candidates must reflect faint AGN activity in bright galaxies of relatively massive halos (analogous to a weak Seyfert activity), rather than faint

AGN activity in small halos as expected in our model. The observed sources might imply a phase in the history of massive halos that corresponds to an additional low–luminosity tail of the quasar lightcurve discussed above. An extended lightcurve would still be consistent with the luminosity function derived from the bright AGN phase and could explain the existence of the embedded AGNs in the HDF. The detection of faint embedded AGNs could also be related to the intrinsic scatter in the distribution of M_{bh}/M_{halo} relation, and reflect objects with unusually small values of this mass ratio.

The longer infrared wavelengths, such as the J and H infrared bands, are better suited for studying the Universe at $z \gtrsim 5$. Forthcoming data on point–sources from NICMOS observations of the HDF [48] could improve the constraints on mini–quasar models. In Figure 3, we show the expected number counts of mini–quasars in the V, I, J, and H bands for the model which is consistent with the optical HDF data. The number of objects predicted in the I, J, and H bands is higher than in the optical HDF. The NICMOS data in these bands would either reveal several high redshift mini–quasars or else place tighter constraints on quasar models than currently possible using V and I data. With the post–reionization feedback on halos imposed by the optical HDF data, $v_{circ} \geq 75$ km s^{-1}, we still expect at least ~ 5 mini-quasars to be found at $z > 3.5$ in the NICMOS J and H bands. A non–detection by NICMOS would translate to a minimum circular velocity of $v_{circ} \gtrsim 100$ km s^{-1}, or a factor of ~ 2 increase in the low–mass cutoff for halos harboring quasars.

WAS THE UNIVERSE REIONIZED BY STARS OR MINI–QUASARS?

Given either the star–formation or quasar black–hole formation histories, we derive the reionization history of the IGM by following the radius of the expanding Strömgren sphere around each source. The reionization history depends on the time–dependent production rate of ionizing photons, their escape fraction, and the recombination rate of the IGM, which are all functions of redshift. The production rate of ionizing photons per quasar follows from the median quasar spectrum [13] and the light–curve we derived. The analogous rate per star follows from the time–dependent composite stellar spectrum, constructed from standard stellar atmosphere atlases [14] and evolutionary tracks [51]. We computed the escape fraction of ionizing photons in each halo, assuming ionization equilibrium inside an isothermal sphere where both the stars and the gas are distributed with a $1/r^2$ profile. For quasars, we assumed that the escape fraction is 100%. During reionization, we also assumed that the formation of low–mass halos is suppressed by photo–ionization heating inside the already ionized cosmological HII regions.

Figure 4 summarizes the resulting reionization histories from stars or quasars in our models with a ΛCDM cosmology. The results for stars are shown in two cases, one with $Z_{IGM} = 10^{-2}Z_{\odot}$ (dashed lines) and the other with $Z_{IGM} = 10^{-3}Z_{\odot}$

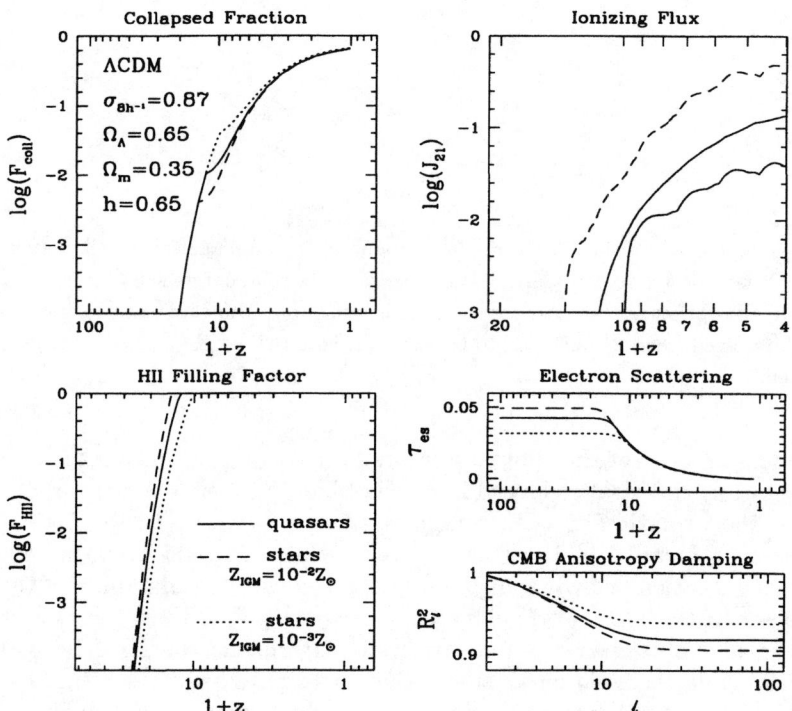

FIGURE 4. Reionization history. Clockwise, the different panels show: (i) the collapsed fraction of baryons; (ii) the background flux at the Lyman limit; (iii) the volume filling factor of ionized hydrogen; and (iv) the optical depth to electron scattering, and the corresponding damping factor for the power–spectrum decomposition of microwave anisotropies as a function of the spherical harmonic index ℓ. The solid curves refer to quasars, while the dotted or dashed curves correspond to stars with a low (2%) or high (20%) normalization for the star formation efficiency.

41

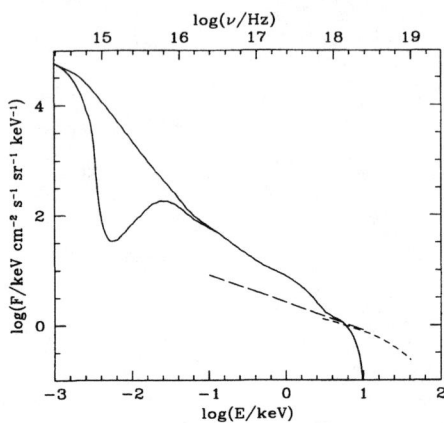

FIGURE 5. Spectrum of the UV/X–ray background in the mini–quasar model, assuming that the median X–ray spectrum of quasars [13] is universal. The solid curves show the spectrum with and without absorption by the Lyα forest at high–redshifts. The short and long dashed lines show the unresolved fraction (assumed to be 25%) of the observed X–ray background spectrum (from [53] and [54]).

(dotted lines), to bracket the allowed IGM metallicity range. The panels in Figure 4 show (clockwise) the total collapsed fraction of baryons available for star or quasar formation; the evolution of the average comoving flux, J_{21} at the local Lyman limit frequency, in units of 10^{-21} erg s^{-1} cm^{-2} Hz^{-1} sr^{-1}; the resulting evolution of the ionized fraction of hydrogen, F_{HII}; and the consequent damping of the CMB anisotropies. In the first two panels, we imposed $v_{\mathrm{circ}} \geq 75$ km s^{-1} at $z < z_{\mathrm{reion}}$. The dashed and dotted curves indicate that stars ionize the IGM by a redshift $9 \lesssim z \lesssim 13$; while the solid curve shows that quasars reionize the IGM at $z \approx 11$. This result can be understood in terms of the total number of ionizing photons produced per unit halo mass; given our normalizations of the efficiencies of star and quasar black hole formation, the relative ratios of this number in the three cases are $1 \div 0.37 \div 0.1$, respectively. A comparison of our quasar and stellar template spectra shows that stars will not reionize HeII, while quasars reionize HeII at essentially the H reionization redshift. Therefore, recent claims that HeII reionization might have been observed at $z \sim 3$ [39] could rule-out the presence of mini-quasars with hard spectra extending to X–rays at high redshifts, if these claims are verified by future observations.

The X–ray background (XRB) [53,54] might provide another useful constraint on the mini–quasar models. In Figure 5 we show the predicted spectrum of the UV to the soft XRB at $z = 0$ in these models (solid lines). In computing the spectrum, we included absorption by neutral H and He in the IGM at $z > z_{\mathrm{reion}} = 11$ and hydrogen absorption [55] by the (extrapolated) Lyα forest at $z \leq 11$. Also shown in this figure is the unresolved 25% fraction of the observed soft XRB [53,54]. The

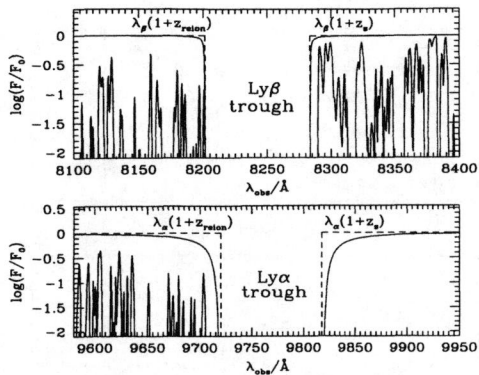

FIGURE 6. Spectrum of a source at $z_s = 7.08$, assuming sudden reionization at a redshift $z_{\text{reion}} = 7$. The solid curves show the spectrum without absorption by the high–redshift Lyα forest, and the dashed lines show the spectrum when the damping wings are also ignored.

dashed lines in figure 5 represent an upper limit on any component of the XRB that could arise from high–redshift quasar activity. As the figure shows, the mini–quasar models overpredict the *unresolved flux* by a factor of $\sim 2-7$ in the 0.1-1 keV range, as they produce a flux comparable to the entire soft XRB flux. If an even larger fraction of the XRB will be resolved into low–redshift AGNs in the future, then the XRB could be used to place stringent constraints on the X-ray spectrum or the abundance of the mini–quasars discussed here.

CAN THE REIONIZATION REDSHIFT BE INFERRED FROM A SOURCE SPECTRUM?

The spectrum of a source at a redshift $z_s > z_{\text{reion}}$ should show a Gunn–Peterson (GP) [56] trough due to absorption by the neutral IGM at wavelengths shorter than the local Lyα resonance at the source, $\lambda_{\text{obs}} < \lambda_\alpha (1 + z_s)$. By itself, the detection of such a trough would not uniquely establish the fact that the source is located beyond z_{reion}, since the lack of any observed flux could be equally caused by: (i) ionized regions with some residual neutral fraction, (ii) individual damped Lyα absorbers, or (iii) line blanketing from lower column density Lyα forest absorbers. On the other hand, for a source located at a redshift z_s beyond but close to reionization, $(1 + z_{\text{reion}}) < (1 + z_s) < \frac{32}{27}(1 + z_{\text{reion}})$, the GP trough splits into disjoint Lyman α, β, and possibly higher Lyman series troughs, with some transmitted flux in between these troughs. Although the transmitted flux is suppressed considerably by the dense Lyα forest after reionization, it is still detectable for sufficiently bright sources, and can be used to infer the reionization redshift.

43

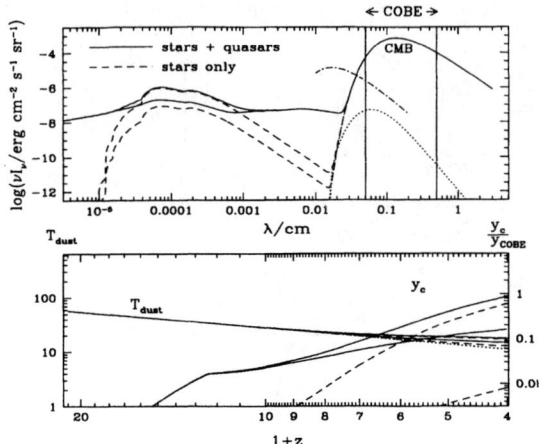

FIGURE 7. Effect of dust on the background flux. The top panel shows the comoving spectra in four different models at $z = 3$, and the bottom panel shows the corresponding evolution of the dust temperature and Compton y-parameter (see text).

As an example, we show in Figure 6 the simulated spectrum around the Lyman α and β GP troughs of a source at redshift $z_s = 7.08$, assuming that reionization occurs suddenly at $z_{reion} = 7$. We have included the extrapolated effects of Lyα absorbers along the lines of sight, whose statistics were chosen so as to obey the redshift dependence and absorption line characteristics of observational data at $z < 4.3$ [57]. Although the continuum flux is strongly suppressed, the spectrum contains numerous transmission features; these features are typically a few Å wide, have a central intensity of a few percent of the underlying continuum, and are separated by ~ 10Å. For sudden reionization, the nominal integration time of about 10 hours for a ~ 10 nJy sensitivity (with $\lambda/\Delta\lambda=100$ and S/N=10) would be sufficient to determine z_{reion} up to a redshift of ~ 7 with a high precision, by determining the location of the short–wavelength edge of the troughs shown in Figure 6. More gradual reionization would smear the edge of the GP trough. Based on the extrapolation of the Lyα forest to higher redshifts, the required sensitivity needs to be one or two orders of magnitude higher if $z_{reion} \sim 8$ or ~ 9 (see [19] for details).

SIGNATURES IMPRINTED ON THE CMB

The free electrons produced by the reionization of the intergalactic medium partially smooth–out the temperature anisotropies of the CMB via Thomson scattering. Given the ionized fraction of hydrogen as a function of redshift, one can readily derive the electron scattering optical depth (τ_{es}), as well as the anisotropy

44

damping factor (R_ℓ^2), as functions of the spherical harmonic index ℓ of the multipole expansion of the anisotropies on the sky [58]. As illustrated in the lower right panel of Figure 4, the amplitude of the anisotropies is reduced by $\sim 6 - 10\%$ on scales below the angular scale of the horizon at reionization ($\ell \gtrsim 10$). Although small, this reduction is within the proposed sensitivities of the future MAP and Planck satellites, provided that both temperature and polarization anisotropy data will be gathered in these experiments (see Table 2 in [59]).

In addition, the dust that is inevitably produced by the first type II supernovae, absorbs the UV emission from early stars and quasars and re-emits this energy at longer wavelengths, where it distorts the CMB spectrum. We have calculated this spectral distortion assuming that each type II supernova yields $0.3 M_\odot$ of dust with the wavelength-dependent opacity of Galactic dust [60]. We have conservatively assumed that similarly to the observed intergalactic mixing of metals, this dust gets uniformly distributed throughout the intergalactic medium. Clumpiness of the dust around UV sources would only enhance our predicted spectral distortion. The top panel of Figure 7 shows the resulting total comoving spectrum of the radiation background (CMB + direct quasar and/or stellar emission + dust emission) at $z = 3$. More distortion could be added between $0 < z < 3$ by dust and radiation from galaxies. For reference, we also show by the dot–dashed lines the recently detected cosmic infrared background (CIB, [61]), and the typical dust peak in our calculations (dotted lines).

Figure 7 shows that in our models the dust emission peaks at a wavelength which is an order of magnitude longer than that of the CIB peak. This is a result of our assumption of a homogeneous dust distribution, and the consequent cold dust temperature. An inhomogeneous distribution of dust would raise its temperature, and could contribute significantly to the observed CIB peak. The deviation from the pure $2.728(1+z)$K blackbody shape is quantified by the Compton y–parameter, whose redshift evolution is shown in the bottom panel. Ignoring the UV flux from quasars, we obtain $1.1 \times 10^{-7} < y_c < 8.2 \times 10^{-6}$ at $z = 3$ (dashed lines), just below the upper limit [62] set by COBE, $y < 1.5 \times 10^{-5}$. Adding the UV flux of quasars increases the y–parameter to $4.1 \times 10^{-6} < y_c < 2 \times 10^{-5}$. The distortion by the intergalactic dust may have been overestimated in our calculation as we ignored the absorption of the UV background by the neutral component of the IGM. Nevertheless, a substantial fraction (~ 10–50%) of the total y–parameter results simply from the direct far-infrared emission by early quasars and could be present even in the absence of any intergalactic dust.

ACKNOWLEDGEMENTS

We are grateful to Martin Rees for sharing his insights during our collaboration on some of the topics described in this review. ZH acknowledges support at Fermilab by the DOE and the NASA grant NAG 5-7092.

REFERENCES

1. Rees, M. J. 1996, preprint astro-ph/9608196
2. Loeb, A. 1998, in Proc. of 34th Liege International Astrophysics Colloquium on the "Next Generation Space Telescope", Belgium, June 1998, preprint astro-ph/9806163
3. Hu, E. M., Cowie, L. L., & McMahon, R. G. 1998, ApJ, 502, L99
4. Dey, A., Spinrad, H., Stern, D., Graham, J. R., & Chaffee, F. H. 1998, ApJ, 498, L93
5. Miralda-Escudé, J. 1998, ApJ, 501, 15
6. Haiman, Z., & Spaans, M. 1998, ApJ, submitted, astro-ph/9809223, see also this proceedings.
7. Carr, B. J., Bond, J. R., & Arnett, W. D. 1984, ApJ, 277, 445
8. Wright, E. L. 1981, ApJ, 250, 1
9. Press, W. H., & Schechter, P. L. 1974, ApJ, 181, 425
10. Tytler, D. *et al.* 1995, in *QSO Absorption Lines*, ed. G. Meylan ed., Springer, p.289
11. Songaila, A., & Cowie, L. L. 1996, AJ, 112, 335
12. Pei, Y. C. 1995, ApJ, 438, 623
13. Ostriker, J. P., & Steinhardt, P. J. 1995, Nature, 377, 600
14. Haiman, Z., Thoul, A., & Loeb, A. 1996, ApJ, 464, 523
15. Haiman, Z., & Loeb, A. 1997a, ApJ, 483, 21
16. Haiman, Z., & Loeb, A. 1997b, in Proceedings of *Science with the Next Generation Space Telescope*, eds. E. Smith & A. Koratkar
17. Loeb, A., & Haiman, Z. 1997, ApJ, 490, 571
18. Haiman, Z., & Loeb, A. 1998a, ApJ, 503, 505
19. Haiman, Z., & Loeb, A. 1998b, ApJ, in press, preprint astro-ph/9807070
20. Haiman, Z., Madau, P., & Loeb, A. 1999, ApJ, in press, preprint astro-ph/9805258
21. Abel, T., Bryan, G., & Norman, M. T. 1998, in proceedings of *Evolution of Large Scale Structure*, Garching, Germany, preprint astro-ph/9810215
22. Gnedin, N. Y., & Ostriker, J. P. 1997, ApJ, 486, 581
23. Bertschinger, E. 1985, ApJS, 58, 39
24. Saslaw, W.C., & Zipoy, D. 1967, Nature, 216, 976
25. Stecher, T. P., & Williams, D. A. 1967, ApJ, 149, L29
26. Haiman, Z., Rees, M. J. R., & Loeb, A. 1997, ApJ, 476, 458
27. Doroshkevich, A.G., Zel'dovich, Ya. B., & Novikov, I. D. 1967, Sov. Astr. A. J., 11, 233.
28. Couchman, H. M. P. & Rees, M. J. 1986, MNRAS, 221, 53
29. Songaila, A. 1997, ApJL, 490, 1
30. Scalo, J. M. 1986, Fund. Cosm. Phys., 11, 1
31. Renzini, A., & Voli, M. 1981, A&A, 94, 175
32. Norman, M. L. 1998, this proceedings
33. Kormendy, J., Bender, R., Magorrian, J., Tremaine, S., Gebhardt, K., Richstone, D., Dressler, A., Faber, S. M., Grillmair, C., & Lauer, T. R. 1997, ApJL, 482, 139
34. Magorrian, J., *et al.* 1998, A&A, 115, 2285
35. van der Marel, R. P. 1998, AJ, in press, astro-ph/9806365
36. Haehnelt, M. G., Natarajan, P., & Rees, M. J. 1998, preprint astro-ph/9712259

37. Eisenstein, D. J., & Loeb, A. 1995, ApJ, 443, 11
38. Loeb, A. 1997, in Proceedings of *Science with the Next Generation Space Telescope*, eds. E. Smith & A. Koratkar, pp. 73-86, astro-ph/9704290
39. see http://www.ngst.nasa.gov
40. Miralda-Escudé, J., & Rees, M. J. 1997, ApJ, 478, L57
41. Woods, E., & Loeb, A. 1998, ApJ, in press, preprint astro-ph/9803249
42. Madau, P., Ferguson, H. C., Dickinson, M. E., Giavalisco, M., Steidel, C. C., & Fruchter, A. 1996, MNRAS, 283, 1388
43. Baugh, C. M., Cole, S., Frenk, C. S. & Lacey, C. G. 1998, ApJ, 498, 504
44. Conti, A., Kennefick, J. D., Martini, P., & Osmer, P. S. 1999, AJ, in press, astro-ph/9808020
45. Thoul, A. A., & Weinberg, D. H. 1996, ApJ, 465, 608
46. Navarro, J. F., & Steinmetz, M. 1997, ApJ, 478, 13
47. Jarvis, R. M., & MacAlpine, G. M. 1998, AJ, in press, preprint astro-ph/9810491
48. Thompson, R., et al. 1998, astro-ph/9810285, see also this proceedings
49. Elvis, M., Wilkes, B. J., McDowell, J. C., Green, R. F., Bechtold, J., Willner, S. P., Oey, M. S., Polomski, E., & Cutri, R. 1994, ApJS, 95, 1
50. Kurucz, R., CD-ROM No. 13, ATLAS9 Stellar Atmosphere Programs (1993)
51. Schaller, G., Schaerer, D., Meynet, G., & Maeder, A. 1992, A&AS, 96, 269
52. Reimers, D., Köhler, S., Wisotzki, L., Groote, D., Rodriguez-Pascual, P., & Wamsteker, W. 1997, A&A, 326, 489
53. Miyaji, T., Ishisaki, Y., Ogasaka, Y., Ueda Y., Freyberg, M. J., Hasinger, G., & Tanaka, Y. 1998, A&A 334, L13
54. Fabian, A. C. & Barcons, X. 1992, ARA&A, 30, 429
55. Madau, P. 1996, ApJ, 441, 18
56. Gunn, J. E., & Peterson, B. A. 1965, ApJ, 142, 1633
57. Fardal, M. A., Giroux, M. L., & Shull, J. M. 1998, AJ, 115, 2206
58. Hu, W., & White, M. 1997, ApJ, 479, 568
59. Zaldarriaga, M., Spergel, D., & Seljak, U. 1997, ApJ, 488, 1
60. Mathis, J. S. 1990, ARA&A, 28, 37
61. Hauser, M. G. 1997, ApJ, in press, astro-ph/9806167
62. Fixsen, D. J., Cheng, E. S., Gales, J. M., Mather, J. C., Shafer, R. A., & Wright, E. L. 1996, ApJ, 473, 576

Formation of Galactic Bulges

Nickolay Y. Gnedin[1], Michael L. Norman[2], and
Jeremiah P. Ostriker[3]

[1] *Center for Astrophysics and Space Astronomy, University of Colorado, Boulder, CO 80309*
[2] *Laboratory for Computational Astrophysics, National Center for Supercomputing Applications,
University of Illinois at Urbana-Champaign, 405 North Matthews Avenue, Urbana, IL 61801*
[3] *Princeton University Observatory, Peyton Hall, Princeton, NJ 08544*

Abstract. We study formation of galactic bulges in a representative CDM-type cosmological model using hydrodynamic numerical simulations. We show that currently
favorable cosmological models predict that galactic bulges form at moderately high
redshifts, $z \sim 4$, with correct density profiles and sizes. On the contrary, cosmological models with more small scale power, like the PIB class of models, would predict
galactic bulges that are two orders of magnitude more dense than the observed ones.

INTRODUCTION

The fact that bulges of ordinary galaxies are very dense is often used as an
argument against the currently favorable CDM-type cosmological models. Really,
since the average density of the universe decreases with time, and the density of a
bound object is directly proportional to the density of the universe at the time when
the object is formed, dense galactic bulges should have formed at very high redshift.
The currently fashionable cosmological models normally do not have enough small
scale power to account for the formation of massive ($10^9 - 10^{10}$ solar masses in
baryons) at $z \gtrsim 10$. (Peebles 1997)

This can be illustrated by the following simple estimate: the average mass density
profile of the Galactic bulge within the range $0.1\,\text{kpc} < r < 3\,\text{kpc}$ can be well fitted
by the following simple formula

$$\rho_{\text{GB}}(r) \approx \frac{1}{\left[1 + (r/1.2\,\text{kpc})^{3/2}\right]^4} \frac{\text{M}_\odot}{\text{pc}^3} \tag{1}$$

assuming mass-to-light ration of 4. The characteristic number density of baryons
in the Galactic bulge within the sphere with the radius of $3\,\text{kpc}$ is thus

CP470, After the Dark Ages: When Galaxies were Young (the Universe at 2 < z < 5),
edited by Stephen S. Holt and Eric P. Smith

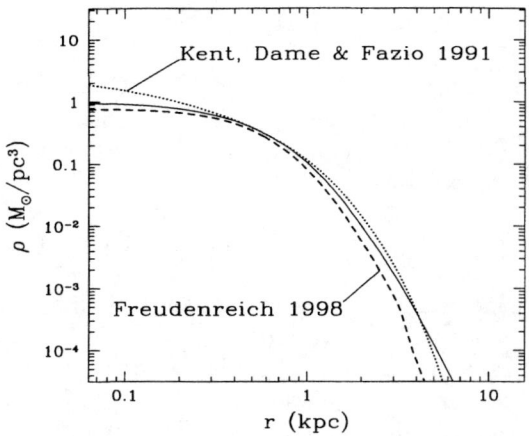

FIGURE 1. Angular averaged mass density profiles for the Galactic bulge from Kent, Dame, & Fazio (1991) (*bold dotted line*) and Freudenreich (1998) (*bold dashed line*). The mass-to-light ration of 4 is assumed in converting luminosity profiles to mass profiles. The thin solid line shows the simple fit from equation 1.

$$n_{\mathrm{GB}} \sim 2\,\mathrm{cm}^{-3}. \qquad (2)$$

In comparison, the average density of the cosmological virialized object formed at redshift z is only

$$n_{\mathrm{TH}} \sim 10^{-4}(1+z)^3\,\mathrm{cm}^{-3} \qquad (3)$$

for $\Omega_b h^2 = 0.04$, and we assume that the average density of the virialized object is 200 times the average density of the universe at the moment of formation. Comparing equations (2) and (3) we can deduce that the bulge of our Galaxy formed at $z_{\mathrm{GB}} \sim 30$. None of the currently acceptable CDM-type models can form a 10^{10} solar mass baryonic objects at $z = 30$ in the numbers even closely comparable to the observed number density of galaxies.

Does this argument imply that the CDM-type models are ruled out? We will try to show in this paper that the answer to this question is no. What this simple argument misses is the ability of baryons to cool and collapse to the densities exceeding that of the dark matter. We present a series of cosmological hydrodynamic simulations which include adequate physical modeling to properly account for formation of bulges of galaxies, and we show that typical bulges form in a realistic CDM-type model at $z \sim 5$ rather than at $z \sim 30$.

SIMULATIONS

We use the SLH cosmological hydrodynamic code (Gnedin 1995, 1996; Gnedin & Bertschinger 1996). Physical modeling included in the code is fully described in Gnedin & Ostriker (1997). We choose a CDM+Λ cosmological model with the following cosmological parameters:

$$\Omega_0 = 0.37, \quad \Omega_L = 0.63, \quad h = 0.70, \quad \Omega_b = 0.049,$$

which is close to the "concordance" model of Ostriker & Steinhardt (1995). We have performed one simulation with 128^3 baryonic resolution elements, the same number of dark matter particles, and a number of stellar particles were formed during the simulation. The simulation box size was fixed to $3h^{-1}$ Mpc, which resulted in the total mass resolution of $1.3 \times 10^6 h^{-1}\,M_\odot$. The spatial resolution was fixed at $1.5h^{-1}$ comoving kiloparsecs. Because of the small box size, this simulation cannot be continued to $z = 0$. Instead, we stopped the simulation at $z = 4$. Another simulation with eight times more resolution elements (256^3) and $6h^{-1}$ Mpc box was also performed. The large simulation thus had the same mass resolution as the small one, and the spatial resolution in the large simulation was fixed at $1.2h^{-1}$ comoving kiloparsecs. However, because the large simulation required a computational expense beyond what was available to us, it was terminated at $z = 9.5$. Thus, we used the large simulation to verify numerical convergence and estimate missing small scale power, but we will use the small (128^3) simulation as the source for scientific results.

By comparing the large and small simulation, we have found that the small simulation included most of the small scale power that was initially present in the baryonic component. Thus, our results are not significantly affected by the finite resolution in the initial conditions (k-space resolution), but they are, of course, subject to finite mass and spatial resolution.

RESULTS

Since we are concerned with the process of formation of galactic bulges, we will focus in this paper on properties of individual objects formed in our simulations. Specifically, we will focus on four most massive objects. Each of those objects contains more than ten thousand particles of each kind (i.e. the dark matter, gas, and stars), and thus they are fully resolved numerically.

Table 5 presents the general properties of the four objects: their total, baryonic, and stellar masses, as well as the ratio of the baryonic to the total mass, and the stellar to the baryonic mass.

A few observations can easily be made from the table. First, three objects contain more baryons than the cosmic average of $\Omega_b/\Omega_0 = 0.13$. Second, they are dominated by gas, as only about 0.2–0.3 of their baryonic mass is turned into stars.

TABLE 1. Four Most Massive Objects at $z = 4$

Object	Total mass ($M_\odot$)	Baryonic mass ($M_\odot$)	Stellar mass ($M_\odot$)	M_b/M_t	M_*/M_b
A	6.9×10^{10}	1.2×10^{10}	2.1×10^9	0.17	0.18
B	4.8×10^{10}	7.4×10^9	1.7×10^9	0.15	0.23
C	3.0×10^{10}	4.7×10^9	1.2×10^9	0.16	0.26
D	2.3×10^{10}	3.1×10^9	0.9×10^9	0.13	0.29

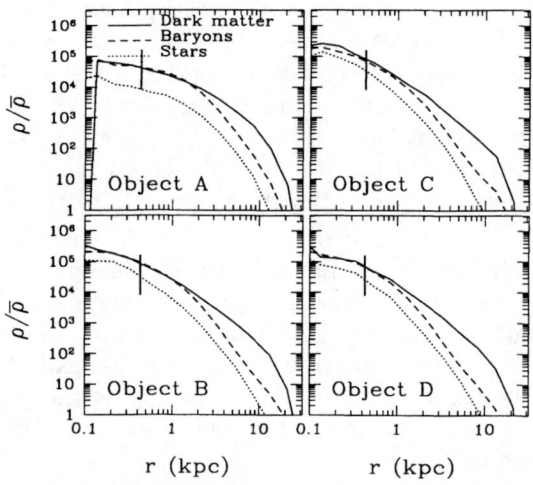

FIGURE 2. Density profiles for the dark matter (*solid lines*), total baryons (gas and stars, *dashed lines*), and stars (*dotted lines*) of four most massive objects at $z = 4$ as a function of radius in physical (not comoving) units. The bold vertical bar marks the spatial resolution of the simulation (430 pc).

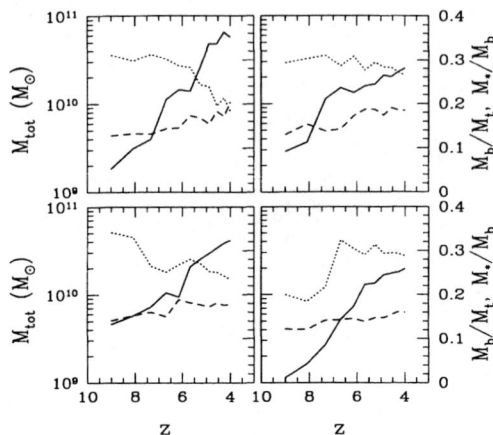

FIGURE 3. Evolution of the total mass (*solid lines*), baryonic fraction (*dashed lines*), and stellar fraction (per unit baryonic mass, *dotted lines*) of four most massive objects.

Density profiles of the four most massive objects are shown in Figure 2. The most massive object, object A, is less dense at the center than other three objects because it experienced a major merger shortly before $z = 4$ and has not fully relaxed yet. Object D has also experienced a major merger at $z \approx 6$, whereas objects B and C have been accreting matter quietly since $z \sim 10$. We point out here that in all four objects baryonic density at the center is similar to the dark matter density, i.e. baryons in all four objects are on the brink of becoming self-gravitating.

Evolution of average properties of these objects is shown in Figure 3. One can immediately see that all four objects are experiencing heavy merging at $z \sim 4 - 6$, increasing their mass by about an order of magnitude. The stellar fraction of object A decreased significantly at $z = 4 - 5$ because of accretion of a large quantity of fresh gas. Also noticeable is the increase in the total baryonic fraction in object A at $z \sim 4$. Since baryons are almost self-gravitating at the center of object A, their efficiently cool and collapse toward the center, leading to the increase in the baryonic fraction of the object.

Figure 4 shows the evolution of the central density for the dark matter, baryonic and stellar components of object A. The central density is defined as the average density within the sphere of two resolution lengths of our simulations ($\sim 900\,\mathrm{pc}$ at $z = 4$). One can see that the dark matter density does not change systematically with time (albeit fluctuating significantly), because the object is close to the virial equilibrium. On the contrary, the baryonic density increases with time because of efficient cooling at the center of the object. The recent merger at $z \sim 5$ triggered a considerable increase in the central density of gas, but this increase has not yet resulted in the burst of star formation. We expect, that if we continued the

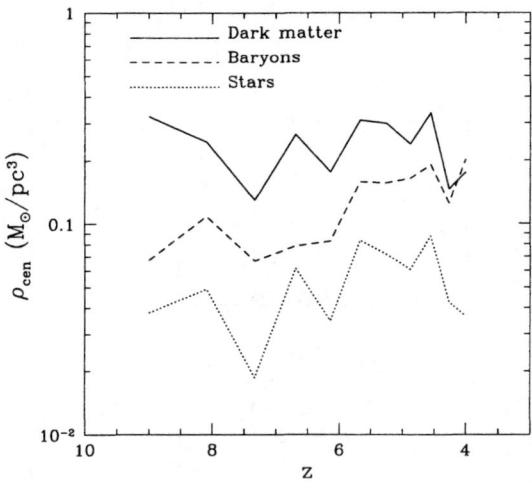

FIGURE 4. Evolution of the central dark matter density (*solid lines*), baryonic density (*dashed lines*), and stellar density (*dotted lines*) for object A.

simulation to lower redshift, object A would experience a burst of star formation at the center, which would transform most of the gas into stars on a rather short time-scale.

Thus, the fact that our objects are still 75% gas, should not mislead the reader. The star formation in those objects is very efficient, and most of this gas will be transformed into stars by $z \sim 2 - 3$.

We are now ready to address the major question of this paper: are those objects formed in the simulation resemble real galactic bulges? In order to answer this question, we show in Fig. 5 the surface density profiles for our four objects. Also, for the total baryonic profile we compute the exponential fits in the form:

$$\Sigma(R) = \Sigma_C e^{1 - R/R_C}. \tag{4}$$

Both, Kent, Dame & Fazio (1991) and Freudenreich (1998) models for the galactic bulge are well fitted by the exponential profile for the range $0.1\,\mathrm{kpc} < r < 3\,\mathrm{kpc}$ (with respective rms errors of 3 and 8 percent respectively).

We put together the parameters of the fits in Table 5. In addition, we list the parameters that describe the Galactic bulge from Kent, Dame, & Fazio (1991) and Freudenreich (1998) for comparison. As one can see, objects that we observe in the simulation are very similar to the bulge of Milky Way, provided, they convert most of their gas into stars.

Why did we then erroneously conclude in the Introduction that the CDM-type models predict too low density bulges? The answer to this puzzle is again illustrated in Fig. 5. The horizontal long-dashed line in that figure shows the central

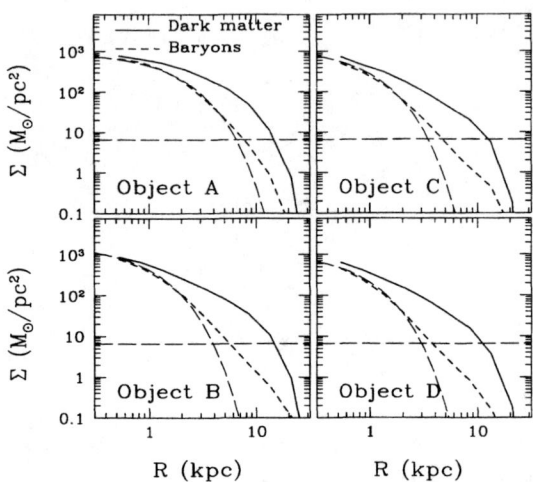

FIGURE 5. Surface mass density profiles for the dark matter (*solid lines*), total baryons (gas and stars, *dashed lines*), and stars (*dotted lines*) of four most massive objects at $z = 4$ as a function of radius in physical (not comoving) units. The profiles are terminated at the resolution limit of the simulation (430 pc). Tilted long-dashed lines show $r^{1/4}$ law for the baryonic profiles, and the horizontal long-dashed line show the surface density of a homogeneous top-hat sphere.

TABLE 2. Fit Parameters for Four Most Massive Objects at $z = 4$

Object	R_C (kpc)	Σ_C ($M_\odot/pc^2$)
A	1.30	350
B	0.70	640
C	0.66	450
D	0.58	430
Galactic bulge[a]	0.58	500
Galactic bulge[b]	0.47	430

[a] Kent, Dame, & Fazio 1991
[b] Freudenreich 1998

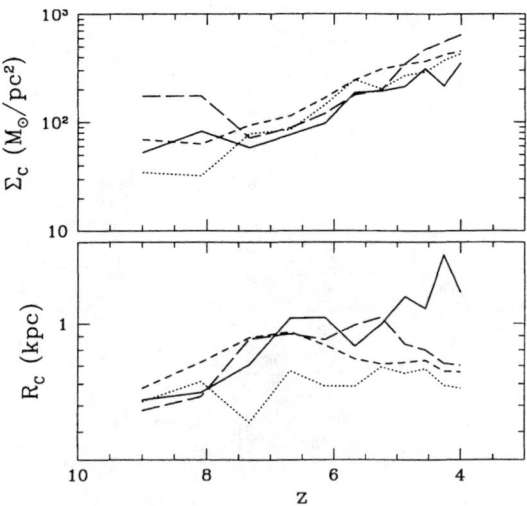

FIGURE 6. Evolution of the characteristic radii (*lower panel*) and densities (*upper panel*) for the four objects: A (*solid lines*), B (*long-dashed lines*), C (*short-dashed lines*), and D (*dotted lines*).

surface density for a homogeneous top-hat sphere, i.e. for a spherical object with the constant density of 200 times the average density of the universe and with a radius equal to the virial radius of object A (all four objects have quite similar virial radii). One can immediately see that the top-hat model underestimates the central density of an object by about two orders of magnitude!

Evolution of the two fit parameters, the characteristic radius and density, is shown in Fig. 6 for the four most massive objects in the simulation. The densities increase steadily as the gas continues to accrete and cool inside the dark matter halos, whereas characteristic radii stay approximately constant for all objects except the object A. This suggests that object A will most likely becomes an elliptical galaxy, a S0 galaxy, or a bulge of a Sa galaxy, rather than a bulge of a Milky Way type spiral.

Finally, if we want to demonstrate that we can form galactic bulges in a realistic cosmological simulation, we should address the question of the bulge shape. The bulge of our Galaxy is spherical, or, at the very least, slightly ellipsoidal. Is this shape also reproduced in the simulation? Figure 7 serves to answer this question. In it we show the axis ratios for the dark matter, gas, and stars for our four objects as a function of radius. One can see that in the central parts the gas and the stellar distributions are quite close to spherical. Shapes at larger radii, $r > 3\,\mathrm{kpc}$, shown shaded in Fig. 7, vary significantly among all objects, but at those large distances the gas is far from equilibrium, and the minute shape of its distribution has little relation to its final state. The fact that the objects we observe in the simulation

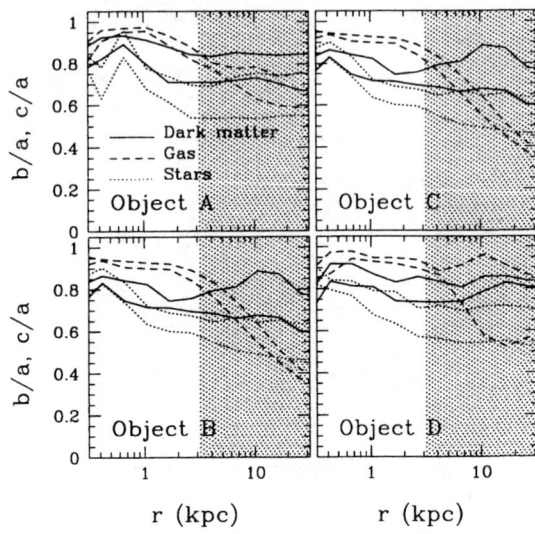

FIGURE 7. Axis ratios for the dark matter (*solid lines*), total baryons (gas and stars, *dashed lines*), and stars (*dotted lines*) of four most massive objects at $z = 4$ as a function of radius in physical (not comoving) units.

are more-or-less spherical, rather than disk-shaped, is due to the fact that at high redshift the slope of the linear power spectrum of the density fluctuations is close to -3. This results in a range of scales becoming nonlinear almost at the same time, which, in turn, leads to heavy merging among objects observed in the simulation. Thus, gaseous disks do not have enough time to form, and the shape of the objects remains quasi-spherical. Only at later times, when the rate of merging falls down, can a gaseous disk form inside an object.

CONCLUSIONS

We have showed that objects that form in a realistic cosmological simulation of a CDM-type cosmological model do look similar to bulges of normal galaxies. Our objects are still 75% gaseous at $z = 4$, but they form stars at a high rate, and when most of the gas in those objects will be converted into stars, they will look like slightly ellipsoidal stellar objects with the density profiles well fit by the exponential profile and with the parameters of the fit similar to the parameters of the Milky Way bulge.

We thus conclude that currently favorable CDM-type cosmological models have no difficulty in reproducing observed properties of galactic bulges. On the contrary, models that have galaxy formation at $z \sim 30$ (Peebles 1997) would form bulges that are two to three orders of magnitude denser than the observed ones. This conclusion is further boosted by the consideration that, due to the limited mass and spatial resolution of our simulation, we can only *underpredict* the densities and masses of cosmological objects. Taking into account, that we found four objects that are similar to the bulge of Milky Way in a comoving volume of only $27h^{-3}\,\mathrm{Mpc}^3$, whereas in the real universe there is only on average one L_* galaxy in this volume, and that we can only underestimate the abundances and central densities of objects due to the finite resolution of our simulations, we can conclude that the cosmological model under consideration likely has too much small scale power, and form galaxies *too early*, rather than too late.

This work was supported in part by the UC Berkeley grant 1-443839-07427. Simulations were performed on the NCSA Origin2000 supercomputer under the grant AST-970006N.

REFERENCES

1. Freudenreich, H. T. 1998, ApJ, 492, 495
2. Gnedin, N. Y. 1995, ApJS, 97, 231
3. Gnedin, N. Y. 1996, ApJ, 456, 1
4. Gnedin, N. Y., & Bertschinger, E. 1996, ApJ, 470, 115
5. Gnedin, N. Y., Ostriker, J. P. 1997, ApJ, 486, 581
6. Kent, S. M., Dame, T. M., & Fazio, G. 1991, ApJ, 378, 131
7. Ostriker, J. P., & Steinhardt, P. J. 1995, Nature, 377, 600
8. Peebles, P. J. E. 1997, ApJ, L1

From Cosmological Initial Conditions to Primordial Protostellar Cloud Cores

Michael L. Norman[1], Tom Abel[2], and Greg Bryan[3]

[1] *Astronomy Department, 1002 W. Green St., Urbana, IL 61801*
[2] *Max-Planck-Institut für Astrophysik, Karl-Schwarzschild Str. 1, 85746 Garching, Germany*
[3] *Physics Department, MIT, Cambridge, MA 02139*

Abstract. We describe results from a 3D adaptive mesh refinement simulation of *first structure formation* in a CDM-dominated universe. A spatial dynamic range of $>$ 250,000 is achieved, allowing us to see the formation of a collapsing primordial protostellar cloud core set within a cosmological framework. The core mass is $\sim 100\ M_\odot$, comparable to the Jeans mass at the densities and temperatures which prevail in the halo center once H_2 cooling has become inefficient.

INTRODUCTION

In a previous paper [3] we presented the first 3D self-consistent cosmological hydrodynamical simulations of *first structure formation* in a cold dark matter-dominated universe. These simulations included a careful treatment of the formation and destruction of H_2—the dominant coolant in low mass halos ($10^4 - 10^7 M_\odot$) which form at high redshifts ($z \sim 30 - 50$). Among the principal findings of that study were: (1) appreciable cooling only occurs in the cores of the high density spherical knots located at the intersection of filaments; (2) good agreement was found with the semi-analytic predictions [1], [11] concerning the minimum halo mass able to cool and collapse to higher densities, (3) only a small fraction (5-8%) of the baryons are able to cool promptly, implying that primordial Pop III star clusters may have very low mass $\sim 50 M_\odot$). Due to the limited spatial resolution of the those simulations (~ 1 kpc *comoving*), we were unable to study the collapse to stellar densities and address the nature of the first stars formed.

In this paper we present new, higher- resolution results using a powerful new numerical technique–adaptive mesh refinement (Norman & Bryan 1998)–which has shed some light on how the cooling gas fragments. Although we are not yet able to resolve individual protostars, we are able to resolve the collapsing protostellar cloud cores which must inevitably form them. We find the cores have typical

CP470, After the Dark Ages: When Galaxies were Young (the Universe at 2 < z < 5),
edited by Stephen S. Holt and Eric P. Smith

masses $\sim 100 M_\odot$, sizes $\sim 1pc$, and number densities $n \geq 10^5$ cm^{-3}–similar to dense molecular cloud cores in the Milky Way.

SIMULATIONS

The three dimensional adaptive mesh refinement calculations presented here use for the hydrodynamic portion an algorithm very similar to the one described by Berger and Collela (1989). The code utilizes an adaptive hierarchy of grid patches at various levels of resolution. Each rectangular grid patch covers some region of space in its parent grid needing higher resolution, and may itself become the parent grid to an even higher resolution child grid. The general implementation of AMR places no restriction on the number of grids at a given level of refinement, or the number of levels of refinement. Additionally the dark matter is followed with methods similar to the ones presented by Couchman (1991). Furthermore, the algorithm of Anninos et al. (1997) to solve the accurate time–dependent chemistry and cooling model for primordial gas of Abel et al. (1997). Detailed description of the code are given in Bryan & Norman (1997,1998), and Norman & Bryan (1998).

The simulations are initialized at redshift 100 with density perturbations of a sCDM model with $\Omega_B = 0.06$, $h = 0.5$, and $\sigma_8 = 0.7$. The abundances of the 9 chemical species (H, H$^+$, H$^-$, He, He$^+$, He^{++} , H$_2$, H$_2^+$, e$^-$) and the temperature are initialized as discussed in Anninos and Norman (1996). After a collapsing high–σ peaks has been identified in a low resolution simulation the simulations is reinitialized with multiple refinement levels on the Langrangian volume of the collapsing structure. The mass resolution in the initial conditions within this Langrangian volume are 0.53 $M_\odot$ in the gas and 8.96 $M_\odot$ for the dark matter component. The refinement criterium ensures the local Jeans length to be resolved by at least 4 grid zones as well as that no cell contains more than 4 times its initial mass of 0.53 $M_\odot$. We limit the refinement to 12 levels within a 64^3 top grid which translates to a maximum dynamical range of $64 \times 2^{12} = 262,144$. As we will show below the simulation is not resolution but physics limited.

RESULTS

Fig. 1 shows the redshift evolution of various quantities calculated for the highest density peak in the simulation. At z=35, a virialized minihalo forms with mass 5×10^3 $M_\odot$. The baryons are unable to cool as the H$_2$ fraction and central baryon density is quite low. As the halo mass increases due to cosmic infall, the central temperature climbs due to adiabatic heating, reaching $\sim 600K$ by z=25. A merger a z=25 doubles the mass of the central object, and shock heating temporarily raises the central temperature to $\sim 800K$. At these temperatures H$_2$ radiates quite efficiently, cooling the gas back down to $\sim 200K$ where H$_2$ cooling becomes inefficient. At $z < 19$, enough gas has collected in the core that it collapses due to

Jeans instability. The Jeans mass at T=200K and $\rho_b = 10^4 \ M_\odot \ \mathrm{cm}^{-3} \sim 300 \ M_\odot$, which is very close to the mass of the collapsing core.

Fig. 2 shows the structure of the core and halo at z=19.1 near the end of the simulation. We plot spherical averages about the baryonic peak. While although the collapsing core is nearly spherically symmetrical, the rest of the halo is not, so these curves must be interpreted with some caution. Nonetheless, one sees in panel a) that the central H_2 fraction has approached 10^{-3}–above the critical threshold discussed by Tegmark $et\,al.$ for cooling. Panel b) shows a comparison of local cooling, freefall, and dynamical times versus radius. For $5 \leq r \leq 60$, the cooling time is shorter than the dynamical time, implying the existence of a H_2 cooling flow, which is further borne out by the temperature profile in panel c). Ordinarily, such conditions lead to a cooling catastrophe. However, this does not occur here for two reasons: (1) the excitation temperature for the lowest rotovib transition in H_2 is about 150K; and (2) above $n \sim 10^4 \ \mathrm{cm}^{-3}$, LTE level populations are established which reduce cooling from quadratic to linear in the gas density. Due to the inefficient cooling of the core, once it becomes Jeans unstable, it collapses quasi-hydrostatically rather than in freefall. Radial mach numbers are found to vary between 0.1 and 0.5 for $0.1 \leq r(pc) \leq 5$.

We acknowledge support from NASA ATP grant NAG5-3923. Simulations were carried out on the Origin2000 computer at the NCSA.

REFERENCES

1. Abel, T. 1995, Thesis, University of Regensburg, Germany
2. Abel, T., Anninos, P., Zhang, Y., Norman, M. 1997a, NewA, 2, 181
3. Abel, T., Anninos, P., Norman, M., Zhang, Y. 1998a, ApJ, 508, 518
4. Anninos, P., Norman, M.L. 1996, ApJ, 460, 556
5. Anninos, P., Zhang, Y., Abel, T., Norman, M.L. 1997, NewA, 2, 209
6. Berger, M.J., Collela, P. 1989, J. Comp. Phys., 82, 64
7. Bryan, G.L., Norman, M.L. 1997, in $Computational Astrophysics$, eds. D.A. Clarke and M. Fall, ASP Conference #123
8. Bryan, G.L., Norman, M.L. 1998, $in preparation$
9. Couchman, H. 1991, ApJL, 368, L23
10. Norman, M.L., Bryan, G.L. 1998, in $Numerical Astrophysics 1998$, eds. S. Miyama & K. Tomisaka
11. Tegmark, M., Silk, J., Rees, M.J., Blanchard, A., Abel, T., Palla, F. 1997, ApJ, 474, 1

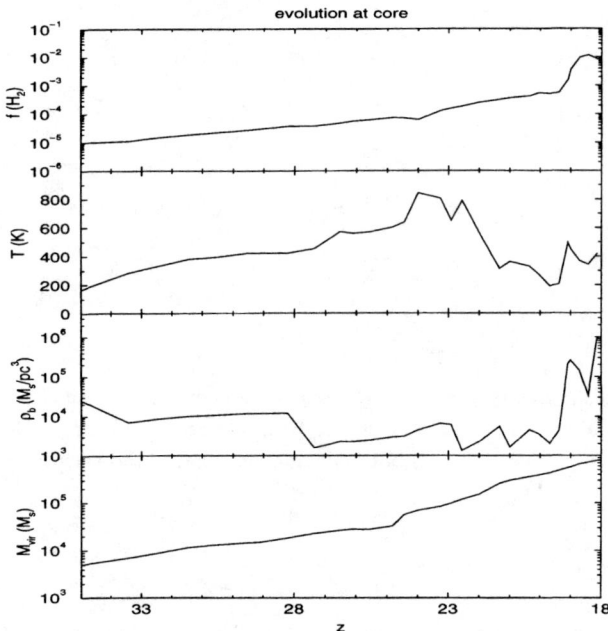

FIGURE 1. Redshift evolution of the densest peak in the simulation. Top to bottom: central H_2 fraction; central temperature; central baryon density; and virial mass. Two collapses are seen: collapse and virialization of the minihalo at $z=35$; and gravitational collapse of the halo core at $z=19$.

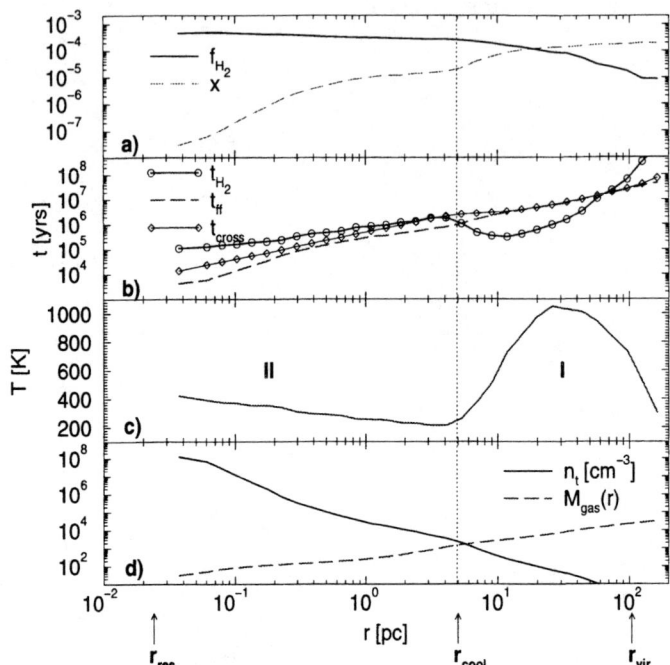

FIGURE 2. Spherically averaged radial profiles of various quantities around the baryonic peak at z=19.1. a) H_2 fraction and ionization fraction; b) a comparison the H_2 cooling time and the freefall time; c) gas temperature; and d) baryonic number density and enclosed mass.

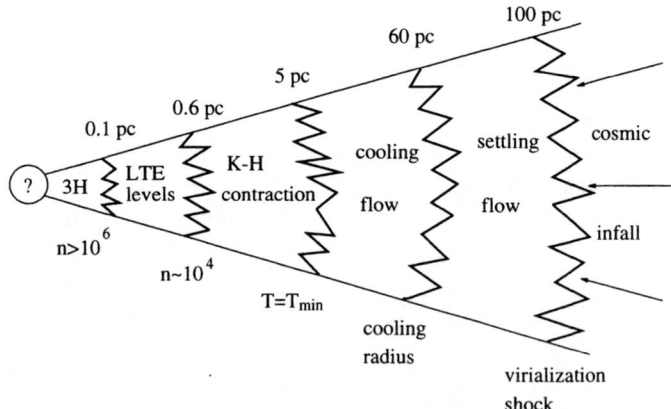

FIGURE 3. A cartoon, in log radius, showing scales and processes in the primordial halo and protostellar cloud core. Since the cartoon is based on spherical averages in what is non-spherically symmetric, transitions (shown as jagged lines) are approximate.

Models for High–Redshift Lyα Emitters

Z. Haiman* and M. Spaans[†]

*Astrophysics Theory Group, Fermi National Accelerator Laboratory, Batavia, IL 60510
[†]Harvard Smithsonian Center for Astrophysics, 60 Garden Street, Cambridge, MA 02138

Abstract. We present models for dusty high–redshift Lyα emitting galaxies by combining the Press–Schechter formalism with a treatment of inhomogeneous dust distribution inside galaxies. These models reproduce the surface density of emitters inferred from recent observations, and also agree with previous non–detections. Although a detailed determination of the individual model parameters is precluded by uncertainties, we find that (i) the dust content of primordial galaxies builds up in no more than $\sim 5 \times 10^8$ yr, (ii) the galactic HII regions are inhomogeneous with a cloud covering factor of order unity, and (iii) the overall star formation efficiency is at least $\sim 5\%$. Future observations should be able to detect Lyα galaxies up to redshifts of $z \sim 8$. If the universe is reionized at $z_r \lesssim 8$, the corresponding decline in the number of Lyα emitters at $z \gtrsim z_r$ could prove to be a useful probe of the reionization epoch.

INTRODUCTION

Until a decade ago, the search for high–redshift Lyα galaxies had enjoyed no compelling successes [1]. With the improved sensitivity on large area telescopes in the last couple of years, these young galaxies are finally being detected [2,3]. Clearly, this population is of great interest to the field of galaxy formation and the early evolution of the universe. Fundamental questions we need to understand are why the earlier surveys have been unsuccessful, how many objects are still expected to be found, and what physical conditions pertain in these early systems so that the Lyα radiation may escape.

In a galactic setting, the emerging luminosity of the Lyα line is strongly modulated by the amount and spatial distribution of stellar dust [4]. Observations have firmly established the strong decrease in Lyα equivalent width with increasing oxygen abundance [5]. Indeed, even a modest amount of dust (observationally traced by oxygen) inside a homogeneous HII region is sufficient to attenuate all of the produced Lyα radiation, because Ly alpha photons are resonantly scattered, and dust absorption significantly increases the effective line optical depths [6]. However, this situation can be alleviated if the medium is inhomogeneous [4]. In a multi–phase

CP470, After the Dark Ages: When Galaxies were Young (the Universe at 2 < z < 5),
edited by Stephen S. Holt and Eric P. Smith

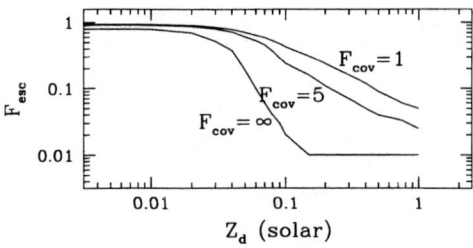

FIGURE 1. *The escape fraction F_{esc} of Lyα photons from an inhomogeneous medium, as a function of dust content. The three curves correspond to three different values of the covering factor of opaque clumps, $F_{cov} = 1, 5,$ and ∞.*

medium where dust resides in opaque clumps, the photons do not penetrate the clumps and spend most of their time in the interclump medium, where the opacity is very small. The result is that the escape of Lyα radiation is significantly enhanced compared to the case of resonant scattering in a homogeneous medium. Alternatively, large enough velocity gradients can provide a second way of efficient escape of Lyα radiation, as suggested by recent observations of 4 nearby galaxies [7].

In a cosmological context, the formation of dust requires the presence of metals. Since the enrichment of the gas with metals is expected to build up only gradually during the star formation history of the universe [8], at high enough redshifts the enrichment level will be low. The correspondingly small dust content of the earliest galaxies would facilitate the escape of their Lyα radiation. Such an interplay between the dust enrichment of the IGM and the galactic environment could be the mechanism that leads to the observed surface density of high–z Lyα emitters [9].

LYα EMISSION FROM INHOMOGENEOUS CLOUDS

The radiative transfer problem which needs to be solved for the Lyα line is well studied [10]. We have modeled individual galaxies with a range of masses for the ionizing stars, dust content, and inhomogeneity, using a numerical Monte Carlo approach [15]. We assumed a Scalo IMF for the spectral types of the central stars in the HII regions, ranging from O5 to B1, with the stars distributed in a statistically homogeneous manner inside a percolating multi–phase medium. We adopted the formalism of Neufeld [4] for a multi–phase medium, and assumed a typical line width of $\Delta v \sim 8$ km/s. We parameterized the multi–phase medium by opaque clumps embedded in an inter–clump medium of negligible opacity. The clump covering factor F_{cov}, i.e. the average number of clumps along a line of sight, fixes the degree of inhomogeneity. The escape fractions F_{esc} were computed on a grid of models with dust contents between $Z_d = 10^{-2} - 1$ solar, and covering factors $F_{cov} = 1, 5, \infty$. Figure 1 shows the escape fraction, and demonstrates its sensitivity to the covering factor. The inhomogeneous percolating slabs are much more transparent than the homogeneous ones; the difference around $Z_d \sim 10^{-1}$ solar is over an order of magnitude.

COSMOLOGICAL ABUNDANCE OF LYα EMITTERS

In order to model the cosmological abundance of high–redshift Lyα emitters, we assumed that the formation of dark matter halos follows the Press–Schechter [12] theory, and every halo with a virial temperature larger than 10^4 K forms a galaxy that goes through a Lyα emitting phase [13]. A fraction $\epsilon_\star$ of the gas was turned into stars, with a constant star formation rate (SFR) over a period of $t_\star$ years. The SFR was related to the intrinsic Lyα luminosity [14], assuming case B recombination. We adopted the simplest assumption, i.e. that $\epsilon_\star$ and $t_\star$ both have the same constant values in each halo. The amount of dust produced and retained in each galaxy, $Z_{d,ISM}$ then increases linearly for $t_\star$ years, after which it reaches the final value of $Z_{d,ISM}(t_\star) = 0.3$ solar. Similarly, we assume that each galaxy deposits dust into the surrounding IGM at a constant rate for $t_\star$ years, and causes an enrichment of the *surrounding regions* within the intergalactic medium to $Z_{d,IGM}(z)$. We normalize $Z_{d,IGM}$ to the cluster metallicity of ~ 0.3 solar at redshift $z \sim 1$. Note that our $Z_{d,IGM}$ is the average metallicity within the polluted regions, and not the universal average dust fraction of the IGM. Given a cosmology, the five parameters $t_\star$, F_{cov}, $\epsilon_\star$, $Z_{d,IGM}(z = 1)$, and $Z_{d,ISM}(t_\star)$ uniquely determine the number density of Lyα emitters at any flux and redshift. Figure 2 shows the resulting surface density of emitters, as a function of redshift, in a flat ΛCDM model with a tilted power spectrum $(\Omega_0, \Omega_\Lambda, \Omega_b, h, \sigma_{8h^{-1}}, n) = (0.35, 0.65, 0.04, 0.65, 0.87, 0.96)$. We also indicate the observed surface density of emitters [2] at the two redshifts $z = 3.4$ and $z = 4.5$.

Figure 2 has several interesting features. It demonstrates that the surface abundance strongly depends on the time over which the dust is produced (t_{star}), and on the ambient inhomogeneity of the HII regions that surround the ionizing OB stars (F_{cov}). Our models indicate that the dust content builds up in $\lesssim 5 \times 10^8$ yr, the galactic HII regions are inhomogeneous with a cloud covering factor of order unity, and the overall star formation efficiency is at least $\sim 5\%$. These numbers should be predicted by more complete and detailed models of galactic evolution, and will be useful discriminators between such models. The surface density of emitters is also a steep function of the detection threshold, a feature that makes our model consistent both with recent detections [2] and earlier upper limits [1]. Our models predict that the surface density changes relatively slowly with redshift to $z \lesssim 8$, and Lyα galaxies will be detectable, around the present flux threshold, up to redshifts as high as $\sim$8. Recent detections of three new Lyα emitters at $z \sim 5.7$ support this conclusion [15]. However, if the universe is reionized at $z_{reion} \lesssim 8$, then the damping wing of Lyα absorption from the neutral IGM would severely damp the Lyα emission line [16]. This would render Lyα emission from $z \gtrsim z_{reion}$ undetectable, and cause a decline in the number of observed emitters beyond z_{reion}. In

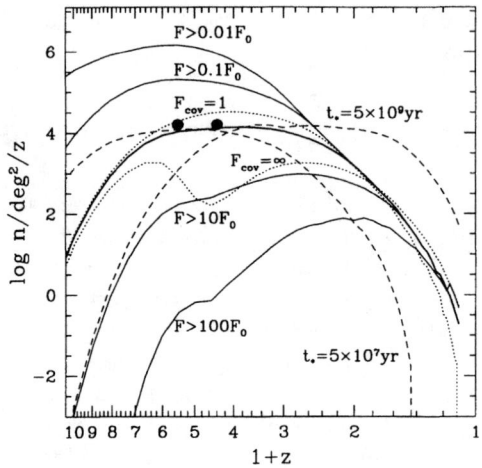

FIGURE 2. *The surface density of Lyα emitters in our standard model (solid lines) with fluxes above different values of the detection threshold. The two data points are taken from [2]. For the fixed threshold $F_0 = 1.5 \times 10^{-17}$ erg cm^{-1} s^{-1}, the dashed lines show the surface density when the star-formation rate is increased or decreased by a factor of 10. Similarly, the dotted lines show the surface density when the covering factor is changed to $F_{cov} = 1$, or ∞.*

addition, the neutral IGM would likely imprint a characteristic asymmetry on the emission line profiles of emitters located near the reionization epoch. Provided this asymmetry is measured, the disappearance of the Lyα emitter population could be useful diagnostic signature of reionization.

ZH was supported at Fermilab by the DOE and the NASA grant NAG 5-7092.

REFERENCES

1. Thompson, D., Djorgovski, S., & Trauger, J. 1995, AJ, 110, 963
2. Hu, E. M., Cowie, L. L., & McMahon, R. G. 1998, ApJ, 502, 99
3. Dey, A., Spinrad, H., Stern, D., Graham, J.R., & Chaffee, F.H. 1998, ApJ, 498, L93
4. Neufeld, D. A. 1991, ApJ, 370, L85
5. Terlevich, E., Diaz,A.I., Terlevich, R., & Garcia Vargas, M.L. 1993, MNRAS, 260, 3
6. Charlot, S., & Fall, S. M. 1993, ApJ, 415, 580
7. Kunth, D., Mas-Hesse, J. M., Terlevich, E., Terlevich, R., Lequeux, J., Fall, S. M. 1998, A&A, 334, 11
8. Haiman, Z., & Loeb, A. 1997, ApJ, 483, 21
9. Haiman, Z., & Spaans, M. 1998, ApJ, submitted, astro-ph/9809223
10. Neufeld, D. A. 1990, ApJ, 350, 216
11. Spaans, M. 1996, A&A, 307, 271
12. Press, W. H., & Schechter, P. L. 1974, ApJ, 181, 425

13. Haiman, Z., Rees, M. J., & Loeb, A. 1997, ApJ, 476, 458
14. Kennicutt, R. C., Jr. 1983, ApJ, 272, 54
15. Cowie, L. L., in these proceedings
16. Miralda-Escudé, J. 1998, ApJ, 501, 15

Do Small-Scale Dark Matter Fluctuations Govern the Fragmentation of Primordial Gas?

Volker Bromm, Paolo Coppi, and Richard Larson

Department of Astronomy, Yale University, New Haven, CT 06520-8101

Abstract. In order to constrain the initial mass function of the first generation of stars (Population III), one has to study the fragmentation properties of primordial gas. We present results from 3D simulations, based on Smoothed Particle Hydrodynamics, which explore the idea that small-scale fluctuations in the (cold) dark matter recreate a filamentary and clumpy structure in the gas component on scales smaller than the initial Jeans mass, where all primordial fluctuations would have been wiped out.

INTRODUCTION

In order to ascertain the influence of the very first generation of stars (Population III) on cosmology, one has to address the problem of how primordial (pure H/He) gas collapses and fragments. Whether the outcome of this process is a super-massive object or a cluster of lower-mass stars, is determined by the Population III initial mass function (IMF), which might differ substantially from the present-day one. The fragmentation properties depend crucially on the adopted initial conditions for the cloud collapse. In principle, these are given by the underlying model of cosmic structure formation. It is straightforward to specify the global properties of the gas cloud in terms of its average temperature, density, and chemical abundances [1].

Much less well-defined are the density and velocity fields inside the cloud. The nature of these initial perturbations could govern the ensuing fragmentation of the gas. There is no direct connection, however, between them and the fundamental physics which determines the primordial power spectrum of density perturbations, since primordial fluctuations in the baryon component below the Jeans length have been wiped out by pressure forces, as described by the small-scale cutoff in the baryon power spectrum:

$$|\delta_{\mathrm{B}}(k)|^2 = \frac{|\delta_{\mathrm{DM}}(k)|^2}{(1 + \gamma k^2)^2} \quad , \tag{1}$$

CP470, After the Dark Ages: When Galaxies were Young (the Universe at 2 < z < 5), edited by Stephen S. Holt and Eric P. Smith

where $|\delta_{\text{B,DM}}(k)|^2$ are the power spectra in the baryon and dark matter (DM) components, respectively, k is the comoving wavenumber, and $\gamma^{\frac{1}{2}} \simeq 1 h^{-1}\text{kpc}$ is the comoving Jeans length (cf. [2]).

In this paper, we investigate whether the small-scale fluctuations in the (cold) dark matter component, which have not been erased, do govern the fragmentation of the baryons, which might start with a completely smooth distribution. At first, the baryons would not be able to follow the growing DM clumping, due to the opposing effect of pressure forces. In the course of a nearly isothermal collapse, however, the Jeans mass decreases as $M_J \propto \rho^{-\frac{1}{2}}$, allowing the baryons to fall into the resulting DM condensations.

In the following, we present first results on how the baryons evolve in this scenario.

SIMULATIONS

For our 3D hydrodynamics/dark matter simulations, we use a variant of TREESPH [3], into which we have incorporated the relevant chemistry of the formation and destruction of molecular hydrogen, which is the main coolant below 10^4 K in the absence of metals. We use an improved version of the H_2 cooling function which takes into account quantum effects at low temperatures ($T < 600$ K) [4]. In order to follow the evolution well into the regime of highly developed clumping, we have devised an algorithm to merge high-density particles (corresponding to excessively small timesteps) into more massive ones, which enables us to follow the evolution beyond the point where otherwise the Courant-condition would force the calculation to a halt. The merging mechanism allows the high-density particles to continually accrete nearby particles, thereby modeling the physics of accretion and merging in an approximate way.

Our starting model consists of a spherical configuration with a total mass of $M = 10^6$ M$_\odot$, a radius of $R = 80$ pc, and a spin parameter of $\lambda = 0.05$. In CDM-like scenarios, these are typical values for a 3σ peak, virializing at $z \simeq 30$ [1]. On top of the originally homogeneous DM mass distribution, we imprint fluctuations by assigning initial velocities to the DM particles, as prescribed by the Zel'dovich approximation [5]. The Zel'dovich velocity field is calculated with a power spectrum $|\delta_{\text{DM}}(k)|^2 = Ak^{-2.9}$, corresponding to the small-scale end of the standard CDM spectrum. The amplitude A is chosen to match $\sigma_{M=10^6 M_\odot} = 0.5$ at $z \simeq 30$, appropriate for a collapsing 3σ peak. Embedded in the DM halo is a homogeneous gas cloud with $\Omega_B = 0.05$, an initially isothermal $T_i = 1000$ K, an H_2 fraction of 10^{-3}, and a free electron abundance of 10^{-4} (cf. [1]). With the exception of the always present SPH shot-noise ($\propto N^{-\frac{1}{2}}$), there are no density fluctuations in the baryonic component.

We have performed the simulations with $N = 65536$ particles in each component, as well as a comparison calculation at low resolution ($N = 8192$). Fig. 1 shows the DM and baryon distribution after one free-fall time. The collapsing dark matter has

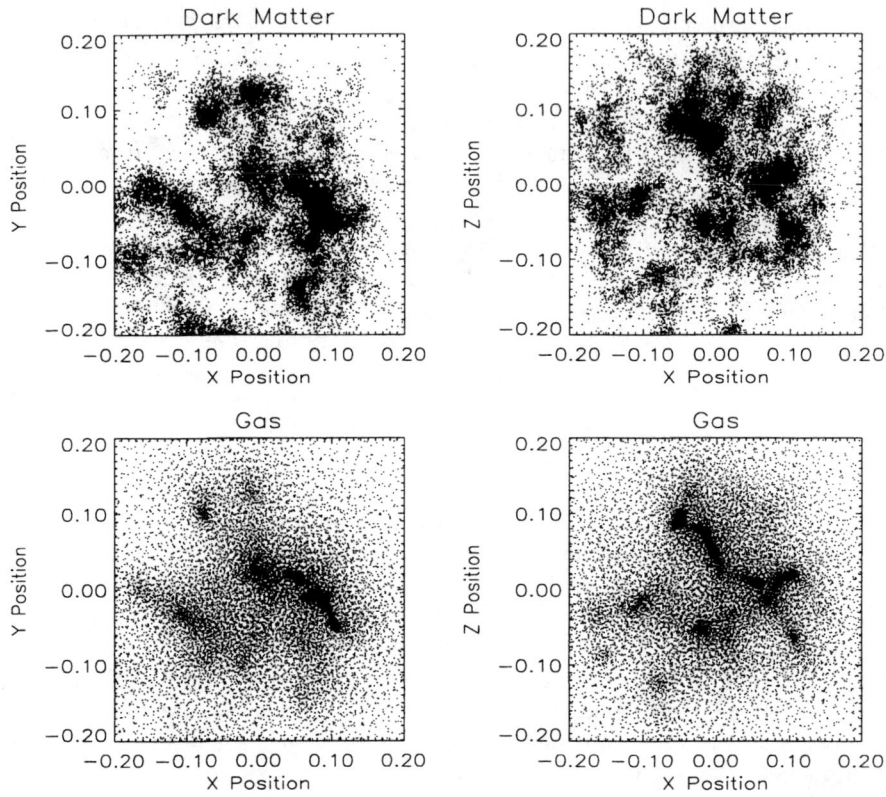

FIGURE 1. *Left column:* Face-on view. *Right column:* Edge-on view. Dark matter and baryon distribution after one free-fall time. $N_{\mathrm{SPH}} = N_{\mathrm{DM}} = 65536$. Shown is a blow-up of the central region, where a dimensionless length of 0.1 corresponds to 8pc. The baryons start to fall into the DM troughs.

been organized into a pronounced structure of filaments and clumps, in response to the initial Zel'dovich velocity field. The cooling baryons have begun to condense into the DM troughs, closely following the DM morphology. Consequently, gas fragmentation is induced before the DM fluctuations are washed out by the process of violent relaxation, which will eventually lead to a smooth (roughly isothermal) mass distribution.

In Fig. 2, we present the distribution of gas at a slightly later time, $t = 1.2t_{\mathrm{ff}}$, for the low-resolution run. Here, most of the gas ($\sim 60\%$) resides in high-density clumps, which are a result of the merging procedure. The resulting mass spectrum of these merged clumps follows a power-law with roughly the Salpeter slope. This result should be taken *cum grano salis*, and its robustness has to be tested in runs with higher resolution and with different initial conditions.

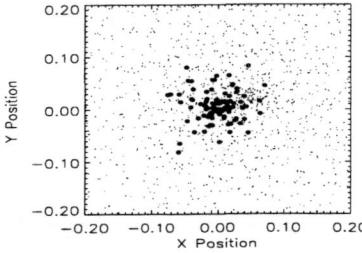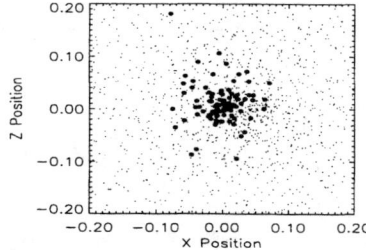

FIGURE 2. Gas distribution after $t = 1.2t_{ff}$. $N_{SPH} = N_{DM} = 8192$. Same units as in Fig. 1. Most of the gas has been incorporated into high-density clumps (*heavy dots*).

OUTLOOK

Although idealized, the appeal of the above scenario lies in the promise of being able to specify the initial conditions of the primordial star formation problem in a non-arbitrary way, directly connected to the underlying model of cosmic structure formation. But there remain many caveats which have to be explored. Does it make sense to treat a high-σ peak in isolation? Is the initial baryon distribution really smooth, or instead structured through complex processes which are difficult to specify without simulating a much larger region of the Universe?

We hope to gain a better understanding of these issues by comparing our results with those from large-scale cosmological simulations. Independent of the particular model proposed above, it makes sense to study the fragmentation of primordial gas in a broad range of circumstances (e.g., using various spectral indices and amplitudes for both the baryon and DM components). Currently, we are undertaking such a comprehensive study, on which we will report elsewhere.

Support from the NASA ATP grant NAG5-7074 is gratefully acknowledged.

REFERENCES

1. Tegmark, M., Silk, J., Rees, M.J., Blanchard, A., Abel, T. & Palla, F. 1997, ApJ, 474, 1
2. Peebles, P.J.E. 1993, Principles of Physical Cosmology (Princeton: Princeton University Press), p. 636
3. Hernquist, L. & Katz, N. 1989, ApJS, 70, 419
4. Galli, D. & Palla, F. 1998, A&A, 335, 403
5. Zel'dovich, Y.B. 1970, A&A, 5, 84

MOND in the Early Universe

Stacy McGaugh

Department of Astronomy, University of Maryland, College Park, MD 20742

Abstract. I explore some consequences of Milgrom's modified dynamics for cosmology. There appear to be two promising tests for distinguishing MOND from CDM: (1) the rate of growth of structure and (2) the baryon fraction. These should be testable with observations of clusters at high redshift and the microwave background, respectively.

STANDARD COSMOLOGY

The standard hot big bang cosmology has many successes; too many to list here. The amount of data constraining cosmic parameters has increased rapidly, until only a small region of parameter space remains viable. This has led to talk of a 'concordant' cosmology with $\Omega_{\mathcal{M}} \approx 0.3$ and $\Omega_\Lambda \approx 0.7$ [1].

This is a rather strange place to end up. The data do not *favor* these parameters so much as they *disfavor* other combinations more. A skeptic might suspect that concordance is merely the corner we've painted ourselves into prior to the final brush stroke.

This is not an idle concern, as there remains one major outstanding problem: dark matter. Something like 90% of the universe is supposedly made of stuff we can not see. There are, to my mind, two ironclad lines of reasoning that *require* the dark matter to be nonbaryonic, cold dark matter (CDM) like WIMPs or axions.

1. $\Omega_{\mathcal{M}} \gg \Omega_b$.

2. Structure does not have time to grow from a smooth microwave background to the rich structure observed today without a mass component whose perturbations can grow early without leaving an imprint on the CMBR.

BUT we have yet to detect WIMPs or axions. Their existence remains an unproven, if well motivated, assumption.

CP470, After the Dark Ages: When Galaxies were Young (the Universe at 2 < z < 5), edited by Stephen S. Holt and Eric P. Smith

IS THERE ANY DARK MATTER?

It is often stated that the evidence for dark matter is overwhelming. This is not quite correct: the evidence for *mass discrepancies* is overwhelming. These might be attributed to *either* dark matter *or* a modification of gravity.

Rotation curves played a key role in establishing the mass discrepancy problem, and remain the best illustration thereof. There are many fine-tuning problems in using dark matter to explain these data. I had hoped that the resolution of these problems would become clear with the acquisition of new data for low surface brightness galaxies. Instead, the problems have become much worse [2].

These recent data are a particular problem for CDM models, which simply do not fit [3]. Tweaking the cosmic parameters can reduce but not eliminate the problems. No model in the concordant range can fit the data unless one invokes some *deus ex machina* to make it so.

There is one theory which not only fits the recent observations, but predicted them [4]. This is the modified dynamics (MOND) hypothesized by Milgrom [5]. The basic idea here is that instead of dark matter, the force law is modified on a small acceleration scale, $a_0 \approx 1.2 \, \text{Å} \, \text{s}^{-2}$. For $a \gg a_0$ everything is normal, but for $a \ll a_0$ the effective force law is $a = \sqrt{a_N a_0}$, where a_N is the usual Newtonian acceleration.

This hypothesis might seem radical, but it has enormous success in rectifying the mass discrepancy. It works exquisitely well in rotating disks where there are few assumptions [6,56,8,9]. MOND also seems to work in other places, like dwarf Spheroidals [4,10–12], galaxy groups [13], and filaments [14]. The only place in which it does not appear to completely remedy the mass discrepancy is in the cores of rich clusters of galaxies [15], a very limited missing mass problem.

It is a real possibility is that MOND is correct, and CDM does not exist. Let us examine the cosmological consequences of this.

SIMPLE MOND COSMOLOGY

There exists no complete, relativistic theory encompassing MOND, so a strictly proper cosmology can not be derived. However, it is possible to obtain a number of heuristic results in the spirit of MOND. The simplest approach is to assume that a_0 does not vary with cosmic time. This need not be the case [16,17], but makes a good starting point. I do not have space to derive anything here, and refer the reader to detailed published work [16–18].

Making this simple assumption, the first thing we encounter is that it is not trivial to derive the expansion history of the universe in MOND [18,19]. This might seem unappealing, but does have advantages. For example, a simple MOND universe will eventually recollapse irrespective of the value of $\Omega_\mathcal{M}$. There is no special value of $\Omega_\mathcal{M}$, so no flatness problem.

Conventional estimates of $\Omega_{\mathcal{M}}$ are overly large in MOND. Instead of $0.2 < \Omega_{\mathcal{M}} < 0.4$, MOND gives $0.01 < \Omega_{\mathcal{M}} < 0.04$. So a MOND universe is very low density, consistent with being composed purely of baryons in the amount required by big bang nucleosynthesis.

This makes some sense. Accelerations in the early universe are too high for MOND to matter. This persists through nucleosynthesis and recombination, so everything is normal then and all the usual results are retained. MOND does not appear to contradict any empirically established cosmological constraint.

The universe as a whole transitions into the MOND regime ($cH_0 \sim a_0$) around $z \sim 3$, depending on $\Omega_{\mathcal{M}}$. Sub-horizon scale bubbles could begin to make this phase transition earlier, providing seeds for the growth of structure and setting the mass scale for galaxies [18]. Nothing can happen until the radiation releases its grip on the baryons ($z \sim 200$), by which time the typical acceleration is quite small. As a result, things subsequently behave as if there were a **lot** of dark matter: perturbations grow fast. This provides a mechanism by which structure grows from a smooth state to a very clumpy one rapidly, without CDM.

Now recall the two ironclad reasons why we must have CDM. In the case of MOND

1. $\Omega_{\mathcal{M}} = \Omega_b \approx 0.02$

2. There is no problem growing structure rapidly from a smooth CMBR to the rich amount seen at $z = 0$ with baryons alone.

PREDICTIONS

The simple MOND scenario makes two predictions which distinguish it from standard CDM models.

1. Structure grows rapidly and to large scales.

2. The universe is made of baryons.

There are indications that at least some galaxies form early, and are already clustered at $z \approx 3$ [20]. At low redshifts, we are continually surprised by the size of the largest cosmic structures. It makes no sense in the conventional context that fractal analyses should work as well as they do [21]. A MOND universe need not be precisely fractal, but if analyzed in this way it naturally produces the observed dimensionality [18]. So there are already a number of hints of MOND-induced behavior in cosmic data.

A strong test may occur for rich clusters. These are rare at $z > 1$ in any CDM cosmology [22]. In the simple MOND universe, clusters form at $z \approx 3$ [18]. Upcoming X-ray missions should be able to detect these [23].

The rapid growth of perturbations in MOND overcomes the usual objections to purely baryonic cosmologies. The baryon fraction makes a tremendous difference to the bumps and wiggles in the CMBR power spectrum (Figure 1) [24]. CDM

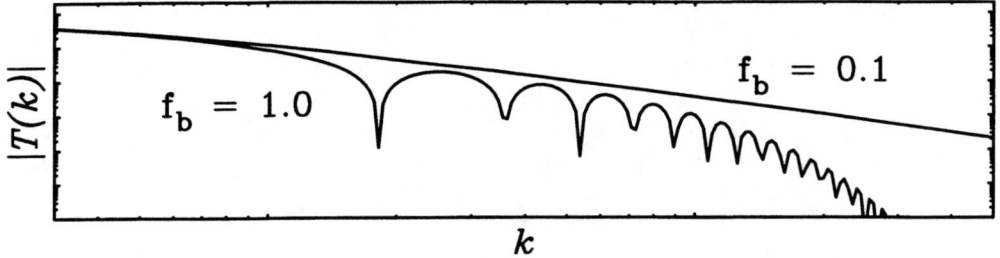

FIGURE 1. Post-recombination transfer functions for low and high baryon fraction models.

smooths out the acoustic oscillations due to baryons in a way which can not happen if $f_b = 1$.

This should leave some distinctive feature in the CMBR that can be measured by upcoming missions like MAP [25]. Unfortunately, it is easy only to predict the spectrum as it emerges shortly after recombination. Since the growth of structure is rapid and nonlinear in MOND, there might be a strong integrated Sachs-Wolfe effect. I would expect this to erase any hint of the oscillations in the $z = 0$ galaxy distribution, but not necessarily in the CMBR. The initial spectrum in the CMBR is sufficiently different in the CDM and MOND cases that there is a good prospect of distinguishing between the two.

REFERENCES

1. Ostriker, J.P., & Steinhardt, P.J. 1995, Nature, 377, 600
2. McGaugh, S.S., & de Blok, W.J.G. 1998, ApJ, 499, 41
3. McGaugh, S.S. 1999, in *Galaxy Dynamics*, eds. Merritt, D., & Sellwood, J., in press
4. McGaugh, S.S., & de Blok, W.J.G. 1998, ApJ, 499, 66
5. Milgrom, M. 1983, ApJ, 270, 371
6. Begeman, K.G., Broeils, A.H. & Sanders, R.H. 1991, MNRAS, 249, 523
7. Sanders, R.H. 1996, ApJ, 473, 117
8. Sanders, R.H., & Verheijen, M.A.W. 1998, ApJ, 503, 97
9. de Blok, W.J.G., & McGaugh, S.S. 1998, ApJ, 508, 132
10. Gerhard, O., & Spergel, D.N. 1992, ApJ, 397, 38
11. Milgrom, M. 1995, ApJ, 455, 439
12. Mateo, M., Olszewski, E.W., Vogt, S.S., Keane, M.J. 1998, AJ, 116, 2315
13. Milgrom, M. 1998, ApJ, 496, 89
14. Milgrom, M. 1997, ApJ, 476, 22
15. Sanders, R.H. 1999, ApJ, in press
16. Milgrom, M. 1989, Comments on Astrophysics, 13, 215
17. Milgrom, M. 1994, Annals of Physics, 229, 384
18. Sanders, R.H. 1998, MNRAS, 296, 1009
19. Felten, J.E. 1984, ApJ, 286, 3

20. Giavalisco, M., Steidel, C.C., Adelberger, K.L., Dickinson, M.E., Pettini, M., Kellogg, M. 1998, ApJ, 503, 543
21. Sylos Labini, F., Montuori, M., & Pietronero, L. 1998, Physics Report, 293, 61
22. Eke, V.R., Cole, S., & Frenk, C.S. 1996, MNRAS, 282, 263
23. Mushotzky, R. 1999, these proceedings
24. Eisenstein, D., & Hu, W. ApJ, 496, 605
25. Spergel, D. 1999, these proceedings

3. First Discrete Structures

Quasar Absorption Line Studies of Galaxies and the Intergalactic Medium at z > 1.5

Lisa J. Storrie-Lombardi

Carnegie Observatories, Pasadena, CA 91101, lisa@ociw.edu

Abstract. The title of this article could of course encompass an entire meeting. I will focus my comments on reviewing of what we know about the most numerous absorption lines, the neutral hydrogen absorbers, and their evolution with redshift. This field of study has undergone a renaissance in last few years driven by observations with the Hubble Space Telescope of low redshift quasar absorption lines, observations of high redshift absorbers with the HIRES instrument on Keck, and cosmological modeling that allows us to make detailed comparisons of lines of sight through simulated universes.

INTRODUCTION

Bright high redshift quasars are particularly valuable as probes of the intervening gas clouds and galaxies superimposed on their spectra in absorption. The galaxies that intercept their lines of sight provide samples selected by gas cross-section, without respect to their surface brightness, luminosity, or star formation rate. Though direct studies of high redshift galaxies are now possible, those selected by the absorption lines they produce in QSO spectra still provide the only means to study in detail their kinematic properties at high resolution. Optical spectroscopy from ground-based telescopes and ultraviolet spectroscopy from space have provided us with a wealth of information about the ionized intergalactic medium and neutral gas and metals in galaxies from very low redshifts out to nearly z=5. Recent advances in cosmological hydrodynamic simulations have provided us real insight into the physical properties of the structures that produce the Lyα forest lines.

In the following sections I briefly review the definitions of the different classes of HI absorption systems, and discuss their evolution in number density with redshift, determinations of their metal abundances, and the interpretations of these results.

CP470, After the Dark Ages: When Galaxies were Young (the Universe at 2 < z < 5),
edited by Stephen S. Holt and Eric P. Smith

TAXONOMY OF NEUTRAL HYDROGEN ABSORBERS

Neutral hydrogen absorption can be detected over a staggering 10 orders of magnitude from the Lyα forest region with the weakest detectable lines having a column density $N_{HI} \sim 10^{12}$ atoms cm^{-2}, up to the damped Lyα absorbers with $N_{HI} \sim 10^{21}$. The rich zoo of these absorbers, in addition to those produced by heavier elements such as carbon, silicon, oxygen, and magnesium, are illuminated along a quasar line-of-sight, leaving their imprint as absorption in the quasar continuum. HI absorbers are typically divided into three classes based on their column density.

- **Lyα Forest:** ($N_{HI} = 10^{12}$ to 10^{16} atoms cm^{-2}) Figure 1 shows a high redshift quasar spectrum. We see the ultraviolet emission lines Lyα (1216Å) and CIV (1549Å) redshifted to $\sim$ 6600Å and 8300Å respectively. All of the absorption structure blueward of the Lyα emission is real, and almost all the lines in this "forest" are neutral hydrogen. Studies of the forest yield a wealth of information about the intergalactic medium, and the background ionizing flux, and structures at high redshifts (see [1] for a complete review of the Lyα forest absorbers). These absorbers are characterized by low metal content, some have been shown to be associated with galaxies, and their number density increases rapidly with increasing redshift.

- **Lyman-Limit Systems:** ($N_{HI} \geq 1.6 \times 10^{17}$ atoms cm^{-2}) As we move to higher column densities, neutral hydrogen becomes optically thick to Lyman continuum radiation at wavelengths below 912Å. The Lyman limit at $z = 4.37$ can be seen in figure 1 at 4900Å where the flux sharply drops to zero. Lyman-limit systems provide a means of directly studying the evolution of galaxies over the redshift range $0.1 < z < 5$ [2–5]. They have been shown to be associated with normal galaxies for $z < 1.6$ [6,7].

- **Damped Lyα Systems:** ($N_{HI} \geq 2 \times 10^{20}$ atoms cm^{-2}) The absorbers detected via the damped Lyα lines they produce show features consistent with an early phase of galactic evolution [8–11]. They are called "damped" because at these high column densities we see radiation damping wings in the line profile. A damped absorber can be seen at 5910Å in figure 1 at a redshift of $z = 3.86$ [12]. The column densities of damped absorbers are comparable to what we see along a typical line of sight through our own galaxy. Though the lower column density forest lines are far more numerous, damped absorbers contain most of the neutral hydrogen in the Universe.

NUMBER DENSITY EVOLUTION WITH REDSHIFT OF HI ABSORBERS

Observations have long shown that as we move to higher redshifts, the number density of lower column density HI absorbers increase at a much faster rate than

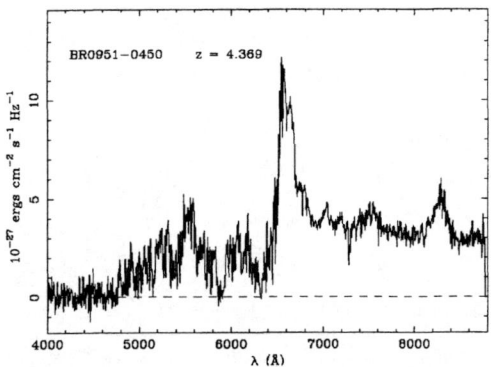

FIGURE 1. The spectrum of the quasar BRI0951-0450. The ultraviolet emission lines Lyα (1216Å) and CIV (1549Å) are redshifted to ~ 6600Å and 8300Å respectively. All of the absorption structure blueward of the Lyα emission is real, and almost all the lines in this "forest" are neutral hydrogen. The sharp drop in flux at 4900Å is due to a Lyman-limit system at $z = 4.37$. A damped Lyα absorber at $z = 3.86$ appears at 5910Å as the strong absorption trough. (Storrie-Lombardi et al. 1996)

for the higher column density systems, and it is clear that the Lyα forest as a whole evolves quite strongly with redshift.

In a standard Friedmann Universe for absorbers with cross-section πR_0^2 and number density Φ_0 per unit comoving volume

$$N(z) = \Phi_0 \pi R_0^2 c H_0^{-1}(1 + z)(1 + 2q_0 z)^{-1/2}. \tag{1}$$

It is customary to represent the number density as a power law of the form

$$N(z) = N_0(1 + z)^\gamma, \tag{2}$$

where $N_0 = \Phi_0 \pi R_0^2 c H_0^{-1}$. This yields $\gamma = 1$ for $q_0 = 0$ and $\gamma = 1/2$ for $q_0 = 1/2$ for the case of no evolution with redshift in the product of the number density and cross-section of the absorbers [13,14].

For the Lyα forest lines (log $N_{HI} < 16$) $N(z)$ is calculated by either counting individual absorption lines [15–19] or by measuring the mean absorption, D_A, caused by the Lyα forest shortward of the QSO Lyα emission line [20–22]. Though there is a large scatter in the values of the exponent γ determined by these different authors, ranging from 1.89 to 2.9 for $2 < z < 4$ and with values of $\gamma > 5$ for redshifts $z > 4$, it is clear that for $z > 2$ the slope of the power law is steeper than would be expected for a non-evolving population, and probably steepens even further at redshifts $z > 4$. This is in contrast to the very flat slope ($\gamma = 0.16 \pm 0.16$) of the power law measured for redshifts $z < 1.5$ [4] from the HST QSO Absorption Line Key Project data. With their complete data set they detect the steepening of the power law at $z \approx 1.7$, which matches well onto the data points determined

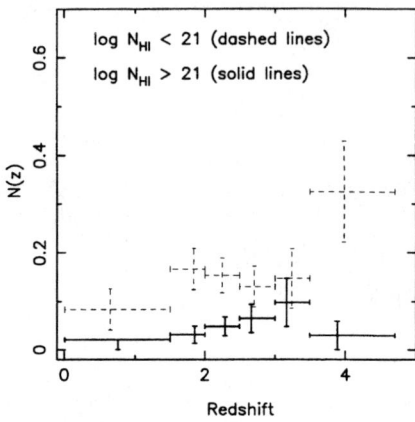

FIGURE 2. The number density per unit redshift, N(z), for damped Lyα absorbers is plotted versus redshift. Differential evolution with column density is evident. The higher column density absorbers (log N_{HI} > 21, plotted as solid lines) disappear at a faster rate from z≈3 to z=0 than the lower column density damped systems (log N_{HI} < 21, plotted as dashed lines). A paucity of very high column density absorbers is evident at z > 4 (Storrie-Lombardi & Wolfe, in preparation).

from ground-based observations. Recent many-body simulations tracing collisionless particles in cold dark matter scenarios ([24], and references therein) reproduce the observations well. By modeling the evolution of the UV background radiation, as well as allowing for shock heating, the simulations indicate that the unshocked material is found primarily in underdense regions while the shocked material is found in condensing regions. The unshocked population evolves more rapidly with increasing redshift, and dominates the absorbers at high redshift. Conversely, the shocked population has a flat evolution and dominates at low redshift.

The Lyman Limit systems (log N_{HI} > 17), on the other hand, are consistent with no evolution with redshift in the product of their number density and cross-section [2,4,5] or mild evolution (for an $\Omega = 1, \lambda = 0$ universe), with $\gamma = 1.55 \pm 0.3$ for redshifts $0.008 < z < 4.7$ [4]. A two power law fit, with $\gamma \approx 1$ for $z < 1$ and $\gamma \approx 2.8$ for $z > 1$ is also consistent with the data but is not required. The damped absorbers (log N_{HI} > 20.2) yield a similar value for the exponent γ (1.11 ± 0.40) but show differential evolution with column density [8,25,26]. The highest column density damped systems (log N_{HI} > 21) disappear at a faster rate from z≈3 to z=0 than the lower column density subset and there is a paucity of very high column density absorbers at z > 4 (figure 2).

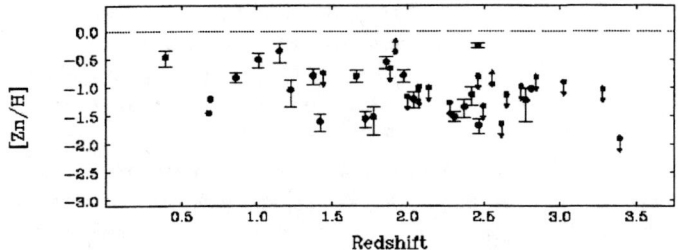

FIGURE 3. The Zn/H abundance for a large sample of damped Lyα absorbers is plotted versus redshift. The damped Lyα systems are metal poor at all redshifts, with a mean abundance of $\langle Zn/H \rangle = -1.13 \pm 0.38$, about 1/13 of solar. There is a large scatter in metallicity at a given redshift and the measured metallicities are lower at higher redshifts (Pettini et al. 1999).

MEASUREMENTS OF METAL ABUNDANCES IN HI ABSORBERS

One of the paradigm shifts that has occurred in QSO absorption studies in recent years is the detection of metals associated with the Lyα forest lines, which were initially thought to be primordial HI clouds. The high resolution/high signal-to-noise spectra that can be obtained with the HIRES instrument [27] on the Keck telescope have made this possible. For HI absorbers with column densities log N_{HI} ≈ 15, most show associated CIV absorption, and for those with $N_{HI} > 3 \times 10^{14}$, about 50 percent show associated CIV lines, with a typical metallicity $Z/Z_\odot = 10^{-2}$ [28,29].

Substantial progress has also been made in measuring metal abundances in the damped Lyα absorbers. A large amount of time on 4m telescopes [30,11,31] and the Keck 1 telescope [32,10] has been applied to this project. The picture that emerges is:

1. Damped Lyα systems are metal poor at all redshifts, with a mean abundance of $\langle Zn/H \rangle = -1.13 \pm 0.38$, about 1/13 of solar.

2. There is a large scatter in metallicity at a given redshift.

3. The measured metallicities are lower at higher redshifts.

4. The damped system abundances don't match the disk star abundances.

5. The velocity profiles from high resolution spectra of low ionization metal lines have multiple, narrow components and are asymmetric in that the component with the strongest absorption tends to lie at one edge of the profile.

Points 1-3 are illustrated in figure 3 from Pettini et al. (1999) [31]. There is still considerable controversy though about exactly how to interpret the results from observations of damped absorbers.

Prochaska & Wolfe (1997a, 1997b) [33,34] have argued that the observationally determined velocity profiles are reflective of rapidly rotating, cold disks. The low ionization metal lines observed in damped systems accurately trace the velocity fields of the neutral gas dominating the baryonic content of the absorbers. Haehnelt, Steinmetz & Rauch (1998) [35] propose a model that reproduces the observed velocity profiles that is consistent with smaller clumps or galactic building blocks, not large disks. Pettini et al. (1997) [11] suggest that damped Lyα systems are drawn from a varied population of galaxies of different morphological types and at different stages of chemical evolution, supporting the idea of a protracted epoch of galaxy formation. When our Galaxy's metal enrichment was at levels typical of damped systems, its kinematics were closer to those of the halo and bulge than a rotationally supported disk. Lu et al. (1996) [10] also found that the chemical evolution history of damped systems is more consistent with the spheroidal component of galaxies or dwarf galaxies.

The interpretation of the elemental ratios measured in damped systems is also the subject of debate. They have been interpreted to be consistent with mild dust depletion, dust depletion combined with supernovae type II enrichment, or supernovae type II enrichment alone. We also do not have definitive evidence regarding what effect dust in foreground damped systems might have on removing these lines of sight from magnitude limited quasar samples [36,9] Though the effect at high redshift is not expected to be large, it could have an important impact in the redshift range $1 < z < 2$. The final word on the nature of damped systems still remains to be determined.

FUTURE WORK

The future holds the potential for substantial new advances in our understanding of the evolution of galaxies using quasar absorption lines due to a number of advances. Large sky surveys, the Sloan Digital Sky Survey (e.g. [37]) and the Two Degree Field (2dF, e.g. [38]), will find substantial numbers of new quasars at low redshifts, as well as the first with redshifts $z > 5$. New surveys for bright quasars ($R < 19.5$) with redshifts $4 < z < 5$ are finding many new objects for detailed follow-up of absorption systems [39,40]. More realistic star formation scenarios and ever faster computer will allow us to do more detailed an realistic simulations of galaxy formation to compare with observations (see [41] for a review of simulations of cosmic structure formation.) Identifying directly the galaxies responsible for damped Lyα absorbers has proved difficult (e.g. [42]) though progress is now being made (e.g. [43–45]. Narrow band Hα searches tuned for a damped absorber redshift may also help answer these questions (Bechtold et al. 1998).

REFERENCES

1. Rauch, M. 1998, ARA&A, 36, 267

2. Sargent, W.L.W., Steidel, C.C., Boksenberg, A. 1989, ApJS, 79, 703
3. Lanzetta, K.M., 1991, ApJ, 375, 1
4. Storrie-Lombardi, L.J., McMahon, R.G., Irwin, M.J., & Hazard C. 1994, ApJ, 427, L13
5. Stengler-Larrea, E.A., Boksenberg, A., Steidel, C.C., Sargent W.L.W., Bahcall, J.N., Bergeron, J., Hartig, G.F., Januzzi, B.T., Kirhakos, S., Savage, B.D., Schneider, D.P., Turnshek, D.A. & Weymann, R.J. 1995, ApJ, 444, 64
6. Steidel, C.C., Dickinson, M. & Persson, S.E. 1994, ApJ, 437, L75
7. Dickinson & Steidel 1995, BAAS, 186, 2509
8. Wolfe, A.M., Lanzetta, K.M., Foltz C.B., & Chaffee F.H., 1995, ApJ, 454, 698
9. Pei, Y.C. & Fall, S.M. 1995, ApJ, 454, 69
10. Lu, L., Sargent, W.L.W., Womble, D.S. & Barlow, T.A. 1996, ApJ, 457, L1
11. Pettini, M., Smith, L.J., King, D.L., Hunstead, R.W. 1997, ApJ, 486, 665
12. Storrie-Lombardi, L.J., McMahon, R.G., Irwin, M.J., & Hazard C. 1996, ApJ, 468, 128
13. Bahcall J.N., Peebles P.J.E. 1969, ApJ, 156, L7
14. Sargent, W.L.W., Young, P.T., Boksenberg, A. & Tytler, D. 1980, ApJS, 42, 41
15. Lu, L., Wolfe, A.M. & Turnshek, D.A. 1991, ApJ, 367, 19
16. Bechtold, J. 1994, ApJS, 91, 1
17. Williger, G.M., Baldwin, J.A., Carswell, R.F., Cooke, A.J., Hazard, C., Irwin, M.J., McMahon, R.G. & Storrie-Lombardi, L.J. 1994, ApJ, 428, 574
18. Cooke, A.J., Espey, B. & Carswell, R.F. 1997, MNRAS, 284, 552
19. Kim, T.-S., Hu, E.M., Cowie, L.L. & Songaila, A. 1997, AJ, 114, 1
20. Press, W.H., Rybicki, G.B. & Schneider, D.P. 1993 ApJ, 414, 64
21. Zuo, L. & Lu, L. 1993, ApJ, 418, 601
22. Fardal, M.A., Giroux, M.L. & Shull, J.M. 1998, AJ, 115, 2206
23. Weymann, R.J., Januzzi, B.T., Lu, L., Bahcall, J.N., Bergeron, J., Boksenberg, A., Hartig, G.F., Kirhakos, S., Sargent, W.L.W., Savage, B.D., Schneider, D.P., Turnshek, D.A. & Wolfe, A.M. 1998, ApJ, 506, 1
24. Riediger, R., Petitjean, P. & Mucket, J. 1998, A&A, 329, 30
25. Storrie-Lombardi, L.J., Irwin, M.J. & McMahon, R.G. 1996, MNRAS, 282, 1330
26. Storrie-Lombardi, L.J. & Wolfe, A.M. 1999, in preparation
27. Vogt, S.S., Allen, S.L., Bigelow, B.C., Bresee, L., Brown, B., Cantrall, T., Conrad, A., Couture, M., Delaney, C., Epps, H.W., Hilyard, D., Hilyard, D.F.; Horn, E., Jern, N., Kanto, D., Keane, M.J., Kibrick, R.I., Lewis, J.W., Osborne, J., Pardeilhan, G.H., Pfister, T., Ricketts, T., Robinson, L.B., Stover, R.J., Tucker, D., WARD, J., & Wei, M.Z. 1994, in proceedings SPIE, eds. Crawford, D.L. & Craine, E.R., 2198, 362
28. Cowie, L.L., Songaila, A. Kim, T.-S. & Hu, E.M. 1995, AJ, 109, 1522
29. Songaila, A. & Cowie, L.L. 1996, AJ, 112, 335
30. Pettini, M., Smith, L.J., Hunstead, R.W. & King, D.L. 1994, ApJ, 426, 79
31. Pettini, M., Ellison, S.L., Steidel, C.C. & Bowen, D.V. 1999, ApJ, in press
32. Wolfe, A.M., Fan, X.-M., Tytler, D., Vogt, S.S., Keane, M.J. & Lanzetta, K.M. 1994, ApJ, 435, 101
33. Prochaska, J.X. & Wolfe, A.M. 1997a, ApJ, 474, 140

34. Prochaska, J.X. & Wolfe, A.M. 1997b, ApJ, 487, 73

35. Haehnelt, M.G., Steinmetz, M. & Rauch, M. 1998, ApJ, 495,647

36. Fall, S.M. & Pei, Y.C. 1993, ApJ, 402, 479

37. Margon, B. 1998, Philosophical Transactions of the Royal Society of London A, in press

38. Croom, S.M., Shanks, T., Boyle, B.J., Smith, R.J., Miller, L., Loaring, N.S., 1999, in proceedings of "Evolution of Large Scale Structure", in press

39. Kennefick, J.D., de Carvalho, R.R., Djorgovski, S.G., Wilber, M.M., Dickson, E.S., Weir, N., Fayyad, U. & Roden, J., 1995, AJ, 110, 78

40. Storrie-Lombardi, L.J., Hook, I.M., Irwin, M.J. & McMahon, R.G. 1999, in preparation

41. Weinberg, D.H., Katz, N., & Hernquist, L. 1998, in *Proceedings of 'Origins'*, eds. Woodward, C.E., Shull, M. & Thronson, H. A., (Astronomical Society of the Pacific:San Francisco), Vol. 148, 47

42. Lowenthal, J.D., Hogan, C.J., Green, R.F., Woodgate, B., Caulet, A., Brown, L. & Bechtold, J. 1995, ApJ, 451, 484

43. Djorgovski, S.G., Pahre, M.A., Bechtold, J. & Elston, R. 1996, Nature, 382 234

44. Le Brun, V., Bergeron, J., Boisse, P. & Deharveng, J.M. 1997, A&A, 321,733

45. Moller, P. & Warren, S.J. 1998, MNRAS, 299, 661

46. Bechtold, J., Elston, R., Yee, H.K.C., Ellingson, E., & Cutri, R.M. 1998, in "The Young Universe: Galaxy Formation and Evolution at Intermediate and High Redshift", eds. S. D'Odorico, A. Fontana, and E. Giallongo, ASP Conference Series, 146, 241

Simulating Galaxy Evolution

Joseph Silk and Rychard Bouwens

Departments of Astronomy and Physics, and Center for Particle Astrophysics, University of California, Berkeley, CA 94720

Abstract. The forwards approach to galaxy formation and evolution is extremely powerful but leaves several questions unanswered. Foremost among these is the origin of disks. A backwards approach is able to provide a more realistic treatment of star formation and feedback and provides a practical guide to eventually complement galaxy formation *ab initio*.

INTRODUCTION

Understanding how galaxies formed is the key to unraveling the mysteries of the high redshift universe. To interpret the deepest images of distant galaxies one has to simulate galaxy evolution. The prescription for such a simulation seems straightforward. Take a gas cloud, massive enough to be self-gravitating, and add a simple prescription for star formation based on the local free-fall time scale. In practice, this approach has yielded star formation histories that appear to match observations.

However the predictive power of this approach is limited. The reason is the following. One has to assume a prescription for star formation. Reasonable guesses can be made, but one has no guarantee that these are valid. There is no way of evaluating the uncertainty in the adopted ansatz for forming stars. This is true for primordial clouds, and equally valid for current star formation. Of course the star formation prescription, once selected, has parameters that can be adjusted, often with little freedom when confronted with the observational data. This approach has been applied to the early universe, commencing with density fluctuations that grow by hierarchical clustering of cold dark matter.

One can try to assess the uncertainties by comparing snapshots of the universe at different redshifts. If one matches the data, one can deduce that one has a working model of galaxy formation, but one cannot expect this to be a useful guide to extreme situations that are not included in the simple algorithm. These might include, for example, the role of active galactic nuclei in primordial and current

CP470, After the Dark Ages: When Galaxies were Young (the Universe at 2 < z < 5), edited by Stephen S. Holt and Eric P. Smith
© 1999 The American Institute of Physics 1-56396-855-X/99/$15.00

epoch star formation. I conclude that it is useful to consider an alternative to "*ab initio*" galaxy formation. In this talk I will describe such an approach that is based on nearby examples of star formation in a global context, that one attempts to run backwards in time. Clearly, forwards and backwards evolution are complementary descriptions of the same fundamental issues that describe galaxy formation.

GALAXY EVOLUTION FROM PRIMORDIAL FLUCTUATIONS

Inflationary cosmology prescribes the initial spectrum of density fluctuations. The horizon scale at matter-radiation equality imprints a scale on the relic fluctuation spectrum: at $L \gg L_{eq} \simeq 12(\Omega h^2)^{-1}$ Mpc, $\delta\rho/\rho \propto M^{-1/2-n/6}$ and $n \approx 1$ whereas at $L \ll L_{eq}$, n_{eff} approaches -3, reflecting scale invariance for fluctuations that entered the horizon during radiation domination. On galaxy scales, $n_{eff} \approx -2$. This leads to a hierarchical formation sequence of structure. Larger and larger structures merge together. Numerical simulations show that some substructure survives.

This is potentially a problem for understanding why galactic disks remain thin if the surrounding dark halos contain even a percent of their mass in massive substructures, characteristic in mass, say, of dwarf galaxies. Dwarf-disk interactions would overheat the disk [19]. This can be partially rectified by gas infall, which certainly helps renew thin disks. The discovery of high velocity hydrogen clouds at the periphery of the halo lends some support to the availability of a gas reservoir today [1].

The properties of dark halos are accounted for by hierarchical clustering. The abundance, mass function, density profile and rotation curve for a typical galaxy halo all agree with empirical estimates. The clustering of galaxies is described by the galaxy correlation function, and simulations of clustering provide a fit over several decades of scale. One accounts for the mass function of galaxy clusters and its evolution with redshift [2,3] by setting $\Omega \approx 0.3\,(\pm 1)$. Interpretation of massive halos as rare peaks accounts for the observed clustering of Lyman break galaxies at $z \sim 3$ [4].

The properties of the intergalactic medium agree with predictions of the hierarchical model. One has to adopt a metagalactic ionizing radiation field. This is taken from the observed quasar luminosity function. The gas distribution from the simulations is exposed to the ionizing radiation field, and the effects of the peculiar velocity field are found to play an important role in reproducing Lyman alpha cloud absorption profiles. One can explain [5] the distribution of observed column densities ranging from damped Lyman alpha clouds with HI column densities in excess of 10^{21} cm^{-2} down to the Lyman alpha forest below 10^{14} cm^{-2}. The gas overdensities range from $\delta\rho/\rho$ of order several hundred for damped clouds to unity for the forest. The structural properties of the Lyman alpha clouds are simply understood. There is some controversy however over the nature of the relatively rare damped clouds. These have been argued to be rotating protodisks [6]. However the observed spread of velocities is not simply a thin disk, but can either be interpreted as a thick disk or as a more incoherent, quasi-spherical halo containing many smaller clouds [7].

More problematic for disk theory is the failure of simulations to reproduce the sizes of galactic disks. Angular momentum conservation of a uniformly collapsing and dissipating cloud of baryons within a dark halo suggests that the disk size is λR_i, where R_i is the halo virialization radius and λ is the critical dimensionless angular momentum. One has $\lambda \approx 0.06$ and R_i is typically about 100 kpc. This argument would actually give the correct disk size. However the clumpy nature of the halo is found to drive efficient angular momentum transfer via dynamical friction. The disk size found in simulations is a factor of five or more smaller than observed disk scale lengths [8]. Evidently feedback from star formation is conspiring to limit the collapse of the gas.

The galaxy luminosity function also represents a challenge for theoretical models, which more naturally specify the galaxy mass function. There are two difficulties. At the low mass end, the predicted slope of the mass function ($dN/dm \backsimeq m^{-2}$) is a poor fit to the power-law tail of the galaxy luminosity function, the slope of which depends on galaxy color selection and varies between $dN/dL \backsimeq L^{-3/2}$ in the blue and $dN/dL \backsimeq L^{-1}$ in the red. One corrects this problem by introducing inefficient star formation in low mass potential wells. The fraction of gas forming stars is assumed to be $(\sigma/\sigma_{cr})^\alpha$, with $\alpha \approx 2$, where $\sigma_{cr} \approx 75$ km s^{-1} denotes the transition velocity dispersion, below which retention of interstellar gas energised by supernova-driven winds becomes suppressed. This assumes that supernovae are effective at disrupting the interstellar gas in the shallow potential wells characteristic of dwarf galaxies [9]. However the efficiency may only be high at masses below $\sim 10^6 M_\odot$, according to a recent analysis of starbursts in dwarf galaxies [10]. This would only flatten the luminosity function at very low luminosities if one counts all gas-retaining galaxies. One difference between dwarfs ($\lesssim 10^8 M_\odot$) and giants is that the supernova ejecta are expelled, so that the residual gas, if retained, is very metal-poor. This might be sufficient to reduce the efficiency of star formation sufficiently so as to produce a population of low surface brightness dwarfs.

At high luminosities the challenge is to explain why the nonlinear clustering mass present is $\sim 10^{14} M_\odot$ whereas the value of L_*, above which the number of galaxies decreases exponentially, is $10^{10} h^{-2} L_\odot$. The corresponding dark halo mass is around $10^{12} M_\odot$. Evidently some physical effect is intervening to limit the luminosity of a galaxy, which does not track the mass of the dark potential wells. The generally accepted resolution is that baryonic cooling is a necessary condition for star formation to occur in a primordial contracting cloud. If the density is too low for gas cooling, the intergalactic gas remains hot and diffuse. Efficient star formation must occur within a dynamical time-scale. This is certainly how monolithic formation of an elliptical must have occurred. In this case, the condition that the gas cooling time be less than a collapse time sets a maximum value on cooled galaxy baryon mass of about $10^{12} M_\odot$.

However gas continues to accrete and cool. The total mass of cooled gas does not provide a distinctive cut-off in the mass function of baryons [11]. One has to vary the efficiency of star formation, reducing it on time-scales longer than a dynamical time, in order to account for L_*. One can appeal to cluster formation to heat up the intergalactic gas, thereby removing the reservoir of cold gas which would potentially be accreted. This would lead one to expect that cluster ellipticals have a relatively homogeneous distribution of formation times, peaked at the epoch of cluster formation. One has to assume a hot gas environment for field ellipticals, associated with galaxy groups, to restrict cold gas infall. However since clustering in the field develops more recently than for rich clusters one expects field ellipticals, on the other hand, to display a much broader range of ages, and reveal, in some cases, signs of recent or current infall. Indications of this effect can be seen in the enhanced scatter in the fundamental plane for field ellipticals relative to cluster ellipticals [12].

For disk galaxies, the comparison of mass and luminosity via the predicted mass function challenges interpretations of the Tully-Fisher correlation between luminosity and maximum rotation velocity [13]. The observed dispersion of fifteen percent in inferred distance [14] may be compared with the dispersion between mass and halo circular velocity in the CDM hierarchy, which is of order 100 percent. Implementation of a prescription for star formation can reduce the dispersion between cooled baryon mass, and hence luminosity, and disk rotational velocity to the observed range by, for example, allowing stars to preferentially form in the more massive disks where the baryons are self-gravitating and dense enough to suppress supernova-driven winds. However there is a price: the Tully-Fisher normalization yields the disk mass-to-luminosity ratio, and the CDM hierarchy inevitably favors a high value relative to the observed value of $M/L \sim 10\,h$ for the baryon-dominated regions of disks.

GALAXY EVOLUTION VIA REVERSE ENGINEERING

A complementary approach to galaxy evolution allows one to circumvent some of these difficulties, although at the risk of introducing other complications. One commences with nearby galaxies, develops a model for star formation, and evolves the galaxies backwards in time. Actual images or idealized models of nearby galaxies are used as the starting point. Suppose one first ignores dynamical evolution. Star formation in disks can be described by an expression of the form

$$\text{SFR} = \epsilon\,\mu_{\text{gas}}\,\Omega(r)\,f(Q)\,.$$

Here μ_{gas} is the surface density of atomic and molecular gas, $\Omega(r)$ is the rotation rate, Q is the Toomre parameter (approximately given for a self-gravitating disk of gas by $\frac{\kappa\sigma}{\pi G \mu_{\text{gas}}}$, where κ is the epicyclic frequency and σ is the gas velocity dispersion) that guarantees gravitational instability to axisymmetric perturbations if $Q < 1$, and ϵ is an efficiency parameter. One needs to generalize the dependence on Q to allow for non-axisymmetric instabilities, such as density waves which are responsible for the growth of molecular clouds and for the gravitational contribution of the stellar component. In general, however, one expects there to be a threshold for local instabilities when the surface density drops below a critical value, for typical disks amounting to about $\mu_{gas} \approx 10\,M_\odot pc^{-2}$. This empirical expression fits global star formation rates in disks remarkably well [15], and ϵ may be interpreted as the fraction of gas converted into stars per dynamical time. Infall is one remaining ingredient that needs to be added.

For individual disks, this model has been exploited to demonstrate that disks form inside out, that disk surface brightness increases by almost a magnitude [16] to $z \sim 1$, and to account for the chemical evolution of old disk stars and of the interstellar medium at high redshift [17]. The model has considerable potential for predicting how galaxies appear to evolve in deep images obtained of the distant universe. In fact, one study [18] has already demonstrated that such a scaling in galaxy size is necessary to reconcile faint galaxies sizes with galaxies at low redshift, this study carefully considering changes in the pixelisation, the PSF, and the surface brightness relative to the noise. Of course, a careful consideration of many of the same effects is important for testing models against the observations. One has to add a disk formation epoch, chosen from an analytical prescription for hierarchical CDM cosmology and some evolution in number density. The latter is required to crudely account for merging and is necessary to reproduce the observed deep galaxy counts. Ellipticals and spheroids must also be incorporated into the model. While these systems do not dominate the number counts, which at faint magnitudes are dominated by disks and their irregular precursors, they are important in the cumulative star formation history of the universe. Approximately half of the mass in stars is in the spheroidal component, and hence mostly in E's and S0's. This is the approximate assessment for the local luminosity function (and is due to the fact that while ~ 30 percent of galaxies are E's and S0's, the associated M/L is about twice as large as for typical spirals). One also reaches an independent verification of this from the cosmic far infrared background. This recently discovered diffuse flux at $100 - 300\,\mu m$ amounts to $\lambda i_\nu \approx 20\,nw/m^2/s^2$, comparable to the diffuse optical light flux when integrated over the HDF and near infrared. Modelling of disk galaxies incorporating dust can reproduce the optical background but only about fifty percent of the FIR background is explained by optically visible systems. The remainder is presumed to be due to dusty ellipticals. Of course if these systems form stars at an early epoch z_E relative to spirals at z_S, then the inferred mass in stars (for the same initial mass function) in dust enshrouded spheroids is equal to $[(1 + z_E)/(1 + z_S)]$ times the contribution from disks. This comparison suggests that $z_E \sim z_S$ though in principle one could have $z_E \gg z_S$.

One might worry that the FIR background could be due to AGN. However modelling of the x-ray background effectively constrains the AGN contribution to diffuse hard photons. Compton self absorption of the x-rays, required to obtain a spectral fit of the XRB, limits the possible contribution to the diffuse FIR background by dust-shrouded, x-ray-emitting AGN to be almost ten percent of the observed background. Direct observations by SCUBA find ultraluminous galaxies at $z = 1 - 3$. Perhaps ten percent of these may be AGN-powered according to the previous argument, and this is consistent with direct spectroscopic signatures.

Disk Parameterisation

There are two major uncertainties in the modelling of the disk star formation rate: infall and efficiency. One can constrain the role of infall by three independent methods that respectively appeal to chemical evolution, disk dynamics, and to the evolution of disk sizes. The best studied is chemical evolution. Infall of metal-poor gas into the early gas is required to account for the paucity of metal poor G dwarfs. The sharp decline in supersolar metallicities of disk stars means that recent metal-poor infall is greatly reduced relative to infall in the first 5 Gyr. Infall of gas-rich clumps is predicted in the CDM model, but these interactions must avoid overheating the disk. Less than 4% of the disk can have fallen in over the past 5 Gyr according to one study [19]. However recent calculations suggest that infalling satellites preferentially tilt rather than heat the disk [20]. The implications for high redshift galaxies is that disks are small at $z > 1$. Without infall, disks would not be sufficiently small, according to one recent analysis, to account for the decrease in faint galaxy angular diameter.

One can only decompose disks from bulges to $z \lesssim 1$, using HST data. Evolution of disk sizes to this redshift is quite model-dependent. Disk size varying as $(1+z)^{\alpha}$, with $\alpha \approx 2$, fits the available data. However selection biases need to be modelled more carefully. One selects earlier type galaxies at high redshift than at low redshift because of surface brightness dimming, and this complicates comparisons.

Disk Physics

The essence of disk formation lies in inefficiency. Galaxies retain a sufficient gas reservoir so as to still be vigorously forming stars at the present epoch. The star formation rate increases dramatically with cosmic epoch, possibly peaking near $z \sim 2$. Hence gas infall drops off dramatically. This also is implicit in models of galactic chemical evolution, where infall of metal-poor gas over the first five or so Gyr helps account for the metal distribution of old disk stars. The inefficiency of star formation must be due, not to the availability of a gas supply but rather arises from being controlled by disk physics.

Feedback of energy and momentum from star formation and death necessarily play an important role. One needs to include such physics to understand disk sizes. One could simultaneously account for gas longevity. Angular momentum transfer is central to such a model. A general class of theories which can successfully reproduce disk profiles is based on contracting viscous self-gravitating disks. The viscosity arises from cloud-cloud collisions, the cold disk being gravitationally unstable to cloud formation. The disk forms as angular momentum is transferred on a viscosity timescale. Since cloud collisions and mergers are assumed to drive star formation, one naturally relates the star formation and viscosity time scales. An exponential surface density profile is naturally generated [21].

Bulge Evolution

Bulges are expected to be prominent in observations of high redshift galaxies, both because of disk evolution and the high bulge surface brightness. Yet the sequence of bulge formation is poorly understood, and this makes it difficult to formulate and test *ab initio* predictions of disk evolution. Consider the following alternatives. Bulges form before disks, either monolithically or in major (i.e. comparable, or at least mass ratio 1:10) mergers. Bulges form simultaneously with disks via satellite mergers. Bulges form after disks, via secular instability of disks, and bar formation followed by dissolution as gas inflow drives bulge formation [22]. Any of these scenarios are possible. Two, or even all three, may be operative. For example, secular evolution can form small bulges but not the massive objects of early type galaxies. Observational evidence that bulge and disk scale lengths are correlated favors a secular evolution origin of bulges for late-type spirals [23]. The ubiquity of bars, which are efficient at torquing accreting gas and driving the gas inwards to form a central bulge, also suggests that secular evolution must have played a significant role in bulge formation. Conversely, massive bulges are most likely formed by mergers. Satellite infall of gas-rich dwarf galaxies is expected to be a common occurrence in hierarchical models and provides a natural mechanism for simultaneously forming the bulges, as the dense stellar cores sink into the center of the galaxy by dynamical friction, and feeding disk growth with gas infall. There are hints of monolithic bulge formation from observations of many compact Lyman break galaxies, which have high star formation rates.

One can try to address this confusing range of bulge formation possibilities by examining the properties of disk galaxies at $z \lesssim 1$, where component separation into bulge and disk is possible at HST resolution. Late-forming bulges are inevitably bluer and smaller than early-forming bulges, at a given redshift. Figure 1 shows a comparison of the model predictions with available data. HST images are shown, in a comparison with the HDF. Analyses of similar images [24] show that only with larger samples at $z \sim 1$ could one be able to distinguish between alternative models of bulge formation.

LOOKING TO THE FUTURE

It will be possible in the not too distant future to greatly refine the observational constraints relevant to galaxy evolution. In Figure 2 we show HST Advanced Camera (2000) and NGST (2007) simulations of the same 85" x 85" field using the secular evolution scheme for bulge formation. The Advanced Camera simulations consider a 150,000-s integration, utilise a pixel size of 0.05 arcsec, and probe the

gri optical bands to $i_{AB} \sim 30.3$, whereas the NGST simulations consider a similar 150,000-s integration, utilise a pixel size of 0.029 arcsec, and probe the 1,3,5-μm wavelength bands to $m_{1\mu m,AB} \sim 31.6$. For comparison, we also show WFPC2 (pixel size is 0.1 arcsec, probes the I_{F814W}, V_{F606W}, and B_{F450W} bands to a limiting magnitude $I_{F814W,AB} \sim 29$) and NICMOS (pixel size is 0.2 arcsec, probes the J_{F110W} and H_{F160W} infrared bands to $H_{F160W,AB} \sim 28.3$) simulations. Since the fiducial secular model for bulge formation breaks down at high redshift, we have included a variation of the Pozzetti, Bruzual, & Zamorani [25] luminosity evolution model at these redshifts. The simulations include both K and evolutionary corrections, cosmological angular size relations and volume elements ($\Omega = 0.15$, $h = 0.5$), appropriate pixelisation, PSFs, and noise (see [18] for a discussion).

Obviously, one of the principal advantages of the Advanced Camera and NGST over WFPC2 and NICMOS are their increases in limiting magnitude, angular resolution, and field of view. Regarding the differing limiting magnitudes, using the Advanced Camera for similar length exposures to those shown here, one could probe to unobscured star formation rates $\sim 0.5 M_\odot/yr$ at $z \sim 5$ whereas with WFPC2, the limiting rate is only $\sim 2 M_\odot/yr$. For higher redshift observations, such as are only possible with NICMOS or NGST, NGST promises to push the sensitivity on unresolved star formation from its current value $\sim 20 M_\odot/yr$ at $z \sim 10$ obtainable with NICMOS exposures down to $\sim 1 M_\odot/yr$.

REFERENCES

1. Blitz, L. et al. 1998, preprint, astro-ph/9803251.
2. Eke, V. et al. 1998, MNRAS, 298, 1145.
3. Bahcall, N. and Fan, X. 1998, ApJ, 504, 1.
4. Giavalisco, M. et al. 1998, ApJ, 503,543.
5. Katz, N. et al. 1996, ApJ, 457, 57.
6. Wolfe, A. M. and Prochaska, J. 1998, ApJ, 507, 113.
7. Haehnelt, M., Steinmetz, M. and Rauch, M. 1998, ApJ, 495, 647.
8. Steinmetz, M. and Muller. E. 1995, MNRAS, 276, 549.
9. Dekel, A. and Silk, J. 1986, ApJ, 303, 39
10. Maclow, M. and Ferrara, A. 1998, preprint, astro-ph/9801237.
11. Thoul, A. and Weinberg, D. 1996, ApJ, 465, 608.
12. Forbes, D., Ponman, T. and Brown, R. 1998, ApJ, 508, L43.
13. Steinmetz, M. and Navarro, J. 1998, preprint, astro-ph/9808076.
14. Willick, J. 1998, preprint, astro-ph/9809160.
15. Kennicutt, R. 1998, ApJ, 498, 541.
16. Bouwens, R., Cayon., L. and Silk, J. 1997, ApJ, 489, L21.
17. Prantzos, N. and Silk, J. 1998, ApJ, 507, 229.
18. Bouwens, R., Broadhurst, and Silk, J. 1998, ApJ, 506, 557.
19. Toth, G. and Ostriker, J. P. 1993, AJ, 389, 5.
20. Huang, S. and Carlberg 1997, ApJ, 480, 503.
21. Lin, D. and Pringle, J. 1987, ApJ, 320, L87.

22. Norman, C., Sellwood, J. and Hasan, H. 1996, ApJ, 462, 114.
23. Courteau, S., R. de Jong and de Broeils, A. 1996, ApJ, 457, L73.
24. Bouwens, R., Cayon., L. and Silk, J. 1999, ApJ, in press, astro-ph/9812193.
25. Pozzetti, L., Bruzual, G., and Zamorani, G. 1996, MNRAS, 274, 832.

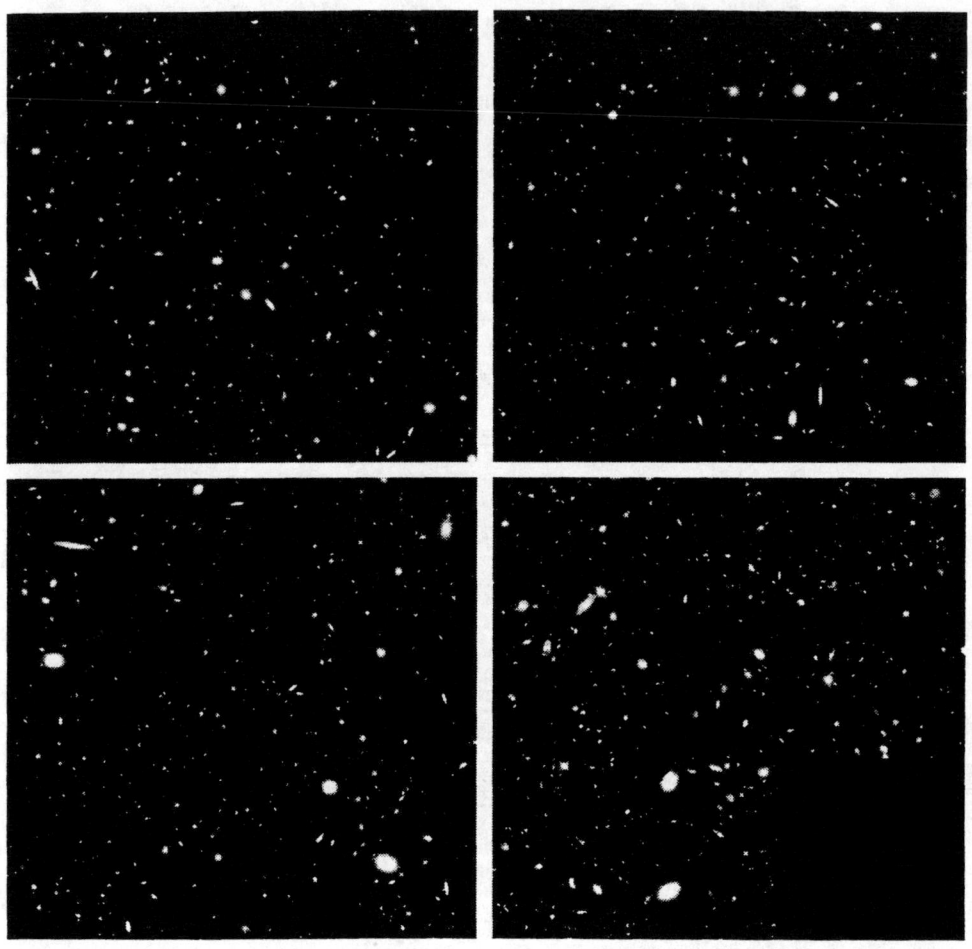

FIGURE 1. Comparison of simulated BVI images of a 2" x 2" patch of the HDF with the observed images (panel d). Panel (a) illustrates our secular evolution model for bulges, panel (b) illustrates our simultaneous formation model, and panel (c) illustrates our early bulge formation model. Calculations are performed using a galaxy-evolution software package written by one of the authors.

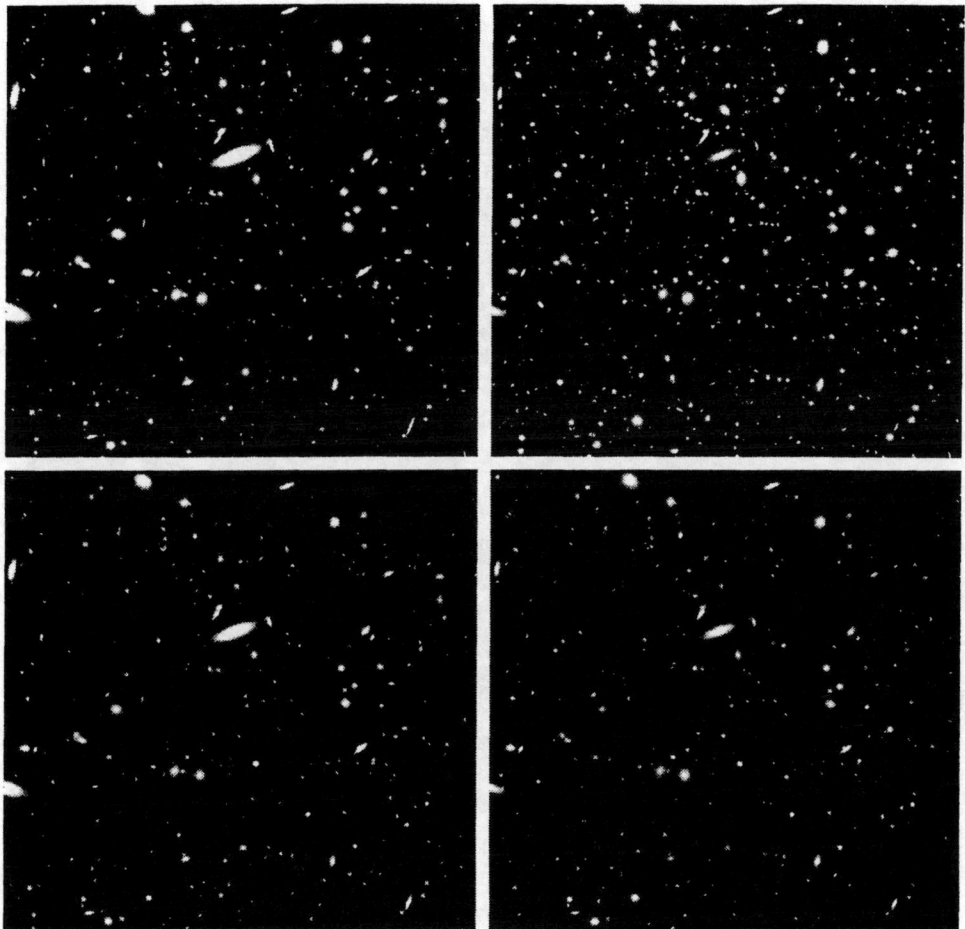

FIGURE 2. Simulated images of a 85" x 85" field using a secular evolution model for disks to $z = 1$ and the Pozzetti, Bruzual, & Zamorani luminosity evolution model for $z > 1$. Shown are 30-orbit *gri* exposures for the HST Advanced Camera (a), 150000-s 1.3.5-μm exposures for NGST (b), 30-orbit *BVI* exposures for HST WFPC2 (c), and 30-orbit *JH* exposures for the HST NIC3 camera (d). Calculations are performed using a galaxy-evolution software package written by one of the authors.

Clustering of Ionizing Sources

Siang Peng Oh

Princeton University Observatory, Princeton, NJ 08544

Abstract. Using existing models for the evolution of clustering we show that the first sources are likely to be highly biased and thus strongly clustered. We compute intensity fluctuations for clustered sources in a uniform IGM. The strong fluctuations imply the universe was reheated and reionised in a highly inhomogeneous fashion.

INTRODUCTION

Fluctuations in the ionizing intensity during reionization result in temperature and ionization fraction fluctuations. This has important implications for the reheating and reionization history of the IGM. There are two contributing components:
- IGM density fluctuations (variable optical depth)
- Spatial distribution of ionizing sources

The second point is usually addressed by computing Poisson fluctuations in the ionizing intensity ([4], [2]). Here we consider a hitherto neglected effect, the clustering of ionizing sources, and show that it is the dominant source of intensity fluctuations.

MODELLING THE IONIZING SOURCES

We work within the framework of the concordance cosmology of [3]:$(\Omega_m, \Omega_\Lambda, \Omega_b, h, \sigma_{8h^{-1}}, n) = (0.35, 0.65, 0.04, 0.65, 0.87, 0.96)$. We parametrise our model of reionization with 3 relevant lengthscales:

•**Mean separation** $n^{-1/3}$ The comoving number density of collapsed objects above a critical mass M_* is given by Press-Schechter theory: $n(z) = \Delta t \int_{M_*}^{\infty} \dot{n}_{PS}(M, z) dM$, where the creation rate of dark matter halos is given in [4]. The critical mass M_* is given by the Jeans mass after reheating, $M_{Jeans} \sim 10^9 (1 + z/10)^{-3/2} M_\odot$. Before reheating, it is given by the virial mass necessary for efficient atomic cooling $M_{cool} \sim 10^8 (1 + z/10)^{-3/2} M_\odot$. The source lifetime is assumed to be $\Delta t \sim 10^7$ yrs, the characteristic timescale for both starbursts and quasars.

CP470, After the Dark Ages: When Galaxies were Young (the Universe at 2 < z < 5),
edited by Stephen S. Holt and Eric P. Smith

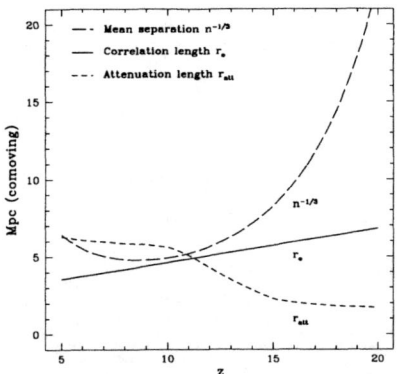

FIGURE 1. The evolution of various lengthscales with redshift. The increasing bias factor with redshift counteracts the fall of the linear growth factor, causing r_o to remain almost constant. The downturn of r_{atten} at low redshift is caused by the increased clumping of the gas, which shortens the recombination time.

- **Attenuation length** r_{atten} This is modelled as $\min(r_{strom}, r_{life}, r_{Lyman})$, where r_{strom} is the Stromgren radius, r_{life} is the radius of the volume ionized by a source during its lifetime, and r_{Lyman} is the mean free path of photons bounded by the covering factor of Lyman limit systems. We assume values for the clumping of the IGM computed in [5]. The covering factor of Lyman limit systems is computed by assuming a mini-halo model (e.g., [6]), and computing their abundance via Press-Schechter theory.

- **Correlation length** r_o The evolution of the matter correlation function with redshift is given by integrating the power spectrum. We assume a linear bias model $\xi_{hh}(r) = b^2 \xi_{mm}(r)$. Given the linear bias factor $b(M)$ as in [7], we then compute $b(M, z)$ by assuming linear theory and thus compute a number weighted bias factor at each epoch $\tilde{b}(z)$:

$$\tilde{b}(z) = \frac{\int_{M_*}^{\infty} b(M, z) n(M, z) dM}{\int_{M_*}^{\infty} n(M, z) dM} \tag{1}$$

The high bias means that the correlation length of halos is large, even though the correlation length of matter falls with redshift. For example, $\tilde{b} \sim 7$ at z=12, which leads to $r_o^{halo} \sim 3.3 \mathrm{Mpc}h^{-1}$ in comoving coordinates, only somewhat weaker than the clustering of objects seen today. We also calculate the evolution of the 3 and 4 point correlation functions, assuming standard hierarchical scaling relations, $\bar{\xi}_N(r) = S_N \bar{\xi}_2^{N-1}(r)$. Note that the S_N, although scale invariant, are a function of bias as well.

To estimate the fluctuations in the radiation field, we conduct a Monte-Carlo simulation, in the spirit of [4]. We generate a mock catalog of ionizing sources in

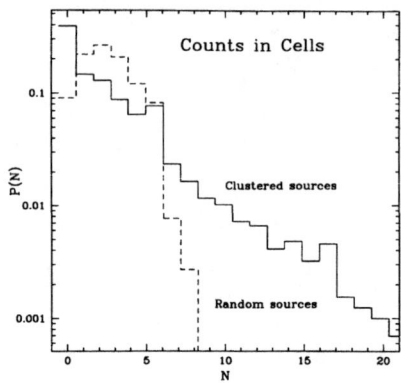

FIGURE 2. The probability distribution function of counts in cells at $z \sim 12$ for clustered and Poisson distributed sources, for a sphere of radius $r_{atten} \sim 4$ Mpc (comoving). The clustered case has a high N tail, where many low luminosity sources can mimic an ultra luminous source. Note also the greatly increased probability for N=0 (voids) in the clustered case.

a box 150 Mpc comoving size at $z_{reion} \sim 12$. A fractal model [8] is used to lay down sources with specified 2, 3 & 4 point correlation functions, with appropriate modelling of the bias factors. This has the chief advantage of speed and convenience over using N-body simulations. It also does not suffer from mass resolutions effects in identifying halos. We verify that the 2 point correlation function and the 2nd, 3rd and 4th moments of counts in cells (which are directly related to the volume-averaged correlation functions, with corrections for shot noise) are correctly recovered. Luminosity is then randomly assigned to each source, assuming a Press-Schechter mass function, and a constant emissivity per unit mass, as given by [9] for their mini-quasar model. The ionizing flux at random points in the box is then computed, assuming each source ionizes out to a radius r_{atten}. We thus compute the distribution function of the ionizing flux, as well as the two point correlation function $\xi_{JJ}(r) = < (J(0) - < J >)(J(r) - < J >) > / < J >^2$.

CONCLUSIONS & OBSERVATIONAL IMPLICATIONS

Clustering of ionizing sources produces a large dispersion in the luminosity of random cells. Clustered sources collectively mimic an ultra-luminous source. In addition, large voids exist where no ionizing sources are present. These are ionized by the percolation of ionizing radiation. Thus, large intensity fluctuations arise in the radiation field, which show coherence on large scales.

This has important observational implications. Emission line searches (e.g. in Lyα and Hα) at high redshift have a stronger signal due to the higher intensity fluctuations. Furthermore, CMB fluctuations due to Thomson scattering off elec-

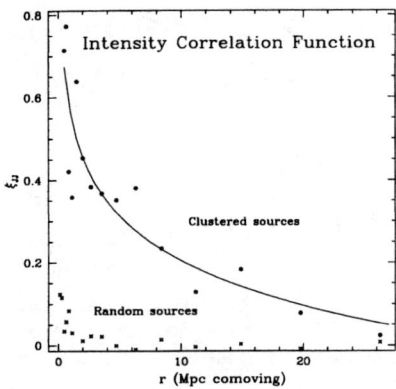

FIGURE 3. The two-point correlation function of ionizing intensity at $z \sim 12$ as a function of separation, assuming a sharp cutoff in ionizing flux at r_{atten}. In the clustered case, the radiation field displays correlations over large lengthscales.

trons in moving ionized patches [10] are enhanced, both due to the larger patch sizes and the spatial correlations of the ionized patches. Finally, uneven reheating provides a possible biasing mechanism.

Acknowledgements I thank my advisor, David Spergel, for his support and encouragement. I also thank Michael Strauss for helpful conversations.

REFERENCES

1. Fardal M.A., & Shull, J.M. 1993, ApJ, 415, 524
2. Zuo L. 1992, MNRAS, 258, 45
3. Ostriker, J.P., & Steinhardt, P. 1995 Nature, 377, 600
4. Sasaki, S., 1994, PASJ, 46, 427
5. Gnedin, N., & Ostriker, J.P. 1997, ApJ, 486, 581
6. Abel, T., & Mo, H.J. 1998, ApJ 494, L151
7. Mo, H.J., & White, S.D.M. 1996, MNRAS 282, 347
8. Soneira, R.M. & Peebles, P.J.E 1978 AJ 83, 845
9. Haiman, Z., & Loeb, A. 1998, ApJ, 503, 505
10. Knox, L., Scoccimarro, R., Dodelson, S., 1998, Phys. Rev. Letters 81,2004

Making Damped Lyman-α Systems in Semi-Analytic Models

Ariyeh H. Maller[1], Rachel S. Somerville[2], Jason X. Prochaska[3] and Joel R. Primack[1]

[1] *Physics Department, University of California, Santa Cruz, CA 95064, USA*
[2] *Racah Institute of Physics, The Hebrew University, Jerusalem, 91904, Israel*
[3] *Observatories of the Carnegie Institution of Washington, Pasadena CA 91101*

Abstract. The velocity profiles of weak metal absorption lines can be used to observationally probe the kinematic state of gas in damped Lyman-α systems. Prochaska and Wolfe [5] have argued that the flat distribution of velocity widths (Δv) combined with the asymmetric line profiles indicate that the DLAS are disks with large rotation velocities ($\sim$200 km/s). An alternative explanation has been proposed by Haehnelt, Steinmetz, and Rauch (HSR) [2], in which the observed large velocity widths and asymmetric profiles can be produced by lines of sight passing through two or more clumps each having relatively small internal velocity dispersions. We investigate the plausibility of this scenario in the context of semi-analytic models based on hierarchical merging trees and including simple treatments of gas dynamics, star formation, supernova feedback, and chemical evolution. We find that all the observed properties of the metal-line systems including the distribution of Δv and the asymmetric profiles, can be reproduced by lines of sight passing through sub-clumps that are bound within larger virialized dark matter halos. In order to produce enough multiple hits, we find that the cold gas must be considerably more extended than the optical radius of the proto-galaxies, perhaps even beyond the tidal radius of the sub-halo. This could occur due to tidal stripping or supernova-driven outflows.

INTRODUCTION

Damped Lyman alpha systems (DLAS), by probing the gas content of the universe at high redshift, are powerful observational probes to study galaxy formation and evolution. Until recently the observations of DLAS included the differential density distribution $f(N)$, its evolution with redshift, and the metalicity of the absorbers [3]. Many authors have shown that hierarchical cosmologies with models of galaxy formation can match this data. Most notable is the work of Kauffmann [4] which uses semi-analytic models (SAMs) to match $f(N)$ and its evolution with redshift, and at the same time reproduces many features of galaxies observed lo-

CP470, *After the Dark Ages: When Galaxies were Young (the Universe at 2 < z < 5)*,
edited by Stephen S. Holt and Eric P. Smith

cally in emission. A general prediction of Kauffmann's models is that the galaxies producing DLAS will typically be smaller then galaxies today.

Prochaska and Wolfe [1,6], hereafter PW, introduced new data by studying the kinematic properties of DLAS as deduced from high resolution spectroscopy of their weak metal lines. They found that DLAS seem to have a wide range of velocity widths, and are asymmetric in velocity space. The only model they tried that could explain this had thick disks with rotational velocities ~ 220 kms^{-1}, which is incompatible with what is expected in models of hierarchal structure formation at the typical redshifts $z \sim 3$ of these observed DLAS. PW excluded the hierarchical single-disk model of Kauffmann.

Recently HSR showed using hydrodynamic simulations and the Press-Schechter approximation that this seeming incompatibility can be reconciled if the DLAS come from merging proto-galactic clumps. Then the DLAS are composed of a few objects, and the large velocity widths come from the motions of these objects within a larger dark matter halo. McDonald and Miralda-Escudé [7] have tested this in an analytic model. We set out to investigate if such a model is feasible in the context of SAMs, which have allowed us to include star formation and feedback, and ultimately to include also a wider range of CDM-type cosmological models.

THE SEMI-ANALYTIC MODELS

SAMs and hydrodynamical simulations have their individual strengths and weaknesses. Hydrodynamical simulations include gas dynamics and gravitation, but this makes them computationally expensive and thus limited in their ability to explore parameter space or even get adequate statistics from the small volume simulated at high resolution. Also because of this it can be difficult to understand in simple terms what is going on in the simulation. Thus while HSR showed that their model is consistent with the PW kinematic data, they only simulated one cosmology and they did not include feedback or metal production.

In contrast, SAMs attempt to capture the most essential aspects of galaxy formation, in a simplified manner. Thus instead of following the gravitational motion of particles a merger history is assigned statistically to a halo. Gas cooling, star formation, and feedback are treated by simple equations. The fact that SAMs are capable of matching many observations suggests that these simpler treatments are successful in reproducing the essential features of galaxy formation. We use the SAMs of Somerville and Primack [10] and Somerville, Primack and Faber (SPF) [9], which are similar to the work of Kauffmann [4].

In particular, we use the fiducial model of SPF, which reproduces the observed number densities of the Lyman break galaxies and $\Omega_{gas}(z)$ and metalicities $Z(z)$ measured from DLAS. In previous work such as PW's analysis of the results of Kauffmann, it was assumed that each observed velocity width arose from the internal rotation velocity of a single disk. The main new feature here is that we explicitly investigate the implications of the substructure in the matter halos for

103

the metal-line kinematics, and different ways of distributing gas within the proto-galaxies. If we are to reproduce the kinematics of DLAS then, the large velocity widths must come from a combination of the rotational motion of the gas and the relative motion between the objects in one halo.

MAKING THE DLAS

We analyze SCDM ($\Omega_{matter} = 1$, $\sigma_8 = 0.67$) with the star formation and feedback described in SFP's fiducial model at a redshift of 3.2. This gives us the number of satellites in a given halo, their distances from the central object, and the amount of cold gas in each object. We position the satellites in the halo randomly in angle along circular orbits. The only quantity not specified by the SAMs is how the cold gas is distributed in each galaxy. We construct five models for the radial distribution of the gas, to explore how the radial profiles affect the observed kinematics (Table 1). We distribute the gas with an exponential of isothermal radial profile N(R), and different normalizations. We compare these models to the data by passing random lines of sight through a given halo, with each halo weighted by the probability of encountering that halo as determined by the Press-Schechter formalism. Those lines of sight that intersect cold gas in excess of $2 \times 10^{20} cm^{-2}$ are labeled as DLAS. Then the kinematics of these systems are investigated by assuming the gas disks have metalicities given by the SAMs and a scale height of one tenth the stellar disks scale length. Spectra are simulated and then analyzed by the same four statistics as the data was in PW, the velocity width ΔV, the mean to median test f_{mm}, the edge leading test f_{edg}, and the two peak test f_{tpk}. We also check to see that we are getting a reasonable distribution of column densities $f(N)$ (Figure 1a). All models but the first and third fit the DLAS $f(N)$ reasonably well.

CONCLUSIONS

Of the models we explore only the last model comes close to matching all of PW's kinematic data, according to the results of a Kolmogorov-Smirnov test (table 1). The radial extent of the gas in this model is larger than the other models
(Figure 1b) and this leads to its producing more multiple hits along a given line of sight than any other model.

In fact the gas extends out beyond the tidal truncation radius (the limit of the fourth model), where it would no longer be gravitationally bound. However, our method of calculating the tidal radius is only approximate, and the large radial extent of the gas can perhaps alternatively be interpreted as due to stripping or outflow. If the gas is stripped away by ram-pressure stripping, or ejected from the small proto-galaxies by supernovae, it may still cover an equivalent amount of area, just not as a disk. We are investigating this more thoroughly.

Our main conclusion is that it is possible to produce the observed kinematic properties of DLAS in SAMs based on CDM-type models by having lines of sight

TABLE 1. Five models of the gas distribution and their KS test results

Model	radial profile	normalization	ΔV	f_{mm}	f_{edg}	f_{tpk}
1	e^{-R}	$R_{gas} = R_{disk}$	0.001	0.008	0.001	0.001
2	e^{-R}	$R_{gas} = 7R_{disk}$	0.001	0.66	0.34	0.058
3	$1/R$	$N = 0.59V_c/R$	0.001	0.004	0.001	0.001
4	$1/R$	$R_{trunc} = R_{tidal}$	0.001	0.350	0.24	0.013
5	$1/R$	$log_{10}(N_{trunc}) = 19.3$	0.51	0.080	0.20	0.66

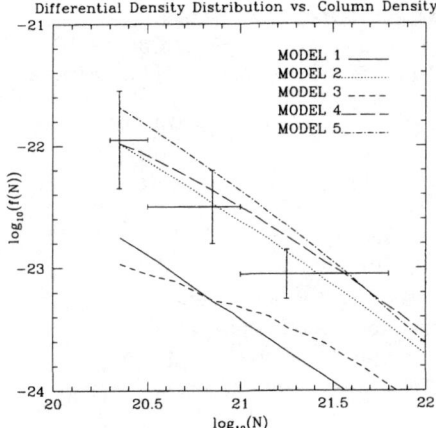
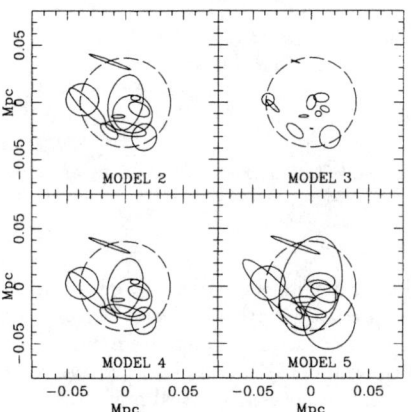

FIGURE 1. a)The left panel shows the five models $f(N)$ distribution, compared to the data from [3]. b)The right panel shows the extent of the gas disks in the models 2-5 for a typical halo with circular velocity of $156 kms^{-1}$, the dashed line is the virial radius for this halo.

pass through multiple objects in the same halo; however, this requires that the cold gas in these proto-galaxies be rather large in radial extent.

REFERENCES

1. Prochaska, J. X., & Wolfe, A. M. 1997, ApJ, 487, 73
2. Haehnelt, M., Steinmetz, M. & Rauch, M. 1998, ApJ, 495, 647
3. Storrie-Lombardi, L., Irwin, M.. & McMahon, R. 1996, MNRAS, 282, 1330
4. Kauffmann, G. 1996, MNRAS, 281, 475
5. Prochaska, J. X., & Wolfe, A. M. 1996, ApJ, 470, 403
6. Prochaska, J. X., & Wolfe, A. M. 1998, ApJ, 507, 113
7. McDonald, P. & Miralda-Esudé, J. 1998, (asto-ph/9809237)
8. Somerville, R. S., & Primack, J. R. 1998, MNRAS, in press (astro-ph/9802268)
9. Somerville, R. S., Primack, J. R., & Faber, S. 1998, MNRAS, in press (astro-ph/9806228)

The Lyman Alpha Forest in Hierarchical Cosmologies

M. Machacek[1], G.L. Bryan[2], P. Anninos[3], A. Meiksin[4], M.L. Norman[5], and Y. Zhang[3]

[1] *Physics Department, Northeastern University, Boston, MA 02115*
[2] *Physics Department, Massachusetts Institute of Technology, Cambridge, MA 02139*
[3] *Laboratory of Computational Astrophysics, National Center for Supercomputing Applications, 405 Matthews Ave, Urbana, IL 61801*
[4] *Institute of Astronomy, University of Edinburgh, Royal Observatory, Blackford Hill, Edinburgh EH9 3HJ, UK*
[5] *Astronomy Department, University of Illinois at Urbana-Champaign, Urbana, IL 61801*

Abstract. The comparison of quasar absorption spectra with numerically simulated spectra from hierarchical cosmological models of structure formation promises to be a valuable tool to discriminate among these models. We present simulation results for the column density, Doppler b parameter, and optical depth probability distributions for five popular cosmological models.

INTRODUCTION

A physical picture of the Lyα forest in hierarchical cosmologies has recently emerged from numerical simulations [1,2] in which the absorbers that give rise to low column density lines ($N_{HI} < 10^{15}$ cm^{-2} at $z \sim 3$) are large, unvirialized objects with sizes of ~ 100 kpc and densities comparable to the cosmic mean. Since the absorbers grow from the primordial density fluctuations through gravitational amplification, statistics of the forest may be used to test various models of structure formation. We discuss the numerical stability of the statistics against changes in simulation box size and spatial resolution in detail elsewhere. [2] We focus here on examples from our model comparison study [3] in which statistics of the Lyα forest are computed in five cosmological models: the standard cold dark matter model (SCDM), a flat cold dark matter model with nonvanishing cosmological constant (LCDM), a low density cold dark matter model (OCDM), a flat cold dark matter model with a tilted power spectrum (TCDM), and a critical model with both cold dark matter and two massive neutrinos (CHDM). The initial fluctuations, assumed to be Gaussian, are normalized using $\sigma_{8h^{-1}}$ to agree with the observed

CP470, *After the Dark Ages: When Galaxies were Young (the Universe at 2 < z < 5)*,
edited by Stephen S. Holt and Eric P. Smith
© 1999 The American Institute of Physics 1-56396-855-X/99/$15.00

distribution of clusters of galaxies, although all but SCDM are also consistent with the COBE measurements of the cosmic microwave background. By varying $\sigma_{8h^{-1}}$ within a given model (SCDM), we also investigate the dependence of the statistics on changes in the fluctuation power spectrum.

The simulation technique uses a particle-mesh algorithm to follow the dark matter and the piecewise parabolic method [4] to simulate gas dynamics. The simulation box length is 9.6 Mpc (comoving) with spatial resolution of $18.75h^{-1}$ $(37.5h^{-1})$ kpc for the model (power) comparison studies, respectively. Nonequilibrium effects are followed [5] for six particle species (HI, HII, HeI, HeII, HeIII, and the electron density). We assume a spatially-constant radiation field computed from the observed QSO distribution [6] which reionizes the universe around $z \sim 6$ and peaks at $z \sim 2$. Synthetic spectra are generated along 300 random lines of sight through the volume, including the effects of peculiar velocity and thermal broadening. The spectra are normalized to give a mean optical depth $< \tau >= 0.3$ at $z = 3$ to agree with observation. We do not include the effects of radiative transfer, self-shielding or star formation and so can not address the physics of the highest column density absorbers $(N_{HI} > 10^{16} \text{ cm}^{-2})$.

FIT DEPENDENT STATISTICS

The synthetic spectra are fit by Voigt profiles at these low column densities to obtain column densities and Doppler widths for each line. The slope of the column density distribution is insensitive to changes in the size of the simulation volume or grid resolution. [2] Fig. 1 confirms analytic work [7] that the slope of the column density distribution depends primarily on the power in the model at scales $\sim 100 - 200$ kpc and steepens for models with lower power. Each of the five cosmologies in our comparison study agrees with the data at the 3σ level, although models (SCDM,LCDM,OCDM) with moderate power at these scales are favored. The Doppler b parameter distributions are determined not only by thermal broadening and peculiar velocities of the absorbers, but also by the Hubble expansion across their width, and require high spatial resolution to be modeled properly. [2,8] The shape of the distribution including the high b tail is well fit by hierarchical models. However, as shown in Fig. 2, observations by Kim, et. al [9] yield median b parameters significantly higher than predicted for the models favored by the column density distributions (LCDM, OCDM, SCDM).

CONCLUSION

Although hierarchical cosmologies reproduce the general characteristics of the Lyα absorption spectra quite well, detailed tests of the models may require new methods of analysis for the simulations and observations. Statistics derived directly from the flux or optical depth distributions without recourse to any line fitting algorithm are particularly interesting. [11] For example, the optical depth probability

107

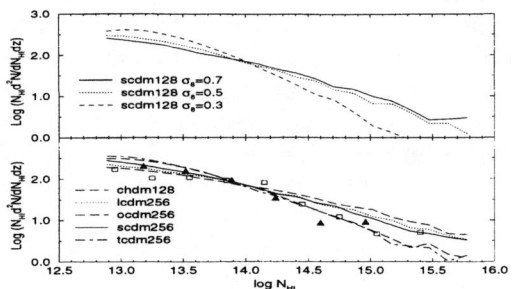

FIGURE 1. HI column density distributions at $z = 3$ for lines with N_{HI} in the range $10^{12.8} - 10^{16}$ cm^{-2}. Results for SCDM varying $\sigma_{8h^{-1}}$ (top); results for the five cosmological models (SCDM, LCDM, OCDM, TCDM, CHDM) are plotted together with observed data from Kim, et. al [9] (open squares, $z = 2.85$) and Kirkman & Tytler [10] (filled triangles, $z = 2.7$)(bottom).

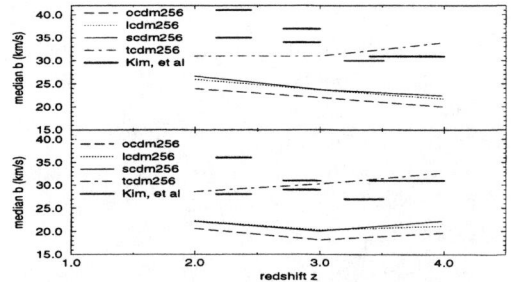

FIGURE 2. The evolution of the median of the b parameter distribution with redshift for SCDM, LCDM, OCDM, & TCDM compared to the observed data by Kim, et. al [9] for lines with N_{HI} in the ranges $10^{13.8} - 10^{16}$ cm^{-2} (top) and $10^{13.1} - 10^{14}$ cm^{-2} (bottom).

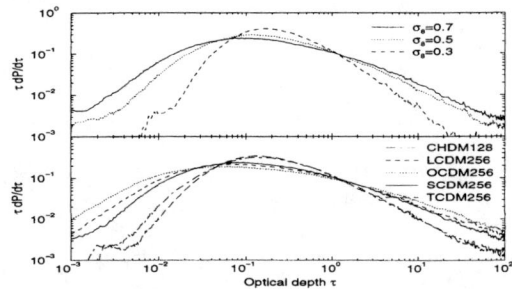

FIGURE 3. Optical depth probability distribution at $z = 3$. Results for SCDM with varying power ($\sigma_{8h^{-1}}$) (top); results for SCDM, LCDM, OCDM, TCDM, & CHDM (bottom).

distribution, like the column density distribution, is stable to changes in simulation spatial resolution. Fig. 3 shows that the distribution narrows for models with lower power and that the shape of the distribution varies significantly for the different models over the range $0.05 < \tau < 4$ accessible to observations. Comparison of high quality observations with high resolution simulations using an ensemble of such statistics may soon clarify the physical properties of the intergalactic medium at intermediate redshifts when galaxies were young.

This work is done under the auspices of the Grand Challenge Cosmology Consortium and supported in part by NSF grant ASC-9318185 and NASA Astrophysics Theory Program grant NAG5-3923.

REFERENCES

1. Cen, R., Miralda-Escude, J., Ostriker, J. P., & Rauch, M. 1994, ApJ, 437, L9; Zhang, Y., Anninos, P., Norman, M. L., & Meiksin, A. 1997, ApJ, 485, 496; Hernquist, L., Katz, N., Weinberg, D., & Miralda-Escude, J. 1996, ApJ, 457, L51

2. Bryan, G. L., Machacek, M., Anninos, P. & Norman, M. L. 1998, ApJ, in press (astro-ph/9805340)

3. Machacek, M., Bryan, G. L., Meiksin, A., Anninos, P., Thayer, D., Norman, M. L., & Zhang, Y. 1998 (in preparation)

4. Bryan, G. L., Norman, M. L., Stone, J. M., Cen, R., & Ostriker, J. P. 1995, Comput. Phys. Comm.,89, 149

5. Anninos, P., Zhang, Y., Abel, T., & Norman, M. L. 1997, New Astronomy, 2, 209

6. Haardt, F. & Madau, P. 1996, ApJ, 461, 20

7. Gnedin, N. Y. 1998, MNRAS, submitted (astro-ph/9706286)

8. Theuns, T. Leonard, A., & Efstathiou, G. 1998 MNRAS, 297, L49

9. Kim, T.-S., Hu, E. M., Cowie, L. L., & Songaila, A. 1997, AJ, 114, 1

10. Kirkman, D. & Tytler, D. 1997, ApJ, 484, 672

11. Rauch, M., Miralda-Escude, J., Sargent, W. L. W., Barlow, T. A., Weinberg, D.H., Hernquist, L., Katz, N., Cen, R., & Ostriker, J. P. 1997, ApJ, 489, 7

4. Galaxy Formation Renaissance

NICMOS Observations of the HDF

Rodger I. Thompson*, Ray J. Weymann†, Lisa Storrie-Lombardi†

*Steward Observatory, University of Arizona, Tucson, AZ 85721
† Carnegie Observatories, 813 Santa Barbara St., Pasadena, CA, 91101-1292

Abstract. This paper presents initial results and performance levels from the Near Infrared Camera and Multi-Object Spectrometer (NICMOS) observations of the Hubble Deep Field (HDF). These observations represent the deepest view of individual objects yet obtained with photometric colors of some objects indicating redshift values greater than 6. These observations add significant value to the previous optical observations of the HDF with the Wide Field and Planetary Camera II (WFPC II).

INTRODUCTION

A major consideration in the instrument design and initial scientific program of NICMOS was the realization that high sensitivity imaging at near infrared wavelengths ($0.8 - 2.5\mu m$) offers a deeper view of distant galaxies than optical measurements that are limited by hydrogen continuum and Lyman line absorption of the galactic light. In particular the wide field NICMOS camera 3 was the best mode for deep observations. The minimum background for the NICMOS instrument occurs at $1.6\mu m$ at the intersection of the zodiacal light reflection that rises toward shorter wavelengths and the thermal emission from the HST mirrors that rises toward longer wavelengths. This minimum background region is exploited by the wide band filter F160W centered on the $1.6\mu m$ wavelength with a spectral width of approximately $0.4\mu m$.

The advent of original HDF observations [7] offered an excellent opportunity to combine deep near infrared and optical imaging over a limited area of the sky. The NICMOS Guaranteed Time Observation (GTO) orbit allocation contributed 127 orbits for the conduct of this program. The majority of the orbits were for imaging but a few orbits utilized the slitless grism mode for wide field spectroscopy. This paper will only consider the imaging mode observations.

One of the main products of the WFPC II HDF observations is a catalog of the photometric properties of all of the detected sources in the region [7]. A similar catalog of the NICMOS sources is now available [4]. Since this publication contains

CP470, After the Dark Ages: When Galaxies were Young (the Universe at 2 < z < 5),
edited by Stephen S. Holt and Eric P. Smith
© 1999 The American Institute of Physics 1-56396-855-X/99/$15.00

the detailed images they will not be repeated here.

OBSERVATIONAL PARAMETERS

The implementation of a high sensitivity, low flux limit imaging program requires careful selection of the observational parameters. These parameters include the camera selection, filters, detector read out modes, and allocation of orbits between imaging and spectroscopic modes. Also included is the exact spatial positioning of the observations.

Camera and Filter Selection

As indicated in the introduction the NICMOS camera 3 was designed with deep imaging as an important priority and is the obvious choice for the observations. Although diffraction limited, cameras 1 and 2 have too limited a field of view for effective areal coverage. The 0.2 arc second pixel size of camera 3 was chosen to be dominated by the natural zodiacal background over the dark current while maintaining a relatively high resolution. Camera 3 was therefore the camera of choice.

Also, as indicated in the introduction, the F160W is designed for low background deep imaging and is therefore an obvious filter choice. A second filter is required to provide color information on high red shift sources that have little or no flux in the optical bands. Without this second filter it is not possible to discriminate between high redshift and heavily extincted galaxies. The exponentially rising thermal flux at longer wavelengths precluded picking a longer wavelength filter. The selected second filter is the wide band F110W filter centered at $1.1\mu m$. Although this filter has significant spectral structure it provides greater signal to noise observations and higher spectral discrimination than any other available filter.

Spatial Distribution of the Observations

Since the field of view the NICMOS camera 3 is significantly less than the WFPC II field there is a choice between coverage of the whole HDF at lower signal to noise or concentration on one area to obtain the deepest possible image. Since our main interest is in high redshift objects beyond the reach of the optical HDF we chose to concentrate on only one area of the HDF. Selection of the particular area was governed in part by the grism observations. These observations require an area free of large bright galaxies whose spectra would overlap large areas of the detector. For this reason we chose our area to be roughly centered on chip 4 of the WFPC image, an area relatively sparse in large galaxies. The orientation of the image was dictated by the scheduling of the observations and available guide stars rather than by scientific criteria.

Once the choice of field is made the exact positioning of the individual images is important. These choices affect the total area covered, the ability to form background images for subtraction and the reduction of artifacts due to pixel to pixel variations in response. Production of the background image requires images separated by enough distance that sources do not predominantly fall on the same group of pixels. High signal to noise on the other hand requires that most of the region is well covered which argues for short steps between images. The number of orbits allocated to the total program must also be in a reasonable range.

Our choice for this program was a 7 X 7 rectangular grid requiring 49 orbits for each filter. Within each orbit there are three integrations each of which is offset from each other by a small step much smaller than the space between grid points. Our orientation placed the field at an approximately 30 degree angle to the original HDF field, therefore, we chose a grid step of 0.918 arc seconds in the X direction and 1.523 arc seconds in the Y direction to produce a slightly rectangular observation grid. Three integrations were taken at each grid point which consumed one orbit. The three integrations were dithered by 0.408 arc seconds in the X and Y direction so that all integrations had each pixel on a different portion of sky. This resulted in 147 integrations per filter and a total integration time per filter of 1.3×10^5 seconds.

A   Areal Coverage

The total area covered by at least one image is a rectangle of 56.91 by 61.39 arc seconds and the area covered by all integrations is 46.85 by 43.01 arc seconds. The image area covered by the listed catalog sources is 49.19 by 48.53 arc seconds. This choice was a subjective judgement on the region containing good signal to noise not compromised by a low number of integrations. Since the quantum efficiency of the NICMOS detectors varies significantly across the array the appropriate area of coverage was not a straightforward calculation.

B   Detector Parameters

From the selection of standard readout patterns we chose the SPARS 64 detector readout for all of the integrations. This pattern has three readouts closely spaced then a readout every 64 seconds. With the number of readouts set to 17 (NSAMP=17) the total integration time was 896 seconds for each image. This choice provided three separate integrations per orbit and enough individual readouts to allow excellent removal of cosmic ray hits. Before the observations it was thought that the integration time would produce a background noise from the zodiacal light roughly equal to the read noise. The zodiacal light at the location of the HDF turned out to be significantly lower than expected (0.55 e/sec), therefore, each integration was read noise limited.

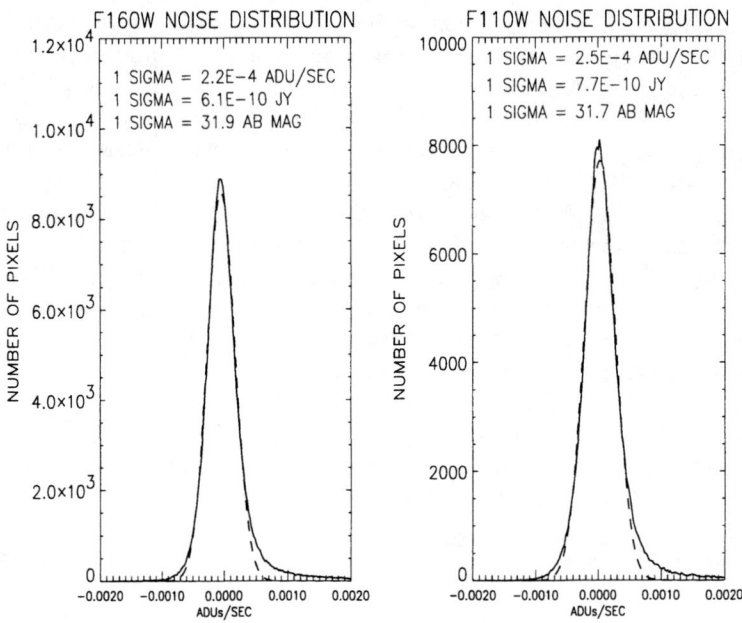

FIGURE 1. Histogram of pixel values for the F160W and F110W images. The 1 sigma values must be multiplied by a factor of 1.9 to account for the correlation of drizzled pixels

I   Performance Levels

The basic result is that the image signal to noise ratio is equal to the ratio calculated from the known read noise, dark current, quantum efficiency and observed sky background. This means that we did not have contributions from previously unknown noise sources or systematic effects that compromised the signal integrity. Aggressive scheduling of the HDF observations by STScI resulted in few integrations that were contaminated by cosmic ray persistence due to recent passage through the South Atlantic Anomaly (SAA). The net usable integrations were 122 in the F160W filter and 118 in the F110W filter.

Figure 1. shows the distribution of flux levels per pixel in ADUs per second for both the F160W and F110W filter. The pixels referred to in this figure are the drizzled [2] pixels of size 0.1 by 0.1 arc seconds instead of the physical pixel size of 0.2 by 0.2 arc seconds. Since the drizzeling process produces correlated noise the numbers in the one sigma numbers in the figure must be multiplied by a factor of 1.9 [7]to produce the proper 1 sigma noise number. The one pixel sigmas are then 1.5×10^{-9} Jy (31.0 AB mag) for the F110W filter and 1.2×10^{-9} Jy (31.2 AB mag)for the F160W filter.

Each of the 896 second integrations that did not have significant residual cosmic

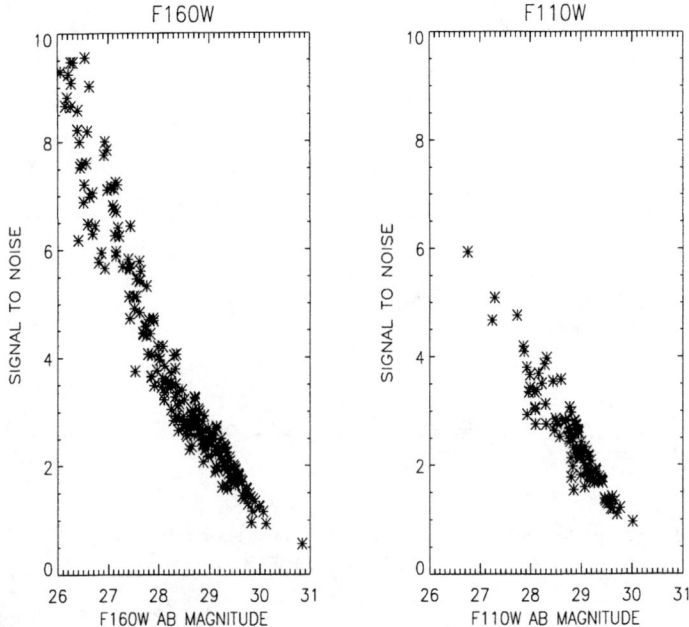

FIGURE 2. Distribution of the signal to noise ratio versus magnitude for the detected sources. Only sources with signal to noise ratios less than 10 are plotted.

ray noise from a SAA passage was used in the final mosaic image. Histograms of the individual images displayed the expected signal to noise ratio from the parameters given above. The final histogram displayed in Fig. 1 had a FWHM reduced by a factor of the square root of the number of integrations from the average individual integration FWHM. This is further indication that systematic errors did not dominate in the final image. It should be noted that the single pixel error is not the expected error level for galaxy detection. Since galaxies are spread over several pixels they will have a higher error level. Figure 2 shows the distribution of signal to noise versus magnitude for the detected sources. Some of the spread in this plot, particularly at higher signal to noise levels is due to sources of the same magnitude spread over a different number of pixels.

II   Results of the Observations

One of the main results of these observations is a catalog of sources with measured signal to noise levels greater or equal to 2.5 [4]. The electronic version of that reference contains an additional catalog of detected sources with signal to noise ratios less 2.5. One of the main purposes of the additional catalog is to provide

117

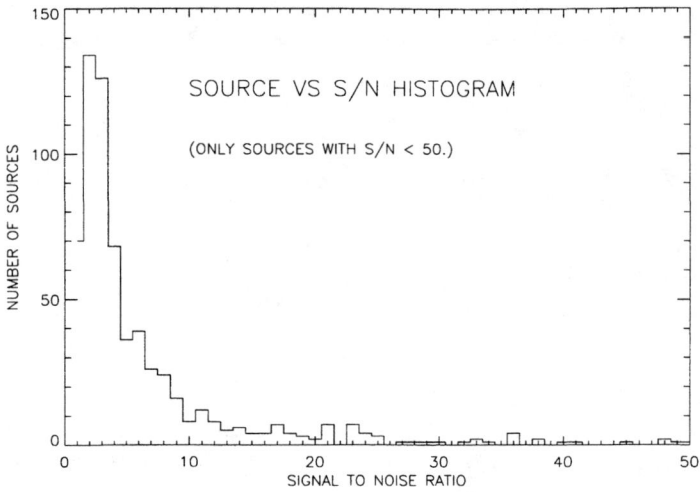

FIGURE 3. Histogram of the number of sources versus signal to noise ratio. The figure has been truncated at a signal to noise level of 50 for clarity.

information of the limits of source level detection for researchers comparing results from other wavelength regions. Figure 3 shows a distribution of sources versus signal to noise in 1.0 sigma bins. The figure is cut at a signal to noise ratio of 50 for clarity, although there are several sources with higher signal to noise ratios.

The NICMOS catalog does contain correspondence information with the WFPC II catalog [7] which gives the identification and distance in arc seconds to the nearest WFPC II source. A NICMOS and WFPC II source are considered to correspond if their separation is less than 0.25 seconds of arc. In some cases differing morphology between the infrared and optical regions results in false noncorrespondance in a source.

In the following results we will refer only to those sources that appear in the catalog of sources with signal to noise ratios greater than or equal to 2.5. This catalog contains 342 separately listed objects.

A   High Redshift Objects

Of the 342 objects in the catalog, 107 objects have no corresponding WFPC II identification. A few of these are due to the differences in infrared and optical morphology mentioned above. Most of these objects are highly reddened galaxies which are too faint at optical wavelengths to be detected in the original HDF observations. A few of these galaxies, however, may be truly high redshift galaxies. These galaxies have little or no flux in the WFPC II F814W filter but have flat or blue color in the NICMOS F110W - F160W bands. It is presumed this flux distribution is due to the Lyman limit falling in the F814W band and the young blue galaxy UV flux falling in the NICMOS F110W and F160W bands. It is possible that at low flux levels this could also be caused by the 4000 Angstrom break falling in the F814W band but in most cases the lack of flux in the WFPC II F606W filter is inconsistent with this interpretation. At present about 10 to 20 objects are high redshift candidates. One of these objects is NICMOS source 184.0 which corresponds to WFPC source 4-473. This source has a spectroscopicly confirmed redshift of 5.61 [6]. It has low F814W flux, but enough to be detected by WFPC, and a blue F110W - F160W color. It is bright enough in the infrared that even if it were dimmed to a redshift of 10 it would still be visible in the NICMOS images. It is strong evidence that if there are sources of this type at redshifts greater than 5.6 they are visible in the NICMOS image. After a redshift of 10 almost all of the flux lies in the F160W band and these sources are indistinguishable from source with very high extinction. Detailed investigation of high redshift candidates is the subject of a paper in preparation [5].

B Small Blue Objects

Another result of inspection of the NICMOS images concerns the nature of the numerous small blue objects that appear in the original WFPC II HDF images. To date these objects have generally assumed to be small galaxies or perhaps fragments of galaxies that will merge at a later date. They can not be extremely distant objects since they have a significant amount of blue flux. Two point correlation studies [1] on the other hand indicate they are spatially more correlated than would be expected for a random distribution of galaxies.

Some clues to the nature of these sources comes from the inspection of WFPC source 4-378. This is clearly a spiral galaxy with bright star formation regions in its spiral arms. This source corresponds to NICMOS source 124.0. In the infrared the star formation regions are dominated by the old stellar population that clearly shows the nucleus of the galaxy, offset from the star formation regions. This source may be a Rosetta stone for the identification of many of the small blue sources. If 4-378 were somewhat fainter only the bright blue star formation regions would be visible in the optical images, similar to the isolated small blue objects. If fainter in the infrared only the dominant red nucleus would still be visible nearby but not coincident with the small blue objects. In fact when the optical and infrared images are overlayed small infrared objects are often found very near individual or groups of small blue objects. The separation distance of the objects are less than the radii of the several $z = 1$ elliptical and spiral galaxies in the field. Since in most cosmologies the angular size of galaxies either reaches a minimum or very slowly decreases past $z = 1$ the grouping of infrared and blue objects fall within the region of a single galaxy. A logical conclusion from these observations is that many of the small blue objects in the optical HDF image may simply be bright star formation regions in a dim galaxy whose nuclear region appears in the infrared image. Details of this analysis will appear in a forthcoming paper [3].

III Conclusions

The NICMOS images of part of the original HDF field gives us the most distant view of individual galaxies in the Universe to date. It provides also provides images of distant galaxies at rest wavelengths in the optical region where galaxies have been well studied. Several high redshift candidates have been identified but spectroscopic follow up will be difficult due to the faintness of the images. In addition we find very suggestive evidence that many of the small blue objects in the WFPC HDF image may be star formation regions in a much larger underlying galaxy.

As with the original HDF image most of the scientific discovery will come with detailed quantitative investigations. Many of these are underway and will come to fruition in the near future. One of the keys to these studies is the combination of the optical and infrared data from HST and data at many different wavelengths from ground based and other space based observatories.

IV Acknowledgements

A very large group of people contributed to the success of the NICMOS HDF observation project. The entire NICMOS Instrument Definition Team contributed both the expertise to make NICMOS a reality and the large number of orbits from the GTO program needed to carry out the observations. We thank Mark Dickinson and Andy Fructher for their help with the reduction and mosaic techniques for the data analysis. Andy Lubenow spent many hours optimizing the observational techniques. Chris Conner and the Lockheed group made an extraodinary effort to keep our pointing accurate during our single guide star integrations. Zolt Levay provided the combined NIMOS and WFPC color image. The work was supported by NASA grant NAG 5-3043 and the observations were obtained with the NASA/ESA Hubble Space Telescope Science Institute managed by the Association of Universities for Research in Astronomy Inc. under NASA contract NAS5-26555.

REFERENCES

1. Colley, W.N., Gnedin, O.Y., Ostriker, J.P., and Rhoads, J.E. 1997, Ap. J., 488, 579.
2. Fruchter, A.S. and Hook, R.N. 1997, in Applications of Digital Image Processing XX, Proc. SPIE, Vol. 3164, ed. A.Tescher, pp 120-125.
3. Storrie-Lombardi, L., Weymann, R.J., and Thompson, R.I. 1999, in preparation.
4. Thompson, R.I., Storrie-Lombardi, L.J., Weymann, R.J., Rieke, M.J., Schneider, S., Stobie, E., and Lytle, D. 1999, A.J., in press.
5. Weymann, R.J., Thompson, R.I., & Storrie-Lombardi, L. 1999, in preparation.

6. Weymann, R.J., Stern, D., Bunker, A., Spinrad, H., Chaffe, F.H., Thompson, R.I., and Storrie-Lombardi, L., 1998, Ap.J. (Letters), 505, L95.

7. Williams, R.E., Blacker, B., Dickinson, M., Dixon, W.V.D., Ferguson, H.C., Fruchter, A.S., Giavalisco, M., Gilliland ,R. L., Heyer, Inge, Katsanis, R., Levay, Z., Lucas, R.A., McElroy, D.B., Petro, L., Postman, M., Adorf, H-M, and Hook, R.N. 1996, A.J. 112, 1335.

A Complete NICMOS Map of the Hubble Deep Field North

Mark Dickinson

Space Telescope Science Institute, Baltimore, MD 21218

Abstract. I present early results from a program to map of the complete Hubble Deep Field North (HDF–N) at 1.1μm amd 1.6μm with NICMOS on HST. The near–infrared data allow us to study galaxies at $1 < z < 3$ at rest–frame optical wavelengths where we are most familiar with the morphological and photometric properties of galaxies nearby. The images are sensitive to emission from redder, evolved stellar populations whose light redshifts out of the optical WFPC2 passbands at $z > 1$. The optical–infrared multicolor parameter space can be used to refine photometric redshift estimates for galaxies, and to search for galaxies at very large redshifts ($z > 5$). I present the observations, briefly discuss the morphological properties of the faint galaxy population, and illustrate one moderately bright object whose extremely unusual colors make it an interesting candidate for having a very large redshift.

INTRODUCTION

The Hubble Deep Fields (North and South, [1,2]) have provided a valuable resource for studying galaxies at high redshift, offering an unprecedentedly deep, clear, and polychromatic view of the distant universe. The original HDF–N data have been used for a variety of research programs, and have been supplemented by additional observations from both the ground and space at almost every accessible wavelength. In nearly every case, the supporting observations represent the state of the art: the deepest radio maps, the faintest redshift surveys, etc. More than 120 spectroscopic redshifts have been gathered (e.g. [3] and others) within this 5 arcmin^2 patch of the sky, extending to $z = 5.6$ [4]. There is little doubt that we will continue to learn from the the HDFs for years to come, and will observe them with the best new instruments as they come on line.

The HDF–N WFPC2 observations provided four–band imaging through the near–UV and optical range (0.3–0.8μm). At redshifts $z > 1$, where many (or even most) of the HDF galaxies lie, the optical rest–frame light where we best understand nearby galaxies is redshifted out of the WFPC2 passbands and into the near–IR. It becomes difficult, therefore, to reliably interpret the morphological and photometric properties of HDF galaxies at $z > 1$ without supplementary observations in the

CP470, After the Dark Ages: When Galaxies were Young (the Universe at 2 < z < 5),
edited by Stephen S. Holt and Eric P. Smith

near–infrared. Until recently, the only data available were ground–based near–IR observations from KPNO [6], Keck [7] and the UH 2.2m and CFHT [8]. However those data do not have nearly the depth nor the resolution of the optical WFPC2 images, resulting in a mismatch which leaves most HDF galaxies without useful infrared photometry or morphological information.

The availability of NICMOS on board HST changed this situation, and two observing programs have targeted the HDF–N for deep NICMOS imaging. NICMOS' small field of view ($52'' \times 52''$ for Camera 3) compared to that of WFPC2 (roughly $150'' \times 150''$) requires care in selecting an observing strategy, and the two projects were designed to be complementary. The NICMOS IDT (see [9] and this conference) observed a single NIC3 field, imaging through the F110W (J) and F160W (H) filters to produce the deepest near–IR view of the universe ever obtained (now supplemented by similar observations in the HDF–S [5]). Our own GO program took a "wide–field" approach, covering the HDF–N with a mosaic of shallower NIC3 pointings. We cover $\sim 8\times$ the solid angle of the GTO field, but at roughly 1/3rd the limiting sensitivity. These data ensure that all HDF galaxies have high angular resolution infrared images available.

THE OBSERVATIONS

The observations were made during the June 1998 NICMOS Camera 3 refocus campaign to ensure the best image quality. We observed a mosaic of 8 NIC3 fields, each dithered through 9 pointings. Each field was observed for 4.5 orbits through F110W and F160W, and the data were mosaiced to produce an image registered with the WFPC2 data. Table 1 summarizes the properties of the NICMOS and WFPC2 data sets. The drizzled NICMOS images have a PSF FWHM$\approx 0''\!.22$, limited by the $0''\!.2$ pixel scale of Camera 3.

Table 1: NICMOS and WFPC2 Observations of the HDF–N

Camera	Filter	λ (μm)	Mean Exposure (ksec)	Limiting AB magnitude[1]
NICMOS:	F160W	1.59μm	12.6	26.5
	F110W	1.10μm	12.6	26.5
WFPC2:	F814W	0.80μm	123.6	27.6
	F606W	0.60μm	109.1	28.2
	F814W	0.46μm	120.6	27.9
	F814W	0.30μm	153.7	27.0

[1]S/N=10 in an 0.2 arcsec^2 aperture

The NICMOS observations are roughly 2.9 and 3.6 magnitudes deeper than the best available ground–based HDF data at J and H, and have much higher angular resolution. Although in terms of limiting AB magnitudes they do not reach as deep as the WFPC2 data, the fact that *most* galaxies are brighter (in f_ν) at longer wavelengths ensures that we detect the majority of the galaxies visible in the WFPC2 images. Our infrared–selected catalog of this data detects ~ 1700 objects, compared to ≈ 3200 objects in the WFPC2 images [1].

GALAXY MORPHOLOGY

From the morphological point of view, the NICMOS data is most interesting for galaxies at $z \gtrsim 1$, where the optical rest–frame redshifts out of the WFPC2 passbands. NICMOS samples $\lambda_{\text{rest}} \leq 4000$Å for all galaxies out to $z \approx 3$, providing a direct means of comparing distant galaxies to local counterparts.

Galaxies at $1 \lesssim z \lesssim 2$

At a casual glance, the NICMOS view of the HDF appears to be generally quite similar to that seen by WFPC2 in the I–band. In detail, many galaxies have interesting and important wavelength dependent morphological differences (figure 1). As might be expected, galaxies become, on average, slightly more regular and centrally concentrated, even when the resolution difference between the WFPC2 and NICMOS images is taken into account. A quantitative asymmetry study of our images by Conselice et al. (in prep., see also this volume) confirms this trend. In many galaxies, red bulges are prominent in the near–IR. Some spirals at $z \approx 1$ develop prominent bars which are invisible at optical wavelengths. This suggests that caution is needed when interpreting recent optical studies which have suggested that the barred galaxy fraction at high redshift is low [11,12], although we have not yet carried out a proper statistical analysis to test this hypothesis. There are also very red galaxies which are quite bright at near–infrared wavelengths: these mostly appear to be early–type (E/S0 – spheroidally dominated) galaxies at $1 < z < 2$, whose old stellar populations dominate the near–IR light, but which fade from view at optical wavelengths. A few irregular red galaxies are visible whose colors and morphologies are probably affected by dust. One such object (shown in figure 1) is a good (albeit rare) example of a galaxy whose appearance is completely transformed as one moves from the optical to the IR – this galaxy was cataloged as several separate objects by Williams et al. (1996), but is clearly a single entity when viewed by NICMOS.

Morphologically Peculiar Galaxies

But in general, the universe does not appear radically different when viewed at twice the wavelength of the reddest WPFC2 bandpasses. Figure 1 includes an example of a "chain galaxy" at $z = 1.355$ whose WFPC2 (i.e. rest–frame UV) structure consists of a roughly co–linear series of bright, star–forming knots. Viewed from U through H, the galaxy hardly changes at all. Figure 2 presents a montage of galaxies selected to have peculiar optical morphologies: a motley collection of objects spanning a wide range of redshifts. While there are many differences in detail, the morphological irregularity of most galaxies persists into the near–IR. This strongly suggests that in general we are *not* being misled about

124

4-550.0 z=1.012 "Grand design" barred spiral

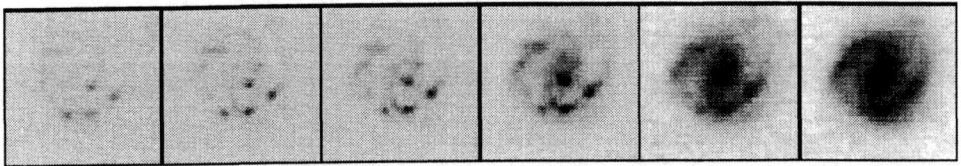

4-752.1 z=1.013 Giant elliptical

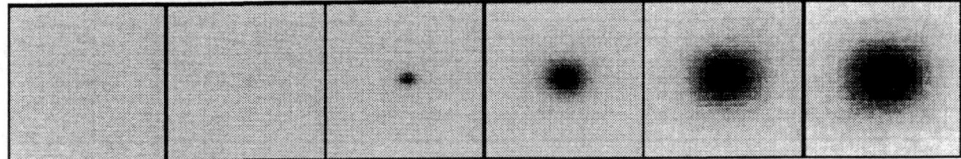

2-736.1 z=1.355 Starbursting "chain galaxy"

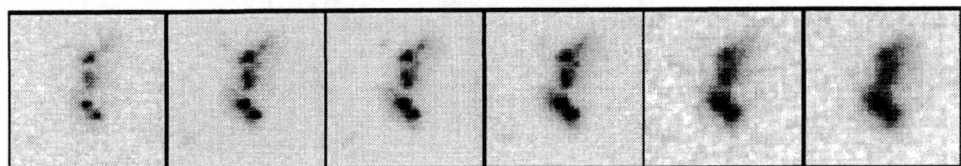

4-948 z ~ 0.6 ? Dusty, possibly interacting disk galaxy

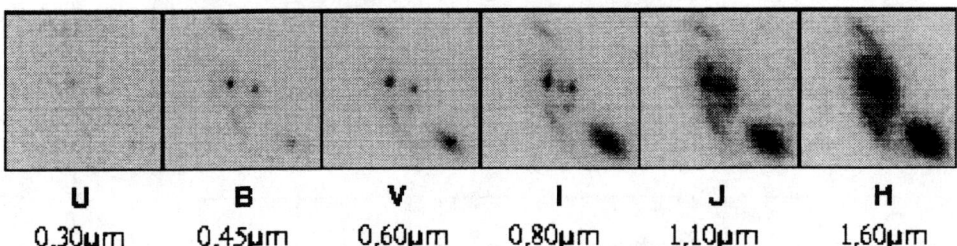

U	B	V	I	J	H
0.30µm	0.45µm	0.60µm	0.80µm	1.10µm	1.60µm

FIGURE 1. Examples of HDF galaxies viewed at wavelengths from 0.3 to 1.6µm. Some objects (e.g. the $z = 1$ spiral at top or the dusty, interacting galaxy at bottom) undergo striking morphological transformations, while others (e.g. the "chain galaxy") do not. Red ellipticals at $z > 1$ are prominent in the NICMOS data: the example shown is perhaps the most luminous galaxy in the HDF, roughly 5 to 8× present–day L^*.

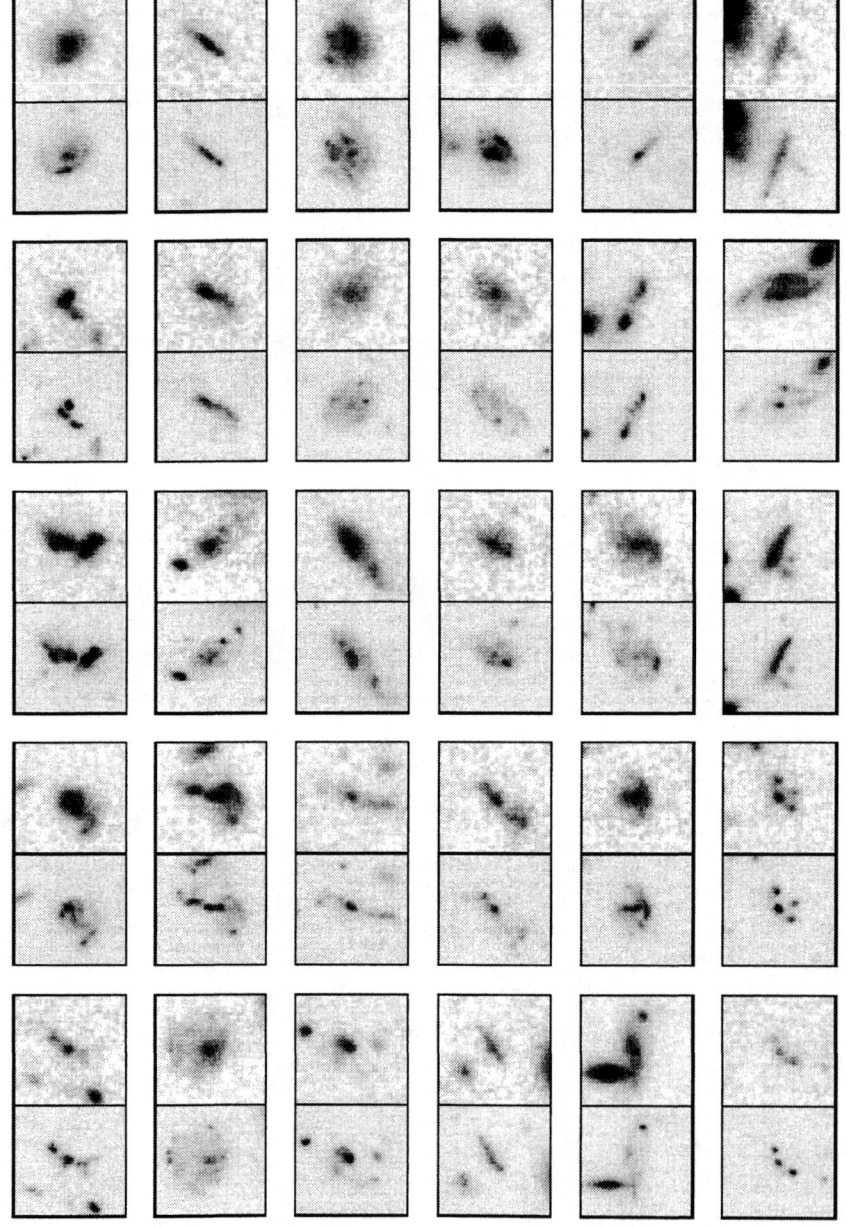

FIGURE 2. A gallery of optical and IR images of HDF galaxies selected to have peculiar WFPC2 morphologies. In all cases the optical image (a BVI composite) is at left, while the near-IR image (IJH composite) is at right.

the prevalence of irregular galaxies in the distant universe [13–15] by the blue and UV rest–frame wavelengths at which we have observed them in the past with WFPC2. The NICMOS data show that, by and large, these are truly *structurally* disturbed systems.

Lyman Break Galaxies

The NICMOS morphologies of Lyman Break selected galaxies [16–18] at $z > 2$ are also generally very similar to their optical WFPC2 counterparts (figure 3). They remain compact, often dominated by several bright knots. In a few cases (e.g. 2–585.1 at $z = 2.008$, 4–52.0 at $z = 2.931$, and 2–604.0 at $z = 3.430$), there are diffuse regions which are optically faint but bright in the near–IR. These may either be regions where older stars dominate the light, or where dust is obscuring ongoing star formation. But the number of such systems is small, and interestingly is mostly limited to the *largest* of the Lyman break galaxies.

While normal spiral and elliptical galaxies are easily identified in the HDF out to $z = 1$ (cf. figure 1), none of the galaxies at $z > 2$ are immediately recognizable as being "classical" Hubble sequence objects. None show regular, symmetric bulge + spiral arm morphologies or look like thin, edge–on disks. We believe that this impression is not entirely the result of surface brightness biases. Figure 4 shows a simulation where a $z = 1$ HDF spiral observed in the J–band is artificially dimmed to show how it would look at $z = 2$ in the NICMOS H–band data. The arms and overall spiral structure remain easily recognizable. If spiral galaxies exist at $z > 2$ they must either be very irregular, or have lower surface brightnesses than their $z \approx 1$ counterparts, or be too small to allow us to resolve their spiral structure with NICMOS. While some Lyman break objects have $R^{1/4}$–law light profiles [17], none of the known or suspected $z > 2$ galaxies in the HDF are as red as "mature" ellipticals at lower redshifts. If they are elliptical progenitors, then by $z \approx 2$ they have not yet, for the most part, built up masses of longer–lived stars sufficient to dominate the rest–frame optical light. The overall impression is that giant galaxies first matured onto the classical Hubble morphological sequence somewhere in the redshift range $1 < z < 2$.

VERY HIGH REDSHIFT CANDIDATES

At $z > 5$, the 1216Å Lyman α forest decrement and the 912Å Lyman limit pass through the reddest WFPC2 bandpasses, suppressing optical flux from very distant objects. With deep, near–infrared data, it should be possible to apply the Lyman break color selection techniques which have proven to be so effective at $2 < z < 4.5$ [16,19] and identify more distant galaxies if they are present. Indeed [4] and [20] have confirmed the redshifts of two galaxies at $z > 5$ previously selected by color [21] as good high redshift candidates.

A complete discussion of such objects in our survey will be presented elsewhere. Here I briefly mention one tantalizing and spectacular candidate, an object which

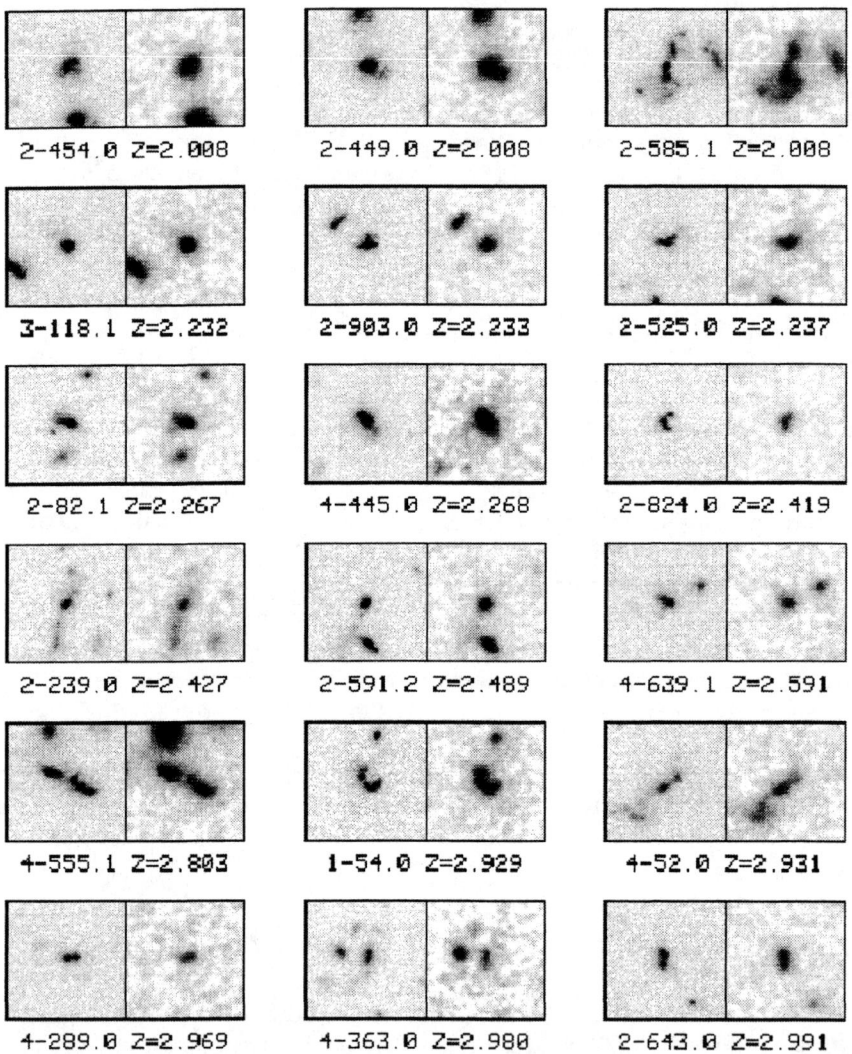

FIGURE 3. A montage of WFPC2 (left) and NICMOS (right) images of Lyman break selected galaxies at $2 < z < 3$. Only galaxies with spectroscopically confirmed redshifts are shown.

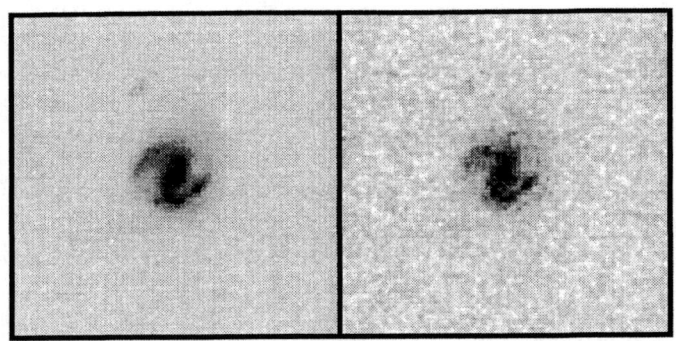

FIGURE 4. Simulation of cosmological surface brightness dimming in the HDF/NICMOS observations. *Left:* the HDF spiral galaxy 4–550.0 ($z = 1.0$) as observed through the F110W filter (rest–frame $\lambda_0 \approx 5500$Å). At $z = 2.0$, the same rest–frame light would shift to the NICMOS F160W filter. *Right:* simulation of the same galaxy at $z \approx 2$ in F160W, including cosmological dimming. The spiral structure, bulge and bar remain visible.

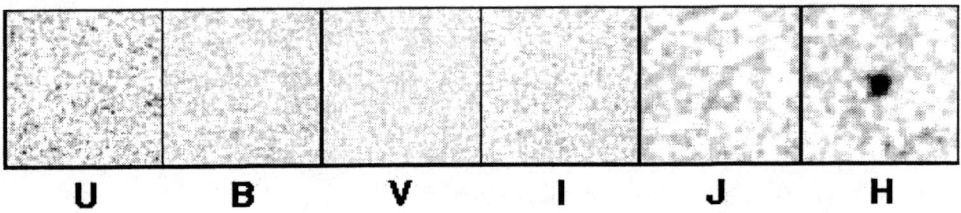

| **U** | **B** | **V** | **I** | **J** | **H** |

FIGURE 5. An extremely unusual and moderately bright "J–band dropout" object, with significant flux detected *only* through 1.6μm NICMOS F160W filter. The object appears to be stellar or nearly so.

is robustly detected *only* in the NICMOS F160W image (figure 5). The object appears to be unresolved or nearly so, and is relatively bright at 1.6μm with $H_{AB} = 25.3$, and with an f_ν flux density decrement of $\gtrsim 10\times$ between 1.6μm and 1.1μm. Extensive checks of the NICMOS data convince us that it is not an artifact, nor a transient object (e.g. a supernova – the J and H–band data were interleaved and taken over a ten day period). The object is probably detected (weakly) in our ground–based K–band images with $K_{AB} \approx 24.3$ at S/N ≈ 5, and in fact was noted by [22] in their list of K–band only candidates in the ground–based data.

We cannot, at this stage, be certain what this object is. Some possibilities include:

- A Lyman break "J–band dropout" object at $z \gtrsim 9$,

- An extremely unusual red object ("EURO"), perhaps a distant, dusty quasar or galaxy,

- An object (AGN? protogalaxy cloud?) with extremely high equivalent width line emission somewhere in the range $1.4 < \lambda < 1.8\mu$m which entirely dominates the H–band, or

- Some kind of exotic galactic star, brown dwarf, or giant extrasolar planet.

The idea that this object may lie at $z \gtrsim 9$ is perhaps the most exciting possibility, but must also be subject to the most rigorous testing. The photometry is somewhat ambiguous: the very red $J - H$ and flatter $H - K$ is suggestive of a spectral break, and the amplitude of the flux discontinuity implicates the Lyman limit or Lyman α forest, but the flux does evidently rise redward of 1.6μm, and dusty objects at lower redshifts are not wholly excluded. The tentative K–band detection may be seen as evidence against the emission line hypothesis (#3 above), unless that line is itself Lyman α or unless there is a second line in the K–band. A reddened starburst with powerful Hα at $z \approx 1.4$ might still fit the colors. Current spectral models for brown dwarfs and super–Jovian planets objects [23] do not predict that stellar or substellar objects should have such colors, but this is no proof of the object's extragalactic origin. The unresolved morphology favors either a stellar or quasi–stellar nature.

Although it is extremely faint, the object may be accessible to near–infrared spectroscopy, particularly (or, realistically, only) if it has strong emission lines (e.g. possibility #3 above), and we will target this object for infrared spectroscopy in the upcoming observing season. If it is in our galaxy, then it may exhibit proper motion which could be detected by reobserving with a revived NICMOS in 2000 after the installation of the NCS cooling system.

DISCUSSION

Our HDF–N/NICMOS data is the largest region of the sky with contiguous, deep NICMOS mapping for studying faint, high redshift galaxies, and is one of very few fields imaged by HST to comparable depths at both infrared and optical wavelengths. Although it covers only 5 arcmin^2, it contains thousands of faint galaxies for which there exists extensive supporting data (spectroscopic, photometric, radio, etc.). Taken together with the smaller but deeper HDF–N GTO field [9] and the HDF–South NICMOS field, these data should be a valuable resource for investigating galaxy evolution in the future.

ACKNOWLEDGEMENTS

I would like to thank my co–investigators and collaborators on this program, particularly Christopher Hanley whose assistance with the data reduction has been invaluable. The other investigators are M. Bershady, A. Connolly, C. Conselice, P. Eisenhardt, R. Elston, H. Ferguson, A. Fruchter, M. Giavalisco, R. Hook, R. Lucas, J. Mack, P. Madau, C. Papovich, M. Postman, S.A Stanford, C. Steidel, and

A. Szalay. Deep thanks are also due to the STScI staff for their considerable efforts scheduling these observations, and to the editors of this volume for their patience.

REFERENCES

1. Williams, R.E., et al. 1996, AJ, 112, 1335
2. Williams, R.E., et al. 1999, in preparation
3. Cohen, J.G., Cowie, L.L., Hogg, D.W., Songaila, A., Blandford, R., Hu, E.M., & Shopbell, P., 1996, ApJ, 471, L5
4. Weymann, R.J., Stern, D., Bunker, A., Spinrad, H., Chaffee F.H., Thompson, R.I., & Storrie–Lombardi, L.J., 1998, ApJ, 505, L95
5. Fruchter, A., et al. 1999, in preparation
6. Dickinson, M., 1998, in *The Hubble Deep Field*, eds. M. Livio, M. Fall & P. Madau, Cambridge Univ. Press, p. 219
7. Hogg, D.W., Neugebauer, G., Armus, L., Matthews, K., & Pahre, M.A., 1997, AJ, 113, 474
8. Barger, A.J., Cowie, L.L., Trentham, N., Fulton, E., Hu, E.M., Songaila, A., & Hall, D., 1999, AJ, 117, 102
9. Thompson, R.I., Storrie–Lombardi, L.J., Weymann, R.J., Rieke, M.J., Schneider, G., Stobie, E., & Lytle, D., 1999, AJ, 117, 17
10. Fruchter, A., & Hook, R., 1998, PASP, submitted
11. Van Den Bergh, S., Abraham, R.G., Ellis, R.S., Tanvir, N., & Santiago, B.X., 1996, AJ, 112, 359
12. Abraham, R.G., Merrifield, M.R., Ellis, R.S., Tanvir, N., & Brinchmann, J., 1998, MNRAS, submitted
13. Glazebrook, K., Ellis, R., Santiago, B., & Griffiths, R., 1995, MNRAS, 275, L19
14. Driver, S.P., Windhorst, R.A., & Griffiths, R.E., 1995, ApJ, 453, 48
15. Abraham, R.G., Tanvir, N., Santiago, B., Ellis, R.S., Glazebrook, K., & Van Den Bergh, S., 1996, MNRAS, 279, L47
16. Steidel, C.C., Giavalisco, M., Pettini, M., Dickinson, M., & Adelberger, K.L., 1996, ApJ, 462, L17
17. Giavalisco, M., Steidel, C.C., & Macchetto, F.D., 1996, ApJ, 470, 189
18. Lowenthal, J.D., Koo, D.C., Guzman, R., Gallego, J., Phillips, A.C., Faber, S.M., Vogt, N.P., Illingworth, G.D., & Gronwall, C., 1997, ApJ, 481, 673
19. Steidel, C.C., Adelberger, K.L., Giavalisco, M., Dickinson, M., & Pettini, M., 1999, ApJ, in press
20. Spinrad, H. Stern, D., Bunker, A., Dey, A., Lanzetta, K., Yahil, A., Pascarelle, S., Fernández–Soto, A., 1998, AJ, in press
21. Fernández–Soto, A., Lanzetta, K., & Yahil, A., 1999, ApJ, in press
22. Lanzetta, K., Yahil, A., 1998, & Fernández–Soto, A., 1998, AJ, 116, 1066
23. Burrows, A., Marley, M., Hubbard, W.B., Lunine, H.I., Guillot, T., Saumon, D., Freedman, R., Sudarsky, D., & Sharp, C., 1997, ApJ, 491, 856

Flux Limited Redshift Surveys in the Optical and Submillimeter.

Lennox. L. Cowie*†, Amy. J. Barger* and Antoinette Songaila*†

*Institute for Astronomy, 2680 Woodlawn Drive, Honolulu, HI96822
†Institute of Astronomy, Madingley Rd., Cambridge CB3 0HA

Abstract. While samples of high redshift galaxies may best be found using color or line selection techniques ultimately only flux limited samples can tell us where the bulk of the galaxies lie and what the history of star formation was. With the new 10m telescopes it is now possible to obtain quite complete spectroscopic redshift samples of optically selected samples to AB magnitudes of near 25 in the B band and 24 in the I band. Despite the fact that samples at these faint magnitudes encompass the majority of the starlight energy density at optical and ultraviolet wavelengths only a very small fraction of galaxies lie above the $z = 2$ boundary of the present conference and more than half of the galaxies lie at $z < 1$. However, a roughly comparable amount of the star formation is now known to be reradiated by dust into the submillimeter and we might hope that the new populations of submillimeter galaxies being found at 850 microns with the SCUBA detector on JCMT might contain high redshift objects since the negative K correction at these wavelengths makes high z objects as easy to detect as those at $z = 1$. Preliminary results suggest that the bulk of these objects lie at $1 < z < 3$.

INTRODUCTION

The cumulative emission from all objects lying beyond the Galaxy, the extra-galactic background light (EBL), provides a direct measure of the integrated star formation history of the Universe if we know the redshift distribution of the galaxies contributing to the light. Recent measurements of both the optical EBL ([5]Bernstein, Freedman & Madore 1999) and the EBL at far-infrared (FIR) and submm wavelengths using data from the *FIRAS* and *DIRBE* experiments on the *COBE* satellite ([33] Puget et al. 1996; [18] Guiderdoni et al. 1997; [34] Schlegel et al. 1998; [17]Fixsen et al. 1998; [21]Hauser et al. 1998; [27] Lagache et al 1999) indicate that the total bolometric energy density from star formation and AGN activity that is absorbed by dust and reradiated into the FIR/submm is comparable to the unobscured bolometric energy density seen in the optical. This implies that unobscured and obscured star formation are responsible for roughly comparable

CP470, After the Dark Ages: When Galaxies were Young (the Universe at 2 < z < 5),
edited by Stephen S. Holt and Eric P. Smith
© 1999 The American Institute of Physics 1-56396-855-X/99/$15.00

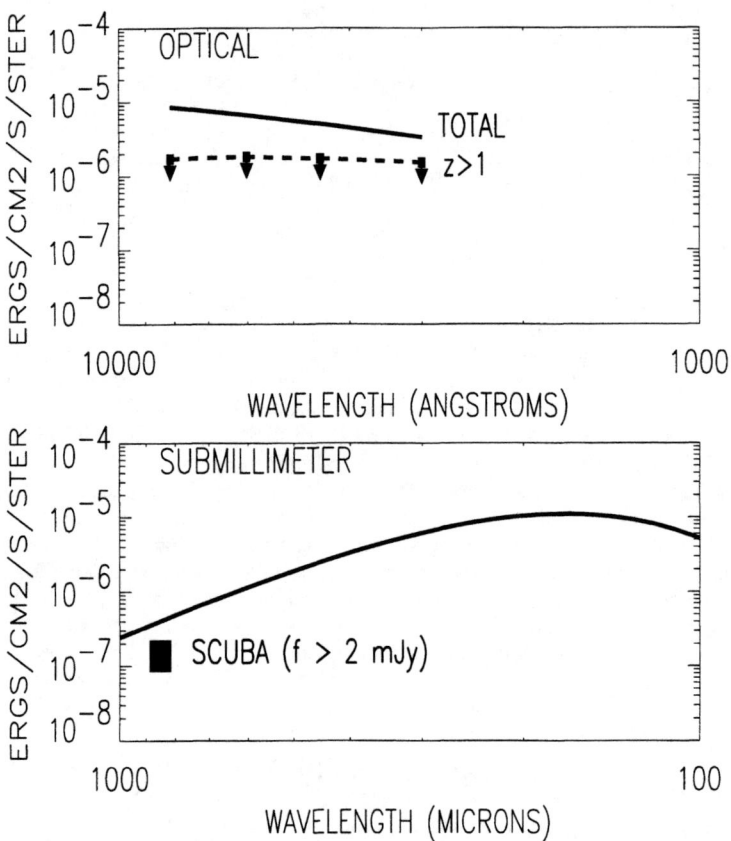

FIGURE 1. OPTICAL and SMM Backgrounds: The upper panel shows the bolometric (νf_ν) extragalactic background light obtained by integrating the HDF counts of Williams et al. (1996), with the dashed line showing the same quantity with known $z < 1$ galaxies excluded. The lower panel shows Fixsen et al.'s (1998) analytic approximation to the submillimeter background light measured by COBE. The two backgrounds are very similar in size. The solid box in the lower panel shows the resolved contribution of the background at 850-μm from sources above 2 mJy directly detected with SCUBA based on the number count determination of Barger et al. 1999.

amounts of the stars and metals seen in the local Universe. We therefore need to determine the redshift distribution of both the optically selected and submillimeter selected populations of galaxies to determine the total star formation history of the universe.

In the absence of information about the redshift distribution, the question of the relative fraction of the star formation producing light in the submillimeter rather than in the optical can be crudely addressed by looking at the total amount of light energy density lying in the two wavelength ranges. The bolometric amount of light reflects the total amount of star formation radiated or reradiated at that wavelength, divided by the $(1 + z)$ factor that reflects the expansion cooling. The bolometric optical light is best determined from the deep counts in the Hubble Deep Field (HDF; Williams et al. 1996). This integrated light is close to the limits and possible measurements on the EBL at these wavelengths (Vogeley et al. 1998; Bernstein et al. 1998). In Figure 1 we show these two backgrounds, with the solid line in the upper panel being the optical light (with the x-axis in Angstroms) and the solid line in the lower panel the submillimeter background (with the x-axis in microns). Since we know the redshifts for many of the HDF galaxies we can further refine the optical EBL that can arise at higher redshifts by excluding known galaxies at $z < 1$. This latter quantity is shown as the dashed line. If most of the galaxies giving rise to the submillimeter background were at $z > 1$, then the submillimeter bolometric light, which considerably exceeds the $z > 1$ limits on the optical light, would map the bulk of the star formation. As we shall discuss here the preliminary evidence suggest that this is indeed the case.

It has been thought in the last year or two that the optically selected star formation history was roughly understood. (At least when dust extinction is neglected as it should be if we also measure the obscured star formation history directly in the submillimeter populations.) The optically visible star formation was thought to have peaked in the $z = 1 - 2$ range and to fall off at higher and lower redshifts. (Madau et al 1998 [31]) Recently both the high and low redshift falloffs have been questioned and it has been suggested that the falloff below $z = 1$ is shallow and that the star formation rate at higher redshifts is relatively constant. (Pascarelle et al. 1998 [32] Cowie et al. 1999 [12], Steidel et al. 1999 [37], Hu et al. 1999 [23]). In particular in Cowie et al. 1999, we have used new faint ultraviolet selected magnitude limited galaxy samples with spectroscopic redshifts together with the local ultraviolet observations of Treyer et al. 1998 [38], to argue that the fall off at $z < 1$ has a rather shallow slope, approximately $(1 + z)$ rather than the very steep evolution of $(1 + z)^{3.8}$ reported by Lilly et al.(1996) [28]. This shallower slope implies that integrated galaxy formation is continuing to rise smoothly to the present time rather than peaking at $z = 1$. However, much of the present formation is taking place in smaller galaxies. As we shall attempt to show here it seems likely that only a fraction of the optically visible star formation is coming from galaxies beyond $z = 2$

The galaxies contributing to the submillimeter background at 850-μm have recently been found using the new camera SCUBA (Submillimeter Common User

Bolometer Array; [22]Holland et al. 1999) on the 15-m James Clerk Maxwell Telescope on Mauna Kea. SCUBA has, for the first time, enabled deep, unbiased surveys to be made of the submm sky (Smail, Ivison & Blain 1997 [35]; Barger et al. 1998, 1999 [2] [3] Hughes et al. 1998 [3]; Eales et al. 1999 [14]). These surveys have uncovered a substantial population of dusty, star-forming galaxies with properties similar to those expected for distant ultraluminous infrared galaxies ([2]Barger et al. 1998; [36]Smail et al. 1998). If the majority of the submm emission in these systems comes from dust-obscured star formation, then their inferred star formation rates are of the order of several hundred solar masses per year. Most importantly for the present talk the deepest submm counts ([8]Blain et al. 1999b), fluctuation analyses (e.g. [3]Hughes et al. 1998), and analyses of deep and wide-area surveys (e.g. [3]Barger et al. 1999) indicate that the bulk of the background emission at 850-μm detected by *COBE* is resolved into discrete sources at a flux limit of around 1 mJy. As is illustrated in Figure 1 roughly 30 percent of the EBL at 850-μm is resolved in the 2 mJy SCUBA samples. Thus, we are now in a position to undertake detailed studies of the population responsible for the majority of the emission in the FIR background just as we have been in the optical. In particular, we can measure the redshift distribution of the submm population and use this to trace the extent and evolution of obscured star formation in the distant Universe. The detailed study of the resolved component of the background also provides the only clear method for determining what fraction of the submm emission originates from AGN activity rather than star formation ([7]Blain et al. 1999a; [19]Guiderdoni et al. 1998).

Most relevantly for the present conference it is conceivable that some fraction of the population responsible for the FIR background lies at high redshift, $z \sim 5\text{--}10$. Submm selection is a powerful technique for locating such distant galaxies. Indeed, the steep thermal dust spectrum, which peaks in the FIR at a rest-frame wavelength of about 100-μm is redshifted into the submm for $z > 1$. The resulting strong negative K-correction for sources out to $z \sim 10$ is sufficient to offset cosmological dimming for $q_0 = 0.5$. Even for low values of q_0 the 850-μm flux density is only expected to decrease by a factor of a few over this redshift range ([6]Blain & Longair 1993; [25]Hughes, Dunlop & Rawlings 1997). However, as we shall discuss below the preliminary evidence we have so far suggests that most of these galaxies lie in the $1 < z < 3$ redshift range.

OPTICALLY SELECTED REDSHIFT SAMPLES

Current spectroscopic redshift samples can probe to galaxy magnitudes below which much of the optical EBL arises. Using the HDF galaxy number counts from Williams et al. [39] roughly 75 percent of the EBL I light arises in galaxies with $I(AB) < 23.75$ around 65 percent of the B light in galaxies with $B(AB) < 24.5$ and 50 percent of the 3400 Å light in galaxies with $AB < 25$. Despite this nearly all the galaxies at these magnitudes lie below $z = 2$ and most below $z = 1$ and the

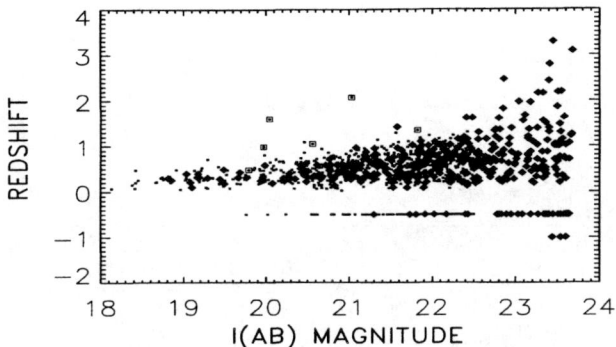

FIGURE 2. Redshift magnitude relation for an $I(AB) = 23.75$ complete magnitude limited sample in the Hawaii survey fields shown as diamonds. Unidentified or unobserved objects are shown at $z = -0.5$ or $z = -1$ respectively. The small dots show the CFRS sample at $I(AB) < 22.5$ where we have distinguished AGN with surrounding boxes.

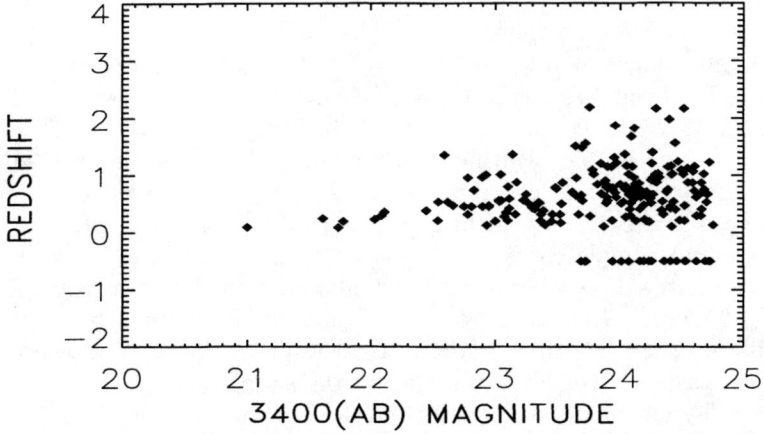

FIGURE 3. Redshift magnitude relation for an $AB = 24.75$ complete magnitude limited selected at 3400 Å in the Hawaii field SSA22 and in an area crossing the Hubble field. Unidentified or unobserved objects are shown at $z = -0.5$ or $z = -1$ respectively.

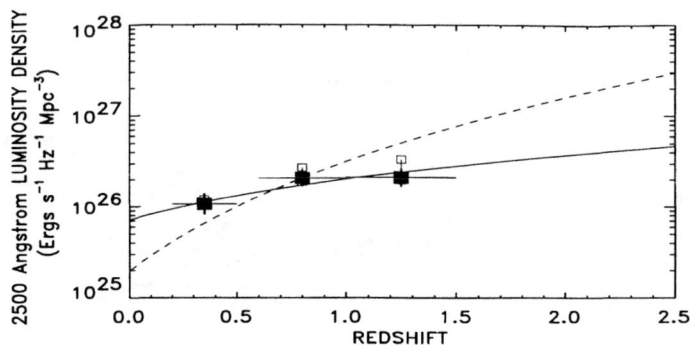

FIGURE 4. Light density at 2500Å from galaxies with absolute AB magnitudes less than -16 as a function of redshift: directly determined in the interval $0.2 < z < 0.5$ and incompleteness-corrected via an $\alpha = -1$ Schechter function for the intervals $0.6 < z < 1.0$ and $1.0 < z < 1.5$. In each redshift interval the light density is shown as a filled square and the redshift range as a thin horizontal line. Statistical and systematic errors (Appendix) are shown as the thick portions of the error bar. Light densities computed using the minimal and maximal possible incompleteness corrections are shown as open squares joined by thin vertical lines. The solid line is a $(1 + z)^{1.5}$ evolution law. The dashed line shows an evolution of $(1 + z)^4$, normalized to the second data point. Their results are calculated for a $q = 0.5$ universe with a Hubble constant of $65 km/s/Mpc$.

majority of the optical and near UV EBL arises in $z < 1$ galaxies as is illustrated in Figure 1 for the HDF sample.

The I sample shown in Figure 2 contains 420 galaxies drawn from complete magnitude limited samples in the Hawaii fields. (The Hubble deep field I selected samples are still somewhat incomplete to be included.) All but 37 of the objects have reliable spectroscopic identifications at this point. The faintest galaxies at $I(AB) > 23.25$ are drawn solely from the SSA22 field where an extremely intensive effort has been made to fully identify the population to the faintest magnitude where substantial completeness is still possible. In the $23.25 < I(AB) < 23.75$ bin there are 57 galaxies with redshifts and 20 unidentified objects and the median redshift is 1.00 with a 1 sigma range $0.47 - 1.49$. In fact many of the unidentified objects will lie in the 'blind spot' redshift range just below $z = 2$ where the OII 3727 line has passed beyond one micron but the short wavelength UV absorption lines and the Lyman break are still below the optical wavelengths. Some of the unidentified objects may also be stars. However, in combination with the small number of galaxies which are actually seen above $z = 2$ we can see that less than 30 percent of the galaxies lie above $z = 2$ and roughly half lie below $z = 1$

The 3400 Å selected sample shown in Figure 3 show this effect even more strongly. Here there are 222 galaxies with $AB < 24.75$ of which 184 have robust spectroscopic IDs. Only 3 have redshifts above 2 and the median redshift of galaxies in the

$AB = 24.25 - 24.75$ bin is 0.871 with a 1 sigma range of 0.54 to 1.02 implying that even at these faint magnitudes most of the galaxies lie below $z = 1$.

The rest frame ultraviolet selected samples which can be constructed from these various color selected samples can be used to construct the ultraviolet luminosity densities as a function of redshift. In Figure 4 we show the evolution of the 2500 Å luminosity density found in Cowie Songaila and Barger 1999 [12] showing the rather slow fall off below $z = 1.5$. If the slope is as shallow as this result suggests then the integrated total of star formation is continuing to grow substantially even at the present time for the stars which are optically visible.

THE SUBMILLIMETER

Barger et al (1999b) [4] have recently carried out a spectroscopic redshift survey of a sample of submm-selected galaxies from the SCUBA cluster lens survey of [35]Smail et al. (1997, 1998). The advantage of this survey is that the clusters magnify any background sources providing enhanced sensitivity in the submm and easing spectroscopic follow-up in the optical. The full 850-μm survey contains 17 sources above 3σ significance over a total surveyed area of 36 arcmin^2 which is quite similar to the areas covered in the optical samples of the previous section.

Unfortunately there is a substantial uncertainty of about 3 arcseconds in the submillimeter source positions and so our ability to conclude whether a particular neighboring galaxy is likely to be the submm source depends strongly on whether the optical spectrum of the galaxy shows any remarkable features, such as particularly strong [OII] $\lambda3727$ or Hα emission lines that indicate a starburst or high excitation lines that show the presence of an AGN. The identification process is therefore somewhat problematic, and while we are capable of robustly identifying some submm sources with striking optical counterparts, (as has been confirmed through the CO detection of two of the sample galaxies), this approach leaves us with a number of ambiguous cases. As a secondary criterion, we can consider the morphologies of the objects and the presence of any merger activity, which can be an indicator of luminous FIR systems in the local Universe. Two of the submm detections in the sample have no visible optical counterparts in very deep imaging, and it is possible that the true counterparts to some other sources in the sample are similarly optically faint. In principal such sources could either be at very high redshift or be so highly obscured that they are emitting their energy almost entirely in the submm ([10] Cimatti et al. 1998 [13] Dey et al. 1999).

The Barger et al. redshift survey has produced highly reliable identifications for four of the submm sources: two central cD galaxies in the lensing clusters , an interacting pair of galaxies at $z = 2.80$, one of which hosts a Seyfert-2 nucleus, and a further pair of galaxies at $z = 2.55$ that show starburst features. Using the spectral features as a guide, there are more subjective identifications for a further two weak AGN sources at $z = 1.06$ and $z = 1.16$. The remaining eight sources have candidate counterparts with redshifts ranging from $z = 0.18$–2.11. The lower

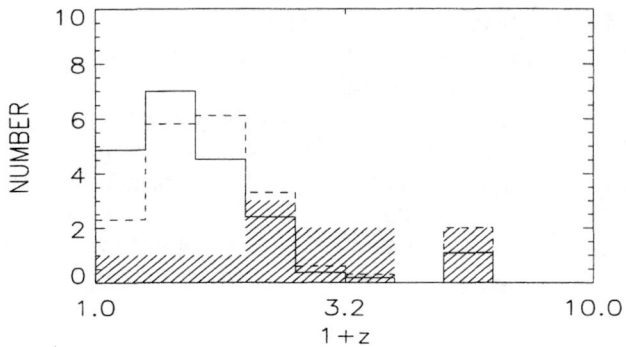

FIGURE 5. Comparison of the redshifts seen in the submillimeter sample shown by the shaded histogram compared with that in the $AB = 24.75$ sample selected at $3400\mathring{A}$ shown by the dashed line. The normalization of the ultraviolet sample has been divided by a factor of 10. Also shown as the solid line is the distribution of 3400 Å light which because the brighter galaxies are at lower redshift peaks at lower redshift than the number distribution. However, both are at substantially lower redshifts than the submillimeter sources. Unidentified objects in both samples are shown at a purely nominal position of $z = 5$

redshift ($z \ll 1$) systems include a pair of interacting galaxies at $z = 0.21$ and two bright spiral galaxies at $z = 0.18$ and $z = 0.33$ The redshift distribution suggests that the majority of the sources outside the clusters themselves are likely to lie at redshifts $z = 1$–3. The lens amplification is not expected to significantly distort the $N(z)$ distribution given the high redshifts of the bulk of the counterparts.

If the current optical identifications are correct then current models formulated to fit the FIR background and various infrared counts using simple analytic descriptions of the evolution of luminous FIR galaxies such as those of [7]Blain et al. (1999a) and [19]Guiderdoni et al. (1998). can be rejected since they predict too high a median redshift. However, Barger et a. 1999b revised the Gaussian model of Blain et al. to include a larger fraction of galaxies at $z = 1$–2 by shifting the peak to lower redshifts. This change can be accommodated without violating any of the observational constraints and provides a better fit to the apparent redshift distribution.

In terms of the dominant energy source for the emission seen in the FIR background, i.e. AGN or starburst, we find that at least three of the fourteen sources surveyed have counterparts with spectral features indicative of AGN activity. These objects may be part of the obscured AGN population predicted to be a major contributor to the X-ray background at energies $> 2\,\text{keV}$ ([30]Madau, Ghisellini, & Fabian 1994; [11]Comastri et al. 1995; [21]Fabian et al. 1998; [20]Gunn & Shanks 1999; [1]Almaini, Lawrence & Boyle 1999). Deep radio, NIR, MIR, and high-resolution X-ray data of fields observed with SCUBA should provide us with improved position estimates and more information on the nature of the sources, making the task of following up the submm detections easier in the future.

CONCLUSION

Ultimately the combination of deep submillimeter and optical redshift samples will allow us to determine the evolution of the total star formation with redshift. The submillimeter redshift samples are still in an early stage but the current best determination of Barger et al. 1999b [4] suggests that most of these sources lie in the $z = 1 - 3$ range. As is illustrated in Figure 5 this is substantially above the distribution seen in ultraviolet selected samples where most of the objects and light in an $AB = 24.75$ sample selected at 3400 Å and accounting for roughly half of the EBL at these wavelengths lies mainly below $z = 1$. It is very tempting to speculate that the very luminous submillimeter sources represent the major merger events giving rise to spheroids (Barger et al 1999b [4], Eales et al. 1999 [14]) as opposed to the more quiescent stages of star formation seen in the optical and ultraviolet samples. In this case the current measurements suggest that spheroids are forming in dust obscured events in the $z = 1 - 3$ redshift range.

ACKNOWLEDGEMENTS

Len Cowie and Toni Songaila would like to thank the IOA for hospitality during the course of this work and Richard Ellis and Richard McMahon for many discussions. We are also extremely grateful to our collaborators Ian Smail, Rob Ivison, Andrew Blain, Jean-Paul Kneib, and Esther Hu whose work is extensively drawn on throughout this talk.

REFERENCES

1. Almaini O., Lawrence, A., Boyle, B., 1999, MNRAS, submitted
2. Barger, A.J., Cowie, L.L., Sanders, D.B., Fulton, E., Taniguchi, Y., Sato, Y., Kawara, K., Okuda, H. 1998, Nature, 394, 248
3. Barger, A.J., Cowie, L.L., Sanders, D.B. 1999, APJl, submitted
4. Barger, A.J., Cowie, L.L., Smail, I., Ivison, R.J., Blain, A.W., and Kneib, J-P. 1999, AJ, submitted
5. Bernstein, R.A., Freedman, W.L., Madore, B.F. 1999, APJ, submitted
6. Blain, A.W., Longair, M.S. 1993, MNRAS, 264, 509
7. Blain, A.W., Smail, I., Ivison, R.J., Kneib, J.-P. 1999a, MNRAS, in press, [astro-ph/9806062]
8. Blain, A.W., Kneib, J.-P., Ivison, R.J., Smail, I. 1999b, APJ, submitted
9. Boyle, B.J., Terlevich, R.J. 1998, MNRAS, 293, L49
10. Cimatti. A., Andreani P., Rottgering H. Tilanus R. 1998, Nature, 392, 895
11. Comastri A., Setti G., Zamorani G., Hasinger G. 1995, A&A, 296, 1

12. Cowie, L.L., Songaila, A., Barger, A.J. 1999, AJ submitted
13. Dey, A., Graham, J.R., Ivison, R.J., Smail, I., Wright, G.S., 1999, APJ, submitted
14. Eales, S., Lilly, S., Gear, W., Dunne, L., Bond, J.R., Hammer, F., Le Fèvre, O., Crampton, D. 1999, APJ, submitted, [astro-ph/9808040]
15. Edge, A.C., Ivison, R.J., Smail, I., Blain, A.W., Kneib, J.-P. 1999, MNRAS, submitted
16. Fabian A.C., Barcons X., Almaini O., Iwasawa K. 1998, MNRAS, 297, 11L
17. Fixsen, D.J., Dwek, E., Mather, J.C., Bennett, C.L., Shafer, R.A. 1998, APJ, 508, 123
18. Guiderdoni, B., Bouchet, F.R., Puget, J.-L., Lagache, G., Hivon, E. 1997, Nature, 390, 257
19. Guiderdoni, B., Hivon, E., Bouchet, F.R., Maffei, B. 1998, MNRAS, 295, 877
20. Gunn, K.F., Shanks, T. 1999, MNRAS, submitted
21. Hauser, M.G. et al. 1998, APJ 508, 25
22. Holland, W.S. et al. 1999, MNRAS, in press, [astro-ph/9809122]
23. Hu E.M., Cowie L.L., McMahon R.,G. 1999 ApJL to be submitted
24. Hughes, D.H. et al. 1998, Nature, 394, 241
25. Hughes, D.H., Dunlop, J.S., Rawlings, S. 1997, MNRAS, 289, 766
26. Ivison, R., Smail, I., Le Borgne, J.-F., Blain, A.W., Kneib, J.-P., Bézecourt, J., Kerr, T.H., Davies, J.K. 1998, MNRAS, 298, 583
27. Lagache, G., Abergerl A., Boulanger, F., Desert, F.X. Puget, J.-L. 1999, AA in press.
28. Lilly, S.J., LeFèvre, O., Hammer, F., Crampton, D. 1996, APJ, 460, L1
29. Lilly, S., Eales, S., Gear, W., Bond, J.R., Dunne, L., Hammer, F., Le Fèvre, O., Crampton, D. 1998, in *NGST: Science and Technological Challenges*, [astro-ph/9807261]
30. Madau, P., Ghisellini, G., Fabian, A.C. 1994, MNRAS, 270, 17
31. Madau, P., Pozzetti, L., Dickinson, M. 1998, APJ, 498, 106
32. Pascarelle, S. M., Lanzetta, K. M. & Fernandez-Soto, A. 1998, ApJ 508, L1.
33. Puget, J.-L., Abergel, A., Bernard, J.-P., Boulanger, F., Burton, W.B., Desert, F.-X., Hartmann, D. 1996, AA, 308, L5
34. Schlegel, D.J., Finkbeiner, D.P., Davis, M. 1998, APJ, 500, 525
35. Smail, I., Ivison, R.J., Blain, A.W. 1997, APJ, 490, L5
36. Smail, I., Ivison, R.J., Blain, A.W., Kneib, J.-P. 1998, APJ, 507, 21L
37. Steidel, C.C., Adelberger, K.L., Giavalisco, M., Dickinson, M., Pettini, M. 1999, APJ, submitted, [astro-ph/9811399]
38. Treyer, M.A., Ellis, R.S., Milliard, B., Donas, J., Bridges, T.J. 1998, MNRAS, 300, 303
39. Williams, R., et al. 1996, AJ 112, 1335

Lyman Break Galaxies at $z \sim 3$: Giants or Dwarfs?

James D. Lowenthal*, L. Simard†, and D. Koo†

*Dept. of Physics and Astronomy, University of Massachusetts, Amherst, MA 01003
† UCO/Lick Obs., University of California, Santa Cruz, CA 95064

Abstract.
We review attempts to constrain masses of Lyman Break Galaxies. Despite the temptation to identify LBGs with precursors of massive galaxies, we find that there is as yet *no direct evidence* that LBGs are drawn from a massive population, or even a uniform one.

Soon after the discovery of LBGs, it was suggested (Steidel *et al.* 1996) [1] that these luminous, star-forming galaxies or protogalaxies at redshifts $z > 2$ were the sites of massive galaxy formation, perhaps the centers of deep potential wells of dark matter that would eventually become massive ellipticals or spheroids of spirals today. Theorists working with semi-analytical cold dark matter models were quick to point out that they had in fact predicted such objects with their simulations(*e.g.*, Baugh *et al.* 1998) [2]). LBGs at $z > 2$ also show significant clustering, apparently supporting the massive galaxy identification (Giavalisco *et al.* 1998) [3].

An alternate viewpoint, however, was proposed by Lowenthal *et al.* (1997) [4], who suggested that the sizes, morphologies, luminosities, and spectra of LBGs were consistent with their being drawn from either a wide range of galaxy types and masses, or even from a predominately low-mass, bursting dwarf population.

To distinguish between those two scenarios, we have undertaken a program of observations with the 10-m Keck telescope to **measure kinematics of LBGs** in the HDF and other deep HST fields. Our goal is to constrain the masses of LBGs in rough analogy to local rotation curves.

While the sample is still small, early results seem to favor our hypothesis, *i.e.* that LBGs are drawn from a wide range of masses. Recent measurements of emission line widths in the observed near-IR by Pettini *et al.* (1998) [5] seem to support this view.

One particularly interesting example is the source known as C4-06 from Steidel et al., at $z = 2.803$. Its elongated, linear structure shows, in the HDF, two major strings of knots extending over some $2''$. Our spectrum of C4-06, obtained with

CP470, After the Dark Ages: When Galaxies were Young (the Universe at 2 < z < 5),
edited by Stephen S. Holt and Eric P. Smith

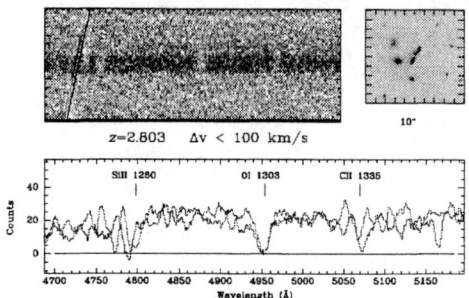

$z=2.803 \quad \Delta v < 100$ km/s

FIGURE 1. (HDF closeup and Keck spectrum of C4-06. Note that the continuum is clearly extended in the 2D spectrum, and that the absorption lines of Si II, O I, and C II are vertical, i.e., no velocity shift is detectable across the system, about 30 kpc in projected size.

the slitlet aligned with the long axis of the galaxy, shows extended emission (Fig. 1 shows a detail), as well as strong absorption lines of Si II 1260, O I 1303, and C II 1335. It is clear that the absorption lines extend over the whole continuum source, and that they are essentially vertical – i.e., there is no detectable rotation or velocity differential over the 20 kpc or so subtended by the object. A cross-correlation of two independent one-dimensional spectra extracted $1''$ apart indicates δv consistent with zero, to within the errors of ~ 50km s^{-1}. Adopting a limit of twice that, or $\delta v < 100$ km s^{-1} over $1''$(or $12\ h_{50}^{-1}$ kpc for $q_0 = 0.05$), implies a gravitational mass $m < 10^{10}$ M$_\odot$, well under the mass of typical L* galaxies today.

Even without calculating masses, we can directly compare these high-redshift star-forming galaxies to more local systems for more insight into their nature. In the half-light-radius/velocity-dispersion (R_e/σ) plane, for example (Fig. 2a, using δv for σ) three of our HDF sources overlap with low-mass spirals at the low end, while one falls above the envelope defined by high-mass spheroids.

The emission line widths of LBGs measured in the near-IR (rest optical, e.g., Hβ) by Pettini et al. (who generally favor a high-mass interpretation) are also consistent with the widths of low-mass local galaxies, including HII galaxies and compact narrow emission line galaxies. Fig. 2b shows those line widths plotted vs. line luminosity

Table 1 summarizes the salient characteristics of LBGs; we have arranged them into those that support a uniform, massive, deep potential well interpretation on the one hand, and a diverse, bursting dwarf interpretation on the other. While these results require confirmation and augmentation, we believe that it is premature to conclude that star-forming galaxies at $2 < z < 4$ selected by their Lyman limit breaks represent a uniform population of high-mass proto-spheroids at the centers of deep, dark matter potential wells. The end products of $U-$ and $B-$band dropout galaxies are more likely to span a range of galaxy types and masses comparable to that *observed locally* in flux- and volume-limited surveys.

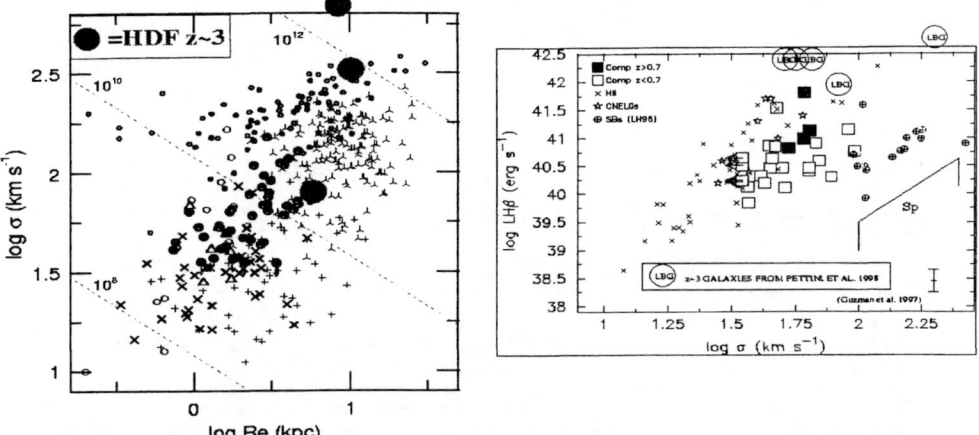

FIGURE 2. (a) Pseudo velocity width *vs.* half-light radius for three LBGs compared to normal local galaxies, from our Keck observations. The LBGs appear to span several orders of magnitude in mass. Original plot from Guzman *et al.* 1996. **(b)** Hβ vs. line width for a sample of Lyman Break Galaxies at $z \sim 3$ (from Pettini *et al.*) compared to local galaxies. Two of the points are scaled from [O III] λ5007 or Hα. Note that the emission line widths and luminosities of LBGs are mostly comparable to those of bright HII galaxies.

PUNCH LINE: Lyman Break Galaxies at $z \sim 3$ are plausibly **diverse, low-mass, starbursting dwarfs** that will wind up in a **wide variety** of end products.

ACKNOWLEDGMENTS

This project is being carried out in conjunction with the DEEP collaboration (see http://www.ucolick.org/ deep/home.html). We thank N. Vogt, A. Phillips, and S. Faber for help on observing runs and D. Kelson for providing the data reduction code.

REFERENCES

1. Steidel, C. C., Giavalisco, M., Pettini, M., Dickinson, M., & Adelberger, K. L. 1996, ApJl, 462, 17L
2. Baugh, C. M., Cole, S., Frenk, C. S., & Lacy, C. G. 1998, ApJ, 498, 504B
3. Giavalisco, M., Steidel, C. C., Adelberger, K. L., Dickinson, M. E., Pettini, M. & Kellogg, M. 1998, ApJ, 503, 543
4. Lowenthal, J. D., Koo, D. C., Guzmán, R. Gallego, J., Phillips, A. C., Faber, S. M., Vogt, N. P., Illingworth, G. D., & Gronwall, C. 1997, ApJ, 481, 673
5. Pettini, M., Kellogg, M., Steidel, C. C., Dickinson, M., Adelberger, K. L, & Giavalisco, M. 1998, ApJ, 508, 539

TABLE 1.

UNIFORM, HIGH-MASS CORES | DIVERSE, LOW-MASS CLUMPS

Morphologies & Sizes

• Uniform Morphologies?	• Diverse Morphologies! > 50% multiple, irregular
• $r^{1/4}$ profiles?	• $(B/T)_{med} \sim 0.1$
• $r_{1/2}$(LBG) $\sim r_{1/2}$(small bulges today)	• $r_{1/2} - M_B$ plot $\rightarrow$ HII galaxy-like

Numbers and Distributions

• $n(I < 25) \simeq n(L > L*)$today	• $n(L > L*) > 4n(L > L*)$today $\rightarrow$ requires merging and/or fading to reconcile with local galaxies (especially in clusters)
• Clustering: $\delta_{Gal} \sim 3.5$ $\rightarrow M_{halo} \sim 10^{12} M_\odot$	• Sub-galactic clumps in halos could explain δ

Luminosities and Star Formation Rates

• SFR up to $100 M_\odot$ yr^{-1} (with dust correction)	• $SFR \sim 10 M_\odot$ yr^{-1} requires 10^{10} yrs to make $10^{11} M_\odot$ $\rightarrow$ conflicts with stellar population constraints
• $L \sim L*$	• Starburst: $L \rightarrow 100 \times L$; $M/L \ll 1$ for HII galaxies

Masses

• Correlation + CDM $\rightarrow 10^{12} M_\odot$; Clumps at *center* of large potential	• Could be clumps *around* large potential
• Absorption line EWs $\rightarrow$ $M_{dyn} > 10^{11} M_\odot$	• $EW \neq f(M)$ (*e.g.*, shocks, SNe...)
• C4-09: $\Delta v \sim 800$km s$^{-1} \rightarrow 10^{12} M_\odot$	• Lyα emission lines narrow $\rightarrow 10^{10} M_\odot$ (caveats: dust, scattering, outflows...)
	• K_{obs} ($\simeq V_{rest}$) $\rightarrow M* < 10^{10} M_\odot$
	• $\sigma_{H\beta} \sim 70$km s$^{-1} \rightarrow$ M $\sim 10^{10} M_\odot$
	• C4-06 kinematics: $M < 10^{10} M_\odot$

Cosmological Implications of Lyman-Break Galaxy Clustering

Risa H. Wechsler[1], Michael A. K. Gross[2], Joel R. Primack[1],
George R. Blumenthal[3], Avishai Dekel[4], and Rachel S. Somerville[4]

[1] *Physics Department, University of California, Santa Cruz, CA 95064*
[2] *NASA/Goddard Space Flight Center, Greenbelt, MD 20771*
[3] *Astronomy & Astrophysics Department, University of California, Santa Cruz, CA 95064*
[4] *Racah Institute of Physics, The Hebrew University, Jerusalem, 91904, Israel*

Abstract. We review our analysis of the clustering properties of "Lyman-break" galaxies (LBGs) at redshift $z \sim 3$, previously discussed in Wechsler *et al.* [6]. We examine the likelihood of spikes found by Steidel *et al.* [2] in the redshift distribution of LBGs, within a suite of models for the evolution of structure in the Universe. Using high-resolution dissipationless N-body simulations, we analyze deep pencil-beam surveys from these models in the same way that they are actually observed, identifying LBGs with the most massive dark matter halos. We find that all the models (with SCDM as a marginal exception) have a substantial probability of producing spikes similar to those observed, because the massive halos are much more clumped than the underlying matter – i.e., they are biased. Therefore, the likelihood of such a spike is not a good discriminator among these models. The LBG correlation functions are less steep than galaxies today ($\gamma \sim 1.4$), but show similar or slightly longer correlation lengths. We have extended this analysis and include a preliminary comparison to the new data presented in Adelberger *et al.* [8]. We also discuss work in progress, in which we use semi-analytic models to identify Lyman-break galaxies within dark-matter halos.

INTRODUCTION

Recently, Steidel *et al.* (S98) [2] discovered that the redshift distribution of "Lyman-break" galaxies (LBGs) in a pencil beam reveals a large "spike" in the LBG distribution near $z \simeq 3.1$. This spike corresponds to a fractional overdensity of LBGs of a few hundred percent over a comoving scale of order $\sim 10 - 20h^{-1}\mathrm{Mpc}$. Since then, they have compiled a sample of more than 600 galaxies in six fields, with measured redshifts between about 2.5 and 3.5 [8], [4].

At a first glance, such a high peak seems surprising, since it suggests substantial nonlinear clustering on rather large scales. In fact, we [6] and other authors [5], [5],

CP470, After the Dark Ages: When Galaxies were Young (the Universe at 2 < z < 5),
edited by Stephen S. Holt and Eric P. Smith
© 1999 The American Institute of Physics 1-56396-855-X/99/$15.00

[7] find from simulations that these spikes arise naturally in a variety of cosmologies, provided that there is substantial galaxy biasing.

Here we compare the clustering properties of the LBGs to those expected in high-resolution simulations of four different cosmological models, by identifying $z \sim 3$ LBGs with the most massive halos in our simulations at that redshift. We are currently working on improving this analysis, both by comparing to the increasing amounts of new data, and by using semi-analytic models to explore different ways of identifying LBGs with halos in N-body simulations.

HOW PROBABLE ARE THE SPIKES?

Our methods are described fully in Wechsler *et al.* (W98) [6], and will be described only briefly here. The data is taken in 9′x9′ fields; the redshift distribution is binned in bins of $z = 0.4$. In determining the probability of spikes, our statistics consider each pixel separately, where a pixel is one z=0.4 redshift bin by the size of an angular field: 9′x18′ (in the case of data from S98), or 9′x9′, (in the case of the newer data from Adelberger *et al.* (A98) [8]). The number of galaxies in each pixel is divided by its selection function; we then consider the galaxy overdensity per pixel: $\delta_g = (N - \bar{N})/\bar{N}$. These data are then compared to N-body simulations by Gross *et al.* [8] of four cosmological models: SCDM ($\Omega_0 = 1, h = .5, \sigma_8 = 0.67$), CHDM ($\Omega_0 = 1, \Omega_{cdm} = 0.8, \Omega_\nu = 0.2, h = .5, \sigma_8 = 0.72$), OCDM ($\Omega_0 = 0.5, h = .6, \sigma_8 = 0.66$), ΛCDM ($\Omega_0 = 0.4, \Omega_\Lambda = 0.6, h = .6, \sigma_8 = 0.77$), in a 75 h^{-1}Mpc box; 57 million CDM particles (+ 113 million HDM particles in CHDM). The halos are identified as virialized regions at $z \sim 3$. To assign LBGs to halos we assume that one LBG resides in each massive halo, then choose a mass cutoff to match the observed number density of candidates and randomly select 40% of these (because S98 find redshifts for $\sim$ 40% of candidates). The mass cutoff ranges from 3×10^{11} for CHDM to 9×10^{11} for SCDM. We then "observe" pixels the size of the observational determined pixels, and compared the statistic δ_g to that calculated from the data. The cumulative probability distribution of this galaxy overdensity statistic is plotted in Figure 1, both for the analysis done in W98, which compared only to the data from the one 9′x18′ field published by S98, and compared to the more complete data set from six 9′x9′ fields published by A98. We find that the probability of observing a spike as large as the largest seen by S98 (corresponding to $\delta_g = 2.6$) is about 37%, 31%, 27%, and 6% for CHDM, OCDM, ΛCDM, and SCDM, respectively. The reason for the small probability for SCDM seems to be due to the shape of the power spectrum [6], [8]. The distribution of the newer data set published in A98 looks relatively well fit by all of the models; a full analysis will be published elsewhere. In future work, we will also take into better account a more accurate selection function for the data.

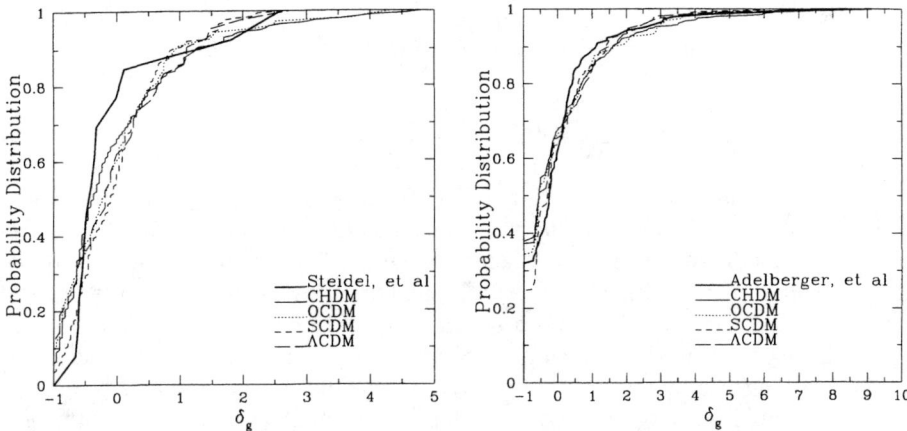

FIGURE 1. (a) The cumulative probability distribution of δ_g for the S98 data as well as for the massive halo distribution in the four models considered here. The distribution reflects the relative excess of halos in each pixel. Spikes the size of the largest spike found by S98 are found with reasonable probability ($\sim 25 - 40\%$) in all models except SCDM, and with marginal probability in SCDM. (b) The cumulative probability distribution of δ_g for the newer data set of A98 as well as for the distribution of massive halos in the four cosmologies considered. The distribution seems reasonably well matched by all of the models.

BIAS & CORRELATION FUNCTION

The clustering of LBGs is found to be biased with respect to the underlying dark matter. The bias factor b, defined as $b = \delta_g/\delta_m$, has an average value of about 2-5 depending on the cosmological model. High-density regions are the most biased. We have measured the correlation function for the LBGs in our simulations. The best fit parameters are: r_0 [h^{-1} Mpc] $= 3.3$, 5.1, 5.0, 7.3, and $\gamma = 1.7$, 1.6, 1.6, 1.5, for SCDM, CHDM, OCDM, and ΛCDM, respectively. A98 have calculated the correlation function using a counts-in-cells method, and find correlation lengths of $\simeq 4 \pm 1, 5 \pm 1$, and $6 \pm 1(h^{-1}\mathrm{Mpc})$ for $\Omega_M = 1$, 0.2 open, and 0.3 flat, assuming that $\gamma = 1.8$ (cf. [11]).

ADDING SEMI-ANALYTIC MODELS

In previous work, we identified LBGs by making the assumption that one object resides in each massive halo above some mass cutoff. Alternatively, we can use semi-analytic models [10], [9], [12] to predict the location of objects luminous enough to be observed as LBGs – they may reside in less massive halos, or there may be more than one per halo. We are currently exploring how different models of galaxy formation affect the clustering properties of the Lyman-break galaxies when identified in this way – which may help to distinguish between galaxy formation models and to determine the nature of the Lyman-break galaxies.

CONCLUSIONS

Large peaks in the observed redshift distribution of LBGs are common at $z \sim 3$ in several cosmological models. Galaxy formation is biased at high-z in all cosmologies we have considered, with a bias factor of $\sim 2 - 5$. The clustering properties will probably not distinguish between different cosmologies, without independent information about the bias. We find a similar correlation length but slightly shallower correlation function slope compared with that observed. We find that the bias and clustering properties are more strongly affected by the shape of the power spectrum than by the mass density; a model with a shallow power spectrum, like SCDM, seems to be somewhat less clustered than the data.

A preliminary analysis of the data published by A98 shows that the original spike [2] is one of the highest in all surveyed fields, and the new data is in fairly good agreement with all four cosmological models. Major improvements to our prior analysis can be made both by comparing to the still-increasing amounts of new data, and by improving our method of assigning LBGs to dark-matter halos. By using semi-analytic models to test and improve upon our method of identifying LBGs within halos from N-body simulations, we may be better able to understand the clustering properties and nature of the Lyman-break galaxies, and may begin to distinguish between galaxy formation scenarios.

REFERENCES

1. Wechsler, R. H., Gross, M. A. K., Primack, J. R., Blumenthal, G. R., & Dekel, A. 1998, ApJ, 506, 19 (W98)
2. Steidel, C.. C., Adelberger, K. L., Dickinson, M., Giavalisco, M., Pettini, M., & Kellogg, M. 1998, ApJ, 492, 428 (S98)
3. Adelberger, K. L., Steidel, C. C., Giavalisco, M., Dickinson, M., Pettini, M., Kellogg, M. 1998, ApJ, 505, 543 (A98)
4. Steidel, C.. C., Adelberger, K. L., Giavalisco, M., Dickinson, M., Pettini, M., & Kellogg, M. 1998 (astro-ph/9804237, astro-ph/9805267)
5. Bagla, J. S., 1998, MNRAS, 297, 251
6. Jing, Y. P., Suto, Y. 1998, ApJL, 494, L5
7. Governato, F., Baugh, C. M., Frenk, C. S., Cole, S., Lacey, C. G., Quinn, T., & Stadel, J. 1998, Nature, 392, 359
8. Gross, M. A. K., Somerville, R. S., Primack, J. R., Holtzman, J., & Klypin, A. A. 1998a, MNRAS, 301, 81
9. Giavalisco, M. Steidel, C. C., Adelberger, K. L., Dickinson, M. E., Pettini, M., Kellog, M. 1998, ApJ, 503, 18
10. Somerville, R. S., Primack, J. R. 1998, MNRAS in press (astro-ph/9802268)

11. Somerville, R. S., Primack, J. R., & Faber, S. M. 1998, MNRAS, in press (astro-ph/9806228)
12. Primack, J. R., Somerville, R. S., Faber, S. M., & Wechsler, R. H., 1998 (astro-ph/9806263)

5. Largest Structures

Distant Large Scale Structures

Richard Mushotzky

LHEA, NASA/GSFC, Greenbelt, MD 20771

Abstract. The existence of massive virialized objects at high redshifts is not expected in most cosmologies. I discuss the evidence for the existence of massive clusters at $z > 0.9$ and conclude that there is strong evidence that such objects exist at $z \sim 1.2$. At larger redshifts the evidence is not conclusive. The number and mass of such objects is difficult to estimate at present because of their optical faintness and the limitations of present data x-ray telescopes.

INTRODUCTION

In most cosmological models the formation of massive, virialized objects occurs at "low" redshifts. Thus "classical" rich relaxed clusters are not expected to exist at large ($z > 2$) redshifts. The relative number of massive clusters at a fixed mass is expected to change rapidly with redshift and their number density and distribution in mass are strong functions of cosmological parameters (in particular Ω and σ_8). In a standard cold dark matter model (with $\sigma_8 = 0.56$, a best fit value based on the cluster abundance) the relative abundance of objects of mass $> 10^{15} M_\odot$ is reduced by over a 1000 from $z = 0$ to $z = 1$ [1]. In an Open cold dark matter model the reduction from $z = 0$ to $z = 2$ is about 10,000 and the mere existence of a massive cluster at $z > 1$ is a strong cosmological test of our world model, numerical simulations and the hierarchical universe calculations. Thus finding even one massive cluster at $z \gg 1$ would be a major discovery. A major concern is determining the mass of the putative high z cluster, since that is what is robustly calculated by theory.

CP470, After the Dark Ages: When Galaxies were Young (the Universe at 2 < z < 5), edited by Stephen S. Holt and Eric P. Smith

WHAT DOES ONE EXPECT?

A. What Do Low ($z < 0.8$) Redshift Clusters "Look" Like?

In the mass range from 10^{14}–$4 \times 10^{15} M_\odot$ appropriate to massive clusters, the observed X-ray temperatures range from $\sim$1–10 keV (10^7–10^8 K), the X-ray luminosities from $L_{\text{bol}} \sim 10^{44}$–$5 \times 10^{45}$ ergs/s . Their optical velocity dispersion 400–1400 km/sec, with optical luminosities $\sim 10^{12-13} L_{\text{sun}} \sim$ 10–100 L_* galaxies . The "size" of the clusters is $R_{\text{virial}} \sim 3h_{50}^{-1}T_8^{1/2}$ Mpc; $\sim$ 30 arcmin at $z = 0.06$. There is a strong correlation of $L_x \sim T_x^3$, $T_x \sim \sigma^2$ and $M_{\text{virial}} \sim T^{3/2}$. The mass to light ratio is $M/L \propto 250h_{50}^{-1}$ and the gas density profiles scales as $\propto r^{-2}$.

There is little or no evidence for evolution ($z < 0.6$) in the relationship between X-ray luminosity and temperature [29], the mean metalicity of clusters the X-ray luminosity function , the size distribution [43] or the temperature function [21]. There is an indication that luminous X-ray clusters were somewhat less common at $z > 0.3$ [43]. Many of these objects look and "behave" as relaxed systems. The lack of evolution in the properties of $z < 0.5$ clusters as has been interpreted as an indication that clusters are "old" and in favor of low Ω universes.

B What do we expect high z clusters to look like?

The general evolution of clusters in a hierarchical universe predicts that **at a fixed mass scale** clusters should be smaller, hotter, denser, and more luminous at higher redshifts [24] with the following rough scaling relationship [15] $T \propto (1 + z)M^{2/3}$; $L \propto (1 + z)^{7/2}$; $R_{\text{virial}} \propto (1 + z)^{2/3}$.

However since in a hierarchical universe the characteristic mass scale changes as $(1 + z)^{-6/(3+n)} \sim (1 + z)^{-3}$ where n is the slope of the power spectrum at the mass scale of clusters, the *average* cluster observed at high redshift should be smaller, cooler, denser and not much less luminous. In other words the evolution of the mass spectrum dominates over the evolution of an object of constant mass. But, there is lots of room to maneuver in varying the cosmological parameters and the simple scaling relations may be poor guides to the detailed calculations. In particular the calculation of the evolution of the X-ray luminosity function [9] Bryan-etal-94 is difficult and complex and strongly dependent on Ω and σ_8. For luminous systems most of the "action" is at high z [9].

Calculation of the optical evolution is much more difficult and depends on understanding galaxy evolution (cf. [25]). At $z \sim 1$ clusters contain "old" red galaxies [13] but the population mix in clusters is varying strongly at $z > 0.2$ [32]. It is not clear what to expect at $z > 2$. However, in a high Ω universe one does not expect massive clusters to exist at all at $z \gg 1$.

III WHAT DO WE NOW KNOW AND HOW DO WE KNOW IT?

To date the 2 main techniques for searching for clusters have been optical imaging and large scale X-ray surveys.

Classical ground based optical techniques [33] (searching for statistically significant projected galaxy overdensities) have great difficulty in finding clusters at $z > 0.8$ for a variety of reasons.

1. massive clusters are dominated by elliptical galaxies which are very "red" . Thus they get dim rapidly with redshift (the "K" correction) in rest frame U,B,V bands as they get redshifted. This can be compensated for somewhat by going to redder optical bands and constructing "optimal" spatial filters [46]

2. the cosmological $(1 + z)^4$ surface brightness diminution makes the galaxies more difficult to detect again the sky brightness limits.

3. foreground contamination increases rapidly with redshift.

4. present technology finds it difficult to obtain large redshift samples of $0.6 < z < 2$ absorption line galaxies.

A major problem with optical techniques is the need for velocity information, since surface density enhancements at high z are often due to projection effects. Thus even when overdensities are found extensive spectroscopic data are needed to confirm the cluster existence and to derive a velocity dispersion to estimate the mass.

X-ray techniques (mainly from Rosat PSPC data) have a similar redshift limit to the optical data.

1. The *ROSAT* PSPC angular resolution (~ 15 arcsec) does not allow easy recognition of clusters as extended at $z > 0.8$, since the angular size (half power radius) of a massive cluster (at least at low redshift) is ~ 15 arcsec at $z \sim 1$ ($q_0 = 0, h = 1/2$) .

2. The $(1 + z)^4$ surface brightness diminution combined with the bright soft X-ray background makes cluster recognition difficult. At low redshift the typical central surface brightness of a massive cluster is ~ 50 times the X-ray background. But at $z \sim 1$ they are roughly equal. Thus even though the *ROSAT* "typical" flux limit of 10^{-14} gives a luminosity limit of $L_x \sim 2.5 \times 10^{43}$ (e.g. 1/50 of Coma) at $z \sim 0.8$ these objects are rather hard to recognize.

3. Most optical data is not adequate to allow cluster IDs: to quote Rosati (1998) [38] *"the fact that very few (clusters) have been discovered so far at $z > 0.85$ is not due to a lack of sensitivity of X-ray searches at these redshifts, but rather reflects the difficulty of carrying out the spectroscopic confirmation with a 4m class telescope."* Thus until X-ray data can uniquely identify and obtain the redshifts of the cluster candidates this will be a fundamental limitation. The

combination of *AXAF* and *XMM* can do this for objects at $z \sim 1$ (*AXAF* will obtain the source image and resolve it and *XMM* will obtain the spectrum and the redshift).

However compared to optical techniques in general the X-ray technique has a major advantage- at low redshift, all luminous extended high latitude X-ray sources are clusters and the X-ray luminosity is strongly correlated with the mass of the system, thus detection of a extended X-ray source at high galactic latitude is almost surely correlated with a massive object. Detailed calculations [15] indicated that $L_x \sim M^{4/3}(1 + z)^{7/2}$.

There have been numerous other techniques:

- *HST* direct imaging [34] and the Lyman-limit redshift searches [42] can find density enhances of galaxies at almost arbitrary limits. Like other optical search techniques redshifts are necessary for confirmation. Near-IR searches (K band) benefit substantially from the redshift of the galaxies, but so far suffer from very small fields of view. Even Keck has difficulty deriving redshifts for objects with K>20, for which $z_{max} \sim 0.8$ [10]

- Looking in and around radio sources: At $z \sim$0.3–0.5 many radio sources are found in dense environments typical of groups and clusters (e.g. [45]) and thus since it is "easier" to find radio sources than high z clusters searches around high z radio sources have proven productive. In addition certain types of radio sources are at low redshift uniquely associated with clusters and groups. In particular radio sources with high Faraday rotation [17] are only found in cooling flow clusters and head tail, wide angle tail sources [6], Hintzen-84 are only found in cluster environments. Thus it is natural to use distant radio sources as guides to the existence of high redshift clusters [4].

All of these techniques have been successful to a greater or lesser degree. A potentially exciting new idea is a "blind" search for Sunyaev-Zeldovich (S-Z) decrements (e.g. [2], Makino-etal-97) – so far there are 2 possible such objects but the solid angle surveyed has been very small. A prime advantage of the S-Z technique is that the signal is constant with redshift (e.g. no cosmological surface brightness dimming) and the predicted evolution of the signal is very different from that of the X-ray and "optical" signals.

IV CLUSTERS AT $Z > 0.95$

Stanford et al (1997) [40] have found a "serendipitous" *ROSAT* counterpart to a "serendipitous" K band selected cluster at $z \sim 1.27$ with $L_x \sim 1.5 \times 10^{44}$ ergs/s and a velocity dispersion of $\sigma = 740 \pm 190$. Using scaling relations found at lower redshifts (corrected for cosmological effects) between X-ray luminosity and optical velocity dispersion imply a total mass of $\sim 5 \times 10^{14} h_{65} M_\odot$ for this object. Dickinson et al (1998) [14] have obtained X-ray images of 6 radio selected objects at $z > 0.8$

and two of them are clearly extended 3C324 at $z = 1.24$ and 3C294 at $z = 1.78$. Optical data for 3C324 (Keck+HST) and the indication of a weak gravitational shear [39] clearly confirm the cluster "nature" of the object. While the mass is difficult to estimate the X-ray luminosity and gravitational shear argue for a massive system with velocity dispersion in the range 900–1500 km/s. A nearby bright star makes optical data for 3C294 difficult. The X-ray spectra of both objects is consistent with originating from a luminous cluster [11]. The "sizes" and luminosity ($L_{\rm bol} \sim 9 \times 10^{14}$) of these objects are consistent with the Coma cluster, but their surface brightness is $\sim 5\times$ higher. Their estimated masses are 1–$10 \times 10^{14} h_{65} M_\odot$ depending strongly on cosmological model and the accuracy of scaling relations between luminosity, size and mass.

Hattori et al (1997) [20] have a X-ray spectrum and image of MG2016+112, a gravitational lens at $z = 1.01$, the X-ray temperature, luminosity and lensing signal argue for a large $> 3 \times 10^{14} h_{50} M_\odot$ mass in this system, however there does not seem to be a significant overdensity of galaxies. Very recent work [3] has unveiled an optically "normal" cluster associated with this object. Hasinger et al 1998 [19] have detected another serendipitous "dark lens" in the deepest $ROSAT$ survey in Lockman hole region. The X-ray source is clearly extended, there is no indication of a galaxy overdensity, but it is associated with a giant arc. While the redshift is not known, general arguments indicate $z \sim 1$ and the combination of the X-ray luminosity and the presence of a gravitationally lensed giant arc is strong evidence for a massive system. The minimum lensing mass is $\sim 2 \times 10^{13} M_\odot$ and the best estimates are 10–50 larger (depending on z) .

There is a recent report [38] of a $z \sim 1.2$ cluster detected serendipitously in $ROSAT$ PSPC data. However at the present time no details are known. At somewhat lower degree of certainty Carilli et al (1998) [8] have an X-ray detection of a high Faraday rotation object at $z = 2.156$. However as of yet there is no data indicating a galaxy overdensity, X-ray extent or temperature. As far as objects without X-ray "confirmation", Deltorn et al (1997) [12] have reasonable indications of a cluster around 3C184 at $z = 0.996$ from galaxy overdensity, the existence of a gravitational arc and galaxy velocity data. Zaritsky et al (1997) [46] have at least one cluster with 3 redshifts at $z \sim 1$.

I V. Optical/UV Data at $z > 2$

Several groups have found evidence for a significant overdensity of Lyman limit galaxies at $z \sim 2.4$ [35], $z \sim 2.57$ [7], $z \sim 3.02$ [41], and of Lyα emission line galaxies and absorption line objects at $z \sim 2.38$ [16]. While it maybe somewhat premature indications from the Steidel et al (1997) [41] work are that such "structures" occur in almost every LRIS Keck field and thus must be fairly common. [42] It is not clear if these are relaxed virialized systems , but these "objects" are large overdensities with scale sizes not inconsistent with rich clusters. Steidel et al argue that these objects represent "proto-clusters" :they have the "right" overdensity and are probably strongly biased. Interpretation of these data are very difficult at present and depend strongly on modeling and theory [44]. However it is clear that significant overdensities on ~10 Mpc scales exists at $z > 2$. Because of the small field of view of HST (1 arcmin $\sim$ 1.3 Mpc $1 < z < 10$ $h = 1/2$, $q_0 = 0$) it is rather difficult to find/study clusters in a serendipitous fashion. Very recently [18] there has been a report of significant enhancements of galaxy counts around quasars at $z > 1.4$. In the "now" standard picture where luminous quasars form in the cores of massive galaxies and that these high density perturbations collapse early one might expect quasars to be tracers of clusters and/or protoclusters. However, these hypothesized clusters do not have measured redshifts and thus estimates of their mass and are extraordinarily uncertain. Similar results have been reported for "extremely red objects" which seem to be a collection of dusty galaxies, AGN and true ellipticals. And several groups have performed K band searches with some success.

VI SUNYAEV-ZELDOVICH EFFECT SEARCHES

There have been 2 recent reports of "blank-field" Sunyaev-Zeldovich (S-Z) decrements. At low redshifts ($z < 0.4$) the S-Z decrements are associated with the scattering of the microwave background off the hot gas in a foreground cluster of galaxies. The net effect is to reduce the brightness of the microwave background at lower frequencies. This has been well measured only in the last 2 years for roughly 15 clusters [31]. The "blank-field" S-Z decrements do not contain an optical cluster or a *ROSAT* source and if real this combination argues for a massive cluster at $z > 0.7$ or 2.8 for the Richard et al (1997) [36] and Jones et al [23] S-Z objects [26] respectively. The Campos et al (1998) [7] high redshift "object" is in the field of a S-Z decrement [36] and may or may not be related. While these results are tentative but potentially very exciting, to paraphrase Kneissl et al [26]: *"Normalized to the present abundance of clusters even $\Omega < 0.25$ models predict only 1 clusters on the whole sky with $M > 2 \times 10^{15}$ at $3 < z < 4$!"*. Clearly future work in this area is potentially exciting.

II   VII Conclusions and Future Prospects

It seems clear that massive clusters exist to $z \sim 1.2$, but their space density is uncertain. However for $\Omega = 1$ models they should "not exist" depending crucially on their "true" mass. Whether their masses and space densities are too high even for lower Ω models is not yet clear. The next generation of X-ray missions will easily detect and resolve "Coma-like" clusters at $z > 2$ (depending on cosmology) and can obtain temperatures out to $z \sim 2$. Thus identification of cluster candidates at high z will be much easier. Optical follow-up work in the $z \sim$1–2 range will require *HST* and 10m telescopes. Near-IR surveys will be much more efficient when the large solid angle detectors come on line. However they will need "optical" spectra to confirm cluster existence. Sunyaev-Zeldovich searches may be very efficient in the near future but need to be confirmed with X-ray and optical follow-ups to determine the origin of the decrements.

REFERENCES

1. Bahcall, N, , Fan, X., Cen, R 1997 ApJ 485, L53
2. Bartlett J.G. , Blanchard A. , Barbosa D. 1998 astro-ph/9808308 Proceedings of the IAP conference *Wide Field Surveys in Cosmology*
3. Benitez, N, Broadhurst, T. , Rosati, P. , Courbin, F. , Squires G. , Lidman C. , Magain P. 1998 astro-ph/9812218
4. Blanton, E. L., Helfand, D. J., Becker, R. H., Gregg, M. D., White, R. L.1997 AAS Meeting, 191, 106.12
5. Bryan, G, Cen, R, Norman, M., L., Ostriker, J. P., Stone, J. 1994 ApJ 428..405
6. Burns, J., O, Honest, R, , White, R, Nelson, E, Morrisette, K,., Moody, J. , 1988 A J 94, 587
7. Campos, A., Yahil, A., Windhorst, R.A., Richards, E.A., Pascaralle, S., Impey, C., Petry, K. MPA/ESO Cosmology Conference "Evolution of Large-Scale Structure: From Recombination to Garching", Garching, Germany, 2-7 August, 1998
8. Carilli, C., Harris, D., Pentericci, L, Rottergering, H., Miley, G, Bremer, M, 1998 Ap J..494 L.143
9. Cen, R and Ostriker, J. P. 1994 Ap J 429 .4
10. Cohen, J. Blandford, R. Hogg, D, Pahre, M.,. Shopbell, P. 1998 astro-ph/9809067
11. Crawford, C. S., Fabian, A. 1996 MNRAS.282.1483
12. Deltorn, J.-M, Le Fevre, O, Crampton, David, Dickinson, M. 1997 Ap. J Lett 483,.L21
13. Dickinson, M. 1997 astro-ph/9703035
14. Dickinson,M.et al 1998 ApJ submitted
15. Eke, V, Navarro, J, and Frenk, C 1998 ApJ 503, 569
16. Francis, P. J., Woodgate, B. E. and Danks, A. C. 1998 in The Young Universe: Galaxy Formation and Evolution at Intermediate and High Redshift. Ed S. D'Odorico, A. Fontana, and E. Giallongo. ASP Conference Series, V 146 496
17. Ge, J and Owen, F. N 1994 AJ. 108, 1523
18. Hall, P, Green, R. F. 1998 Ap J 507,558
19. Hasinger G. ,Giacconi R. Gunn J. , Lehmann I., Schmidt M., Schneider D.P. Truemper J., Wambsganss J., Woods D., and Zamorani G. 1998 astro-ph/9810347
20. Hattori, M., Ikebe, Y.,Asaoka, I., Takeshima, T., Boehringer, H., Mihara, T., Neumann, D. M., Schindler, S., Tsuru, T and Tamura, T. 1997 Nature .388..146
21. Henry, J. P. 1997ApJ 489 L 1
22. Hintzen, P. 1984 Ap J Supp 55, 533
23. Jones ,M., Saunders,R. Baker,J., Cotter, G. Edge,A. Grainger,K., Haynes,T., Lasenby,A., Pooley,G., Rottgering,H. 1997 Ap J Lett 479,L1
24. Kaiser, N. 1991 Ap J 383, 104
25. Kauffmann, G. and Charlot S. 1998 astro-ph/9810031
26. Kneissl, R, Sunyaev, R, White, S.D 1998 . MNRAS 297, L29

27. Makino , N. Shin S, Suto, Y. 1997 astro-ph/9710344

28.

29. Mushotzky, R. F. and Scharf, C. A. 1997 ApJ 482 L 13

30. Mushotzky, R. F and Loewenstein, M. 1997ApJ 481L..63

31. Myers, S. T., Baker, J. ,Readhead, A. , Leitch, E., Herbig, T. 1997 Ap. J. .485,.1

32. Oemler, A, Jr.Dressler, A and Butcher, H, 1997ApJ 474..561

33. Oke, J. B., Postman, M, Lubin, L M.1998 AJ 116, 549-559.

34. Ostrander, E. J., Nichol, R. C.,.Ratnatunga, K. U, Griffiths, R. E. 1998 astro-ph/9808304

35. Pascarelle, S. M., Windhorst, R. A., Driver, S. P., Ostrander, E. J and Keel, W. C. 1996 ApJ 456 L..21

36. Richards, E. A.; Fomalont, E. B.; Kellerman, K. I.; Partridge, R. B.;

37. Windhorst, R. A. 1997 A. J. 113,

38. Rosati, P. 1998 astro-ph/9810054

39. Smail, Ian, Dickinson, Mark 1995 ApJ 455, L99

40. Stanford, S.A., Elston, R. , Eisenhardt, P, Spinrad, H, Stern, D., Dey, A. 1997 A. J. .114,.2232

41. Steidel, C, Adelberger, K, Dickinson, M, Giavalisco, M, Pettini, M, Kellogg,M, 1998 Ap J 492,428

42. Steidel,C. 1998 in "Evolution of Large-Scale Structure: From Recombination to Garching"

43. Vikhlinin, A., Mcnamara, B. R., Forman, W., Jones, C., Quintana, H., Hornstrup, A. 1998 Ap. J .498,L21

44. Wechsler, R, Gross, M, Primack, J, Blumenthal, G,. and Dekel, Ap J 506,19

45. Yee, H. K. C, Ellingson, E.1993 Ap J411, 43

46. Zaritsky, D, Nelson, A., Dalcanton, J., and Gonzalez, A. 1997 ApJ 480, L..91

Results from the CASTLES Survey of Gravitational Lenses

C.S. Kochanek[*], E.E. Falco[*], C.D. Impey[†], J. Lehár[*], B.A. McLeod[*] & H.-W. Rix[†]

[*] Center for Astrophysics, 60 Garden St., Cambridge, MA 02138
[†] Steward Observatory, Univ. of Arizona, Tucson, AZ 85721

Abstract. We show that most gravitational lenses lie on the passively evolving fundamental plane for early-type galaxies. For burst star formation models (1 Gyr of star formation, then quiescence) in low Ω_0 cosmologies, the stellar populations of the lens galaxies must have formed at $z_f \gtrsim 2$. Typical lens galaxies contain modest amounts of patchy extinction, with a median differential extinction for the optical (radio) selected lenses of $\Delta E(B - V) = 0.04$ (0.07) mag. The dust can be used to determine both extinction laws and lens redshifts. For example, the $z_l = 0.96$ elliptical lens in MG 0414+0534 has an $R_V = 1.7 \pm 0.1$ mean extinction law. Arc and ring images of the quasar and AGN source host galaxies are commonly seen in NICMOS H band observations. The hosts are typically blue, $L \lesssim L_*$ galaxies.

1. INTRODUCTION

In the last few years the number of known gravitational lenses has exploded to a total of over 40 systems[1]. With such a large statistical sample the lenses become excellent tools for studying the structure and evolution of the lens galaxies, the luminosity function of lens galaxies, dust in the lens galaxies, and cosmology. Many of these applications, particularly the evolution and luminosity function studies, depend on possessing accurate surface photometry of the lens galaxies. When we attempted the first survey of galaxy evolution and structure using lenses (Keeton et al. 1998), we discovered that the accumulated photometric data were inadequate to the task. Individual groups had observed individual lenses in a remarkable array of filters, frequently using short snapshots which were adequate to verify lens candidates but inadequate for detailed surface photometry of the lens galaxies.

The goal of the CfA/Arizona Space Telescope Lens Survey (CASTLES) is to remove these limitations and to fully enable the use of lenses as precision tools in astronomy. To date we have obtained NICMOS H images of 37 lenses and

---

[1] see http://cfa-www.harvard.edu/castles.

CP470, After the Dark Ages: When Galaxies were Young (the Universe at 2 < z < 5),
edited by Stephen S. Holt and Eric P. Smith

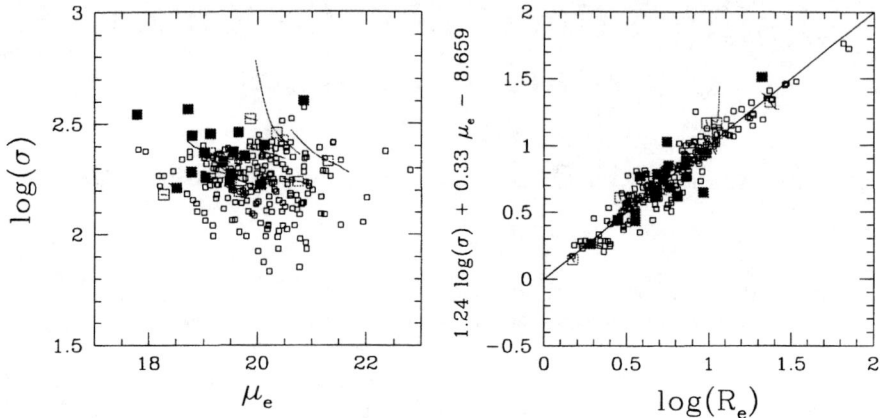

FIGURE 1. The Fundamental Plane of lens galaxies. The right panel shows an edge-on view of the FP, and the left panel shows the distribution of galaxies in surface brightness μ_e and velocity dispersion σ_c. The small open squares are the galaxies from the local sample of Jorgensen et al. (1996). The large filled (open) squares are the lenses with known (unknown) lens redshifts. The curves through the open squares show the parameter uncertainties created by the remaining redshift uncertainty.

binary quasars, with complementary WFPC2 V and I images scheduled for 1999. Further observations proposed for Cycle 8 would expand the database to 60 lenses and binary quasars. The data are homogeneously reduced with typical astrometric accuracies of 3 mas or better (checked on VLBI lenses), and initially fit with a series of standard photometric and gravitational lens models (e.g. Lehár et al. 1998).

Rather than discuss well known applications of gravitational lenses such as determinations of H_0 (e.g. Barkana 1998, Impey et al. 1998), the cosmological model (Kochanek 1996, Falco et al. 1998), or the mass distribution of the lenses (e.g. Keeton et al. 1997), we will discuss three new applications. First, we will study the fundamental plane of lens galaxies and its evolution, along with a few preliminary observations about the luminosity function of lens galaxies. Second, we will study extinction and extinction laws in lens galaxies. Third, and finally, we will quickly survey the properties of the host galaxies of the source quasars and AGN.

2. THE FUNDAMENTAL PLANE OF LENSES

Djorgovski & Davis (1987) and Dressler et al. (1987) discovered that early type galaxies show a tight correlation between the effective radius R_e, mean surface brightness μ_e and central velocity dispersion σ_c which is now known as the fundamental plane (FP). The FP is clearly related to the virial theorem, and the differences between the FP and the relation expected from the virial theorem are usually interpreted as variations in the mass-to-light ratio with luminosity. Van Dokkum et al. (1998) and Pahre (1998) have used the fundamental plane of early-type galaxies in rich clusters out to $z \sim 1$ to study the evolution of the mass-to-light ratio of the early-type galaxies and to demonstrate that they follow the predictions for passively evolving stellar populations formed at $z_f \gtrsim 2$. There is no comparable study of early-type galaxies in low density environments due to observational limitations, although some models of galaxy formation (e.g. Kauffmann & Charlot 1998) would predict significantly different star formation histories for field early-types.

Most lens galaxies are predicted (e.g. Fukugita & Turner 1991) and observed (e.g. Keeton et al. 1998) to be early-type (E and S0) galaxies in low density environments. Many lenses are in groups and very few are in poor clusters. To use the FP we need not only R_e and μ_e, which we can obtain from the CASTLES photometry, but also the central velocity dispersion. While we know the mass enclosed by the lensed images with extraordinary accuracy compared to that obtainable from stellar dynamical studies of even nearby galaxies, we have few direct measurements of the central velocity dispersion. We can, however, indirectly estimate σ_c from the separation of the images $\Delta\theta$. Both lens models (e.g. Kochanek 1995) and modern stellar dynamical models (e.g. Rix et al. 1997) of early-type galaxies favor an overall mass distribution corresponding to a flat rotation curve. In these models, the image separation depends only on the dispersion of the dark matter $\Delta\theta \propto \sigma_{DM}^2$, and stellar dynamical models of local early-type galaxies using the same mass model show that $\sigma_{DM} = \sigma_c$ to remarkable accuracy (Kochanek 1994). Thus we estimate the velocity dispersion by $\sigma_c = 225 \left[(\Delta\theta/2\overset{\prime\prime}{.}91)(D_{OS}/D_{LS})\right]^{1/2}$ km s^{-1} where D_{LS} and D_{OS} are the lens-source and observer-source distances.

We compare the lenses to the local FP by mapping the properties of the lenses to the redshift of Coma, and then comparing the predicted properties of the lenses to the sample of early-type galaxies in nearby clusters studied by Jorgensen et al. (1996). To make the transformation we must select a cosmological model (we will use the currently popular $\Omega_0 = 0.3$, $\Lambda_0 = 0.7$ model) and a stellar population evolution model. We used burst models in which star formation starts at a formation redshift z_f and lasts for 1 Gyr, which provides a good match to the colors of the lens galaxies for $z_f \gtrsim 2$.

Figure 1 shows the FP at Coma and the predicted properties of the lens galaxies. *With no adjustable parameters other than the star formation epoch, the vast majority of lenses with known redshifts lie on the fundamental plane with little more scatter than seen in the local samples!* The success of the comparison again emphasizes that most lens galaxies are early-type galaxies, and that the dark matter model we have used to estimate the central velocity dispersion is making accurate dynamical predictions. Changes in the cosmological model tend to make the galaxies slide along the FP, while changes in the star formation history tend to move the galaxies perpendicular to the FP. Figure 2 shows the relative probabilities of the star formation onset redshift z_f for the burst models. For this particular model the preferred epoch is $z_f \simeq 3$, rather similar to the results for early-type galaxies

165

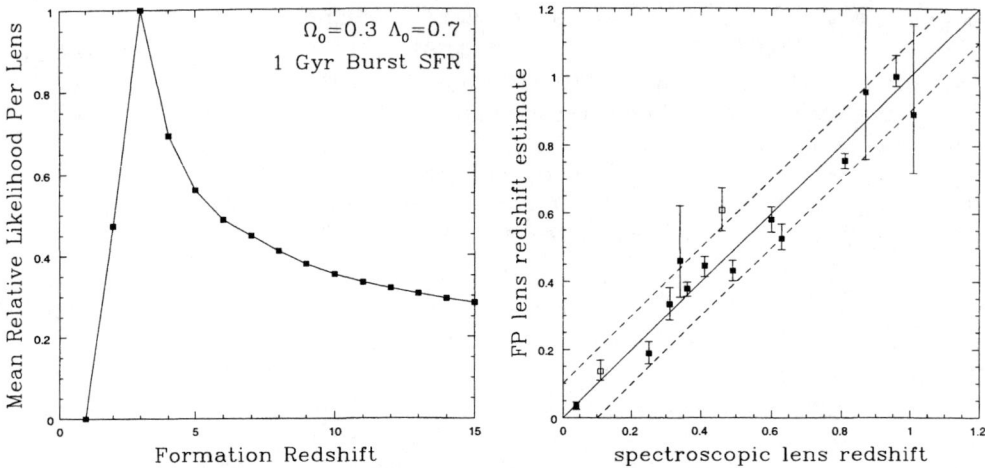

FIGURE 2. The relative probabilities of different star formation epochs z_f for $\Omega_0 = 0.3$, $\Lambda_0 = 0.7$, $H_0 = 65$ km s^{-1} Mpc^{-1} and a 1 Gyr burst star formation history.

FIGURE 3. Spectroscopic redshifts versus redshifts estimated from the FP. Filled (open) squares are lenses with known (unknown) source redshifts. HST 14113+5211, the most discrepant case, is the open square at $z_l = 0.46$.

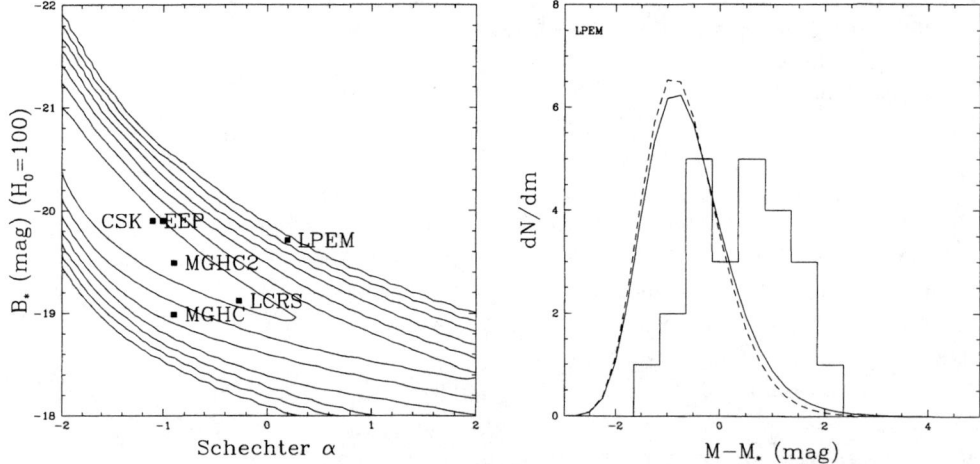

FIGURE 4. Left: Likelihood contours spaced by a factor of 10 for the Schechter function parameters α and B_*. The points show the standard model of Kochanek (1996, labeled CSK) and the five LF models used by Chiba & Yoshii (1998, labeled EEP, MGHC, MGHC2, LCRS, and LPEM). Chiba & Yoshii (1998) used the two most discrepant points (LPEM and MGHC) to revise the cosmological limits. Right: An illustration of why the LPEM model fails. The observed (histogram) and LPEM predicted (curves) luminosity functions of the lenses for an $\Omega_0 = 0.3$ flat cosmology. The dashed (solid) curves include (exclude) a model for the selection effects due to finite angular resolution in lens surveys.

in rich clusters despite the vast difference in environmental densities.

Even though the lens galaxies fall on the present day FP, they are not a random subset of it. As the left panel of Figure 1 shows, the lenses are concentrated toward high velocity dispersion. We see little difference in the effective radius and surface brightness distribution of the lens galaxies and the Jorgensen et al. (1996) local sample. Note, however, the selection effect that the lenses with spectroscopic redshifts tend to have higher surface brightnesses than the lenses with only estimated redshifts. The concentration of the lens galaxies at high velocity dispersions compared to local samples is expected from the strong velocity dispersion dependence of the probability that a galaxy will be a lens ($\propto \sigma_{DM}^4 \propto \sigma_c^4$).

Many redshifts for optically selected lenses remain unmeasured because the quasar/galaxy contrast makes the observations technically challenging, and the continuing redshift incompleteness is a severe limitation on using the lenses to determine the cosmological model. We can use the constraint that a lens lies on the FP as a means of estimating unmeasured lens redshifts because the predicted physical properties of the lens at Coma as a function of lens redshift follow a trajectory that is largely perpendicular to the FP. Thus, it is only at or near the true lens redshift that the galaxy properties will lie on the FP. Figure 3 compares the spectroscopic and FP estimates for the redshifts of 15 lenses. The rms redshift difference is only 0.06. The least accurate estimate is for HST 14113+5211 (Fischer et al. 1998), where the lens galaxy is in a cluster. The cluster potential boosts the image separation, leading to an overestimated galaxy velocity dispersion. Sometimes the trajectory for a particular filter moves along the FP for some redshift region, causing the larger uncertainties seen for two lenses near $z = 1$ with only H band data. Multicolor data breaks the degeneracy.

Finally, if we know the redshifts of the lens galaxies we can also estimate the mean luminosity function (LF) of the lenses and compare it to the predictions from local estimates of the LF. The LF of the lens galaxies differs from the LF of all galaxies because the lens cross section rises with luminosity, and we must include the appropriate cross section weighting of the LF when we make comparisons. As emphasized by Kochanek (1996) and Falco et al. (1998), the uncertainties in the luminosity function of galaxies by type contributes as much to the uncertainties in the cosmological limits derived from lens statistics as the Poisson errors arising from the small size of the samples. When parameterized by a Schechter function, $dn/dL = (n_*/L_*)(L/L_*)^\alpha \exp(-L/L_*)$, different local surveys (e.g. EEP (Efstathiou et al. 1988), LPEM (Loveday et al. 1992), MGHC (Marzke et al. 1994), LCRS (Lin et al. 1996)) find mutually discrepant values for the faint end slope α, break luminosity L_*, and number density of early-type galaxies.

Figure 4 shows the likelihood of fitting the observed LF of the lens galaxies and the separation distribution of the images as a function of the Schechter function parameters α and B_* (the absolute B magnitude corresponding to L_*) in an $\Omega_0 = 0.3$, $\Lambda_0 = 0.7$ cosmology. We mark the central point of the standard model used by Kochanek (1996) and Falco et al. (1998) and five alternative models used by Chiba & Yoshii (1998). The value of α is that from the original LF surveys (listed above), while the value of B_* is an estimate by Chiba & Yoshii (1998) after converting from the photometric band of the original survey. The two models selected by Chiba & Yoshii (1998) to revise the cosmological limits (LPEM and MGHC) are the two models most discrepant with the lens data, probably because of problems in the estimate of B_*. Figure 4 reveals why the LPEM model fails. The likelihood contours for the lens data show a degeneracy between B_* and α which is very similar to the apparent degeneracy that links most of the local LF estimates. These LF comparisons are extremely preliminary, but with the full sample it should be possible to constrain B_* and α more accurately, while measuring the changes in the comoving density n_* with redshift.

3. EXTINCTION

We possess little direct information on extinction in galaxies outside the Local Group and almost none on early-type galaxies (see reviews by Mathis 1990, Fitzpatrick 1998). Accurate extinction estimates almost always depend on knowing the intrinsic spectrum of the reddened object, which is generally true only of stars. Once inferences about extinction depend on modeling the fluxes of stellar populations mixed with dust, the accuracy drops dramatically (e.g. Witt et al. 1992). Extinction laws are measured almost exclusively in the Galaxy, the LMC and the SMC. No accurate extinction curve is measured in an early-type galaxy, although several studies (e.g. Warren-Smith & Berry 1983) suggest that dust in early-type galaxies may be quite different from "standard" Galactic dust. Moreover, since both the mean metallicity and star formation rates are strong functions of redshift, it would be surprising if the mean extinction curve failed to evolve with redshift.

We can use the lenses to determine the differential extinction between the lensed images from the variation in the flux ratios with wavelength. If there is sufficient dust and wavelength coverage, the extinction law can be determined (Nadeau et al. 1991), and it may be possible to determine the redshift of the dust (Jean & Surdej 1998). The magnitude difference between two images i and j as a function of wavelength λ is

$$m_i(\lambda) - m_j(\lambda) = -2.5 \log\left(M_i/M_j\right) + (E_i - E_j) R\left(\lambda/(1 + z_d)\right) \tag{1}$$

where M_i/M_j is the magnification ratio, $E_i - E_j$ is the extinction difference ($\Delta E(B-V)$), and $R(\lambda/(1+z_d))$ is the extinction law in the rest frame of the dust. Systematic errors arise if the magnification ratio depends on wavelength or temporal variations by the source mimic a wavelength dependence.

Figure 5 presents a histogram of differential extinctions on 37 lines of sight in 23 lens galaxies. The median rest frame differential extinction of the optically selected lenses is $\Delta E(B - V) = 0.04$ mag and the median for the radio selected lenses is 0.07 mag. The distributions for optical and radio selected lenses are similar except for the two radio-selected lenses with high differential extinctions. Both B 0218+357 and PKS 1830–211 are face-on spiral galaxies (Lehár et al. 1998)

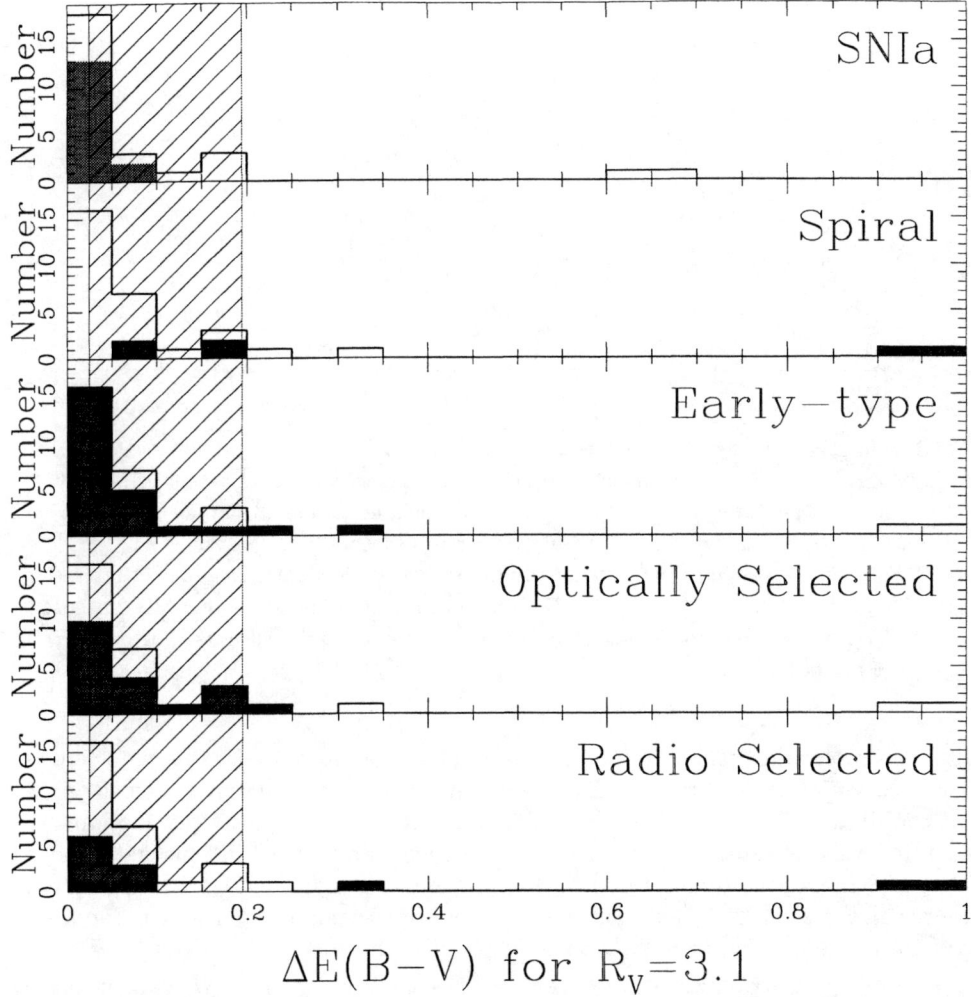

FIGURE 5. Histograms of differential extinctions. The solid histogram is the total sample, and the shaded histogram shows the distribution of the (from bottom to top) radio-selected, optically-selected, early-type and late-type subsamples. The hatched region shows the mean extinction estimated from a comparison of the statistics of lensed quasars and radio sources by Falco et al. (1998). The two objects in the high extinction bin actually have $\Delta E = 1.0$ and 3.0 mag. The top panel shows the extinction distribution of Type Ia supernovae from Riess et al. (1998) for low redshift (open) and high redshift (shaded). Note, however, that 11 of the 15 high redshift supernovae have negative extinctions which have been reset to zero.

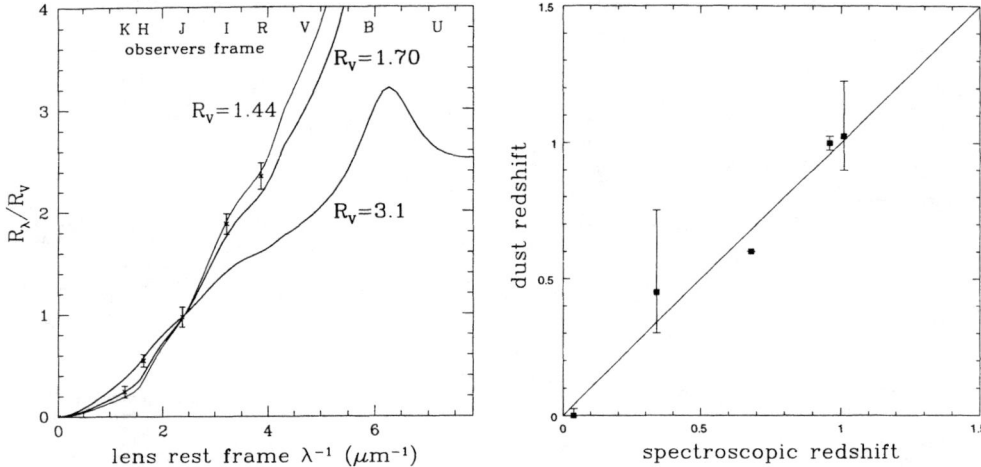

FIGURE 6. The extinction law in the $z_l = 0.96$ elliptical lens galaxy MG 0414+0534. The non-parametric extinction curve is shown by the points along with the Cardelli et al. (1989) model with the same $R_V = 1.44 \pm 0.09$. The best Cardelli et al. (1989) parametric model has $R_V = 1.7 \pm 0.1$. The standard $R_V = 3.1$ Galactic curve is shown for comparison.

FIGURE 7. Spectroscopic versus dust redshifts for five lens systems. In order of increasing redshift they are Q 2237+0305, B 1422+231, B 0218+357, MG 0414+0534, and MG 2016+112. Using $\Delta\chi^2 = 1$ appears to underestimate the uncertainties in the dust redshifts. The mean error is $\langle z_{dust} - z_l \rangle = 0.01 \pm 0.07$.

with high molecular gas content (e.g. Wiklind & Combes 1995, 1996). There is no correlation of the differential extinction with impact parameter, which suggests that the diffuse dust is patchy. For comparison to the differential extinction, we had previously obtained an estimate of the mean extinction in lens galaxies of $A_B = 0.58 \pm 0.45$ mag by comparing the statistics of radio and optically selected lenses (Falco et al. 1998). Thus the mean extinction is comparable to the differential extinction, also consistent with a patchy dust distribution.

For systems with sufficient extinction and wavelength coverage we can estimate the extinction law. Figure 6 shows the extinction law of the $z_l = 0.96$ elliptical lens in MG 0414+0534 derived both parametrically using the Cardelli et al. (1989) models and non-parametrically. The values of $R_V = 1.44 \pm 0.09$ in the non-parametric models and $R_V = 1.7 \pm 0.1$ in the parametric models are well below the standard Galactic value of $R_V = 3.1$. In fact, standard Galactic dust is ruled out at a $\Delta\chi^2 = 56$, and we confirm the local indications that dust in early-type galaxies may have a different mean extinction law than that of the Galaxy.

Following Jean & Surdej (1998) we can also estimate a dust redshift for the systems using the dependence of the extinction at observed wavelength λ on the extinction curve at the rest wavelength in the dust $R(\lambda/(1 + z_d))$. The deter-

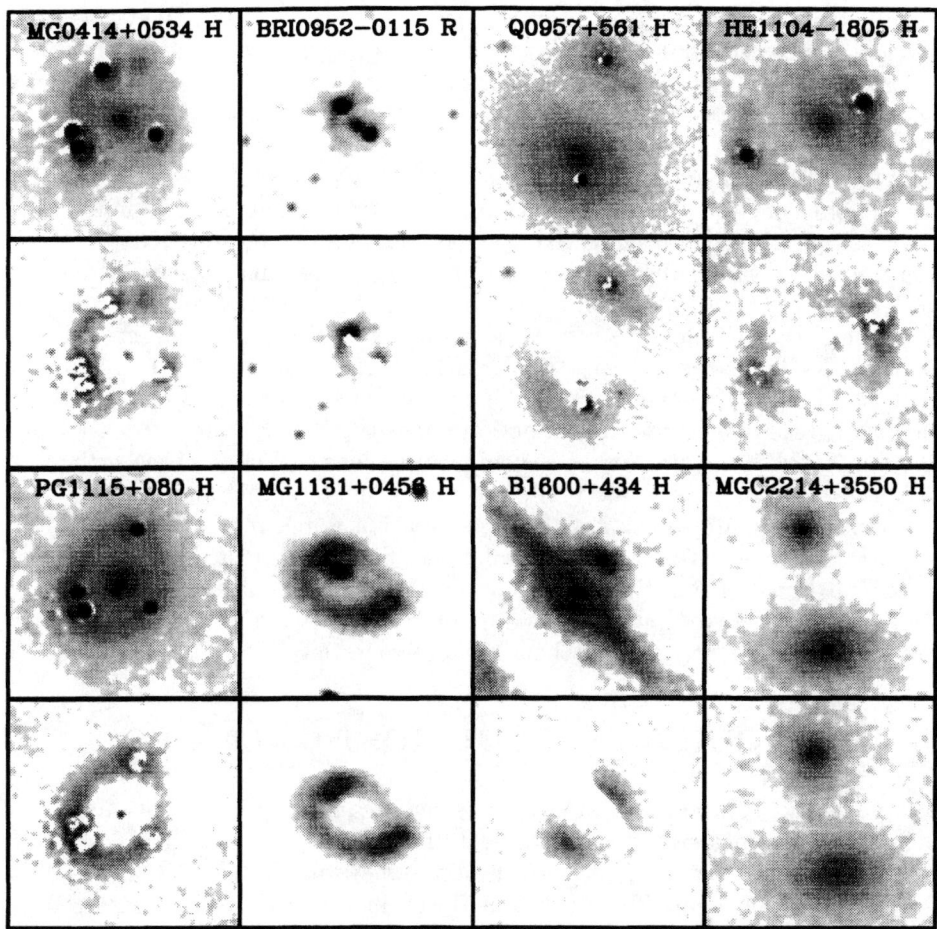

FIGURE 8. Host galaxies. The top image shows the full image, and the bottom image shows the host after subtracting the lens and the active nucleus. All images are 5″.8 square except for Q 0957+561 which is 11″.5 square.

mination of dust redshifts requires better data than determining the extinction curve because it depends on detecting the deviations of the extinction curve from a self-similar power law $R_\lambda \propto \lambda^{1.7\pm0.1}$ (e.g. Mathis 1990). For five cases we can compare the spectroscopic lens redshift to the dust redshift (see Figure 7), including MG 0414+0534. The agreement is remarkably good. Moreover, the fact that the anomalous extinction law predicts the observed redshift of the MG 0414+0534 lens strongly suggests that the result is not a consequence of systematic errors.

We are measuring extinctions and extinction laws at redshifts and impact parameters very similar to the those of the Type Ia supernovae used by Perlmutter et al. (1997) and Riess et al. (1998) to determine the cosmological model. It is critical to the cosmological determination that the supernova fluxes be accurately corrected for extinction. We can make four observations about the supernovae from the lens samples. First, if even early-type galaxies at these redshifts contain dust, it would be surprising if the supernova samples showed no dust. Second, from the radio lenses like MG 0414+0534, we can rule out the existence of dust which is gray in the optical. Where we can measure the extinction curves in lens galaxies they resemble the Cardelli et al. (1989) parameterized forms. Third, galaxies at high redshift, like galaxies at low redshift, show a range of extinction curves. It is dangerous to assume that the extinction curve will match a mythical standard extinction law. Fourth, we could not compare the lens and supernovae extinction distributions. Perlmutter et al. (1997) do not estimate extinctions. While Riess et al. (1998) estimate extinctions, the distribution is peculiar because 11 of 15 supernovae have negative estimated extinctions. Simple statistical tests show that the preponderance of negative extinctions in the sample is inconsistent with the assumption that they are produced by random photometric errors with the stated uncertainties at a slightly greater than 2-σ confidence level.

4. A QUICK TOUR OF HOST GALAXIES

A large fraction of the lens systems now show arc and ring images of the quasar or AGN host galaxies either in the optical or in the infrared. Figure 8 shows a sample of the hosts. Seven of the eight hosts are new discoveries. MG 0414+0534 ($z_s = 2.64$) is the system used to determine the extinction law in §3. BRI 0952−0115 ($z_s = 4.5$) is the highest redshift detection of a host galaxy. Q 0957+561 ($z_s = 1.41$) shows two enormous arc images of the host whose morphologies essentially rule out the popular Grogin & Narayan (1996) models for the system. These models predict a radially stretched image of the Northern host relative to the lens galaxy, while we observe a tangentially stretched image. HE 1104−1805 ($z_s = 2.32$) shows two arc images of the host. PG 1115+080 ($z_s = 1.72$) has an Einstein ring image of the host galaxy which could be used to break the degeneracies in the estimates of H_0 for the system if NICMOS is repaired (Impey et al. 1998). B 1600+434 ($z_s = 1.59$) shows two images of the host straddling the bulge of the lens. MGC 2214+3550 ($z_s = 0.88$) is a beautiful example of a binary quasar. The lower source is a radio

172

loud quasar and the upper is not, with a radio flux ratio of > 80. Both H band sources are perfectly modeled by a point source for the quasar at the center of a de Vaucouleurs profile host. MG 1131+0456 ($z_s =$?) has a spectacular infrared ring image of the host (Kochanek et al. 1998). The ring is 4–5 times brighter than the $z_l \simeq 0.85$ lens galaxy in the infrared and virtually invisible in the optical. The red colors of the system are due to the flux from the stars in the host galaxy rather than to dust in the lens.

We have started to estimate the properties of the host galaxies, and the two generic statements seem to be that many correspond to sub-L_* galaxies for their redshifts and that they are relatively blue. In many cases the host galaxies are actually bluer in their I–H colors than the early-type lens galaxies. While the very luminous hosts of radio-loud objects (e.g. MG 1131+0456) have been seen previously, our lensing results constitute the first secure detection for a sample of radio-quiet hosts. While in nearby quasars the host luminosities do not depend on the radio properties, we find that at $z > 1$, the radio loud hosts are on average 2 mag brighter than the radio quiet ones. One possible interpretation is that the radio flux as well as the rest-optical luminosity get boosted during star-burst phases.

5. SUMMARY

The next ten years will be the period when gravitational lenses make their most dramatic scientific impact. As recently as two years ago, there were too few lenses to attack many of the most interesting scientific problems, and ten years from now there will be several hundred lenses and progress will again slow. Today, with 40–50 lens systems and 4–5 time delay determinations we are at the cusp where science using gravitational lenses will advance most rapidly. The CASTLES project in combination with archival HST observations of gravitational lenses now has about half of the final data set for the 47 currently known lenses, but new lenses are being discovered almost as fast as HST is observing the old lenses.

We can use this explosion in data to dramatically expand the range of scientific problems that gravitational lenses can attack. In this short review we have illustrated only three new examples of gravitational lenses as tools. The fundamental plane of lens galaxies shows that most lenses are normal early-type galaxies, that early-type galaxies in low density environments are very similar to those in the centers of rich clusters, and that the stellar populations of the early-type lenses must have formed at $z_f \gtrsim 2$. The differential extinction in lens galaxies shows that early-type galaxies contain modest amounts of diffuse dust. The amount of dust is sufficient to bias cosmological limits based on the statistics of lensed quasars, as already discovered by Falco et al. (1997) in their comparison of the statistics of lensed quasars and radio sources. The dust in the lenses can be used to determine both extinction laws and lens redshifts. Finally, HST images of lenses, particularly infrared images, commonly show arc and ring images of the quasar or AGN host galaxies. The lens magnification pulls the host galaxy out from under the bright central point source and makes it significantly easier to detect and model the properties of the host galaxies.

Acknowledgements: Support for the CASTLES project was provided by NASA through grant numbers GO-7495 and GO-7887 from the Space Telescope Science Institute, which is operated by the Association of Universities for Research in Astronomy, Inc. CSK was also supported by the NASA Astrophysics Theory Program grant NAG5-4062. HWR is also supported by the Alfred P. Sloan Foundation.

REFERENCES

1. Barkana, R., Lehár, J., Falco, E.E., et al., 1998, astro-ph/9808096
2. Cardelli, J.A., Clayton, G.C., & Mathis, J.S., 1989, ApJ, 345, 245
3. Chiba, M., & Yoshii, Y., 1998, astro-ph/9808321
4. Djorgovski, S., & Davis, M., 1987, ApJ, 313, 59
5. Dressler, A., Lynden-Bell, D., Burstein, D., et al., 1987, ApJ, 313, 42
6. Efstathiou, G., Ellis, R.S. & Peterson, B.A., 1988, MNRAS, 232, 431
7. Falco, E.E., Kochanek, C.S., & Muñoz, J.A., 1998, ApJ, 494, 47
8. Fischer, P., Schade, D., & Barientos, F., 1998, astro-ph/9806273
9. Fitzpatrick, E.L, 1998, astro-ph/9809387
10. Fukugita, M., & Turner, E.L., 1991, MNRAS, 253, 99
11. Grogin, N.A., & Narayan, R., 1996, ApJ, 564 92 (erratum: ApJ, 473, 570)
12. Impey, C., Falco, E., Kochanek, C., et al., 1998, ApJ in press, astro-ph/9809371
13. Jean, C., & Surdej, J., 1998, astro-ph/9810218
14. Jorgensen, I., Franx, M., & Kjaergaard, P., 1996, MNRAS, 280, 167
15. Kauffmann, G., & Charlot, S., 1998, astro-ph/9810031
16. Keeton, C.R., Kochanek, C.S., & Seljak, U., 1997, ApJ, 482, 604
17. Keeton, C.R., Kochanek, C.S., & Falco, E.E., 1998, ApJ in press, astro-ph/9708161
18. Kochanek, C.S., 1994, 436, 56
19. Kochanek, C.S., 1995, 445, 559
20. Kochanek, C.S., 1996, ApJ, 466, 638
21. Lehár, J., Falco, E.E., Impey, C., et al., 1998, in preparation
22. Lin, H., Kirshner, R.P., Schechtman, et al., 1996, ApJ, 464, 60
23. Loveday, J., Peterson, B.A., Efstathiou, G., & Maddox, S.J., 1992, ApJ, 390, 338
24. Marzke, R.O., Geller, M.J., Huchra, J.P., & Corwin, H.G., 1994, AJ, 108 437
25. Mathis, J.S., 1990, ARA&A, 28, 37
26. Nadeau, D., Yee, H.K.C., Forrest, W.J., et al., 1991, ApJ, 376, 430
27. Pahre, M., 1998, Caltech PhD thesis
28. Perlmutter, S., Gabi, S., Goldhaber, G., et al., 1997, ApJ, 483, 565
29. Riess, A.G., Filippenko, A.V., Challis, P., et al., 1998, AJ, 116, 1009
30. Rix, H.-W., de Zeeuw, P.T., Cretton, N., et al., 1997, ApJ, 488, 702
31. van Dokkum, P., Franx, M., Kelson, D., & Illingworth, G., 1998, ApJL, 504, 17L
32. Warren-Smith, R.F., & Berry, D.S., 1983, MNRAS, 205, 889
33. Wiklind, T., & Combes, F., 1995, A&A, 299, 382
34. Wiklind, T., & Combes, F., 1996, Nature, 379, 139

35. Witt, A., Thronson, H., & Capuano, J., 1992, ApJ, 393, 611

Two-point Correlation Functions of High-redshift Objects on a Light-cone

Kazuhiro Yamamoto* and Yasushi Suto[†]

*Department of Physics, Hiroshima University, Higashi-Hiroshima 739-8526, Japan.
[†]Department of Physics and Research Center for the Early Universe (RESCEU)
School of Science, University of Tokyo, Tokyo 113-0033, Japan.

Abstract. A theoretical formulation for a two-point correlation function of high-redshift objects is investigated. While all the cosmological observations are carried out on a light-cone, the null hypersurface of an observer at $z = 0$, the clustering statistics has been properly defined only on the constant-time hypersurface. We develop the theoretical formulation for the two-point correlation function on the light-cone, and derive a simple approximate expression relevant to the discussion of clustering of high-redshift objects. As an example, we present a prediction of the two-point correlation function for the SDSS quasar catalogue.

INTRODUCTION

Clustering of high-redshift objects is one of the current topics in the fields of cosmology and astrophysics, e.g., Lyman-break galaxies and the 2dF QSO survey (e.g., [1]). The clustering statistics of such high-z objects provides several important pieces of cosmological information. In discussing the clustering of the high-redshift objects, an important but less often discussed point is the light-cone effect, that is, such cosmological observations are feasible only on the light-cone hypersurface defined by the current observer. The light-cone effect makes it ambiguous and difficult to distinguish the scale-dependence of clustering in the survey volume from the intrinsic redshift evolution (e.g., change of bias, the clustering amplitude, and the mean number density of the objects considered).

Some aspects of the light-cone effect have been discussed in [2–4]. In those papers rather intuitive formulas for the two-point correlation function were used to incorporate the light-cone effect. The primary purpose of this paper is to present a rigorous theoretical formulation to define and compute a two-point correlation function of cosmological objects which explicitly takes into account the light-cone effect [5]. We propose an expression (eq.[9]) as a practically useful formula relevant

CP470, After the Dark Ages: When Galaxies were Young (the Universe at 2 < z < 5),
edited by Stephen S. Holt and Eric P. Smith

to the discussion of clustering of high-redshift objects at large separations. Here we use the units $c = 1$.

FORMULATION

In this section we derive a theoretical formula for the two-point correlation function in the spatially-flat Friedmann-Robertson-Walker universe, whose line element is written in terms of the conformal time η as

$$ds^2 = a^2(\eta) \left[-d\eta^2 + dr^2 + r^2 d\Omega_{(2)}^2 \right]. \tag{1}$$

Since our fiducial observer is located at the origin of the coordinates ($\eta = \eta_0$, $r = 0$), an object at r and η on the light-cone hypersurface of the observer satisfies a simple relation of $r = \eta_0 - \eta$. Denote the comoving number density of observed objects at η and $\mathbf{x} = (r, \vec{\gamma})$ by $n(\eta, r, \vec{\gamma})$, then the corresponding number density defined on the light-cone is written as $n^{\mathrm{LC}}(r, \vec{\gamma}) = n(\eta_0 - r, r, \vec{\gamma})$. If we introduce the mean *observed* number density (comoving) and the density fluctuation at η, $n_0(\eta)$ and $\Delta(\eta, \mathbf{x})$, on the constant-time hypersurface: $n(\eta, \mathbf{x}) = n_0(\eta) \left[1 + \Delta(\eta, \mathbf{x}) \right]$, then $n^{\mathrm{LC}}(r, \vec{\gamma})$ is rewritten as

$$n^{\mathrm{LC}}(r, \vec{\gamma}) = n_0(\eta_0 - r) \left[1 + \Delta(\eta_0 - r, r, \vec{\gamma}) \right]. \tag{2}$$

When the observed density field of objects on the light-cone, $n^{\mathrm{LC}}(r, \vec{\gamma})$ is given, one may compute the following two-point statistics:

$$\mathcal{X}(R) = \frac{1}{V^{\mathrm{LC}}} \int \frac{d\Omega_{\hat{\mathbf{R}}}}{4\pi} \int dr_1 r_1^2 d\Omega_{\vec{\gamma}_1} \int dr_2 r_2^2 d\Omega_{\vec{\gamma}_2} n^{\mathrm{LC}}(r_1, \vec{\gamma}_1) n^{\mathrm{LC}}(r_2, \vec{\gamma}_2) \delta^{(3)}(\mathbf{x}_1 - \mathbf{x}_2 - \mathbf{R}),$$

$$\tag{3}$$

where $\mathbf{x}_1 = (r_1, r_1 \vec{\gamma}_1)$ and $\mathbf{x}_2 = (r_2, r_2 \vec{\gamma}_2)$ and $R = |\mathbf{R}|$, $\hat{\mathbf{R}} = \mathbf{R}/R$, and V^{LC} is the comoving survey volume of the data catalogue:

$$V^{\mathrm{LC}} = \int_{r_{\min}}^{r_{\max}} r^2 dr \int d\Omega_{\vec{\gamma}} = \frac{4\pi}{3} (r_{\max}{}^3 - r_{\min}{}^3), \tag{4}$$

with $r_{\max} = r(z_{\max})$ and $r_{\min} = r(z_{\min})$ being the boundaries of the survey volume.

Substituting equation (2) into (3), we can show that the ensemble average of the estimator $\mathcal{X}(R)$ is explicitly written as [5]

$$\langle \mathcal{X}(R) \rangle = \mathcal{U}(R) + \mathcal{W}(R), \tag{5}$$

where

$$W(R) = \frac{1}{V^{\text{LC}}} \frac{1}{\pi R} \int \int_{\mathcal{S}} dr_1 dr_2 r_1 r_2 \prod_{j=1}^{2} [n_0(\eta_0 - r_j) D_1(\eta_0 - r_j)]$$

$$\times \int dk k^2 P(k) j_0(kR) b(k; \eta_0 - r_1) b(k; \eta_0 - r_2), \qquad (6)$$

$$\mathcal{U}(R) = \frac{1}{V^{\text{LC}}} \frac{2\pi}{R} \int \int_{\mathcal{S}} dr_1 dr_2 r_1 r_2 n_0(\eta_0 - r_1) n_0(\eta_0 - r_2), \qquad (7)$$

where $D_1(\eta)$ is the linear growth rate normalized to unity at present (η_0), $P(k)$ is the power spectrum of the mass fluctuations at η_0, $b(k; \eta)$ is the k-dependent linear bias factor, $j_0(x)$ is the spherical Bessel function of the 0-th order, and $\mathcal{S}$ denotes the region $|r_1 - r_2| \leq R \leq r_1 + r_2$.

The remaining task is to define the two-point correlation function on the light-cone. We propose the following definition:

$$\xi_A^{\text{LC}}(R) \equiv \frac{\langle \mathcal{X}(R) \rangle - \mathcal{U}(R)}{\mathcal{U}(R)} = \frac{W(R)}{\mathcal{U}(R)} . \qquad (8)$$

In the case of $R \ll r_{\text{max}}$, we can derive the approximate expression:

$$\xi_A^{\text{LC}}(R) = \frac{\int_{r_{\text{min}}}^{r_{\text{max}}} dr r^2 n_0(\eta_0 - r)^2 \xi(R; \eta_0 - r)_{\text{Source}}}{\int_{r_{\text{min}}}^{r_{\text{max}}} dr r^2 n_0(\eta_0 - r)^2} , \qquad (9)$$

where $\xi(R; \eta)_{\text{Source}} = (2\pi^2)^{-1} \int k^2 dk P(k) j_0(kR) [b(k; \eta) D_1(\eta)]^2$. The estimator of equation (8) can be computed from a given catalogue of object and we propose it as the best estimator of the two-point correlation function on a light-cone [5].

DISCUSSION

We apply our formula, $\xi_A^{\text{LC}}(R)$, to predict the correlation functions for SDSS QSO catalogue as a demonstration. For that purpose we assume (1) the B-band quasar luminosity function according to Wallington & Narayan [6], (2) the effective bias $b_{\text{eff}}(z)$ integrated over the halo mass larger than M_{min} [2], and (3) the standard cold dark matter (SCDM) model (see ref. [5] for details). In figure 1 we show our $\xi_A^{\text{LC}}(R)$ for the SDSS QSO sample together with the nonlinear mass two-point correlation functions.

In summary we developed a theoretical formulation which properly takes account of the light-cone effect. This effect becomes very important either as a contamination of the real signal or as a cosmological probe for on-going wide and deep surveys like SDSS and 2dF QSO surveys. We propose equation (9) as a well-defined and practically useful expression for the light-cone correlation function. Strictly speaking we were able to derive equation (9), only in linear theory. However the nonlinearity becomes important only on $R \lesssim 1 h^{-1}\text{Mpc}$ (see figure 1), which implies that

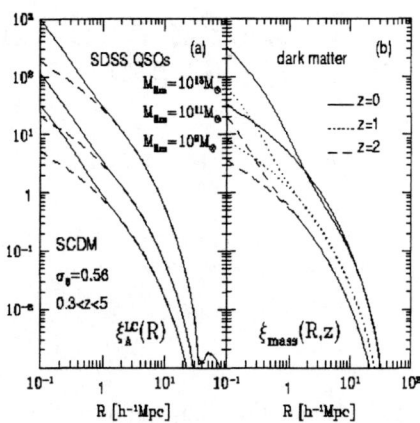

FIGURE 1. Two-point correlation functions of QSOs defined on the light-cone hypersurface in the standard CDM models. (a) $\xi_A^{LC}(R)$ for the SDSS QSO catalogue for the threshold mass $M_{lim} = 10^9$, 10^{11} and $10^{13}h^{-1}M_\odot$. Nonlinear mass correlation function by Peacock & Dodds [7] is used for solid lines, while mass correlation function in linear theory is used for dashed lines;(b) linear (lower curves) and nonlinear (upper curves) mass correlation functions defined on constant-time hypersurfaces $z = 0$, 1 and 2.

one can safely ignore the nonlinear effect on scales which are probed by (sparse) QSO samples in general.

There are several issues which remain to be solved before making theoretical predictions. The QSO luminosity function and the evolution of bias play central roles in confronting observations and predictions of the QSO correlation functions. The redshift-space distortion to the peculiar velocity of the objects is crucial in the quantitative comparison to high precision because an observational map is obtained in the redshift space. These effects will be discussed elsewhere [8].

REFERENCES

1. Boyle, B.J., et al. 1998, Phil.Trans.R.Soc.Lond.A, in press (astro-ph/9805140).
2. Matarrese, S., Coles, P., Lucchin, F., & Moscardini, L. 1997, MNRAS, 286, 115
3. Matsubara, T. , Suto, Y., & Szapudi 1997, ApJ, 491, L1
4. Nakamura, T.T., Matsubara, T. & Suto, Y. 1998, ApJ, 494, 13
5. Yamamoto, K. & Suto, Y. 1998, submitted to ApJ
6. Wallington, S., & Narayan, R. 1993, ApJ, 403, 517
7. Peacock, J.A. & Dodds, S.J. 1996, MNRAS, 280, L19
8. Nishioka, H. & Yamamoto, K. 1999 ; Yamamoto, K. & Suto, Y. 1999, in preparation

Gravitational Lensing and the Hubble Deep Field

Asantha R. Cooray, Jean M. Quashnock, M. Coleman Miller

Dept. of Astronomy and Astrophysics, University of Chicago, Chicago IL 60637.
E-mail: asante@hyde.uchicago.edu

Abstract.
We calculate the expected number of multiply-imaged galaxies in the Hubble Deep Field (HDF), using photometric redshift information for galaxies with $m_I < 27$ that were detected in all four HDF passbands. A comparison of these expectations with the observed number of strongly lensed galaxies constrains the current value of $\Omega_m - \Omega_\Lambda$, where Ω_m is the mean mass density of the universe and Ω_Λ is the normalized cosmological constant. Based on current estimates of the HDF luminosity function and associated uncertainties in individual parameters, our 95% confidence lower limit on $\Omega_m - \Omega_\Lambda$ ranges between -0.44, if there are no strongly lensed galaxies in the HDF, and -0.73, if there are two strongly lensed galaxies in the HDF. If the only lensed galaxy in the HDF is the one presently viable candidate, then, in a flat universe ($\Omega_m + \Omega_\Lambda = 1$), $\Omega_\Lambda < 0.79$ (95% C.L.). These limits are compatible with estimates based on high-redshift supernovae and with previous limits based on gravitational lensing.

INTRODUCTION

The Hubble Deep Field (HDF; [1]) is the deepest optical survey that has been performed to date, allowing detailed studies of the galaxy redshift distribution and the global star formation history. Galaxies in the HDF have redshifts which are estimated to range from 0.1 to 5, with a large portion having redshifts between 2 and 4. Such galaxies have a significant probability of being strongly lensed.

The combination of high resolution and deep exposures in multiple colors provides a rich ground for gravitational lens searches, and it was expected that the HDF would contain between 3 to 10 lensed galaxies, based on the number of lensed quasars and radio sources in other surveys [2]. Instead, a careful analysis of the HDF [3] has revealed a surprising dearth of candidates for lensed sources. In fact, the best estimate is either 0 or 1 lensed sources in the entire field, although very faint images with small angular separations may have escaped current analyses. This lack of lensing has led to suggestions (e.g., Ref. [3]) that the HDF data may

CP470, After the Dark Ages: When Galaxies were Young (the Universe at 2 < z < 5),
edited by Stephen S. Holt and Eric P. Smith

be incompatible with the high probability of lensing expected in a universe with a large cosmological constant.

Here, we calculate the expected number of detectable, multiply-imaged galaxies in the HDF for different cosmological parameters, and we constrain these parameters by comparing the expectations with the observations. Further details of our calculation can be found in Ref. [4].

EXPECTED NUMBER OF LENSED GALAXIES

In order to calculate the number of lensed galaxies in the HDF, we model the lensing galaxies as singular isothermal spheres (SIS) and use the analytical filled-beam approximation (e.g., Ref. [5]). We also include the effects of "magnification bias" due to a magnitude limit of point-source detection in the I-band. To describe the foreground lensing galaxies, we assume that the brightness distribution of galaxies at any given redshift is described by a Schechter function. We use the redshift-dependent luminosity function given in Ref. [6] to describe the foreground galaxies.

The background galaxies are described by magnitude and redshift distributions given in two photometric redshift catalogs [6,7]. These catalogs are complete to an I-band limiting magnitude of 27 and contain a total of 848 galaxies. In Figure 1, we show the estimated photometric redshifts of the HDF galaxies according to Ref. [6] (*left panel*) and Ref. [7] (*right panel*) versus I-band magnitude. Even though the redshift distribution is different for the two catalogs — because redshifts are estimated using two different techniques — our constraints on cosmological parameters are almost the same for either catalog.

CONSTRAINTS ON COSMOLOGICAL PARAMETERS

In Figure 2, we show the expected number, $\bar{N}$, of strongly lensed sources in the HDF as a function of Ω_m and Ω_Λ, using the photometric redshift catalog of Ref. [6]. A universe dominated with Ω_Λ has a higher number of multiply-imaged sources than the number in a universe dominated with a large Ω_m. As shown in Fig. 2, $\bar{N}$ is essentially a function of the combined quantity $\Omega_m - \Omega_\Lambda$. This degeneracy in the lensing probability permits us to constrain $\Omega_m - \Omega_\Lambda$ rather than Ω_m or Ω_Λ separately.

We constrain the quantity $\Omega_m - \Omega_\Lambda$ by comparing the observed and expected number of lensed galaxies in the HDF and by using a Bayesian likelihood approach [4]. We consider cases in which either 0, 1 (the best estimate from observations), or 2 lensed sources are found in the HDF, for lens search programs that have been carried out to an I-band limiting magnitude of $m_{\text{lim}} = 28.5$.

If there are no lensed galaxies in the HDF, then at the 95% confidence level $\Omega_m - \Omega_\Lambda > -0.44$, so that in a flat universe $\Omega_\Lambda < 0.72$. If there is one lensed galaxy in the HDF, our constraint depends only slightly on the galaxy redshift,

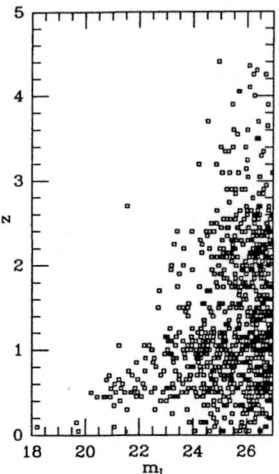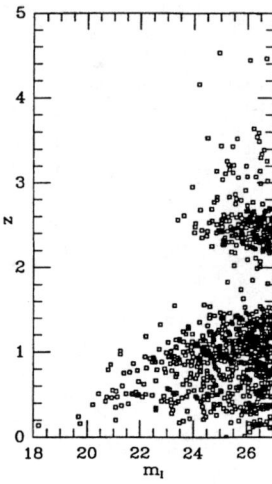

FIGURE 1. Redshift-magnitude distribution of 848 galaxies with I-band magnitude < 27 in the HDF. The plot shows the estimated photometric redshifts in the catalogs of Ref. [6] (*left panel*) and Ref. [7] (*right panel*) versus I-band magnitude. Both catalogs appear to trace the same redshift distribution, with two peaks ($z \sim 0.6$ and 2.3). However, there is a lack of galaxies in the Ref. [7] catalog between $z \sim 1.5$ to 2.2. This is the same range in redshift where no spectroscopic redshifts are currently available for the HDF.

estimated to be between 1 and 2.5 [3]. If the galaxy redshift is 1, then $\Omega_m - \Omega_\Lambda > -0.52$, implying $\Omega_\Lambda < 0.76$ in a flat universe. If instead the galaxy redshift is 2.5, then $\Omega_m - \Omega_\Lambda > -0.58$, and hence $\Omega_\Lambda < 0.79$ in a flat universe. If 2 strongly lensed galaxies are present in the HDF, $\Omega_m - \Omega_\Lambda > -0.73$ at the 95% confidence level. These limits are compatible with estimates based on high-redshift type Ia supernovae [8] and with previous limits on the cosmological constant based on gravitational lensing [9,10].

In Ref. [4], we discussed some of the systematic uncertainties in our calculation, the largest of which are due to effects of reddening and extinction on (optical) lens search programs. Such effects tend to reduce the number of observed lenses, and hence could lead to a systematic underestimate of the upper bound on the cosmological constant. We allowed m_{lim} to vary by as much as 1 magnitude to indicate the possible scope of such an effect: we find that our upper bound on Ω_Λ increases by about 0.06.

We find that using photometric redshifts from two different catalogs [6,7] yields results that are almost indistinguishable. Thus, we have shown that photometric redshifts can be used to estimate the expected number of lensed galaxies in the HDF with reasonable accuracy. This bodes well for the upcoming Southern Hubble Deep Field redshift catalog that is expected in the near-future. The Southern HDF will double the number of high redshift galaxies and will increase the expected number

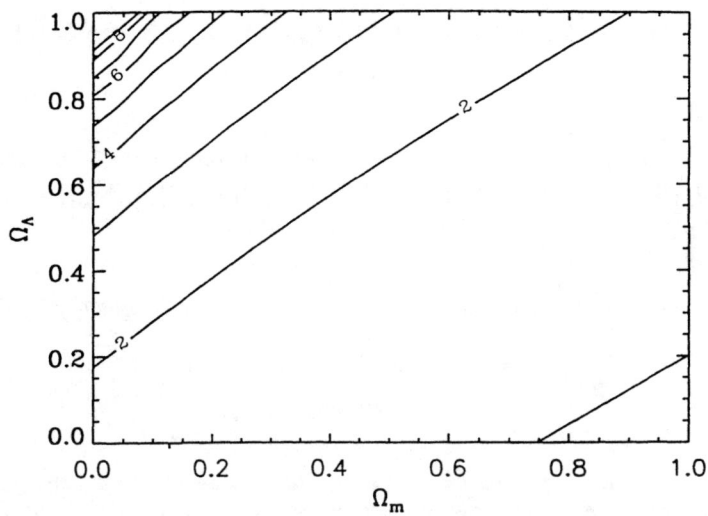

FIGURE 2. Expected number of multiply-imaged galaxies, $\bar{N}$, in the HDF, as a function of Ω_m and Ω_Λ. $\bar{N}$ is constant along lines of constant $\Omega_m - \Omega_\Lambda$, allowing for direct constraints on this quantity. Shown here is the expected number based on the redshift catalog of Ref. [6], and for lens search programs that have been carried out to an I-band limiting magnitude of $m_{lim} = 28.5$.

of gravitationally lensed galaxies. The actual number of lensed sources will lead to stronger constraints on $\Omega_m - \Omega_\Lambda$.

REFERENCES

1. Williams, R. E., et al. 1996, AJ, 112, 1335.
2. Hogg, D. W., et al. 1996, ApJ, 467, L73.
3. Zepf, S. E., et al. 1997, ApJ, 474, L1.
4. Cooray, A. R., Quashnock, J. M., & Miller, M. C. 1999, ApJ, 511, in press.
5. Fukugita, M., Futamase, T., Kasai, M., & Turner, E. L. 1992, ApJ, 393, 3.
6. Sawicki, M. J., Lin, H., & Yee, H. K. C. 1997, AJ, 113, 1.
7. Wang, Y., Bahcall, N., Turner, E. L. 1998, AJ, 116, 2081.
8. Riess, A. G., et al. 1998, AJ, 116, 1009.
9. Kochanek, C. S. 1996, ApJ, 466, 638.
10. Falco, E. E., Kochanek, C. S., & Munoz, J. A. 1998, ApJ, 494, 47.

Sunyaev-Zel'dovich Effect in Galaxy Clusters

Asantha R. Cooray*, John E. Carlstrom*, Laura Grego*, Gilbert P. Holder*, William L. Holzapfel*, Marshall Joy†, Sandeep K. Patel†, and Erik Reese*

*Dept. of Astronomy and Astrophysics, U. of Chicago, Chicago IL 60637. E-mail: asante@hyde.uchicago.edu
†Space Science Laboratory, NASA MSFC, Huntsville AL 35812.

Abstract. We review recent results of Sunyaev–Zel'dovich-effect (SZE) observations toward galaxy clusters. Using cm-wave receivers mounted on the OVRO and BIMA mm-wave arrays, we have obtained high signal-to-noise images of the effect for more than 25 clusters. Over 90% of these clusters are scheduled to be observed with *AXAF* during the first year of its observations. We present current estimates of cosmological parameters H_0 and Ω_m based on the SZE in galaxy clusters.

INTRODUCTION

Over the last few years there has been a tremendous increase in the study of galaxy clusters as cosmological probes, initially through the use of X-ray emission observations and, more recently, through the use of the Sunyaev–Zel'dovich-effect (SZE). Briefly, the SZE is a distortion of the cosmic microwave background (CMB) radiation by inverse-Compton scattering of thermal electrons within the hot intercluster medium [1]; see Ref. [2] for a recent review. The change in the CMB brightness temperature observed is:

$$\frac{\Delta T}{T_{\mathrm{CMB}}} = \left[\frac{x(e^x + 1)}{e^x - 1} - 4 \right] \int \left(\frac{k_B T_e}{m_e c^2} \right) n_e \sigma_T dl, \tag{1}$$

where $x = h\nu/k_B T_{\mathrm{CMB}}$, and n_e, T_e and σ_T are the electron density, electron temperature and the cross section for Thomson scattering. The integral is performed along the line of sight through the cluster.

The other important observable of the hot intercluster gas is the thermal Bremsstrahlung X-ray emission, whose surface brightness S_X can be written as:

CP470, After the Dark Ages: When Galaxies were Young (the Universe at 2 < z < 5), edited by Stephen S. Holt and Eric P. Smith

$$S_X = \frac{1}{4\pi(1+z)^3} \int n_e^2 \Lambda_e dl, \qquad (2)$$

where z is the redshift and $\Lambda_e(\Delta E, T_e)$ is the X-ray spectral emissivity of the cluster gas due to thermal Bremsstrahlung within a certain energy band ΔE. By combining the intensity of the SZE and the X-ray emission observations, and knowing the cluster gas temperature T_e, the angular diameter distance to the cluster can be derived due to the different dependence of the X-ray emission and SZE on the electron density, n_e. Combining such distance measurements with redshift allows a determination of the Hubble constant, H_0, as a function of certain cosmological parameters (e.g., [3]). If distance measurements for a sample of clusters exist, then the angular diameter distance with redshift relation can be used to put constraints on the cosmological models, similar to current supernovae constraints at high redshift.

INTERFEROMETRIC OBSERVATIONS OF THE SZE

We have imaged the SZE by outfitting the OVRO and BIMA mm-wave arrays, with low-noise cm-wave receivers. One of the key advantages of our system is the ability to use interferometric techniques to produce two-dimensional images of the SZE with sensitivity to large angular scales (up to 2.5′). The system as installed at OVRO, and the first images obtained are discussed in Ref. [4]. In Ref. [5], we presented the observed cluster sample at OVRO and BIMA during the summers of 1995 to 1997 and detections of radio sources in galaxy clusters at 28.5 GHz. One of the main problems of SZE observations at cm-wavelengths is the presence of bright radio sources towards galaxy clusters, and catalogs of such sources are important for future SZE and CMB anisotropy observations. In Ref. [6], we reviewed current progress on the determination of cosmological parameters based on galaxy cluster SZE.

In Figure 1 we present SZE images towards a sample of 12 clusters. The rms noise of individual cluster SZE image is typically 15 to 30μK, and was observed for an average of 40 to 50 hours.

COSMOLOGICAL PARAMETER CONSTRAINTS

Table 1 in [6] lists published Hubble constant measurements that have been obtained by combining SZE and X-ray emission observations. In recent years, several studies have questioned the reliability of H_0 measurements based on SZ/X-ray route. This is primarily due to various systematic effects involved with this method, which include the nonisothermality of the electron temperature for cluster gas, gas clumping, asphericity of the cluster gas distribution, and radio source contamination and gravitational lensing effects (see Ref. [2]). It is likely that deep AXAF observations will produce reliable cluster electron temperature profiles and

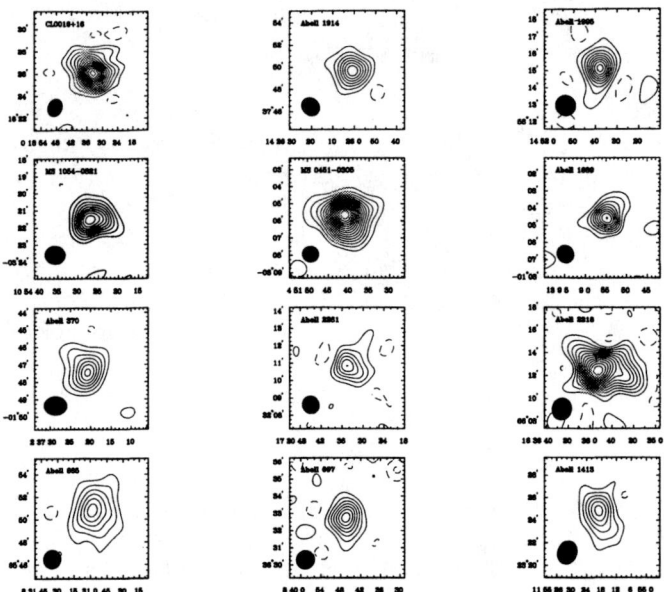

FIGURE 1. A sample of SZE images from OVRO and BIMA. Contours are multiples of -2σ. The beam FWHM is shown by the filled black symbol in the left corner of each panel.

constrain the amount of gas clumping. Systematic changes in H_0 due to aspherical gas distribution can be treated in a statistical manner for a large sample of clusters, and such effects can also be understood based on other cluster properties, such as gravitational lensing [7,8].

In Figure 2 we show the current distance measurements to galaxy clusters as a function of cluster redshifts based on SZE/X-ray route. The vertical dotted lines depict clusters for which we have obtained high signal-to-noise SZE data and for which AXAF observations are planned to be performed during the first year of observations. When combined, the SZE/X-ray data will not only enable the determination of H_0, but also measurements of the cosmological parameters Ω_m and Ω_Λ based on the angular diameter distance relation with redshift. We note that the current SZE/X-ray data are consistent with $q_0 > -0.73$ (90% C.I.).

The SZE is a measurement of the integrated gas (baryonic) mass along the line of slight through the cluster. The total (including non-baryonic) mass of a cluster can be derived based on three methods: gas temperature, gravitational (strong and weak) lensing, and velocity dispersion measurements. The baryonic mass fraction, when compared to the primordial nucleoynthesis determined value for Ω_b allows constraints on Ω_m assuming the cluster baryonic fraction is the same as Ω_b/Ω_m (based on hierarchical clustering models, where clusters represent the composition of the universe). The present limits on Ω_m are: $\Omega_m h < 0.3$ [9,10], based on SZE measurements, and $\Omega_m h^{2/3} < 0.28$ [11], based on X-ray measurements.

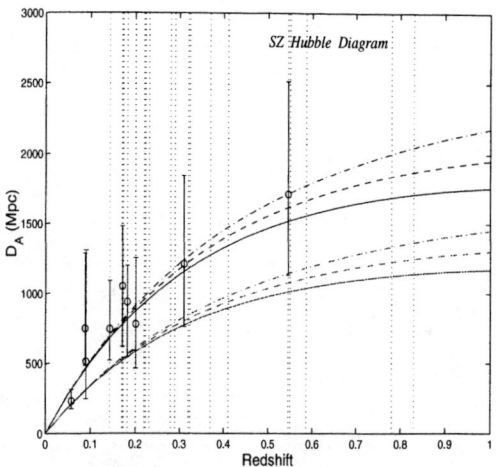

FIGURE 2. Angular diameter distance relation with redshift. The two sets of curves corresponds to an H_0 of 75 (*lower*) and 50 (*upper*) km s^{-1} Mpc^{-1} and Ω_m of 1 (*solid*), 0.5 (*dashed*), and 0.1 (*dot-dashed*). The plotted points are clusters for which distance measurements are available from existing SZE/X-ray data (see [6]). The dotted vertical lines correspond to clusters with interferometric SZE data and which are planned to be obeserved with AXAF.

In addition to individual cluster gas mass fraction measurements as an estimate on Ω_m, the evolution of the gas mass fraction with redshift can be used to determine the cosmological parameters Ω_m and Ω_Λ [12]. The current 90% confidence upper limits on Ω_m based on only the X-ray determined gas mass fractions are $\Omega_m < 0.6$ in a flat universe ($\Omega_m + \Omega_\Lambda = 1$) and $\Omega_m < 0.7$ in an open universe ($\Omega_\Lambda = 0$). When combined with SZE determined gas mass fractions, it is likely that tighter constraints on these two parameters would be possible.

REFERENCES

1. R. A. Sunyaev and Ya. B. Zel'dovich, ARAA 18 (1980) 537.
2. M. Birkinshaw, Phys. Rep. (1998) submitted (astro-ph/9808050).
3. J. P. Hughes and M. Birkinshaw, ApJ in press (1998), (astro-ph/9801183).
4. J. E. Carlstrom, M. Joy and L. Grego, ApJ 456 (1996) L75.
5. A. R. Cooray, L. Grego, W. L. Holzapfel, et al., Astron. J. 115 (1998) 1388.
6. A. R. Cooray, J. E. Carlstrom, M. Joy, et al., in *Dark Matter 98*, (astro-ph/9804149).
7. A. R. Cooray, A&A 339 (1998) 623.
8. M. Bartelmann and T. S. Kolatt (1998) MNRAS submitted, (astro-ph/9706184).
9. S. T. Myers, J. E. Baker, A. C. S. Readhead, et al., ApJ 485 (1997) 1.
10. L. Grego, J. E. Carlstrom, M. Joy, et al., (1998) in preparation.
11. A. E. Evrard, MNRAS 292 (1977) 289.

6. Galaxy Formation and Mergers

Mergers and Galaxy Assembly

Joshua E. Barnes

Institute for Astronomy, University of Hawai'i
2680 Woodlawn Drive, Honolulu, Hawai'i, 96822, USA

Abstract. Theoretical considerations and observational data support the idea that mergers were more frequent in the past. At redshifts $z = 2$ to 5, violent interactions and mergers may be implicated by observations of Lyman-break galaxies, sub-mm starbursts, and active galactic nuclei. Most stars in cluster ellipticals probably formed at such redshifts, as did most of the halo and globular clusters of the Milky Way; these events may all be connected with mergers. But what *kind* of galaxies merged at high redshifts, and are present-epoch mergers useful guides to these early collisions? I will approach these questions by describing ideas for the formation of the Milky Way, elliptical galaxies, and systems of globular clusters.

INTRODUCTION

Why is it so plausible that galactic mergers and tidal interactions were more frequent in the past? Several theoretical reasons come to mind:

- Hierarchical clustering, in which small objects are progressively incorporated into larger structures [1], is common to many accounts of galaxy formation. In the "core-halo" picture [2], clustering of dark matter creates galaxy halos which subsequently accumulate cores of baryons, forming visible galaxies.

- Tidal encounters generate short-lived features; a population of binary galaxies with highly eccentric orbits is required to explain the peculiar galaxies observed today [3]. If these binaries have a flat distribution of binding energies, their merger rate has declined with time as $t^{-5/3}$, and the 10 or so merging galaxies in the NGC catalog are but the most recent additions to a population of about 750 remnants [4].

- The CDM model [5] provides a concrete example of galaxy formation in which merging of dark halos is easily calculated and clearly important [6].

Observations, though not always reaching the redshift range emphasized in this meeting, also imply rapid merging at high redshift:

CP470, After the Dark Ages: When Galaxies were Young (the Universe at 2 < z < 5),
edited by Stephen S. Holt and Eric P. Smith

- Various counting strategies indicate that the pair density grows like $(1 + z)^m$, where $m \simeq 3 \pm 1$ [7,3].

- Peculiar morphology becomes more common with increasing redshift [9]. For example, the fraction of irregular galaxies in the CFRS survey increases from about 10% at $z \sim 0.4$ to a third at $z \sim 0.8$ [10].

Thus, both theory and observation support the notion that there was "a great deal of merging of sizable bits and pieces (including quite a few lesser galaxies) early in the career of every major galaxy" [4]. But the *nature* of these early mergers is not so clear; were the objects involved dominated by dark matter, by gas, or by stars? And can we learn anything about early mergers by studying present-epoch examples?

SIGNPOSTS OF HIGH-REDSHIFT MERGERS

Merging is hard to prove at redshifts $z \gtrsim 1.5$; cosmological dimming renders tidal tails nearly invisible, while bandshifting effects complicate interpretation of the observations [11]. But circumstantial evidence implicates merging in various high-z objects.

Starburst Galaxies

The most extensive and unbiased sample of high-redshift galaxies are the "Lyman-break" objects at $z \sim 3$, which have rest-frame UV luminosities consistent with star formation rates of $\sim 10^1 \, \mathrm{M}_\odot \, \mathrm{yr}^{-1}$ [12]. The actual rates could be several times higher, since much of the UV emitted by young stars may be absorbed by dust (eg. [15]). Spectra show gas outflows with velocities of $\sim 500 \, \mathrm{km \, sec}^{-1}$ [14], atypical of quiescent galaxies but fairly normal for starburst systems. Heavily obscured high-z starbursts have been detected at sub-mm wavelengths [15,16]. These have IR spectral energy distributions similar to ultra-luminous starburst galaxies like Arp 220 and appear to be forming stars at rates of $\sim 10^2 \, \mathrm{M}_\odot \, \mathrm{yr}^{-1}$.

At low redshifts, luminous starbursts are often triggered by mergers of gas-rich galaxies [17]. The gas in such systems is highly concentrated; H_2 surface densities of 10^3 to $10^5 \, \mathrm{M}_\odot \, \mathrm{pc}^{-2}$ are typical of nearby starbursts [18], and similar surface densities are indicated in high-z starbursts [15]. In the potential of an axisymmetric galaxy, gas becomes "hung up" in a disk several kpc in radius (Frenk, these proceedings) instead of flowing inward. Violently changing potentials in merging galaxies enable gas to shed its angular momentum and collapse to as little as $\sim 1\%$ of its initial radius [19].

But models based on mergers of low-z disk galaxies may not apply to high-redshift starbursts [20]. First, bar instabilities in isolated galaxies can drive rapid gas inflows without external triggers [21]. Second, disks forming at higher redshifts are more compact [22] and thus may already have the surface densities associated with starbursts. Third, the starbursts in Lyman-break galaxies occur on scales of several kpc (Weedman, these proceedings), whereas inflows concentrate gas into much smaller regions. Nonetheless, these objects also have irregular morphologies suggestive of mergers, and deep HDF images reveal faint asymmetric features which may be due to tidal interactions [9,23]. Mergers seem to be the "best bet" for high-z starbursts, but something more than naive extrapolation from low-z is needed to test this conjecture.

Radio Galaxies

At low redshifts, powerful radio sources are often associated with merger remnants; some 30% exhibit tails, fans, shells, or other signatures of recent collisions [24]. But at redshifts $z \gtrsim 0.6$ the most striking morphological feature of powerful radio sources is a near-ubiquitous alignment between the radio lobes and continuum optical emission [25,26]. This "alignment effect" seems at odds with the merger morphologies seen at low redshift; one explanation invokes jet-induced star formation (eg. [25]).

Recent observations suggest the alignment effect is compatible with mergers [27]. Strong polarization is found in several $z \gtrsim 2$ radio galaxies, implying that the aligned emission is scattered light from an obscured AGN (eg. [28]); in several cases there is good evidence that dust is the primary scattering agent [29,30]. HST imaging of the radio galaxy 0406–244 at $z = 2.44$ reveals a double nucleus and what appear to be tidal debris illuminated by an AGN [30].

From a theoretical perspective, merging may even be *necessary* to form powerful radio sources. The most plausible engines for such galaxies are rapidly spinning black holes (Blandford, these proceedings). Accretion from a disk can't spin up a black hole unless the accretion phase lasts $\sim 0.1\,\mathrm{Gyr}$; on the other hand, two black holes of comparable mass can coalesce to produce a rapidly-spinning hole [31].

Quasars

Evidence that low-redshift quasars frequently occur in interacting systems has been accumulating for two decades [27]. Early claims that quasars have close companions are supported by recent studies out to redshifts $z \sim 1$ [32–34]. Even more telling are the tidal tails and other signs of violent interactions in nearby cases [35–39].

The very nature of these interactions makes their detection difficult at higher redshifts – tidal tails and other signs are hidden by cosmological dimming and quasar glare. Nor does the low-z evidence preclude the possibility that high-redshift quasars may have nothing to do with mergers. However, the peak in quasar activity at $z \sim 2$ to 3 seems to broadly coincide with other indications of extensive merging activity reviewed above. Given the observational difficulties, a compelling case that this high-z activity is driven by mergers probably awaits a theory for the formation of supermassive black holes.

ASSEMBLING THE MILKY WAY

Complementing the data gathered by looking back to high redshift is information gleaned by "archeological" studies of objects at $z \sim 0$. The oldest components of the Milky Way provide evidence that mergers of small galaxies played an important role [60]:

1. A "second parameter" – which may not be age [41] – is required to account for variations in the stellar content of globular clusters.

2. This second parameter is correlated with orbital direction; clusters with retrograde orbits have Oosterhoff class I variables [42].

3. Halo stars with [Fe/H] ~ -1 have a large range of [α/Fe] values [43,57].

4. The outer halo exhibits retrograde rotation with respect to the rest of the galaxy [1].

5. The halo is not completely well-mixed, as indicated by observations of star streams and moving groups [1,46,47].

Items 1–3 indicate that different parts of the halo have different enrichment histories, items 2 & 4 imply that some part of the halo fell in on a retrograde orbit, and item 5 is direct evidence for the gradual dissolution of fragments after merging.

Halo accretion is clearly an ongoing process, as shown by the discovery of the Sgr I dwarf galaxy [48] and by observations of high-latitude A stars [49]. But two different arguments suggest that the *bulk* of the halo fell into place long ago.

First, halo stars are old. The halo as a whole shows a well-defined turn-off at $B - V \sim 0.4$, corresponding to ages $\gtrsim 10$ Gyr; only $\sim 10\%$ of the stars appear younger [50]. To be sure, this does not rule out recent accretions of objects containing only old stars, but most dwarf galaxies in the local group contain intermediate-age stars as well. Thus, unless the accreted galaxies were unlike those we observe today, most fell in more than 10 Gyr ago.

Second, galactic disks are dynamically fragile; accretion of satellite galaxies can easily ruin a stellar disk. Analytic estimates limit the mass accreted by the Milky Way to less than 4% in the past 5 Gyr [51]. N-body experiments show less disk heating than the analytic work predicts; dark halos absorb much of the damage, and disks may tilt as well as thicken [52–54]. Still, accretion events of any size increase the disk's vertical dispersion, σ_z. Significant structure is seen in the σ_z–age relation; most striking is the jump from $\sigma_z \simeq 20$ to 40 km sec^{-1} which marks the transition to the ~ 10 Gyr-old thick disk [55].

In sum, the Milky Way last suffered a significant merger at least 10 Gyr ago; relics of this event include the outer stellar halo and possibly the thick disk. Presumably, the Milky Way's dark halo was largely in place at this time, since a major merger would have disrupted even the thick disk.

ASSEMBLING CLUSTER ELLIPTICALS

Galaxy clusters are old in two distinct respects: first, cluster galaxies probably collapsed early; second, dynamical processes run faster in proportion to $\sqrt{\rho}$. Thus clusters should contain remnants of many high-redshift mergers. Archeological evidence from nearby clusters provides important clues to these mergers.

Merger Formation

After some controversy, it's generally accepted that elliptical galaxies can be formed by fairly *recent* mergers of disk galaxies. Support for this position includes:

- Studies of proto-elliptical merger remnants like NGC 7252 [56] and models of disk galaxy mergers reproducing such objects [57,58].

- $H\beta$ line strengths in some ellipticals indicating recent star formation [59].

- "Fine structures" in elliptical galaxies correlating with residuals in luminosity–color and luminosity–line strength relations [60,61].

These results enable us to trace the gradual assimilation of recent merger remnants into the larger population of field ellipticals. But such evidence is not available for cluster ellipticals, which seem to be a more homogeneous population (eg. [62]). Studies of the fundamental plane out to $z \simeq 0.8$ indicate that cluster ellipticals evolve passively and probably formed the bulk of their stars at $z \gtrsim 2$ [63]. Thus cluster ellipticals are unlikely to show the signs which betray aging merger remnants in the field.

Counter-rotating or otherwise decoupled "cores" are probably the clearest signs that cluster ellipticals were formed by ancient mergers [64,65]. High-resolution imaging shows that kinematically distinct nuclear components are usually *disks* [64,66]. Such disks typically have high metal abundances [67] and low velocity dispersions [68]. These properties indicate that they formed dissipationally during major mergers [69,70]; merger simulations producing counter-rotating nuclear gas disks back up this hypothesis [71].

The nature of the mergers which formed cluster ellipticals is unknown; often invoked are highly dissipative encounters of gaseous fragments. But the existence of counter-rotating disks indicates that the penultimate participants can't have been very numerous or very gassy. If many small objects coalesced, the law of averages would make counter-rotation extremely rare. And counter-rotation is unlikely to arise in essentially gaseous mergers since gas flows can't interpenetrate.

Once formed, kinematically distinct disks would be easily disrupted by dissipationless mergers [72]. Thus observations of such structures in cluster galaxies imply that few mergers occur once a cluster has virialized. This is entirely plausible on dynamical grounds since encounters at speeds higher than about twice a galaxy's internal velocity dispersion don't result in mergers [73].

Abundance Ratios

In elliptical galaxies, α-process elements are more abundant with respect to Fe than they are in the disk of the Milky Way [74]. This may constrain the timescale for star formation in ellipticals, since α-process elements are produced in SN II, which explode on a short timescale, while Fe is also produced in SN Ia, which explode after $\sim 1\,$Gyr. Indeed, [Mg/Fe] $\simeq 0.5$ for the nuclear disks in cluster ellipticals [64,65]. High α-process abundances indicate that SN Ia played little role in enriching these galaxies; on the face of it, they also imply that cluster ellipticals formed on timescales $\lesssim 1\,$Gyr (eg. [12]).

High abundances of α-process elements with respect to Fe are also seen in X-ray observations of the diffuse gas in galaxy clusters (eg. [76], but see [77]). The large amounts of metals in cluster gas require remarkably high SN rates which may not be possible with a Salpeter IMF [78]. These results undermine the argument that high α/Fe ratios imply short enrichment timescales, since abundances in the cluster gas presumably represent integrated metal production over $\sim 10\,$Gyr. The abundance patterns of cluster ellipticals are clearly inconsistent with mergers of present-day spirals, but do not preclude mergers of moderately gas-rich galaxies containing substantial stellar disks.

Globular Clusters

Young star clusters are observed in star-forming galaxies like the LMC [79] and in intense starburst galaxies [80,81]. These clusters have half-light radii of less than $5\,$pc, masses of 10^4 to $10^7\,\mathrm{M_\odot}$, and metal abundances comparable to their parent starbursts. Their luminosity functions follow power laws with slopes of -1.6 to -2, intriguingly close to the mass function of giant molecular clouds [82]. However, it's not entirely clear that cluster luminosity is a good indicator of mass since some range of cluster ages is usually present.

Evidence is accumulating that the globular cluster systems of field ellipticals are partly due to cluster formation in merger-induced starbursts:

- Ongoing and recent mergers (eg., NGC 4038/9, NGC 7252, NGC 3921) have populations of blue luminous clusters with ages of less than 1 Gyr [81,83,84].

- Older remnants (eg., NGC 3610) have redder and fainter clusters with ages of a few Gyr [85].

- Predicted specific frequencies[1] in merger remnants increase to $S_N \simeq 2$ or 3 over $\sim 10\,\mathrm{Gyr}$ as the stellar populations fade [83,84].

- Globulars in elliptical galaxies have bimodal color (metalicity) distributions.

These findings imply that metal-rich star clusters form during mergers and are gradually assimilated into existing globular cluster populations [86]. However, the large populations of metal-*poor* globulars found in cluster ellipticals are *not* consistent with mergers of field spirals [87]; predicted specific frequencies of metal-poor clusters are $S_N^P \simeq 1$, while in fact $S_N^P \simeq 4$. This problem is even worse for cluster systems in cD galaxies, which have $S_N^P \simeq 10$; obviously, no amount of merging between metal-rich systems will produce metal-poor clusters!

The question of high-S_N in cluster ellipticals boils down to this: fewer stars, or more globulars? One way to get fewer stars is to merge galaxies *after* their metal-poor globulars have formed but before they build up substantial disks. For example, the Milky Way as it was $\sim 10\,\mathrm{Gyr}$ ago could serve as a building-block for cluster ellipticals; the halo of our galaxy, considered alone, has $S_N^P \simeq 4$. However, mergers of Milky Way halos (or dwarf elliptical galaxies [88]) still fall short of the high S_N^P values of cD galaxies. Another way to end up with fewer stars is to eject most of the gas after the initial epoch of cluster formation; the problem here is that the ejection efficiency must be *higher* in cD galaxies, which have the deeper potential wells and should be better at retaining gas [89].

Alternately, the production of globular clusters may have been more efficient in high-redshift starbursts. Even at low-z, about 20% of the UV emitted by starbursts comes from knots identified with young clusters [80]; if all these clusters survive, the specific frequency for a pure starburst population is $S_N \simeq 60$. Moreover, these clusters are concentrated where the surface densities are highest; it's likely that net yields of star clusters increase rapidly with increasing surface density.

If so, then globular cluster systems reflect the starburst histories of their parent galaxies: Large populations of metal-poor globulars are due to efficient cluster production in early starbursts, while predominantly metal-rich systems (eg., NGC 5846) formed in more recent starbursts. Metallicity distributions for cluster systems support this idea; giant elliptical galaxies have a range of distributions with multiple peaks between [Fe/H] $\simeq -1.2$ and 0.2 [90]. Such variety seems hard to explain in a picture where internal events determine the timing of cluster formation (eg., [87]); on the other hand, it's easy to imagine that different distributions result from the different merging histories of individual galaxies.

CONCLUSIONS

Circumstantial evidence suggests that merging played an important role in galactic evolution long before the present epoch. The key points of the argument can

---

[1] The specific frequency S_N is defined as the number of globular clusters divided by the galaxy luminosity in units of $M_V = -15$.

be summed up as follows:

1. Starbursts and AGN are signposts of high-redshift mergers; the high incidence of such objects at $z \simeq 2$ to 4 reflects frequent merging of juvenile galaxies.

2. The bulk of the Milky Way's halo merged more than 10 Gyr ago as part of this activity.

3. Cluster ellipticals merged before $z \simeq 2$; their immediate progenitors were few and only moderately gassy.

4. The metal-rich globular cluster systems of these ellipticals are relics of their final mergers.

Finally, direct observations of high-redshift events are complemented by archeological investigation of nearby systems. Both approaches are needed to discover what happened at redshifts $z = 2$ to 5.

I thank Alex Stephens and Hector Velázquez for communicating results in advance of publication. I also thank Jun Makino and the University of Tokyo for hospitality while I prepared this article. This research made use of NASA's Astrophysics Data System Abstract Service. Travel to the conference was covered by air miles accumulated while following the Grateful Dead.

REFERENCES

1. Layzer, D. 1954, AJ, 59, 170
2. White, S.D.M. & Rees, M.J. 1978, MNRAS, 183, 341
3. Toomre, A. & Toomre, J. 1972, ApJ, 178, 623
4. Toomre, A. 1977, in The Evolution of Galaxies and of Stellar Populations, eds. B.M. Tinsley & R.B. Larson (Yale Observatory, New Haven), p. 401
5. Blumenthal, G.R., Faber, S.M., Primack, J.R., Rees, M.J. 1984, Nature, 311, 517
6. Lacey, C. & Cole, S. 1993, MNRAS, 262, 627
7. Zepf, S.E. & Koo, D.C. 1989, ApJ, 337, 34
8. Abraham, R.G. 1999, in Galaxy Interactions at Low and High Redshifts, eds J.E. Barnes & D.B. Sanders (Kluwer, Dordrecht), p. 11
9. van den Bergh, S., Abraham, R.G., Ellis, R.S., Tanvir, N.R., Santiago, B.X., & Glazebrook, K.G. 1996, AJ, 112, 359
10. Brinchmann, J., Abraham, R., Schade, D., Tresse, L., Ellis, R. S., Lilly, S., Le Fevre, O., Glazebrook, K., Hammer, F., Colless, M., Crampton, D., & Broadhurst, T. 1998, ApJ, 499, 112
11. Hibbard, J.E. & Vacca, W.D. 1997, AJ, 114, 1741
12. Steidel, C.C., Giavalisco, M., Pettini, M., Dickinson, M., & Adelberger, K.L. 1996, ApJ, 462, L17
13. Heckman, T.M. 1998, astro-ph/9801155

14. Pettini, M., Kellog, M., Steidel, C.C., Dickinson, M., Adelberger, K.L., & Giavalisco, M. 1998, astro-ph/9806291
15. Hughes, D.H., Serjeant, S., Dunlop, J., Rowan-Robinson, M., Blain, A., Mann, R.G., Ivison, R., Peacock, J., Efstathiou, A., Gear, W., Oliver, S., Lawrence, A., Longair, M., Goldschmidt, P., & Jenness, T. 1998, Nature, 394, 241
16. Barger, A.J., Cowie, L.L., Sanders, D.B., Fulton, E., Taniguchi, Y., Sato, Y., Kawara, K., & Okuda, H. 1998, Nature, 394, 248
17. Sanders, D.B. & Mirabel, I.F. 1996, ARAA, 34, 749
18. Kennicutt, R. 1998, ApJ, 498, 541
19. Barnes, J.E. & Hernquist, L. 1996, ApJ, 471, 115
20. Somerville, R.S., Primack, J.R., Faber, S.M. 1998, astro-ph/9806228
21. Schwarz, M.P. 1981, ApJ, 247, 77
22. Mo, H.J., Mao, S., & White, S.D.M. 1998, MNRAS, 295, 319
23. Steidel, C.C., Giavalisco, M., Dickinson, M., & Adelberger, K.L. 1996, AJ, 112, 352
24. Heckman, T.M., Smith, E.P., Baum, S.A., van Breugel, W.J., Miley, G.K., Illingworth, G.D., Bothun, G.D., & Balick, B. 1986, ApJ, 311, 526
25. McCarthy, P.J., van Breugel, W.J.M., Spinrad, H., & Djorgovski, S.G. 1987, ApJ, 321, L29
26. Chambers, K.C., Miley, G.K., & van Breugel, W.J.M. 1987, Nature, 329, 604
27. Stockton, A. 1999, in Galaxy Interactions at Low and High Redshift, eds. J.E. Barnes & D.B. Sanders, (Kluwer, Dordrecht), p. 311
28. Tadhunter, C.N., Fosbury, K.R.A.E., & di Serego Alighieri, S. 1989, in BL Lac Objects, eds. L. Maraschi, T. Maccaro, & M.H. Ulrich (Berlin: Springer), p. 79
29. Knopp, G.P. & Chambers, K.C. 1997, ApJ, 487, 644
30. Rush, B., McCarthy, P.J., Athreya, R.M., & Persson, S.E. 1997, ApJ, 484, 163
31. Wilson, A.S. & Colbert E.J.M. 1995, ApJ, 438, 62
32. Disney, M.J., Boyce, P.J., Blades, J.C., Boksenberg, A., Cane, P., Deharveng, J.M., Macchetto, F., Mackay, C.D., Sparks, W.B., & Phillipps, S. 1995, Nature, 376, 150
33. Fisher, K.B., Bahcall, J.N., Kirhakos, S., & Schneider, D.P. 1996, ApJ, 468, 469
34. Stockton, A. & Ridgway, S.E. 1998, ApJ, 115, 1340
35. Stockton, A. & Mackenty, J.W., 1983, Nature, 305, 678
36. Stockton, A. & Ridgway, S.E. 1991,
37. Bahcall, J.N., Kirhakos, S., & Schneider, D.P. 1995, ApJ, 447, L1
38. Boyce, P.J., Disney, M.J., Blades, J.C., Boksenberg, A., Crane, P., Deharveng, J.M., Macchetto, F.D., Mackay, C.D., & Sparks, W.B. 1996, ApJ, 473, 760
39. Stockton, A., Canalizo, G., & Close, L.M. 1998, ApJ, 500, L121
40. Searle, L. & Zinn, R. 1978, ApJ, 225, 357
41. Stetson, P.B., VandenBerg, D.A., & Bolte, M. 1996, PASP, 108, 560
42. van den Bergh, S. 1993, AJ, 105, 971
43. Gilmore, G. & Wyse, R.F.G. 1998, astro-ph/9805144
44. Stephens, A. 1998, ApJ, submitted
45. Majewski, S.R. 1996, ApJ, 459, L73
46. Eggen, O.J. 1987, in The Galaxy, eds. G. Gilmore & B. Carswell, (Reidel, Dordrecht), p. 211
47. Lynden-Bell, D. & Lynden-Bell, R.M. 1995, MNRAS, 275, 429

48. Ibata, R.A., Gillmore, G., & Irwin, M.J. 1995, MNRAS, 277, 781

49. Preston, G.W., Beers, T.C., & Schectman, S.A. 1994, AJ, 108, 538

50. Unavane, M., Wyse, R.F.G., & Gilmore, G. 1996, MNRAS, 278, 727

51. Tóth, G. & Ostriker, J.P. 1992, ApJ, 389, 5

52. Walker, I.R., Mihos, J.C., & Hernquist, L. 1996, ApJ, 460, 121

53. Huang, S. & Carlberg, R.G. 1997, ApJ, 480, 503

54. Velázquez, H. & White, S.D.M. 1998, MNRAS, submitted

55. Freeman, K.C. 1993, in Galaxy Evolution: The Milky Way Perspective, ed. S.R. Majewski (ASP, San Francisco), p. 125

56. Schweizer, F. 1982, ApJ, 252, 455

57. Barnes, J.E. 1988, ApJ, 331, 699

58. Hibbard, J.E. & Mihos, J.C. 1995, AJ, 110, 140

59. Faber, S.M., Trager, S.C., Gonzalez, J.J., & Worthey, G. 1994, in Stellar Populations, eds. P.C. van der Kruit & G. Gilmore (Kluwer, Dordrecht), p. 249

60. Schweizer, F., Seitzer, P., Faber, S.M., Burstein, D., Dalle Ore, C.M., & Gonzalez, J.J. 1990, ApJ, 364, L33

61. Schweizer, F. & Seitzer, P. 1992, AJ, 104, 1039

62. de Carvalho, R.R. & Djorgovski, S. 1992, ApJ, 389, L49

63. van Dokkum, P.G., Franx, M., Kelson, D.D., & Illingworth, G.D. 1998, 504, L17

64. Surma, P. & Bender, R. 1995, AA, 298, 405

65. Mehlert, D., Saglia, R.P., Bender, R., & Wegner, G. 1998, AA, 332, 33

66. Carollo, M., Franx, M., Illingworth, G.D., & Forbes, D.A. 1997, ApJ, 481, 710

67. Bender, R. & Surma, P. 1992, AA, 258, 250

68. Rix, H.-W. & White, S.D.M. 1992, MNRAS, 254, 389

69. Franx, M. & Illingworth, G.D. 1988, ApJ, 327, L55

70. Schweizer, F. 1990, in Dynamics and Interactions of Galaxies, ed. R. Wielen (Springer: Berlin), p. 60

71. Hernquist, L. & Barnes, J.E. 1991, Nature, 354, 210

72. Schweizer, F. 1998, in Galaxies: Interactions and Induced Star Formation, eds. D. Friedli, L. Martinet, & D. Pfenniger (Springer, Berlin), p. 105

73. Makino, J. & Hut, P. 1997, ApJ, 481, 83

74. Worthey, G., Faber, S.M., & Gonzalez, J.J. 1992, ApJ, 398, 69

75. Bender, R. 1997, in The Nature of Elliptical Galaxies, eds. M. Arnaboldi, G.S. Da Costa, & P. Saha (ASP, San Francisco)

76. Mushotzky, R., Loewenstein, M., Arnaud, K.A., Tamura, T., Fukazawa, Y., Matsushita, K., Kikuchi, K., & Hatsukade, I. 1996, ApJ, 466, 686

77. Ishimaru, Y. & Arimoto, N. 1997, PASJ, 49, 1

78. Renzini, A., Ciotti, L., D'Ercole, A., & Pellegrini, S. 1993, ApJ, 419, 52

79. Elson, R.A. & Fall, S.M. 1985, PASP, 97, 692

80. Meurer, G.R., Heckman, T.M., Leitherer, C., Kinney, A., Robert, C., & Garnett, D.R. 1995, AJ, 110, 2665

81. Whitmore, B.C. & Schweizer, F. 1995, AJ, 109, 960

82. Harris, W.E. & Pudritz, R.E. 1994, ApJ, 429, 177

83. Schweizer, F., Miller, B.W., Whitmore, B.C., & Fall, S.M. 1996, AJ, 112, 1839

84. Miller, B.W., Whitmore, B.C., Schweizer, F., & Fall, S.M. 1997, AJ, 114, 2381

85. Whitmore, B.C., Miller, B.W., Schweizer, F., & Fall, S.M. 1997, AJ, 114, 1797

86. Ashman, K.M. & Zepf, S.E. 1992, ApJ, 384, 50

87. Forbes, D.A., Brodie, J.P., & Grillmair, C.J. 1997, AJ, 113, 1652

88. Miller, B.W., Lotz, J.M., Ferguson, H.C., Stiavelli, M., & Whitmore, B.C. 1998, astro-ph/9809400

89. Harris, W.E., Harris, G.L.H., & McLaughlin, D.E. 1998, AJ, 115, 1801

90. Harris, W.E. 1994, in Stellar Populations, eds. P.C. van der Kruit & G. Gilmore (Kluwer, Dordrecht), p. 85

Clues from Deep HST Images to Galaxy Formation and the Role of Mergers

Rogier A. Windhorst, Seth H. Cohen, & Ian Waddington

Dept. of Physics & Astron., Arizona State Univ., Tempe, AZ 85287

Abstract. We review recent clues from deep HST images on the formation and evolution of galaxies, and the role that mergers have played in this process. First, we first review the evidence from deep ground-based and HST studies for an epoch dependent merger rate, which was higher in the past by $\sim (1+z)^m$ with $m \simeq 1.5$–3 out to $z \lesssim 1$–2. Next, we review clues from deep HST broad-band images to the role of (minor) mergers. The galaxy counts as a function of restframe type show that E/S0's and Sabc's are only marginally evolving, but show that Sd/Irr's are the dominant population for $B \lesssim 27$ mag, or $1 \lesssim z \lesssim 2$. The scale-lengths of all galaxy types decrease steadily towards the faintest flux limits. A significant fraction of the faint blue irregulars have close companions and/or tidal features indicating merging at $z \gtrsim 1.5$–2, implying an epoch of merger-induced star-formation for $z \sim 1.5$–2, which also coincides with the peak in the star-formation rate [44]. This suggests a Universe at $z \gtrsim 1.5$ dominated by hierarchical merging of star-forming (irregular) "sub-galactic" sized objects. We then summarize the clues that HST medium-band searches have played in finding these "sub-galactic" building blocks, as well as searches with similar techniques from the ground. Next, we review high resolution PC images, as well as other data, on the faint radio galaxy 53W002 at z=2.390, and suggest how and when merging would have taken place in this object to produce its most likely counterpart today — a giant elliptical galaxy. Last, we present recent HST/NICMOS images of microJansky radio sources without clear optical counterparts in the HDF/Flanking Fields, and suggest how the process of (early) major mergers could have resulted in a non-negligible fraction of the radio-selected star-formation rate to be optically enshrouded by dust.

THE EPOCH-DEPENDENT MERGER RATE

In the last decades, studies of the nature and evolution of faint field galaxies have concentrated on the numerous faint blue galaxies ("FBG's") seen in deep ground based CCD images (e.g., [23], [24], [36], [38], [47], [66]), and more recently with HST (e.g., [18], [19], [20], [29], [49], [70]). Recent studies generally found an increased fraction of close companions for $19 \lesssim I \lesssim 24$ mag, suggesting that the galaxy merger

CP470, *After the Dark Ages: When Galaxies were Young (the Universe at 2 < z < 5)*,
edited by Stephen S. Holt and Eric P. Smith

the galaxy pair-fraction or companion-rate is visible in Fig. 1, which is a complete compilation of all galaxies in the HDF as function of spectroscopic or photometric redshift [21]. Other studies like [79] found little evidence for evolution in the merger rate from galaxy pair-counts. The problems with analyzing the galaxy pair-counts are, amongst others: (a) how well can the survey find companions near the flux limit, especially pairs of greatly unequal luminosity, and how does this affect the ability to find minor mergers? (equal-luminosity merger candidates are the easiest to recognize); (b) at how large a separation can one reliably announce a physical companion without running into background contamination; (c) without complete redshift information, how can one be sure that the two objects are at the same redshift and will in fact merge?; (d) even with complete redshift information, how can one be sure the two objects will merge, given that one does not know the tangential velocities, nor the geometry of the configuration, nor the dark matter components that will allow mergers to occur at a larger velocity difference; and (e) how does one convert the observed pair fraction as function of cosmic epoch into a merger rate? (One can only take the time derivative of $(1+z)^m$ if all physical pairs result into mergers). A systematic study of pair-counts combined with spectroscopic data from the CFRS suggest that the merger rate was higher in the past by $(1+z)^3$ [40]. This work suffers from the same problems as the previous pair-count studies, except that identification of merger candidates is easier when both objects have redshifts, and that the dependence on epoch can be traced directly, instead of using photometric redshifts (Fig. 2). A review of high-z mergers is given by [2].

Another clue to the relevance of mergers may be obtained from the *micro*Jansky radio source population. The mJy population is dominated by AGN (giant ellipticals and quasars [37]), but the μJy population consists mostly of interacting, peculiar, or merging galaxies [76], and a minority of early-type galaxies [30]. There is good evidence that the μJy population is dominated by star-forming and post-starburst galaxies [56]. The sub-mJy population is a mixture of AGN and star-bursting objects [5], [71]. The cosmological evolution of mJy, sub-mJy and μJy sources is approximately $\propto (1+z)^e$ with $e \simeq 3$ at least for $z \lesssim 1$ (e.g., [68], [69], [71], [78], and Fig. 3 here). Hence, we suggest that both the cosmological evolution of the faint radio galaxy population, as well as the process of galaxy formation itself, are driven by the same epoch dependent rate of (minor) mergers, with both $\propto (1+z)^3$. We will argue below that both peak around $z \simeq 1$–2, remarkably close to the peak in the optically selected star-formation rate in the universe [44]. We adopt $H_0 = 50\ km\ s^{-1}\ Mpc^{-1}$ & $q_0 = 0.1$ throughout.

DEEP HST IMAGES AND THE ROLE OF MERGERS

Ground based studies of the FBG excess are always limited by atmospheric seeing because the *median* scale length of faint field galaxies is $\sim 0\rlap{.}''3$ ([49] and Fig. 4]. The superb resolution of the HST WFPC2 allowed various groups to study the sub-kpc morphology, light-profiles and color gradients of faint field galaxies out to substantial redshifts (e.g., [18], [19], [29], [49], [1], [70]). The results from these HST projects are summarized here, because they provide important clues to the nature and evolution of faint field galaxies, as likely driven by the epoch dependent rate of (minor) mergers. For the complete HDF sample with $I \lesssim 26$, [21] estimated the luminosity and type-dependent redshift distribution N(z, mag, type) in Fig. 2. They used spectroscopic redshifts where available, and otherwise photometric redshift estimates from UBRIJHK filters [25]. The morphology was derived from the restframe ANNs of [49], [50]. The following are the main HST results:

(a) Luminous bulge-dominated galaxies (E/S0's) have evolved passively since $z \lesssim 1$, and early to mid-type spirals (Sabc's) have undergone some evolution for $0.5 \lesssim z \lesssim 2$ ([18], [21], [49]; Fig. 1 & 2). The reddest objects are mostly classified in the I-band as E/S0's and Sabc's, either by eye or using the restframe Automated Neural Network classifiers (ANNs) of [49]. Their counts do not increase rapidly for $24 \lesssim B_J \lesssim 27$ mag, suggesting that they have been assembled largely before $z \gtrsim 1$. Recent ground-based spectroscopic surveys (CFRS, Keck) show similar trends for $B \lesssim 24.5$ mag [6], [13], [41]. Careful morphological studies with HST can push these results 2–3 mags fainter than can be done spectroscopically routinely from the ground.

(b) The majority of the FBG counts is made up of late-type/Irregular galaxies, which had the strongest evolution for $1 \lesssim z \lesssim 3$, but largely disappeared for $z \lesssim 1$ (Fig. 1 & 2). The Sd/Irr population must have both a steep local luminosity function ("LF", with Schechter slope $\alpha \simeq 1.5–1.8$; [18], [19]) and/or largely have escaped detection from nearby surveys due to surface brightness ("SB") constraints. A significant fraction must have undergone substantial evolution in luminosity and/or space density since $z \lesssim 1$ [21], [49]. Fig. 2 shows that non-evolving models are inadequate to explain the high B-band counts for the Sd/Irr+M population. The FBG excess is dominated by late-type galaxies, with a non-negligible merger fraction ($\sim$35%), which rises strongly for $B_{450} \gtrsim 23$ mag [49]. It is possible that galaxies move back-and-forth between panels in Fig. 2, i.e., Irr's may merge to become E/S0's, and/or turn into Sabc's later, if any remaining neutral HI-gas settles back into disks, surrounding the bulges which resulted from the earlier mergers [31].

(c) A significant fraction of very compact galaxies with scale lengths $r_{hl} \lesssim 0.1–0.2"$ was found by HST. The median WFPC2 scale-length at $B \simeq 27$ mag is $r_{hl} \simeq 0".25–0".3$ [49] (see Fig. 4), which corresponds to $\sim$1.2–2.5 kpc for the redshift range $z \sim 0.5–2.5$ (and for $q_0 = 0.1–0.5$). This is smaller than the characteristic scale-lengths of mid to late-type galaxies measured locally, as well as those measured with HST for $B \lesssim 23-24$ mag [42], [45], [58]. A possible explanation is that the faint galaxy population becomes progressively more dominated by lower luminosity — and therefore smaller — late-type objects at fainter fluxes [18], [19].

Together, (a)–(c) are very important results that may herald the formation of Sabc's and E/S0's from the gradual hierarchical merging of compact (Irr) objects in the epoch $1 \lesssim z \lesssim 3$ (cf. [21], [52]). Complete spectroscopic redshifts, and more accurate photometric redshifts beyond the level where spectroscopic redshifts can be measured routinely, will further constrain the evolution of the different Hubble types with cosmic time, and help trace the physical cause of this evolution at z≃1–4, *e.g.*, a possibly merger induced migration between Hubble types with time (*cf.* Fig. 2). Larger and deeper HST samples are needed to study this (HDF-S & ACS).

Fig. 3 shows the redshift distribution of the faint radio source population in three different flux intervals. The bulk of the μJy –mJy radio source population is at intermediate redshifts ($z_{med} \simeq 0.5$–1 [78]). The global SFR of optically selected field galaxies peaks at z≃1.5–2 [44], with an apparent decline for z≳2, which is possibly caused by dust. Since faint radio sources: (a) are nearly completely

optically identified for R≲26 mag [69], [78]; (b) have an redshift distribution with a median of only z_{med} ≃0.7 and a strong decline for z≳1.5 (Fig. 3); (c) are not affected by dust; (d) have a steep source count, so that they are likely strongly evolving [74], it is therefore possible that the apparent peak in the global SFR at z≃1–2 as seen in *optically* selected surveys is not entirely due to dust at z≳2, but may truly indicate the formation epoch of the bulk of faint (radio) galaxies at z∼1–2. Since the bulk of μJy radio sources are caused by starbursts [5], [26], [56], [76], its redshift distribution may reflect the (radio selected) star-formation history of galaxies, which in the radio apparently peaks at z∼1.

SEARCHES FOR SUB-GALACTIC OBJECTS

HST and Ground-Based Medium-Band Searches: Recent medium-band searches for high-redshift, weak compact Lyα emitters have been remarkably successful. We here first discuss how these objects were found, what their relationship is to galaxy formation, and how their redshift distribution and (group-) velocity dispersion may contribute to mergers at high redshifts. Using the WFPC2 medium-band filter F410M (λ_c ∼4100Å), [52] discovered a group of 17 faint Lyα emitting candidates at $z \simeq 2.4$ surrounding the radio galaxy 53W002 at z=2.390 [72]. Three long WFPC2 F410M parallel fields yielded another 18 $z \simeq 2.4$ candidates [53]. Most of the $z \simeq 2.4$ candidates have rather blue colors and are likely not very reddened, as argued below. Eight of the 18 $z \simeq 2.4$ candidates were spectroscopically confirmed to B_J ≲24.5 mag with the MMT and KPNO [52], [53], and several more to B≲26.5 mag with Keck [3]. The reliability of this method to find $z \simeq 2.4$ candidates is thus likely ≳75%. [11] used similarly a F433M filter to search for Lyα objects surrounding two known radio quasars at z≃2.56 [76] close to a CBR decrement in the $13^h+43°$ field [55], and confirmed three more objects at z≃2.56, possibly part of the actual proto-cluster responsible for the CBR decrement.

Fig. 5a shows that the luminosities of the $z \simeq 2.4$ candidates are -22≲M_B ≲-18 mag, after subtraction of a few AGN contributions [*cf.* 73]. The evolving value of M^* at z=2.4 is M_B ≃-22 mag. Hence, most of these compact z=2.4 galaxies have luminosities of 1/6 L^* or less, and possibly only ≲10^{10} $M_\odot$ in stars at z=2.4 [72]. These $z \simeq 2.4$ candidates are very compact (r_{hl} ≲0″15 or≲1.1 kpc, see Fig. 4–5), and their average light-profiles follow an $r^{1/4}$-law over ≲5 mag in SB [52].

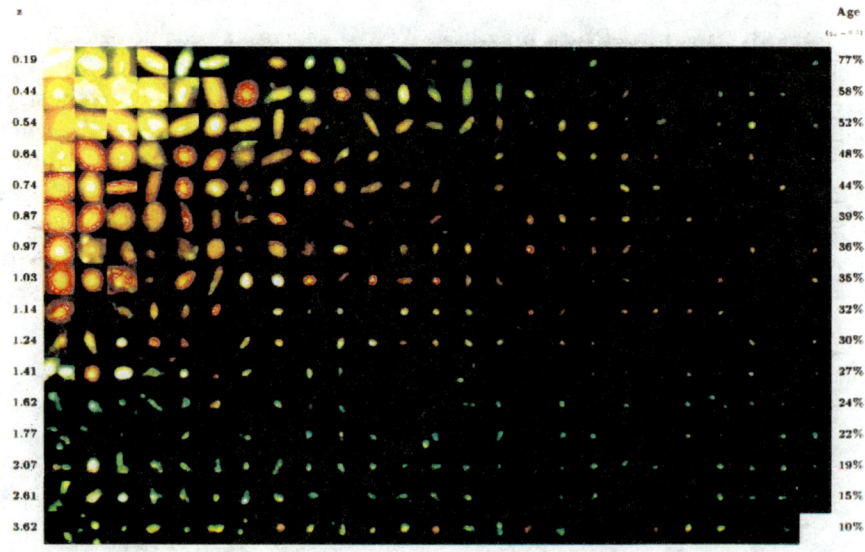

FIGURE 1. The entire HDF sample for I ≲ 26 mag from [21], classified into 16 photometric redshift bins [25] — using spectroscopic redshifts where available (for ∼25 % of the sample) — and sorted towards fainter I-mag in each redshift bin, and hence roughly towards fainter absolute magnitude in each redshift bin [21]. The progression down the rows qualitatively reflects the process of galaxy evolution, although it does not include K-corrections and redshift dependent selection windows. Objects classified as high-redshift Irregulars are frequently seen with close companions and/or tidal features indicating merging at z ≳ 1.5–2, implying an epoch of merger-induced star-formation for z∼1.5–2 (coincident with the peak in the universal star-formation rate [44]). This suggests a Universe at z ≳ 1.5 dominated by hierarchical merging of star-forming (irregular) sub-galactic sized objects [52].

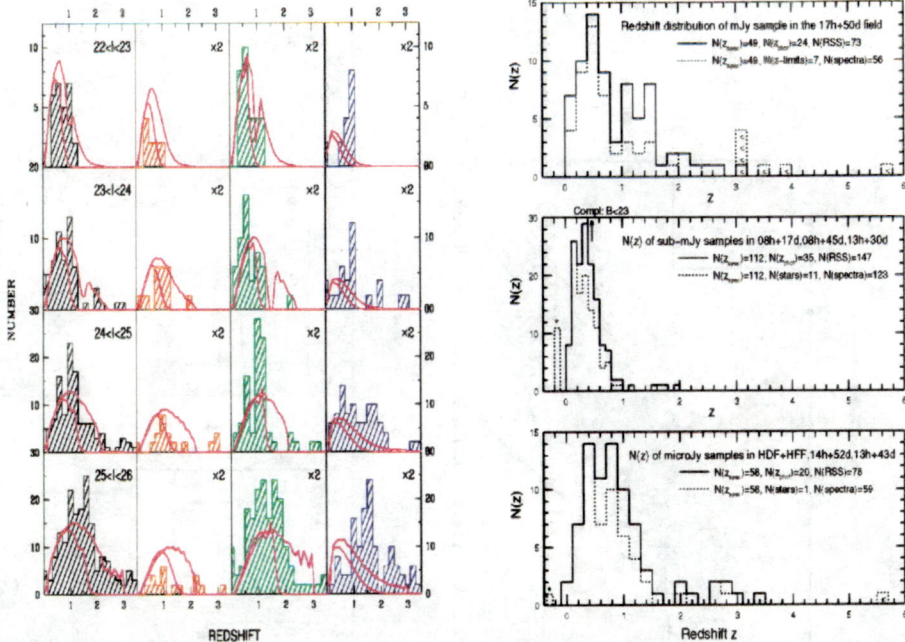

FIGURE 2. [Left]. Redshift distributions N(z) in the HDF as a function of HST morphological type and I-band magnitude, following [21]. Spectroscopic redshifts were used where available, and otherwise photometric redshifts. The WFPC2 morphological types from the restframe Artificial Neural Networks of [49] were used. N(z) is shown for four magnitude intervals (top to bottom rows) and for four different morphological types: all galaxies (left panels), E+S0's (middle left), Sabc's (middle right), and Sd/Irr's (right panels). No-evolution (lower line) and passive evolution (upper line) models are shown. The diagrams shows evidence for *differential* evolution as function of Hubble type: Sd/Irr's show the largest excess for $1 \lesssim z \lesssim 3$, and E/S0 a deficit, possibly related to the formation of early type galaxies from merging between compact Irr's in the epoch $1 \lesssim z \lesssim 3$.

FIGURE 3. [Right]. The redshift distribution for mJy, sub-mJy and μJy radio source samples [78]. Dotted lines show sources with measured spectroscopic redshifts, or upper limits to z from inconclusive spectra. Full-drawn lines show the complete radio sample, with photometric redshift estimates for those sources without a spectroscopic redshift [68]. The mJy and μJy samples are $\gtrsim 70\%$ complete, but the sub-mJy sample is not complete for $B \gtrsim 23$ or $z \gtrsim 0.4$ (see arrow). The μJy population is dominated by starbursting objects (driven by mergers [76]), and its redshift distribution reflects the radio selected star-formation history of galaxies, which apparently peaks at $z \lesssim 1$.

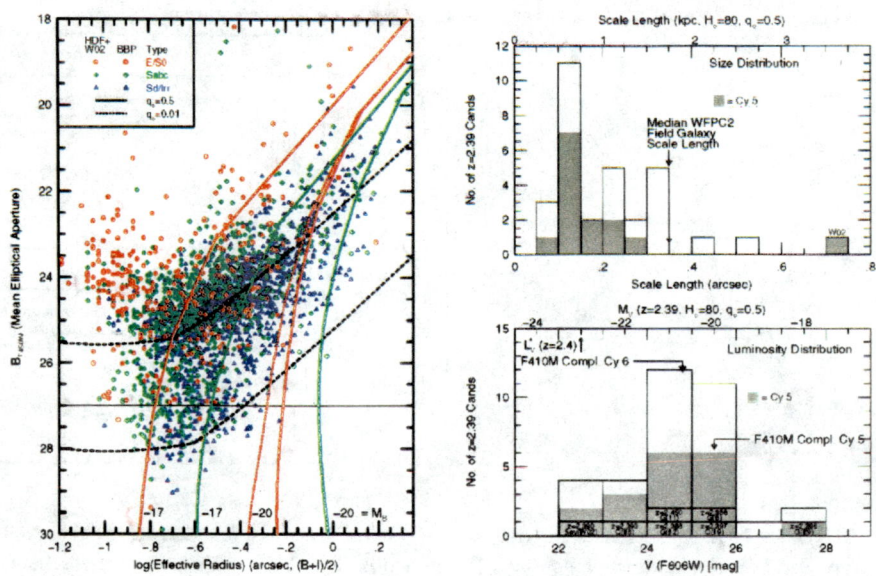

FIGURE 4. [Left]: The B_{450} magnitude vs. half-light radius r_{hl} for all classified galaxies in 33 WFPC2 fields (parallel fields, plus the deeper W02 field and the HDF). Symbols indicate E/S0's, Sabc's, or Sd/Irr+M's classified by the *restframe* ANN [49]. The solid almost-horizontal curves indicates the B_{450}-detection limits. The almost-vertical curves indicate how the median scale-length of RC3 galaxies of given Hubble type and M_B decline towards fainter magnitude [49]. Galaxies classified as E/S0 have on average smaller observed scale-lengths than Sabc's, which are generally smaller than Sd/Irr galaxies. The observed scale-lengths reach a median of $r_{hl} \simeq 0\rlap.{''}25$–$0\rlap.{''}3$ at $B \simeq 27$ mag; [14] & [49] show color versions of this Figure.

FIGURE 5. [Right]: *(a) Top:* Histogram of WFPC2 continuum scale lengths averaged over BVI for the significant $z \simeq 2.4$ candidates in different WFPC2 F410M fields (shaded [52]; white [53]): *(b) Bottom:* Histogram of V-band luminosities for all significant $z{\simeq}2.4$ Lyα emitting candidates. The upper axis indicates their absolute magnitudes, and the evolving L^* value at z=2.39 is indicated. AGN contributions were subtracted in three cases as indicated (*cf.* [73], [77]). Spectroscopically confirmed $z{\simeq}2.4$ objects of [52] are indicated.

Their scale lengths are smaller than — or at most comparable in size to — those of *bulges* in local late-type spirals, which range from 0.2–4 kpc with a median of ~1 kpc for S0–Sbc's [16]. The $z \simeq 2.39$ candidates may be young (blue) spheroids, possibly the bulges of young galaxies that have not (yet) developed significant disks around them, and/or disks that are reduced in brightness in the HST images by the $(1+z)^4$ SB-dimming. Such objects likely existed throughout the entire redshift range z≃1–4 and beyond [21], [49]. According to the CDM models [46], these small "sub-galactic" units could have grown into the luminous giant galaxies seen today, through the process of repeated hierarchical merging, with an epoch dependent merger rate that was likely much higher in the past. A merger rate proportional to $(1 + z)^{2.5}$ would result in a time integral of ~10–20 mergers from z≃3.5 to z=0. It has been suggested that disks can be regenerated (or generated) after such mergers [31], [77]. If this scenario is true, this could have important consequences to theories of galaxy formation (*e.g.*, CDM), and the evolution of large scale structure in general (*e.g.*, the epoch dependent merger rate).

Other systematic searches have been made for high redshift Lyα emitting candidates. We mention, *e.g.*, the work from Palomar in the optical [63] and the CADIS survey in the optical and near-IR [63], [64], and references therein. The number of high redshift Lyα emitters detected in these surveys has been rather small, which [64] explain as possibly due to the fact that any such Lyα emitting objects may have formed hierarchically. Their stringent survey limits just permit the hierarchical models, but are ruling out other models. The large numbers of *faint, compact* Lyα emitters at z≃2.4 found by [52], [53], and the successfully search for faint Lyα emitters at higher redshifts (z≃4–5.6) by [33] using the same method, appear to be consistent with the earlier ground-based upper limits.

Possible Effects from Dust on Lyman-α Searches: There have been other recent reports of high redshift star-forming galaxies with possibly somewhat higher luminosities (*e.g.*, [27], [28], [32], [43], [61], [65]). Among 41 objects in these combined Keck samples of [43], [61] and [65] with measured redshifts $2.5 \lesssim z \lesssim 4$, 24 (or ~60%) have Lyα in emission with W_λ large enough that they would have been significantly detected at z~2.4 in the WFPC2 F410M surveys. Hence, possibly as much as 50% of these high redshift samples are weak Lyα emitters, while another 40% are clear Lyα absorbers, and the remainder show no Lyα at all. Even small amounts of dust, if properly distributed, can effectively quench Lyα and UV continuum emission [35]. However, the (sub-galactic) Lyα emitters of [27], [32], [52], & [80] have rather low luminosity, and may have had fewer generations of O stars, and so fewer supernovae to produce significant dust to absorb Lyα . Or, gas and dust may have been blown out of these lower luminosity objects by the first generation of supernovae [4].

Constraints to their Luminosity Function at z=2–3: [53] show the LF of the z≃2.4 candidates in all four WFPC2 fields together, binned according to their B-band luminosities following [57]. Given the F410M completeness limit, the true luminosity distribution of the z≃2.4 candidates is likely steep for $M_B \gtrsim -19$ mag, and the (luminous) mass spectrum at $z \simeq 2.4$ could be quite steep. [57] show a similar LF from *photometric* redshifts in the HDF, showing both a brightening and steepening with increasing redshift up to $z \simeq 3$, in agreement with measured galaxy redshift surveys at $z \sim 1$ (*cf.* [41]). A steep faint-end of the LF is expected in hierarchical models of galaxy formation, in which merging is important.

Constraints from their Redshift Distribution and (group) Velocity Dispersion: The confirmed $z \simeq 2.4$ objects around 53W002 show a remarkably small group velocity dispersion ($z \simeq 2.391 \pm 0.004$ with $\sigma_v \simeq 286$ km s^{-1} — corrected to $z=0$; [3], [52]), despite the fact that Lyα emission could have been detected in F410M from objects in the entire range $z \simeq 2.28$–2.45. This implies that these "sub-galactic" objects may have existed to some extent in groups or proto-clusters at high redshift. Recent deep KPNO 4m imaging in F410M and spectroscopy [34] shows that the 53W002 "cluster" stretches over $7'$ ($\simeq 4$ Mpc), with two more spectroscopic confirmations at $z \sim 2.391$, and may be part of some larger-scale structure. Similar structures have been seen out to $z \lesssim 1$ [7], [39], and with Keck for $z \gtrsim 1$ [13] and at $z \simeq 3.0$ by [62]. The "spikes" or "frothiness" observed in the redshift distribution of small fields out to $z \simeq 1$ may already have existed at some level at $z \simeq 1$–3, although the clustering amplitude at higher redshift would have to be lower according to most CDM models. Possible frothiness in N(z) (*cf.* the models of [54] may be another reason why narrow-band searches of high redshift Lyα emitters may not be as efficient as medium-band searches: the narrow bands may be looking largely in between any such structures or sheets.

FORMATION AND EVOLUTION OF 53W002

Here we summarize clues regarding the formation and evolution of the mJy radio galaxy 53W002 at $z=2.39$ from deep HST/PC images [77]. The surrounding 17 sub-galactic sized candidates at $z \simeq 2.4$ [52] play an important role in the subsequent future evolution of this object, and the role of mergers therein. The following main components are visible in the PC images of this young radio galaxy: (1) the maximum possible central point-source or AGN contribution is 20–25% in BVI; (2) a remaining dominant $r^{1/4}$-like profile is seen with $r_e \sim 0\farcs27$, although an early-type galaxy with a bulge-to-disk ratio $\gtrsim 3$–5 cannot be ruled out from the PC data. Following spectral evolution models [75], the colors of the symmetric component of 53W002 — if interpreted as coming from stars — suggests a stellar population with age$\sim$0.4 Gyr, of the same order as its dynamical time scale [67]. The lack of a discernible color gradient does not allow to distinguish whether 53W002 formed through a sudden global halo collapse [22] or through rapid merging of many sub-galactic sized units [52], [60]; (3) there is a two-sided blue cloud, the largest side of which is quite extended and vaguely triangular [77], with a brighter "arc" dominated by Lyα line emission $\sim 0\farcs6$ from the core. These clouds are likely a combination of: (a) AGN-light shining through a reflection cone, which implies the existence of a substantial amount of gas and/or dust beyond the optical extent of this galaxy; and (b) a starbursting region induced by the radio jet.

A recent interferometric OVRO image in redshifted CO has 3″ FWHM and provides an important clue to the nature of 53W002. CO was detected ~2–3″ away on both sides of 53W002's AGN, and in the same direction as both blue HST clouds and the extended 8.4 GHz radio source, but not perpendicular to this direction. Since the CO extends further in both directions than the two aligned blue clouds and the currently visible extended radio source (Fig. 3 of [59]), the CO was likely deposited there by physical processes related to the jet. Since carbon and oxygen had to be formed in massive stars, jet-induced star-formation thus likely played a role at some stage in the evolution of 53W002. Its overall $r^{1/4}$-like stellar population is extended in the same direction as the radio source, so that the jet possibly triggered a non-negligible fraction of 53W002's mass to form stars in these two directions. As long as this all happened within a few$\times 10^8$ years, there would have been just enough time for the stellar population to settle into a $r^{1/4}$-like profile. Following [77], the total stellar mass of 53W002 — integrated over its assumed exponentially declining SFR — is $\sim 1.8 \times 10^{11}$ $M_\odot$.

The CO-flux implies $\sim 2.1 \times 10^{11}$ $M_\odot$ in gas around 53W002 alone. The CO velocity widths are ~250 $km\ s^{-1}$ (HWHM), possibly indicating a forming rotation curve [59], and implying an enclosed Keplerian mass of 1.5–3.8×10^{11} $M_\odot$ — consistent with its total stellar mass above [72]. If all this gas settled into disk stars within a few free-fall times (~1 Gyr), 53W002 could evolve into an mid-type spiral galaxy (with B/D-ratio ~0.5), or into an earlier-type galaxy ($B/D \gtrsim 1$) if most of the gas was used up during the initial starburst, and/or if a substantial fraction of the gas remained neutral (as seen in some nearby ellipticals and merger remnants [31]). The small velocity dispersion ($\lesssim 300$ $km\ s^{-1}$) in the group of $z \simeq 2.4$ objects with measured redshifts [52], and the small area ($\lesssim 1$ Mpc^2) over which these 17 $z \simeq 2.4$ candidates are seen — together with assumed significant dark-matter halos — implies that many of these objects will likely merge into a few larger galaxies during the next half Hubble time after $z = 2.4$. Hence, while 53W002 may have formed as a $r^{1/4}$-dominated galaxy during a relatively quick and sudden collapse that started at $z \simeq 3$ (or ~0.4 Gyr before $z = 2.4$) — possibly induced by star-formation along its radio jet — it appears to be also developing a massive disk at $z \simeq 2.4$. This disk may completely settle ~1–2 Gyrs later (or at $z \simeq 1.5$; see [21] and Fig. 1 & 2 here), but possibly be destroyed again during future mergers (at $z \lesssim 1.5$) with the surrounding sub-galactic sized objects [52], so that 53W002 may end up as a gE galaxy today.

NICMOS IMAGING OF MICROJANSKY SOURCES

A remaining important question is how large a fraction of the star-formation in the universe induced by (major) mergers is hidden by dust. At high redshifts, such objects may not show up on deep optical images, since these sample the restframe UV which is likely the most obscured by dust, or in any case severely depressed below the Lyman break. However, they may show up in the near-IR. Important

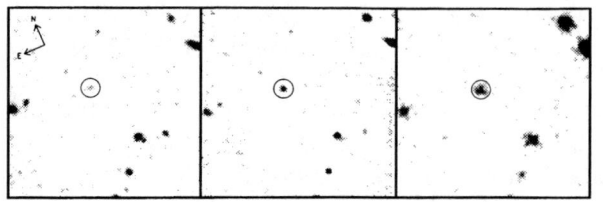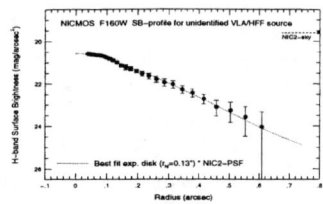

FIGURE 6. [LEFT]: WFPC2 I-band [Left], 6-orbit NIC2 H_{F160W}-band [middle], and ground-based K-band image [right], of a microJy radio source in the HFF. Each image is 20″.5 on a side, and the circular aperture is of radius 1″. The I- and K-band data have been re-sampled to the NIC2 resolution of 0″.075; all images have been smoothed over a 5×5-pixel box. The NIC2 H-band sky is 19.6 mag arcsec^{-2} .

FIGURE 7. [RIGHT]: The NIC2 H-band light-profile of this objects has as best fit an exponential disk with r_{hl} =0″.13.

clues may be obtained from the fraction of μJy (and therefore starbursting [56]) radio sources that remain unidentified for R$\gtrsim$27–29 mag. The complete sample in Fig. 3c and [78] suggests that the fraction of such sources is small, although perhaps not entirely negligible. For I$\lesssim$22-23 mag, the Hubble relation for mJy and even μJy sources is rather well defined, and has a relatively small dispersion, especially for mJy sources [68], [78]. If the spectral evolution models [9] that best fit these observed relations may be extrapolated to higher redshifts, they suggest that most unidentified mJy and μJy radio galaxies with R$\gtrsim$26.5 are likely at z$\gtrsim$3, and quite possibly at z$\gtrsim$4–10. Given that massive early-type galaxies may have assembled very rapidly from the merging of many smaller sub-clumps [52], such luminous objects could have already existed at z$\sim$5–10. We here discuss one such candidate.

Fig. 6a shows the WFPC2 image of a radio source in one of the HDF flanking fields that remained essentially unidentified at I$\gtrsim$25.5 mag. A 9-orbit CVZ image with NICMOS — with 6 orbits in H-band — clearly shows an object with H$\sim$22.7 mag (Fig. 6b), and provided a limit of J$\gtrsim$24.4 mag. This object was also seen in deep ground–based near-IR images in the K-band with K$\sim$21.1 mag (Fig. 6c).

Hence, the object is very red with (I–K)$\gtrsim$4.4, (J–K)$\gtrsim$3.3, (J–H)$\gtrsim$1.7 and (H–K)$\simeq$1.6 mag. Its NIC2 light-profile is (much) better fit with an exponential disk than an $r^{1/4}$ profile (see Fig. 7 & [69]). Given the distribution of brighter radio galaxies in the Hubble and color-redshift diagrams of [68], [78], this object is most likely intrinsically luminous, and so possibly a dusty starburst at z$\sim$3–6 with a disk that had just formed, and with a (major, first?) starburst that is still mostly obscured in the optical (=restframe UV). This would then suggest a recent dissipational collapse. Several more such unidentified objects have been found in the HDF/VLA survey [56], and it would be of extreme interest to get deep high resolution near-IR images of these other objects, as well as spectra in the far-red or near-IR, to see how large a fraction of the radio selected star-forming objects is hiding optically. Such objects would most likely be major mergers, experiencing a (first) major phase of star-formation that is heavily enshrouded by dust [51].

Acknowledgments: We thank the STScI staff for their continuous help in these projects. We thank our collaborators Ken Kellermann, Bruce Partridge, and Eric Richards for allowing us to quote a few unpublished results. This work was supported by NSF grant AST-9802963 and NASA grants GO-5308.01-93A, GO-5985.01-94A, GO-6609.01-95A, GO-6610.01-95A, & GO-7452.01-96A from STScI under NASA contract NAS5-26555.

REFERENCES

1. Abraham, R., et al. 1996, MNRAS, 279, L47
2. Abraham, R. 1998, in IAU Symposium No. 186, "Galaxy Interactions at Low and High Redshifts", Ed. D. Sanders & J. Barnes (Dordrecht: Kluwer), in press.
3. Armus, L., Scoville, N., Pascarelle, S., & Windhorst, R. 1999, ApJ, in prep.
4. Babul, A., & Rees, M. J. 1992, MNRAS, 255, 346
5. Benn, C. R., et al. 1993, MNRAS, 263, 98
6. Brinchmann, J. et al. 1998, ApJ, 499, 112 (astro-ph/9712060)
7. Broadhurst, T., Ellis, R., Koo, D., & Szalay, A. 1990, Nature, 343, 726
8. Broadhurst, T. J., Ellis, R. S., & Glazebrook, K. 1992, Nature, 355, 55
9. Bruzual A., G., & Charlot S. 1993, ApJ, 405, 538
10. Burkey, J., Keel, W., Windhorst, R., & Franklin, B. 1994, ApJL, 429, L13
11. Campos, A., et al. 1999, ApJL, in press (astro-ph/9809146)
12. Carlberg R. G., Pritchet, C. J., Infante L. 1994, ApJ, 435, 540
13. Cohen, J. G., et al. 1996, ApJL, 471, L5
14. Cohen, S. H., Windhorst, R. A., Odewahn, S. C., et al. 1999, AJ, in prep.
15. Colless, M., Schade, D., Broadhurst, T. J., & Ellis, R. S. 1994, MNRAS, 267, 1108
16. Courteau, S., deJong, R. S., & Broeils, A. H. 1996, ApJ, 457, L73
17. Cowie, L. L., Hu, E. M., & Songaila, A. 1995b, AJ, 110, 1576
    M. 1994, MNRAS, 266, 155
18. Driver, S., Windhorst, R., Ostrander, E., Keel, W., et al. 1995a, ApJL, 449, L23
19. Driver, S. P., Windhorst, R. A., & Griffiths, R. E. 1995b, ApJ, 453, 48
20. Driver, S., Windhorst, R., Phillipps, S., & Bristow, P. 1996, ApJ, 461, 525
21. Driver, S. P., Fernandez-Soto, A., Couch, W. J., Odewahn, S. C., Windhorst, R. A., Phillipps, S., Lanzetta, K., & Yahil, A. 1998, ApJL, 496, L93
22. Eggen, O. J., Lynden-Bell, D., & Sandage, A. 1962, ApJ, 136, 748
23. Ellis, R. S. 1997, ARA&A, 35, 389 (astro-ph/97040)
24. Ellis, R., et al. 1996, MNRAS, 280, 235
25. Fernandez-Soto, A., Lanzetta, K., Yahil, A. 1999, ApJ, in press (astro-ph/9809126)
26. Fomalont, E. B., et al. 1997, ApJL, 493, L5
27. Francis, P. J., Woodgate, B. E., & Danks, A. C. 1997, ApJL, 482, L25
28. Giavalisco, M., Steidel, C. C., & Szalay, A. S. 1994, ApJL, 425, L5
29. Glazebrook, K., Ellis, R. S., Santiago, B. & Griffiths, R. 1995, MNRAS, 275, L19
30. Hammer, F., et al. 1995, MNRAS, 276, 1085

31. Hibbard, J. E. & Mihos, J. C. 1995, AJ, 110, 140
32. Hu, E. M. & McMahon, R. G. 1996, Nature, 382, 281
33. Hu, E. M., Cowie, L. L., & McMahon, R. G. 1998, ApJL, 502, L99 (astro-ph/9803011)
34. Keel, W. C., Cohen, S. H., & Windhorst, R. A. 1999, ApJ, in preparation
35. Kinney, A. L., et al. 1996, ApJ, 467, 38
36. Koo, D. C., & Kron, R. G. 1992, ARA&A, 30, 613
37. Kron, R. G., Koo, D. C., & Windhorst, R. A. 1985, A&A, 146, 38
38. Lacey, C. G., Guiderdoni B., Rocca-Volmerange B., & Silk J. 1993, ApJ, 402, 15
39. Le Fevre, O., et al. 1994, ApJ, 423, L89
40. Le Fevre, O., et al. 1997, in "The Hubble Space Telescope and the High Redshift Universe", Eds. N. Tanvir et al. al (Singapore: World Scientific), p. 88.
41. Lilly, S. J., et al. 1995, ApJ, 455, 108
42. Lilly, S. J., et al. 1998, ApJ, 500, 75 (astro-ph/9712061)
43. Lowenthal, J. D. et al. 1997, ApJ, 481, 673
44. Madau, P., et al. 1996, MNRAS, 283, 1388
45. Mutz, S. B., et al. 1994, ApJL, 434, L55
46. Navarro, J. F. & White, S. D. 1994, MNRAS, 267, 401
47. Neuschaefer, L. W., & Windhorst, R. A. 1995a, ApJ, 439, 14
48. Neuschaefer, L. W., Griffiths, R. E., Ratnatunga, K. U. 1997, ApJ, 480, 59
49. Odewahn, S., Windhorst, R., Driver, S., & Keel, W. 1996, ApJL, 472, L13
50. Odewahn, S. C., Burstein, D., & Windhorst, R. A. 1997, AJ, 114, 2219
51. Ostriker, J. P., & Heisler, J. 1984, ApJ, 278, 1
52. Pascarelle, S., Windhorst, R., Keel, W., Odewahn, S. 1996, Nature, 383, 45
53. Pascarelle, S. M., Windhorst, R. A., & Keel, W. C. 1998, AJ, 116, 2659
54. Rauch, M., Haehnelt, M. G., & Steinmetz, M. 1997, ApJ, 481, 601
55. Richards, E. A., et al. 1997, AJ, 113, 1475
56. Richards, E. A., et al. 1998, AJ, 116, 103
57. Sawicki, M. J., Lin, H., & Yee, H. K. C. 1997, AJ, 113, 1
58. Schade, D., et al. 1995, ApJL, 451, L1
59. Scoville, N., Yun, M., Windhorst, R., Keel, W., Armus, L. 1997, ApJL, 485, L21
60. Searle, L. & Zinn, R. 1978, ApJ, 225, 357
61. Steidel, C., Giavalisco, M., Dickinson, M., & Adelberger, K. 1996b, AJ, 112, 352
62. Steidel, C., et al. 1998, ApJ, 492, 428
63. Thompson, D., Djorgovski, S. G., & Trauger, J. 1995, AJ, 110, 963
64. Thompson, D., Manucci, F., & Beckwith, S. V. W. 1996, AJ, 112, 1794
65. Trager, S. C., Faber, S. M., Dressler, A., & Oemler, A. 1997, ApJ, 485, 92
66. Tyson, J. A. 1988, AJ, 96, 1
67. van Albada, T. S. 1982, MNRAS, 201, 939
68. Waddington, I. 1998, Ph.D. thesis, (Univ. of Edinburgh)
69. Waddington, I., Windhorst, R. A., et al. 1999, ApJ, in preparation
70. Williams, R. E. et al. 1996, AJ, 112, 1335
71. Windhorst, R., Miley, G., Owen, F., Kron, R., Koo, D. 1985, ApJ, 289, 494
72. Windhorst, R. A., et al. 1991, ApJ, 380, 362
73. Windhorst, R. A., Mathis, D. F., & Keel, W. C. 1992, ApJL, 400, L1

74. Windhorst, R. A., et al. 1993, ApJ, 405, 498
75. Windhorst, R. A., et al. 1994b, ApJ, 435, 577
76. Windhorst, R. A., et al. 1995, Nature, 375, 471
77. Windhorst, R. A., Keel, W. C. & Pascarelle, S. M. 1998, ApJL, 494, L27
78. Windhorst, R. A., Waddington, I. 1999, in "The Birth of Galaxies: Proceedings of the Xth Rencontres de Blois", Eds. B. Guiderdoni et al. , in press
79. Woods, D., Fahlman, G. G., & Richer, H. B. 1995, ApJ, 454, 32
80. Yee, H. K. C. & Ellingson, E. 1995, ApJ, 445, 37

Lyman Break Galaxies as Collision-Driven Starbursts

J.S. Bullock[1], T.S. Kolatt[1,2], R.S. Somerville[2], Y. Sigad[2], A.V. Kravtsov[3], A.A. Klypin[3], J.R. Primack[1,2], A. Dekel[2]

[1] Physics Department, University of California, Santa Cruz, CA 95064
[2] Racah Institute of Physics, The Hebrew University, Jerusalem, 91904, Israel
[3] Astronomy Department, New Mexico State University, Box 30001, Dept. 4500, Las Cruces, NM 88003-0001

Abstract. Explaining the nature of the Lyman Break Galaxies (LBGs) recently discovered [1,2] at redshift $z \sim 3$, is an exciting challenge for the paradigm of hierarchical structure formation. These galaxies are forming stars at a rate comparable to locally rare "starburst" galaxies [3], but are as luminous and numerous as local bright galaxies. In addition, the brightest LBGs have small emission line-widths [4], indicating virial masses of $\sim 1 - 5 \times 10^{10} M_\odot$, however LBGs exhibit strong clustering, similar to the properties expected of the most massive ($\sim 10^{12} M_\odot$) dark matter halos [5–8] at this redshift. We explore a possible solution to these apparent paradoxes: that LBGs are a population of collision-driven starburst galaxies which are abundant due to an increased collision rate at high redshift [2,9]. We use high-resolution cosmological N-body simulations and a hierarchical halo finder to estimate the galaxy collision rate as a function of time in a popular cosmological model (ΛCDM). We find that appropriate collisions are frequent enough, and the ensuing bursts are plausibly bright enough, to account for most of the LBGs. Although many of the simulated collisions have relatively small masses ($\sim 10^{10} M_\odot$), they tend to cluster about large-mass halos. They therefore exhibit strong clustering, similar to that observed [1,11,8] and stronger than that of halos. The collision-induced starburst scenario [2,9] thus appears to explain the key observed properties of the high-z galaxies. This picture can be further tested observationally, and distinguished from other scenarios, by more detailed studies of the evolution of the number density of LBGs with redshift and the dependence of their clustering on scale and environment.

INTRODUCTION

The Hubble Deep Field [10] (HDF) and deep ground-based surveys [13], combined with spectroscopy at the Keck Telescope [1,2], have led to the discovery of a population of $z \sim 3$ Lyman-Break Galaxies (LBGs). Determining the nature of these galaxies is an intriguing challenge. Observations indicate that LBGs are

CP470, After the Dark Ages: When Galaxies were Young (the Universe at 2 < z < 5),
edited by Stephen S. Holt and Eric P. Smith

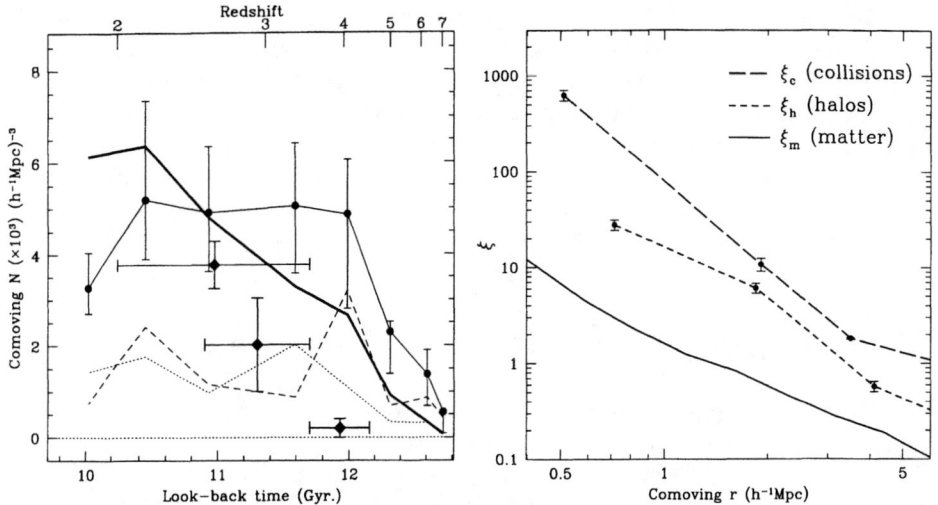

FIGURE 1. (a) The comoving number density N of bursts brighter than $\mathcal{R} < 25.5$. The thick solid line is the halo collisions converted into a magnitude-limited burst number density ($\mathcal{R} < 25.5$). The diamonds depict observational estimates of the number density of LBGs ($\mathcal{R} < 25.5$) from the HDF ($z \sim 2.75$, $z \sim 4$) and ground-based observations ($z \sim 3$). The filled circles connected by a thin solid line refer to halo collisions with no imposed luminosity cut, and an assumed burst duration of $\tau_{\text{vis}} = 100$ Myr. Errors are estimated from the collision-identification procedure. The dashed and dotted curves are the partial contributions from unbound collisions and collisions involving at least one sub-halo, respectively. (b) The long-dashed line shows two-point auto-correlation function of DM halos of $M > 7 \times 10^9 h^{-1} M_\odot$ that experienced collisions in the redshift interval $3.9 > z > 2.9$. The lower curve is for the underlying dark matter. The starburst scenario predicts that LBGs should be strongly clustered, even more than massive halos.

1) highly abundant, with a comoving number density comparable to present-day bright ($\geq L_*$) field galaxies [8,11]; and 2) strongly clustered, with clustering properties similar to that those of the highest mass ($\sim 10^{12} h^{-1} M_\odot$) dark matter halos at $z \sim 3$ [5–8]. These properties are consistent with identifying LBGs as the central objects of massive halos, with a roughly monotonic relationship between the LBG luminosity and the dark matter halo mass [1,7,8].

While this "massive halos" scenario does well in reproducing the clustering properties of LBGs, when confronted with other data, there are some puzzling difficulties. For example, data on a few of the brightest LBGs indicates that they have small emission line-widths [4], consistent with small virial masses of $\sim 1 - 5 \times 10^{10} M_\odot$. In addition, the massive halos picture predicts a much smaller comoving number density at $z \sim 4$ than at $z \sim 3$, which appears to be inconsistent with observations [10,9,14]. Finally, if the LBGs are massive, quiescently star-forming objects, then it is difficult to explain the very high star-formation rates

[2,10,3,13] and small sizes/high surface brightnesses observed [2].

Here, we review the results of Kolatt et al. [12], where we have investigated the validity of an alternate idea which may prove to explain the intriguing characteristics mentioned above: that most LBGs correspond to relatively low-mass galaxy collisions, which trigger bright bursts of star formation [2,9]. We test this hypothesis by finding the collisions between dark matter halos in a ΛCDM N-body simulation. We use semi-analytic estimates of the luminosity of the ensuing bursts in order to directly compare the clustering properties and number densities of these events with the observed properties of LBGs.

DETERMINING THE COLLISION RATE

Our analysis uses the output from Adaptive Refinement Tree (ART) dissipationless simulations [15] to determine the clustering properties of galaxy collisions and the collision rate as a function of lookback time between $z \approx 2 - 5$. The ART simulations make use of a successive refinements of grid cells in order to achieve extremely high force resolution. Our cosmology is assumed to be ΛCDM, with $\Omega_0 = 1 - \Omega_\Lambda = 0.3$, $h = 0.7$, and $\sigma_8 = 1.0$, and a box size of $L_{box} = 30h^{-1}h^{-1}$Mpc. We track the evolution of $N_p = 256^3 \approx 1.7 \times 10^7$ DM particles, and have a mass resolution of $m_p = 1.3 \times 10^8 h^{-1} M_\odot$.

We identify DM halos and collisions between DM halos using techniques described elsewhere [12,16,17]. We assign a luminosity to each collision as described [12]. The thick solid line in Figure 1 shows our estimated magnitude-limited ($\mathcal{R} < 25.5$) starburst burst number density as a function of z. The diamonds are observational estimates of the number density of LBGs ($\mathcal{R} < 25.5$) from the HDF [14] ($z \sim 2.75$, $z \sim 4$) and ground-based observations [8] ($z \sim 3$). Note that the predicted number densities are somewhat larger than those observed, perhaps due to dust extinction and other selection effects.

The two-point auto-correlation function DM halo collisions is shown by the long dashed line in Figure 1b. In the range $1 - 5h^{-1}$Mpc, it can be approximated by a power law $\xi_c(r) \simeq (r/r_0)^{-\gamma}$, with $r_0 \simeq 5h^{-1}$Mpc and $\gamma \simeq 2.6$ Shown for comparison is $\xi_h(r) \simeq (r/3.5h^{-1}\,\text{Mpc})^{-2.2}$ for halos more massive than $10^{12}h^{-1}M_\odot$; these are what the massive-halo scenario would identify with LBGs. The lower curve is for the underlying dark matter. The starburst scenario thus predicts that LBGs should be strongly clustered, more than the halos themselves. Given the current uncertainties, both the halo and collision correlation functions are consistent with the parameters derived from observations [8,11] for the simulated cosmology: $r_0 \simeq 6h^{-1}$Mpc and $\gamma \simeq 2$.

CONCLUSIONS

We have found that starburst galaxies associated with collisions of relatively low mass halos (in a ΛCDM cosmology) can account for the observed number density and clustering of bright LBGs at $z \sim 2.5 - 4.5$. This picture can be further tested observationally, and distinguished from other scenarios, by more detailed studies of the evolution of the number density of LBGs with redshift and the dependence of their clustering on scale and environment. In addition, new instruments may provide better measurements of the virial masses of these objects.

REFERENCES

1. Steidel, C.C. , Giavalisco, M. , Pettini, M. , Dickinson, M. & Adelberger, K.L. 1996, ApJ, 462, L17
2. Lowenthal, J., Koo, D.C., Guzman, R., Gallego, J., Phillips, A.C., Faber, S.M., Vogt, N.P., & Illingworth, G. 1997, ApJ, 481, 673
3. Heckman, T.M., Robert, C., Leitherer, C., Garnett, D.R. & van der Rydt, F. 1998, ApJ, 503, 646
4. Pettini, M., Kellogg, M., Steidel, C.C., Dickinson, M., Adelberger, K.L. & Giavalisco, M. 1998, stro-ph/9806219
5. Jing, Y. P., Suto, Y. 1998, ApJL, 494, L5
6. Wechsler, R. H., Gross, M. A. K., Primack, J. R., Blumenthal, G. R., & Dekel, A. 1998, ApJ, 506, 19
7. Governato, F., Baugh, C. M., Frenk, C. S., Cole, S., Lacey, C. G., Quinn, T., & Stadel, J. 1998, Nature, 392, 359
8. Adelberger, K. L., Steidel, C. C., Giavalisco, M., Dickinson, M., Pettini, M., Kellogg, M. 1998, ApJ, 505, 543
9. Somerville, R. S., Primack, J. R., & Faber, S. M. 1998, MNRAS, in press (astro-ph/9806228)
10. Dickinson, M. 1998, in "The Hubble Deep Field", Ed. M. Livio, S.M. Fall & P. Madau
11. Giavalisco, M. Steidel, C. C., Adelberger, K. L., Dickinson, M. E., Pettini, M., Kellog, M. 1998, ApJ, 503, 18
12. Kolatt, T.S. et al. 1998, Nature, submitted
13. Steidel, C., Giavalisco, M., Dickinson, M., & Adelberger, K. 1996, AJ, 112, (352)
14. Pozzetti, L. , Madau, P. , Zamorani, G. , Ferguson, H.C. & Bruzual, A. 1998, MNRAS, submitted.
15. Kravtsov, A. V. , Klypin, A. A. & Khokhlov, A. M. 1997, ApJ Suppl., 111, 73
16. Bullock, J.S. , Kolatt, T.S. , Sigad, Y. , Somverville, R.S., Primack, J.R. , Dekel, A. , Kravtsov, A.V. & Klypin, A.A. 1998, in prep.
17. Kolatt, T.S., Bullock, J.S., Sigad, Y., Primack, J.R., Dekel, A., Kravtsov, A.V. & Klypin, A.A. 1998, in prep.
18. Mihos, J.C. & Hernquist, L. 1994, ApJ, 424, (L13); Mihos, J.C. & Hernquist, L. 1996, ApJ, 464, 641

Evidence for Multiple Mergers Among Ultraluminous IR Galaxies

Kirk D. Borne*, H.Bushouse†, L.Colina†, and R.A.Lucas†

*Raytheon ITSS, NASA GSFC Code 631, Greenbelt, MD 20771
†Space Telescope Science Institute, Baltimore, MD 21218

Abstract. Using an HST imaging survey, we have devised a morphological classification scheme for ULIRGs. We have used the results to uncover evidence for a multiple-merger origin for a significant fraction of the ULIRG population.

RESULTS FROM AN HST IMAGING SURVEY

We have used the Hubble Space Telescope (HST) to study a large sample of Ultra-Luminous IR Galaxies (ULIRGs), the most luminous galaxies in the local universe, with $L_{FIR}[8–1000\mu m] > 10^{12}L_\odot$. They are believed to be powered by massive starbursts that are induced by violent gas-rich galaxy-galaxy collisions and they may be the missing link in the chain of evolution from QSOs at high redshift $(2 < z < 5)$ to normal quiescent galaxies seen today $(z \lesssim 0.5)$ ([1], [2]). Furthermore, ULIRGs at $2 < z < 5$ may actually be the dominant source of the IR background ([3], [4]). Several discoveries have been made through our HST imaging survey of this unique collection of violently starbursting systems ([5], [6], [7]); see Fig. 1 for some representative images). We present some new results here.

Morphological Classification of ULIRGs

We have examined a complete subsample of ULIRG images and have identified 4 main morphological classes, plus 2 additional sub-classes (which are included in the main classes for statistical counting purposes). Figure 1 depicts 6 representative ULIRGs, one from each of the following classes:

1 — Strongly Disturbed Single Galaxy (Fig. 1a)
2 — Dominant AGN/QSO Nucleus (Fig. 1b)
3 — Strongly Interacting Multiple-Galaxy System (Fig. 1c)
4 — Weakly Interacting Compact Groupings of Galaxies (Fig. 1f)
5 — Collisional Ring Galaxy (Fig. 1d)

CP470, After the Dark Ages: When Galaxies were Young (the Universe at 2 < z < 5),
edited by Stephen S. Holt and Eric P. Smith

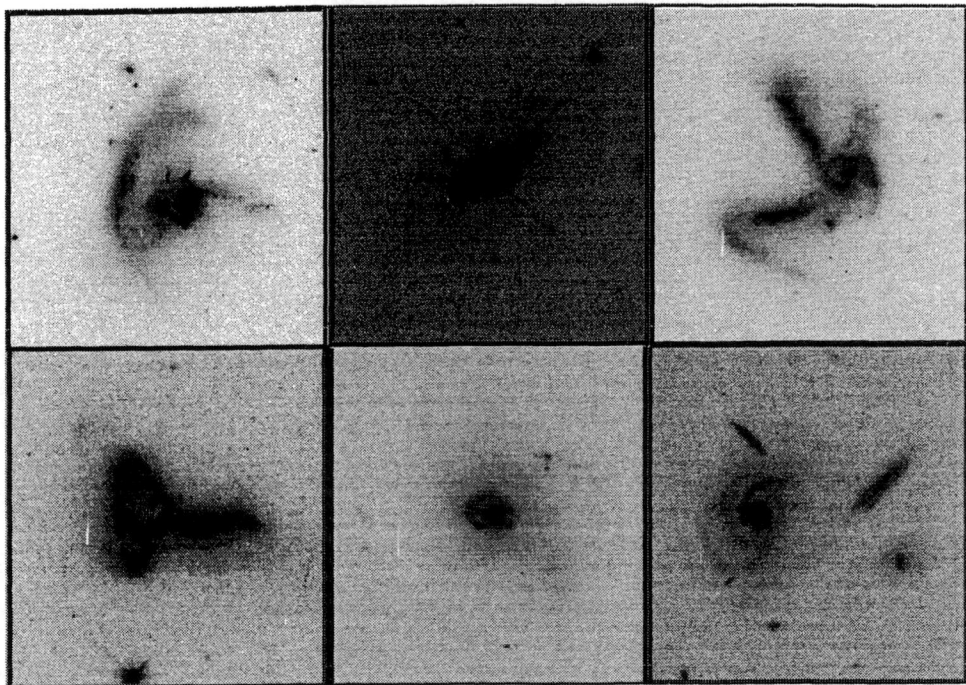

FIGURE 1. HST WFPC2 I-band images of 6 representative ULIRGs. **(a) Upper Left** – *IR 12112+0305* – A strongly disturbed ULIRG, showing evidence for only a single remnant galaxy (presumably the remnant of a major merger event). In many of these cases, massive star formation is seen on all scales in the HST images, including super star clusters of the type seen in HST images of other colliding galaxies, indicating that a giant starburst is the dominant power source in most ULIRGs. **(b) Upper Middle** – *IR 13349+2438* – A star–like nucleus, as seen in 15% of the ULIRGs, for which the dominant power source may be a dust–enshrouded AGN/QSO. **(c) Upper Right** – *IR 16007+3743* – A strongly disturbed ULIRG, showing evidence for two or more galaxies. **(d) Bottom Left** – *IR 21130−4446* – A classical collisional ring galaxy, one of a few in the sample, similar to the Cartwheel ring galaxy imaged by the HST. **(e) Bottom Middle** – *IR 09425+1751* – A ULIRG previously classified as non-interacting from ground-based images, now showing here clear evidence of merging (a second nucleus) and interaction (e.g., tidal tails). **(f) Bottom Right** –*IR 13342+3932* – A ULIRG showing several physically associated companion galaxies, which may be related to the collision, merger, and subsequent burst of star formation. In these cases, the signs of interaction (e.g., distortions) are usually weak.

6 — Previously Classified Non-Interacting Galaxy (Fig. 1e)

The following table shows the distribution of ULIRGs among our six morphological classes. We find no luminosity dependence on class and that there is the same number of single ULIRGs (classes 1 and 2) as multiple ULIRGs (classes 3 and 4).

Class	Number	Fraction	$< logL_{IR}/L_\odot >$	Notes
Disturbed Singles	30	34%	11.85	morph. class 1
AGN/QSO Nucleus	13	15%	11.74	morph. class 2
Interacting Multiples	29	33%	11.81	morph. class 3
Compact Groupings	14	16%	11.94	morph. class 4
Collisional Rings	1-3	~1-3%	...	re-classified
"Non-Interacting"	~5	~5%	...	re-classified

Interaction / Merger Fraction

Given the high angular resolution (~0.1–0.2″) of our HST images, some ULIRGs that were previously classified as "non-interacting" have now revealed secondary nuclei at their centers (remnants from a merger event?) and additional tidal features (tails). An example of one such system is shown in Figure 1e. It now appears that the fraction of ULIRGs showing evidence for interaction is nearly 100%. Observational estimates of this number have varied from 30% to 100% over the past 10 years, but it now seems to be converging to a value significantly above 90%.

Evidence for Multiple Mergers

It is not obvious that there is a well-defined point during a merger at which the ULIRG phase develops, nor is it clear what the duration of the ultraluminous phase is. Our new HST imaging surveys indicate that the mergers are well developed (with full coalescence) for some ULIRGs. Others show clear evidence for 2 (or more) nuclei, while others (~5%) can best be described as wide binaries, still a long way from coalescence. One possible explanation for this *dynamical diversity* is a multiple-merger model for ULIRGs ([8]). In this scenario, the existence of double nuclei is taken as evidence of a second merger, following the creation of the current starburst nuclei from a prior set of mergers. In fact, this would indicate for some systems (with double AGN or double starburst nuclei) that the currently observed merger is the third (at least) in the evolutionary sequence for that galaxy. This may be reasonable if these particular ULIRGs are *remnants of previous compact groups of galaxies*. Compact groups are known to be strongly unstable to merging, and yet examples are seen in the local (aged) universe. These may be the tail of a distribution of dynamically evolving galaxy groups. Similarly, the ULIRGs are presumed to be at the tail end of a distribution of major gas-rich mergers. A connection between the two populations, if only in a few cases, is therefore possible. Figure 2 presents images of 12 ULIRGs from our HST sample that appear to have

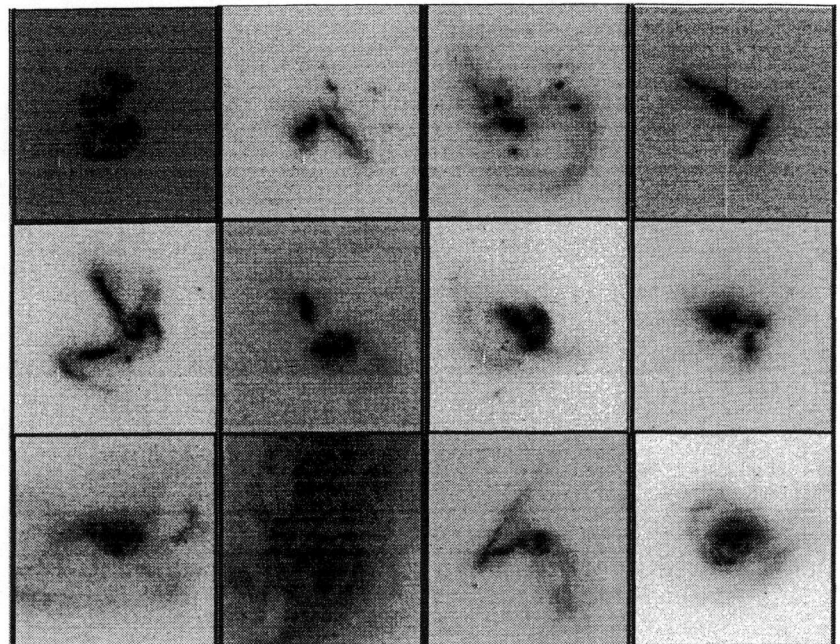

FIGURE 2. Sample of ULIRGs whose morphologies appear to be derived from > 1 merger.

evolved from multiple mergers. The evidence for this includes: >2 remnant nuclei, or >2 galaxies, or an overly complex system of tidal tails and loops.

Acknowledgments

Support for this work was provided by NASA through grant numbers GO–6346.01–95A and GO–7896.01–96A from the Space Telescope Science Institute, which is operated by AURA, Inc., under NASA contract NAS5–26555.

REFERENCES

1. Sanders, D.B., & Mirabel, I.F. 1996, ARAA, 34, 749
2. Genzel, R., et al. 1998, ApJ, 498, 579
3. Barger, A. J., et al. 1998, Nature, 394, 241
4. Hughes, D. H., et al. 1998, Nature, 394, 248
5. Borne, K.D., et al. 1997a, in "Star Formation, Near and Far", Woodbury: AIP, 295
6. Borne, K.D., et al. 1997b, in "Extragalactic Astronomy in the Infrared", Paris: Editions Frontieres, 277
7. Borne, K.D., et al. 1997c, in "The Ultraviolet Universe at Low and High Redshift", Woodbury: AIP, 423

8. Taniguchi, Y., & Shioya, Y. 1998, ApJ, 501, L167

The Symmetry, Color and Morphology of Galaxies in the Hubble Deep Field

Christopher J. Conselice* and Matthew A. Bershady[†]

Department of Astronomy, University of Wisconsin-Madison
475 N. Charter St. Madison WI, 53706
* chris@astro.wisc.edu
[†] mab@mingus.astro.wisc.edu

Abstract.

We present a new method of utilizing the color and asymmetry values for galaxies in the Hubble Deep Field to determine both their morphological features and physical parameters. By using a color-asymmetry diagram, we show that various types of star-forming galaxies (e.g. irregular versus interacting, peculiar galaxies) can be distinguished in local samples. We apply the same methods to the F814W images of the Hubble Deep Field, and show preliminary results indicating that galaxy mergers and interactions are the dominate process responsible for creating asymmetries in the HDF galaxies.

INTRODUCTION

One of the main purposes of the *Hubble Deep Field* (HDF) is to provide an unprecedented opportunity to examine the morphologies of distant galaxies. However, for a large fraction of the galaxies in the HDF (40%), no meaningful morphological indicator of type can be assigned. A problem coupled with the morphology of HDF galaxies is the question of whether the "peculiar" galaxies are merging systems, or if these galaxies are just undergoing an intense episode of star-formation. In this paper, we present a method for determining the morphologies of galaxies in the HDF based on both galaxy asymmetry and rest-frame UBV colors. We use a nearby galaxy sample to simulate the appearance of HDF galaxies, as well as to develop methods that insure reliable comparisons can be made between nearby galaxies and the more distant HDF sample. Our preliminary results indicate that a large portion of the 'peculiar' galaxies in the HDF are probably undergoing a merger or interaction, based on their asymmetry and color values.

CP470, After the Dark Ages: When Galaxies were Young (the Universe at 2 < z < 5),
edited by Stephen S. Holt and Eric P. Smith

MORPHOLOGY USING ASYMMETRY

The first uses of asymmetry for distant galaxy classification [1], [2] used asymmetry as a rough morphological parameter, with image concentration as a second classification parameter [2]. Asymmetry and concentration, while useful for segregating irregular or 'peculiar' galaxies, do not alone distinguish between irregular morphology due to interactions, or simply non-uniform star formation in turbulent environments (e.g. gas rich, dwarf irregulars). Conselice [3] tested the use of asymmetry as a general morphological parameter for nearby galaxies, finding a strong correlation between asymmetry and color.

Here we reformulate earlier methods ([2], [3]) to derive a consistent approach of measuring galaxy asymmetry applicable at both high and low redshifts. Simulations of nearby galaxies degraded in resolution and S/N show significant changes in the asymmetry. It is therefore necessary to correct for these effects, which are common in high-redshift galaxy images. In this preliminary presentation we have corrected for noise, but not for image degradation. Another critical facet of our new formulation concerns finding a self-consistent center about which to rotate the galaxy image. Both [2] and [3] defined the center of rotation by the brightest pixel centroid value. This is not a reliable method for finding the center: galaxy images have brightest pixels which change considerably with image degradation, yet very small changes in the center pixel values change the measured asymmetry considerably. To avoid this problem, we compute the asymmetry for a grid of rotation points centered on an initial best guess. We search for the center yielding the minimum asymmetry value, and iterate the search as necessary until a true minimum is found. This allows a robust method of finding the asymmetry to be computed which is fairly insensitive (in this regard) to resolution.

For galaxies in the HDF with high S/N $> (50)$ in I_{814} which span the redshift range from $(0<z<4.5)$ we compute the asymmetry parameter using the corrections mentioned above. We also compute rest frame B and V colors for each galaxy based on either spectroscopic redshifts, or from photometric ones based on F300W, F414W, F606W, F814W, J,H, and K magnitudes [5]. The k-corrections are computed in the empirical and interpolative manner as described in [6]. As such, the rest-frame UBVI colors at higher redshifts, while accurate, have lower precision because they rely increasingly on the near-infrared observed bands which are at lower S/N than the optical data. This should improve with the addition of photometry from deep NICMOS imaging of the HDF.

COLOR-ASYMMETRY DIAGRAM AT $Z = 0$

The color-asymmetry diagram can give a good morphological and physical indication of the present physical state of a galaxy. For the nearby sample of Frei et. al. [4] there is a strong correlation between the asymmetry of a galaxy and the color index for non-interacting face-on systems (Figure 1, left panel). The trend is, as might be expected: blue galaxies are asymmetric while red galaxies are symmetric. However, if we plot the entire Frei sample, which contains galaxies considered 'irregular' or 'peculiar' as well as edge-on disk galaxy systems, we obtain features which do not lie along the normal color-asymmetry sequence (Figure 1, labeled sources in left panel). The fact that these objects do not coincide with the face-on normal galaxies gives a method for deciphering these objects from normal face-on high-surface brightness galaxies at high-redshift. In contrast, the objects at the bluest-asymmetric end of the normal galaxy sequence are true irregulars – that is: the asymmetries are caused by star-formation, and not from projection effects (inclination), or from interactions.

By simply plotting asymmetry and color for a sample of galaxies, one cannot immediately disentangle which galaxies falling outside the normal sequence are inclined or interacting. This requires additional morphological information obtained by direct image inspection. Since, for the most part, inclined systems have a high axis ratio, they can be distinguished easily from interacting systems.

Asymmetries of the Hubble Deep Field galaxies

To first order the HDF sample *avoids* the trend of asymmetry with color as seen in the local sample of Frei et al. The majority of galaxies in the HDF are

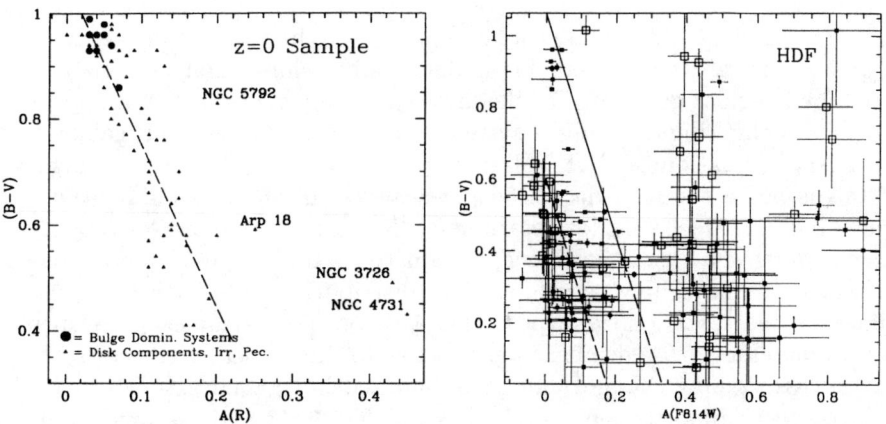

FIGURE 1. Asymmetry-Color diagrams for a local $z = 0$ sample (left panel), as well as for the high S/N galaxies in the HDF (right pabel). A tight correlation between color and asymmetry can be seen in the local sample for most objects with A<0.2 (solid line, both panels). The objects not on this line, labeled with NGC numbers, are galaxies generally regarded as undergoing an interaction or merger. In the HDF sample ($z > 1$, open symbols; $z < 1$, filled symbols), we see a strong bimodal pattern, with most galaxies either very asymmetric or symmetric for their blue colors.

either too asymmetric or too symmetric for their colors (compared with the local sample). However, the local sample includes almost entirely bright ($\sim$L*) galaxies (presumably as does the HDF for galaxies at the largest redshifts), while for $z < 1$, many of the HDF galaxies are low luminosity (sub-L*). Indeed, the colors of the HDF galaxies are blue compared to galaxies in the local Frei et al. sample, with very few galaxies having colors with (B-V) >0.80 (e.g. the colors of un-evolved ellipticals).

It is likely that the highly asymmetric HDF galaxies are the result of interactions. Normal star-formation processes as defined by the local sample would still lead galaxies to lie along the local sequence, albeit at the extreme blue, asymmetric end. Moreover our visual inspection of the asymmetric, blue HDF galaxies (A>0.3) reveals that the majority (80%) are not highly inclined. The asymmetric 'objects' in the HDF cannot be star-bursting regions embedded in a largely hidden galaxy, since in this scenario the bursting region would likely be blue and symmetric.

A class of galaxy not seen in the local sample but present in profusion in the HDF are blue, symmetric objects. These systems might be related to the blue-nucleated objects similar to those found in z $\approx$ 0.5 surveys (e.g. [1]), although Jangren et al. [7] find these specific objects to have A>0.2. The appearance of such a 'new' class of objects could be due either to increased, nucleated (symmetric) star-formation, or to resolution effects – an issue we are exploring. We emphasize, however, that in our asymmetry formulation, resolution effects will only tend to lower A. Hence the highly asymmetric galaxies discussed previously cannot be an artifact of the analysis.

Our preliminary results indicate that a large portion of the galaxies in the HDF, while extremely blue, are not undergoing a burst of star formation in a manner similar to nearby irregular galaxies. For the most part these galaxies do not appear morphologically as thin, highly inclined systems, but appear as 'peculiar' galaxies. Another substantial fraction of HDF galaxies appear to be highly blue and symmetric. If this is not a result of decreased physical resolution, we would surmise these systems have enhanced nuclear starbursts. Roughly 40%, however, appear too asymmetric for their blue color. Such asymmetry is indicative of interactions or mergers which are disturbing the global light distribution. While these results are preliminary, if the deviation from the normal-galaxy color-asymmetry sequence is confirmed to increase with redshift as we have found, this indicates that *merging* is a *critical* process shaping the morphology of high redshift galaxies.

This research was supported by NASA LTSA NAG5-6043 and research funds from the UW Graduate School.

REFERENCES

1. Schade et al., *ApJ* **451**, 1L (1995)
2. Abraham et al., *ApJS* **471**, 694 (1996)
3. Conselice C., *PASP* **109**, 1251 (1997)

4. Frei et al. *ApJ* **111**, 174 (1996)
5. Fernandez-Soto et al., *ApJ* in Press (1999)
6. Bershady M., *AJ* **109**, 87 (1995)
7. Jangren, Bershady, Conselice, Guzman, & Koo *AJ* in prep. (1999)

7. QSOs, AGN, and the CXRB

Surveys for High-redshift Quasars

Donald P. Schneider

The Pennsylvania State University, University Park, PA 16802

Abstract. It has been just over a decade since quasars first opened the $z > 4$ universe for detailed study. These distant beacons have cast a revealing light on the conditions present in the universe's first billion years, on topics ranging from the formation of massive structures to the properties of intergalactic gas. This paper reviews the techniques used to discover distant quasars, the current state of the field (which is based on but a few dozen objects), and some predictions for what awaits us in the next ten years.

INTRODUCTION

The most striking aspect of this conference is that after a span of 35 years galaxies have reclaimed their role as our primary probe of the distant universe. For the past generation the frontiers of the universe were set (with the exception of the cosmic background radiation) by our ability to identify distant quasars; from 1963 to 1991 the volume encompassed by the highest-known redshift increased by a factor of approximately 40!

This paper presents a brief historical review of high-redshift quasars, with an emphasis on the work since the discovery of the first quasar with a redshift larger than four. At the time of this writing (Fall 1998), there are 65 quasars with redshifts larger than four that have appeared in journals (a number of additional $z > 4$ quasars can be found on various web pages across the globe, but throughout this paper I will only discuss the published objects). This is a particularly appropriate time for a review as it is expected that new surveys planned in the next few years will produce an order of magnitude increase in the number of known $z > 4$ quasars.

Historical Perspective

Quasars were first identified as a discrete class of objects in 1960 by Matthews and Sandage [1]. Quasars' combination of strong radio emission, substantial brightness variations, and unusual optical spectra (a series of broad, unidentified emission

CP470, *After the Dark Ages: When Galaxies were Young (the Universe at 2 < z < 5)*,
edited by Stephen S. Holt and Eric P. Smith

lines superposed on a smooth continuum) was unprecedented; these "radio stars" were thought to represent an exotic stage of stellar evolution. Within a year of Maarten Schmidt's decoding of the spectrum of 3C 273 in 1963 [2], the largest known redshift passed from the radio galaxy 3C 295 ($z = 0.46$, [3]) to the quasar 3C 147 ($z = 0.55$, [4]). For more than 30 years the largest redshift belonged to quasars; their reign as the most distant known objects ended in 1997 with the Hubble Space Telescope-aided discovery of a $z = 4.92$ gravitationally-lensed galaxy [5].

Table 1 lists the objects that have at one time possessed the largest known quasar redshift. It is clear that the study of the distant universe has not proceeded at a steady rate; there was a rapid increase in the mid-1960s (note the spectacular discovery of 3C 9), a hiatus of 9 years from 1973-1982, and a spate of discoveries of $z > 4$ quasars in the late 1980s. A particularly striking aspect is the change in the discovery technique for high-redshift quasars; before 1985 all the "highest-redshift" quasars identifications were initially based upon radio detections, while all subsequent listings in the table were found purely from optical observations. The definition of what constitutes a high-redshift quasar obviously changes with time; for the past decade this term has been applied to quasars with redshifts larger than four, and will be adopted for the rest of this article. Physical quantities in this review are calculated assuming $H_0 = 100$ km s^{-1} Mpc^{-1} and $q_0 = 0.5$.

SURVEY TECHNIQUES

Fig. 1 displays the spectrum of PC1247+3406, the quasar with the highest known redshift ($z =4.90$; [23]); this spectrum is a typical example of high-redshift quasars. The emission lines seen in low redshift quasars are all present, with the Lyman α feature being particularly prominent. The continuum redward of the Lyman α line is relatively smooth, but as one traverses the line to the blue, there is a dramatic decrease in the flux produced by absorption by the Lyman α forest [24]. At a redshift of four, approximately 50% of the radiation is removed by intervening material; this fraction rises to nearly 80% at redshifts near five [25,26]. Another common feature in high-redshift quasars, a strong Lyman-limit system near the emission-line redshift of the quasar [23], is also clearly seen in the figure at 5400 Å.

The identification of high-redshift quasars poses a difficult challenge, as is demonstrated from the nearly quarter-century lapse between the initial quasar redshift and the discovery of the first $z > 4$ source. Attempts to extend radio observations beyond a redshift a four were not successful (until quite recently); to discover high-redshift quasars with optical telescopes, a number of obstacles must be overcome.

• High-redshift quasars are rare. On average there is one $z > 4$ quasar brighter than an R magnitude of 20 for every ≈ 15 square degrees of sky. Even in the regions of lowest stellar density (near the Galactic poles), this amount of sky would contain approximately 10,000 stars of similar brightness.

• High-redshift quasars are red. At redshifts above four, most of a quasar's light below 6000 Å is absorbed by intervening material. This property required the development of red-sensitive detectors, and moving into a wavelength region where the atmosphere's spectrum is bright and complex.

• High-redshift quasars are faint. Only a small fraction of $z > 4$ quasars can be detected in the Palomar Sky Survey, which has served as the primary optical data base for the second half of this century.

The explosion in the level of computational ability available to individual investigators in the 1980s led to a revolution in high-redshift quasar research. Using data obtained from either red-sensitive photographic plates or electronic detectors such as CCDs, it became possible to digitize large scale surveys, find and characterize all objects using automated software packages, and sift through the myriads of stars for the 1-in-10,000 source. An excellent review of the initial $z > 4$ discoveries is given by Warren and Hewett [27].

The vast majority of high-redshift quasars have been identified with optical telescopes. These objects have been identified via a number of techniques:

• **Multicolor Imaging.** The colors of stars fall along a well-defined locus of points in multidimensional color-space; in these diagram high-redshift quasars, with their prominent emission and absorption features, lie significantly (often more than a

TABLE 1. Highest Known Quasar Redshift

Quasar	z	Month-Year [a]	Discovery	Ref.
3C 273	0.16	03-1963	Radio	[2]
3C 48	0.37	03-1963	Radio	[6]
3C 147	0.55	02-1964	Radio	[4]
3C 9	2.01	04-1965	Radio	[7]
PKS 0106+01	2.11	02-1966	Radio	[8]
PKS 1116+12	2.12	04-1966	Radio	[9,10]
PKS 0237−23	2.22	02-1967	Radio	[11]
4C 25.05	2.36	06-1968	Radio	[12]
5C 02.56	2.39	12-1968	Radio	[13]
4C 05.34	2.88	05-1970	Radio	[14]
OH 471	3.40	04-1973	Radio	[15]
OQ 172	3.53	06-1973	Radio	[16]
PKS 2000−330	3.78	09-1982	Radio	[17]
Q 1208+1011	3.80	07-1986	Optical[b]	[18]
Q 0046−293	4.01	01-1987	Optical[c]	[19]
PC 0910+5625	4.04	10-1987	Optical[b]	[20]
Q 0051−279	4.43	12-1987	Optical[c]	[21]
PC 1158+4635	4.73	12-1989	Optical[b]	[22]
PC 1247+3406	4.90	09-1991	Optical[c]	[23]

[a] Date of publication of discovery
[b] Slitless Spectroscopy
[c] Multicolor

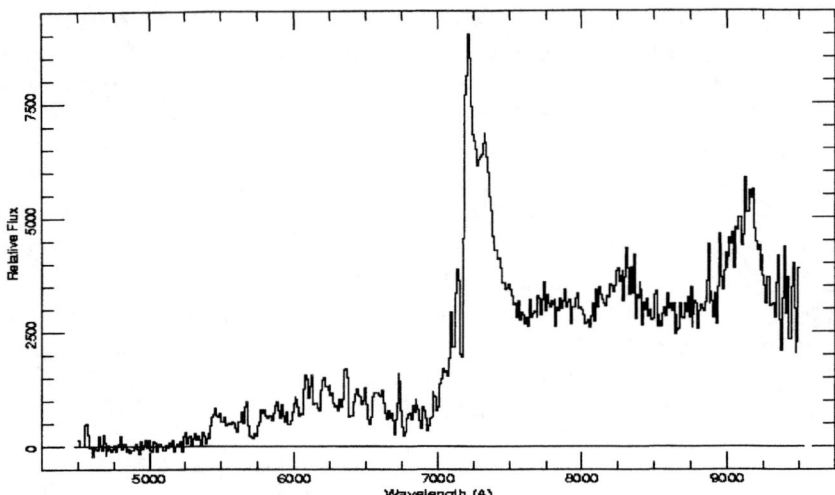

FIGURE 1. The spectrum of PC 1247+3406, the quasar with the highest known redshift ($z = 4.90$; [23]). The strong emission lines are labelled. The sharp drop in the continuum at wavelengths below the Lyman α emission line is absorption produced by intervening material.

magnitude) away from the stars. For example, in a BRI survey, the Lyman α emission line in a $z = 4.4$ quasar falls near the center of the R filter. The $(B - R)$ color of the quasar will be extremely red (line-enhanced R emission compared to the Lyman α forest/Lyman-limit system depressed B flux) while the $(R - I)$ color will be moderately blue. Stephen Warren and his collaborators [19] employed a multicolor technique to identify Q0046–293, the first $z > 4$ quasar; Fig. 2 is a reproduction of the plot used to identify Q0046–293. To date about 60% of the $z > 4$ quasars have been found using multicolor photometry, including PC 1247+3406.

• **Slitless Spectroscopy.** Observing a large region of sky through a dispersive element (commonly referred to as "objective prism" technique) produces spectra of all sources in the field; the spectral resolution is defined by the instrumental configuration and the atmospheric seeing. This has the advantage over mulitcolor imaging in that all of the data are obtained at once as well as at considerably higher (≈ 50) spectral resolution at the cost of reduced sensitivity and higher source confusion (the spectrum from an individual object covers a much larger region of sky than does an image). Fig. 3 displays the data used to identify the $z = 4.73$ quasar PC 1158+4635 [22]. Approximately one-sixth of the known high-redshift quasars have been found via this technique.

• **Serendipity.** In the summer of 1987 Patrick McCarthy and collaborators discovered a $z = 4.41$ quasar (then the second largest redshift) while obtaining a spectrum of a radio galaxy [28]. This supposedly once-in-a-lifetime event was repeated a few years later with the identification of a $z = 4.21$ quasar [29].

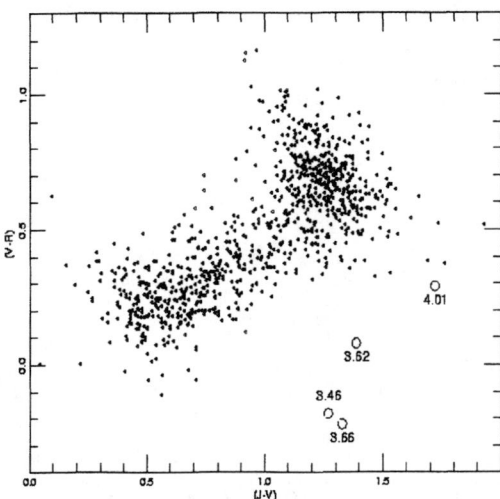

FIGURE 2. Multicolor diagram that led to the discovery of the first $z > 4$ object [19]. The open circles are four quasars that stand out from the stellar locus; the redshift for each quasar is given. (Only 5% of the stars are plotted.)

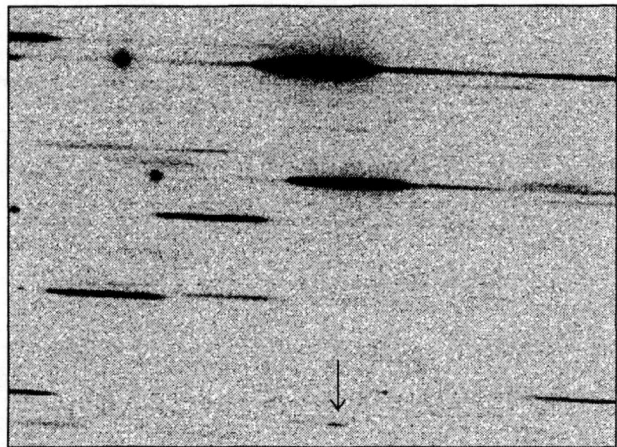

FIGURE 3. The grism data of the field containing PC 1158+4635 ($z = 4.73$; [22]). The points are the zero order images of the objects. The quasar's Lyman α emission line is indicated by the arrow; Lyman α forest absorption removes most of the signal blueward of the emission line.

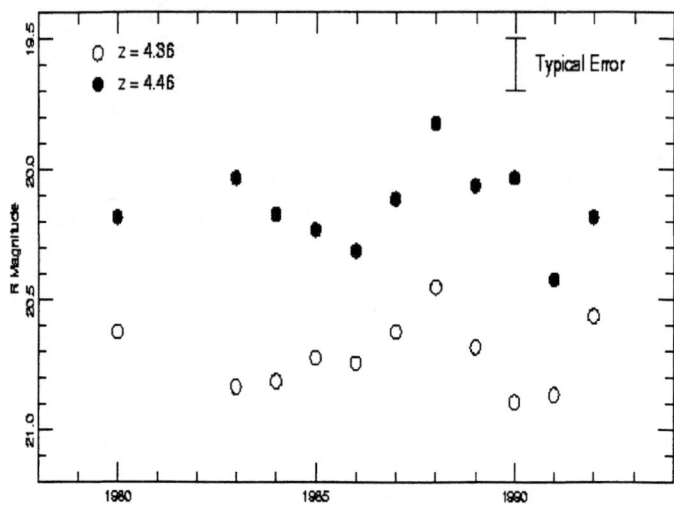

FIGURE 4. The R magnitude as a function of time of two $z > 4$ quasars. The measurements were made using Schmidt plates; these are the two highest redshift objects identified by variability [36].

- **X-ray.** Patrick Henry and collaborators discovered the first $z > 4$ quasar identified in a waveband other than the optical (a ROSAT source, [30]) in 1994. Earlier this year the X-ray discovery redshift limit was raised to 4.45 with the identification of an $R \approx 23$ quasar [31]; this observation was made possible because of the high accuracy $(2'')$ of the X-ray position.
- **Radio.** Radio astronomy, the prime source of distant quasars in the 25 years following the identification of the first quasar redshift, has for the most part contributed very little to the discovery of $z > 4$ quasars; only 5–10% of optically selected $z > 4$ quasars are radio-loud [32]. The first radio discoveries of $z > 4$ quasars were made in 1995 [33,34]. In 1997 a $z = 4.72$ quasar was found in a radio survey [35]; this object is also a strong X-ray source [36].
- **Variability.** It has been known since the discovery of the first quasars that they could exhibit spectacular variations in brightness; it is now believed that virtually all quasars display fluctuations on the 10–30% level on a timescale of years. Several high-redshift quasars have been identified by long-term monitoring programs [37]; Fig. 4 shows the light curves used to identify two high-redshift quasars.

Even when a high-redshift candidate has been identified, the chances that it is a high-redshift quasar are small; the typical success rate (number of $z > 4$ quasars per candidate) of the various techniques to date is 5-10%. Thus in addition to the large efforts required for the survey observations and data processing, the investigation of the individual candidates is an exhausting process. Table 2 presents a summary of the published $z > 4$ discoveries. For each of the six detection methods are listed the date of the first $z > 4$ discovery, the number of high-redshift quasars found,

TABLE 2. Detection Techniques

Technique	Date	Quasars	z_{max}
Multicolor Photometry	1987	43	4.90
Slitless Spectroscopy	1987	10	4.73
Serendipity	1988	2	4.41
X-Ray	1994	2	4.45
Radio	1995	4	4.72
Variability	1996	4	4.46

and the maximum redshift.

A summary of the properties of the 65 $z > 4$ quasars is presented in Fig. 5. The upper right panel, which displays the redshift distribution of the $z > 4$ quasars, reveals that the vast majority of these objects lie at redshifts below 4.5; only a handful of quasars have $z > 4.6$. The bulk of the sample lies between 18th and 20th magnitude, with a few objects in a faint tail extending down to 23rd magnitude. The quasars are fairly luminous; in this cosmology the "dividing" line between AGNs and quasars is $M_B = -21.5$ and 3C 273 has a luminosity of $M_B = -25.5$.

SCIENTIFIC PROGRAMS

Before 1987, our only direct observational information of the $z > 4$ universe came from the Cosmic Microwave Background Radiation field. In the past twelve years high-redshift quasars have produced a dramatic increase in our understanding of the first 10% of cosmic time; this section presents a brief summary of the highlights.

• **Constraints on Cosmological Models.** Soon after the discovery of the $z > 4$ population, George Efstathiou, Martin Rees, and Edwin Turner [38–40] pointed out the consequences of the existence of billion solar mass black holes within a billion years of the Big Bang. These objects provide one of the key constraints on theories of galaxy formation.

• **Quasar Evolution.** Within a few years of the discovery of quasars, it was shown that they were a rapidly evolving population of objects; their co-moving number density out to a redshift of one was proportional to $\approx (1 + z)^6$ [41] (or, from a chronological viewpoint, the number density of quasars has dropped by a factor of 15–20 since the universe was half of its current age). In 1982 Patrick Osmer published a study, based on objective prism photography, suggesting that this rise with redshift did not continue indefinitely; it appeared that beyond a redshift of 3.5 the space density of quasars actually declined [42]. A number of recent investigations, using both photographic plates and CCDs and a variety of detection methods, indicate that the number density of luminous quasars peaked near redshift three [43,33,45,37] (see Fig. 6).

239

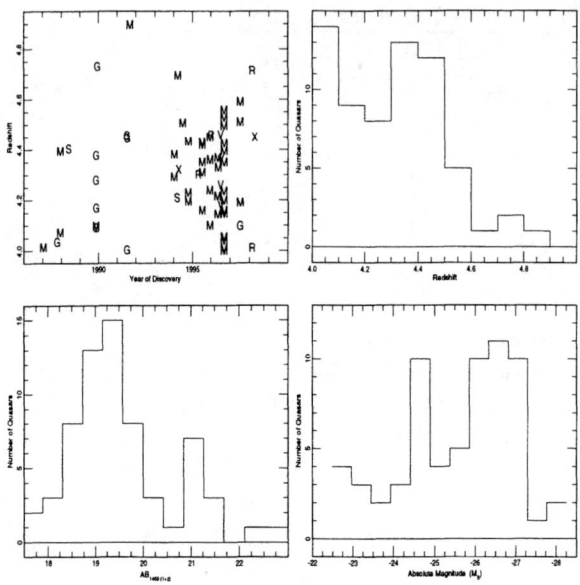

FIGURE 5. Summary of the properties of known $z > 4$ quasars. Upper left: Redshift vs publication date; objects are coded by detection technique [M:multicolor, G:grism (slitless spectroscopy), S:serendipity, X:X-ray, R:radio, V:variability]. Upper right, lower left, lower right: histograms of the number of objects as a function of redshift, $AB_{1450(1+z)}$ (the AB magnitude at rest wavelength 1450 Å [25]), and the absolute B magnitude (calculated assuming an optical spectral energy index of -0.5.)

- **The Properties of Intervening Material.** One of the most important (if not the most important) areas of quasar research does not involve the properties of the quasars themselves but the ability to investigate a wide variety of types of non-luminous matter through absorption features in the quasar spectrum produced by material that lies along the line-of-sight. High-redshift quasars have extended our ability to investigate, in some detail, this material out to redshifts of nearly five; we can now trace the properties of a number of types of features (*e.g.*, the Lyman α forest, haloes and disks of galaxies) throughout nearly the entire history of the universe [47,48,26].

- **The BAL Phenomenon.** The broad absorption lines seen in the spectra of 5-10% of quasars is thought to be material within ≈ 1 kpc of the central engine that is experiencing a high velocity outflow [49]. One of the first $z > 4$ quasars discovered was a spectacular BAL [21]; several more have been identified [47] including a possible member of the rare "Mini-BAL" class [50].

- **Evolution of Abundances of the Elements.** Quasars possess prominent emission lines of elements produced only in stellar nucleosynthesis. One might expect that the metal abundances of quasars would decrease with increasing redshift, but the observations do not show any redshift dependence in the strengths or profiles of the emission lines [25]. Quasar abundance determinations are notoriously difficult, but estimates suggest that even in the highest redshift quasars the metal abundances are several times the solar value [51].

• **Clustering of Quasars.** The current sizes of high-redshift quasar samples are too small to perform detailed studies of their spatial distribution, but there are indications that high-redshift quasars are clustered with a correlation length of order 20 Mpc [52,53].

• **Detection of Distant Clusters of Galaxies.** If high-redshift quasars mark the sites of substantial density perturbations, as seems plausible, it may be possible, with substantial efforts of observing time, to detect "normal" galaxies that are physically associated with the quasar. Some preliminary studies have already yielded intriguing results (*e.g.*, [54]).

• **Gravitational Lenses.** The longer the path length to an object, the greater its chance of being gravitationally lensed, hence high-redshift quasars are attractive candidates for gravitational lenses. If a multiply-imaged high-redshift quasar is found where the lens can be identified and characterized (especially if the quasar exhibits significant brightness variations), one will have a test of General Relativity on the largest imaginable scale! Gravitational lenses have already been identified at redshifts larger than 3.75 [55,56].

• **Spectral Energy Distribution.** Less than 10% of the known $z > 4$ objects have been found in non-optical surveys, but the radio properties of optically-selected

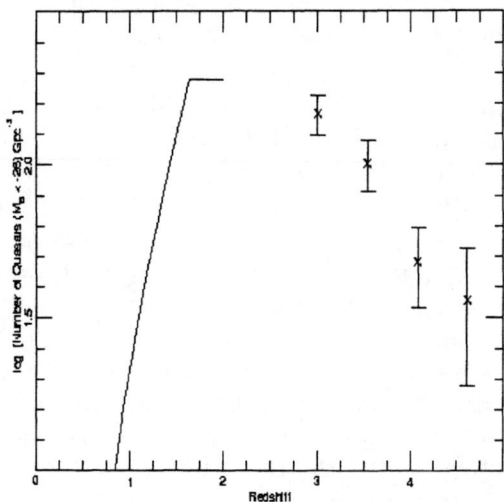

FIGURE 6. The number density of luminous quasars as a function of redshift based on a low-redshift study containing several hundred quasars (line, $z < 2$; [45]) and a slitless survey with 90 quasars with $z > 2.7$ (points; [43]).

high-redshift quasars have been investigated in some detail [32]. A number of optically (and radio) selected high-redshift quasars have been detected in X-rays, but no systematic X-ray study of this population has been published.

FUTURE SURVEYS

The relentless march of technology made possible the detection of the first $z > 4$ quasars; new instrumentation now becoming operational is expected to increase the number of $z > 4$ quasars by more than an order of magnitude in the next decade. Some avenues of investigation:

• **Large Scale Photographic Surveys.** The second Palomar Sky Survey has already yielded an impressive number of $z > 4$ quasars and provided a measurement of the high-redshift quasar luminosity function [45] based on but a small fraction of the survey area.

• **The Faint End of the Quasar Luminosity Function.** Both ground-based [57] and HST [58] observations have been used to determine the $z > 4$ quasar luminosity function at brightnesses that are considerably fainter than the $\approx$ 20-21 mag limit of current surveys.

• **Radio Surveys.** In the past few years four high-redshift quasars have been identified in radio surveys; this pace should dramatically quicken when the superb data base provided by the FIRST survey [59] is combined with the material produced by large-scale optical surveys.

• **X-ray Surveys.** To date X-ray surveys have not been particularly effective at identifying high-redshift quasars, but the recent discovery of a $z = 4.45$, $R = 23$ X-ray quasar [31] demonstrates what is possible with accurate X-ray positions. The impressive positional accuracy of the next generation of X-ray satellites (AXAF and XMM [60,61]) should enable the construction of sizable samples of X-ray selected $z > 4$ quasars.

• **The Sloan Digital Sky Survey (SDSS).** The SDSS [62,63], will perform photometry in five optical bands of 10,000 sq deg of the North Polar Cap. One of the primary goals of the project is to produce a sample of 100,000 quasars brighter than $B \approx 20$. First light for the project occurred in May 1998, and the first SDSS quasar discoveries were recently announced [64]. The SDSS filter system should be able to identify quasars beyond a redshift of 5.5; extrapolations of the measured quasar luminosity function suggest that the SDSS should identify of order 1,000 quasars with $z > 4.5$!

I would like to thank Maarten Schmidt, Michael Strauss, and Stephen Warren for their comments on the manuscript. This work was partially supported by NSF grant AST95-09919.

REFERENCES

1. Matthews, T.A., and Sandage, A.R. 1963, ApJ, 138, 30
2. Schmidt, M. 1963, Nature, 197, 1040
3. Minkowski, R. 1960, ApJ, 132, 908
4. Schmidt, M., and Matthews, T.A. 1964, ApJ, 139, 781
5. Franx, M., et al. 1997, ApJL, 486, 75
6. Greenstein, J.L., and Matthews, T.A. 1963, Nature, 197, 1040
7. Schmidt, M. 1965, ApJ, 141, 1295
8. Burbidge, E.M. 1966, ApJ, 143, 612
9. Schmidt, M. 1966, ApJ, 144, 443
10. Lynds, C.R., and Stockton, A.N. 1966, ApJ, 144, 446
11. Arp, H.C., Bolton, J.G., and Kinman, T.D. 1967, ApJ, 147, 840
12. Schmidt, M., and Olson, E.T. 1968, AJS, 73, 117
13. Burbidge, E.M. 1968, ApJL, 154, 109
14. Lynds, C.R., and Wills, D. 1970, Nature, 226, 532
15. Carswell, R.F., and Strittmatter, P.A. 1973, Nature, 242, 394
16. Wampler, E.J., Robinson, L.B., and Baldwin, J.A. 1973, Nature, 243, 336
17. Peterson, B.A., et al. 1982, ApJL, 260, 27
18. Hazard, C., McMahon, R.G., and Sargent, W.L.W. 1986, Nature, 322, 38
19. Warren, S.J., et al. 1987, Nature, 325, 131
20. Schmidt, M., Schneider, D.P., and Gunn, J.E. 1987, ApJL, 321, 7
21. Warren, S.J., et al. 1987, Nature, 330, 453
22. Schneider, D.P., Schmidt, M., and Gunn, J.E. 1989, AJ, 98, 1951
23. Schneider, D.P., Schmidt, M., and Gunn, J.E. 1991, AJ, 102, 837
24. Lynds, C.R. 1971, ApJL, 164, 73
25. Schneider, D.P., Schmidt, M., and Gunn, J.E. 1991, AJ, 101, 2004
26. Storrie-Lombardi, L.J. 1999, this conference
27. Warren, S.J., and Hewett, P.C. 1990, Rep.Prog.Phys., 53, 1095
28. McCarthy, P.J., et al. 1988, ApJL, 328, 29
29. Schneider, D.P., Schmidt, M., and Gunn, J.E. 1994, AJ, 107, 880
30. Henry, J.P., et al., 1995, AJ, 107, 1720
31. Schneider, D.P., et al. 1998, AJ, 115, 1230
32. Schmidt, M., et al. 1995, 109, 473
33. Hook, I.M., et al. 1995, MNRAS, 273, L63
34. Shaver, P.A., Wall, J.V., and Kellermann, K.I. 1996, MNRAS, 278, L11
35. Hook, I.M., and McMahon, R.G. 1998, MNRAS, 294, L7
36. Fabian, A.C., et al. 1997, MNRAS, 291, L5
37. Hawkins, M.R.S., and Véron, P. 1996, MNRAS, 281, 348
38. Efstathiou, G., and Rees, M. 1988, MNRAS, 230, 5P
39. Turner, E.L. 1990, ApJL, 365, 43
40. Turner, E.L. 1991, AJ, 101, 5
41. Schmidt, M. 1968, ApJ, 151, 393
42. Osmer, P.S. 1982, ApJ, 253, 28
43. Warren, S.J., Hewett, P.C., and Osmer, P.S. 1994, ApJ, 421, 412

44. Schmidt, M., Schneider, D.P., and Gunn, J.E. 1995, AJ, 110, 68
45. Kennefick, J.D., Djorgovski, S.G., and de Carvalho, R.R. 1995, AJ, 110, 2553
46. Hewett, P.C., Foltz, C.B., and Chaffee, F.H. 1993, ApJL, 406, 43
47. Storrie-Lombardi L.J., et al. 1996, ApJ, 468, 121
48. Lu, L., Sargent, et al. 1996, ApJ, 472, 509
49. Weymann, R.J., et al. 1991, ApJ, 373, 23
50. Churchill, C.W., et al. 1999, AJ, in press
51. Hamann, F., and Ferland, G. 1993, ApJ, 418, 11
52. Kundic, T. 1997, ApJ, 482, 631
53. Stephens, A.W., et al. 1997, AJ, 114, 41
54. Giallongo, E., et al. 1998, AJ, 115, 2169
55. Maoz, D., et al. 1992, ApJL, 386, 1
56. Kochanek, C. 1999, this volume
57. Schneider, D.P., Schmidt, M., and Gunn, J.E. 1999, AJ, in press
58. Conti, A., et al. 1999, AJ, in press
59. Becker, R.H., White, R.L., and Helfand, D.J. 1995, ApJ, 450, 559
60. Weisskopf, M.C., O'Dell, S.L., and van Speybroeck, L. 1996, Proc. SPIE, 2805, 2
61. Mason, K.O., et al. 1995, Adv. Space. Res., 16(3), 41
62. Gunn, J.E., and Weinberg, D. H. 1995, The Sloan Digital Sky Survey. In *Wide-Field Spectroscopy and the Distant Universe*, ed. S.J. Maddox and A. Aragón-Salamanca (Singapore: World Scientific), 3
63. Knapp, G.R. 1997, *Sky & Telescope*, 94, 40
64. Fan, X., et al. 1999, this volume

X-ray Observations of AGN at Intermediate to High Redshift

K. A. Weaver*

*NASA/GSFC, Code 662, Greenbelt, MD 20771

Abstract.
  The cores of active galactic nuclei (AGN) harbor some of the most extreme conditions of matter and energy in the Universe. One of the major goals of high-energy astrophysics is to probe these extreme environments in the vicinity of supermassive black holes, which are intimately linked to the mechanisms that produce the continuum emission in AGN. X-ray studies seek to understand the physics responsible for the continuum emission, its point of origin, how nuclear activity is fueled, and how supermassive black holes evolve. The key to finding answers to these questions lies in measuring the intrinsic luminosities and spectral shapes, the relation of these properties to other wavebands, and how the source properties change with redshift. This article reviews X-ray observations of AGN from redshifts of $\sim 0.1 - 3$ with the goal of summarizing our current knowledge of their X-ray spectral characteristics. Results are evaluated in terms of their robustness and are examined in the light of current theoretical predictions of energy release via processes associated with the accretion mechanism. A possible evolutionary scenario is discussed, along with the importance of AGN studies at high redshift as they relate to the total energetics of the Universe.

INTRODUCTION

Since their discovery, active galactic nuclei (AGN) have stood out as uniquely luminous objects in the Universe. We are fairly confident that their ultimate power source is the release of gravitational energy sustained by an accretion disk, which feeds matter directly onto a supermassive black hole. Evidence for the accretion mechanism is found in X-ray-bright Seyfert galaxies, which have broad Fe Kα lines indicative of radiatively efficient, geometrically thin accretion disks extending down to the radius of marginal stability [1,2]. However, the mechanisms to produce electromagnetic radiation from the accreting material and the manner in which this material actually reaches the black hole are still unclear. Understanding the accretion mechanism is important because nuclear activity in galaxies is common (perhaps more common than previously thought) and therefore accretion is likely to play a fundamental role in the energetics of the Universe.

CP470, After the Dark Ages: When Galaxies were Young (the Universe at 2 < z < 5),
edited by Stephen S. Holt and Eric P. Smith
1999 The American Institute of Physics 1-56396-855-X

The Universe is populated with many classes of AGN, the most powerful of which are QSOs ($L_X \sim 10^{46-48}$ erg s^{-1}). Their high luminosities make them observable at great distances, thereby providing ways to obtain fundamental information about the formation and subsequent evolution of galaxies. As we look toward the early Universe, studies of QSOs can help answer these important questions:

• Do all galaxies contain massive black holes and what roles do AGN play in the formation and evolution of galaxies?

• How does the accreting material make its way into the surroundings of the black hole, and how is this material fed directly into the black hole?

• Is there a change in the accretion efficiency or accretion rate with z?

• Are we seeing the flaring of short lived QSO events in many nuclei or a slow decline in a few nuclei that have been QSOs from the start?

Or, in observational terms: *In what ways do AGN exhibit spectral evolution?*

X-ray observations are of particular value because they provide a powerful diagnostic of the environs of the accretion flow and a powerful means for tracing evolution. Variability studies show that the X-ray continuum emission in AGN originates on the spatial scales we are most interested in – close to the black hole. X-rays can also penetrate large amounts of gas and dust in which some active nuclei are embedded. Moreover, X-ray emission appears to be a universal property of QSOs [3], which allows us to trace their properties out to high redshift. The major limitation of X-ray studies is the need for different instruments to cover the entire X-ray continuum range. Unbiased and statistically valid X-ray samples of QSOs have been difficult to obtain, but this will change with the next generation of X-ray observatories, beginning with the launch of *Chandra* and *XMM* in 1999. This article evaluates the current status of X-ray spectroscopy studies of QSOs, with emphasis on objects whose spectra appear to be dominated by accretion mechanisms rather than jet/beaming mechanisms.

X-RAY CONTINUUM PROPERTIES OF QSOS

The X-ray spectral characteristics of QSOs for redshifts up to z $\sim$ 3 are summarized in Table 1. All results derive from spectral modeling techniques, which assume that the QSO spectrum in a given energy band can be modeled with an absorbed power law having a photon index Γ ($N(E) \propto E^{-\Gamma}$). For cases where the line-of-sight absorption is consistent with that due to our Galaxy (N_{Hgal}), the fits have N_H fixed at N_{Hgal}. For cases where N_H is significantly larger than N_{Hgal}, N_H is left as a free parameter. In this article, the term "soft X-ray" is loosely applied

246

TABLE 1. The X-ray Spectral Properties of QSOs

Radio class	Sample size	z^a	Energy [keV]	Instr.	$<\Gamma_X>$	Intrinsic [b] N_H?	Comments [c]	Ref.
RQQ	42	0.12 ± 0.05	$0.1 - 2.4$	*Rosat*	$2.56^{+0.10}_{-0.11}$	no	[d]	[6]
	9	0.3 ± 0.03	$0.1 - 2.4$	*Rosat*	2.47 ± 0.33	no	[d]	[6]
	19	0.19 ± 0.08	$0.1 - 2.4$	*Rosat*	2.72 ± 0.09	no	[d]	[7]
	390	$0 \rightarrow 2.5$	$0.1 - 2.4$	*Rosat*	$2.58 \rightarrow 2.22$	—	[d]	[8]
	16	0.18 ± 0.21	$0.3 - 3.5$	*Einstein*	$1.91^{+0.67}_{-0.36}$	no	s. excess (7)	[9]
	12	0.076 ± 0.04	$0.1 - 10$	*Exosat*	2.18 ± 0.35	no	s. excess (5)	[10]
	9	0.54 ± 0.47	$0.5 - 10$	*ASCA*	1.93 ± 0.06	yes (3)	Fe K (5) [d]	[11]
	5	2.1 ± 0.13	$2 - 10$	*ASCA*	1.68 ± 0.09	no	FeK (1) [d]	[4]
	7	0.13 ± 0.08	$2 - 20$	*Ginga*	1.90 ± 0.38	—		[12]
RLQ	65	$0.08 \rightarrow 2.3$	$0.1 - 2.4$	*Rosat*	$2.52 \rightarrow 1.87$	no	[d]	[6]
	4	0.27 ± 0.11	$0.1 - 2.4$	*Rosat*	2.15 ± 0.14	no	[d]	[7]
	4	3.16 ± 0.23	$0.1 - 2.4$	*Rosat*	1.71 ± 0.08	yes (3)	[d]	[13]
	9	2.56 ± 0.8	$0.1 - 2.4$	*Rosat*	1.53 ± 0.06	yes (2)	[d]	[14]
	17	0.34 ± 0.19	$0.3 - 4.5$	*Einstein*	$1.48^{+0.63}_{-0.36}$	no		[9]
	5	0.27 ± 0.22	$0.1 - 10$	*Exosat*	1.79 ± 0.19	no	s. excess (1)	[10]
	3	2.3 ± 0.83	$0.5 - 10$	*ASCA*	1.67 ± 0.20	yes (1)	[d]	[15]
	15	2.42 ± 1.29	$0.5 - 10$	*ASCA*	1.63 ± 0.04	yes (9)	Fe K (2) [d]	[11]
	9	2.56 ± 0.8	$0.5 - 10$	*ASCA*	1.61 ± 0.04	yes (6)	[d]	[14]
	6	0.38 ± 0.3	$2 - 20$	*Ginga*	1.71 ± 0.16	—	[d]	[12]

[a] Mean redshift and standard deviation except for cases with arrows,
which indicate the range of z

[b] Indicates whether absorption in excess of the Galactic value is present.
Parenthesis contain the number of objects.

[c] Indicates cases for which excess soft X-ray emission or Fe Kα emission is detected.
Parenthesis contain the number of objects.

[d] Denotes data that are plotted in Figure 1

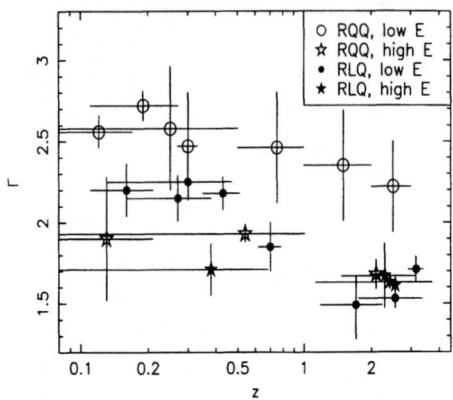

FIGURE 1. Observed photon index in the soft X-ray band (0.1 − 2.4 keV; *Rosat*) and hard X-ray band (∼ 0.5 − 20 keV; *Ginga* and *ASCA*) vs. redshift.

to photon energies between 0.1 and 2.4 keV and the term "hard X-ray" is loosely applied to photon energies between 2 and 10 keV. The energy bands of the experiments listed in Table 1 overlap and so only the photon indices that are strongly weighted toward soft or hard energies are discussed.

When photon index is plotted against redshift (Figure 1), it is apparent that Γ decreases with increasing z. For RQQs, Γ(soft) ranges from ∼ 2.6 at low z to ∼ 2.2 at high z ($\Delta\Gamma = 0.4$ from $z = 0.1 \to 2$) while Γ(hard) changes only slightly with z, from ∼ 1.9 to ∼ 1.7. For RLQs, Γ(soft) decreases by a larger amount from ∼ 2.5 at low z to ∼ 1.7 at high z ($\Delta\Gamma = 0.8$ from $z = 0.1 \to 3$), while Γ(hard) remains fairly constant with z. In addition, the soft X-ray index is related to radio loudness in the sense that RLQs have systematically smaller values of Γ(soft) than RQQs. At high energies, the spectral shapes are similar implying a common emission mechanism and minimal spectral evolution.

What about possible selection biases? The set of observations that most likely contain a selection bias is the *ASCA* sample of high-z RQQs [4]. For a given optical luminosity, RQQs are ∼ 3 times less luminous in X-rays than RLQs [5] and so only the most luminous RQQs have reliable hard X-ray data, especially at high z. For a given distribution of Γ(hard), *ASCA* may be biased towards detecting objects with small Γ(hard). On the other hand, these objects still possess fairly steep soft X-ray slopes [8]. Since these objects have both steep low-energy spectra *and* flat high-energy spectra, the trend seen with *ASCA* probably does represent RQQs at high z.

The dichotomy of spectral indices is robust and provides strong evidence for two distinct emission mechanisms, one at low energies that dominates the spectra up to $z = 1 - 2$, and one at high energies that is approximately independent of z. A distinct energy for this spectral "break" toward low energies (obviously a larger effect in RQQs than in RLQs) has not been found, but it is probably somewhere between 0.5 and 1 keV for low-z objects [10]. To explain the spectral changes with z, RLQs must have their soft component shifted out of the *Rosat* band by $z = 2$, beyond which Γ reaches its redshift-independent value [16]. RQQs, on the other hand, have not yet displayed the point at which the soft component is shifted out of the *Rosat* band, and so the spectral break must occur at a higher energy compared to RLQs (Figure 1). Depending on how the soft and hard X-ray components are normalized, either the soft X-ray emission is enhanced in RQQs relative to RLQs or the hard component is enhanced in RLQs relative to RQQs. The fact that RLQs are the stronger X-ray sources implies the latter.

PHOTOELECTRIC ABSORPTION

Characteristics for QSOs at $z > 2$ that have high-quality X-ray data are listed in Table 2[1]. More than 1/2 of the RLQs possess absorption in excess of the Galactic value.[2] In contrast, RQQs and low z quasars lack significant amounts of intrinsic absorption, with a handful of RQQs, such as PG 1114+445 [18] showing evidence for ionized absorption. The absorption in RLQs can be as much as $\sim 5 \times 10^{22}$ cm^{-2} in the quasar frame, which is similar to that seen in intermediate-type Seyfert galaxies such as NGC 4151. However, the physical properties of the absorbers are not well known because the X-ray data are ambiguous. In some cases, data at other wavebands support the contention that the absorption is physically associated with the quasar [19], while in other cases, the data favor absorption at low z [20]. If the absorber is intrinsic to the quasar, it could be nuclear material as in low-z, low-L_x objects or it could exist on a larger scale such as the host galaxy (or protogalaxy).

Because of the uncertainties associated with the properties of the absorbing material in high-z RLQs, some general spectral trends are not clear. For example, does the shape of the intrinsic spectrum depend on the luminosity of the source? Other QSO studies have not found a significant correlation between Γ and L_X except for the general trend that RLQs have smaller X-ray indices than RQQs [11,12,14]. Table 2 also shows no correlation between Γ and L_X, but Γ depends on how the absorption is handled in the spectral modeling. Having N_H "wrong" can make Γ artificially large or small and so the true values of Γ are somewhat uncertain.

The data also do not require that *only* high-z RLQs possess intrinsic absorption. Indeed, if only the hard X-ray component is absorbed, this underlying absorption could remain "hidden" in other QSOs. In RQQs, the stronger soft component might

---

[1] Not necessarily a complete sample.
[2] A strong case for absorption is made by Fiore et al. (1998) [17]. An alternative but less favored explanation is downward curvature in the intrinsic spectrum, such as that resulting from emission dominated by synchrotron losses.

TABLE 2. X-ray Properties of a Representative Sample at z > 2

Quasar	z	Γ_x	N_H^a [10^{20} cm^{-2}]	N_{Hgal} [10^{20} cm^{-2}]	$\log L_x$ [ergs s^{-1}]	Radio Class	Fe Kα EW^b [eV]	Ref.
S5 0014 + 81	3.38	1.7 ± 0.07	27.4 ± 4 554^{+196}_{-170} (qf)	13.9	47.8	RLQ	< 70	[14]
0040 + 0034	2.00	$1.79^{+0.15}_{-0.14}$	$9.7^{+5.5}_{-5.2}$	2.45	46.4	RQQ	—	[4]
PKS 0237 − 233	2.22	1.68 ± 0.06	122^{+62}_{-54} (qf)	2.39	46.8	RLQ	< 32.2	[11]
0300 − 4342	2.30	$1.69^{+0.24}_{-0.13}$	< 6.8	1.83	46.0	RQQ	—	[4]
Q 0420 − 388	3.12	2.24 ± 0.48	34 ± 33 (qf)	1.91	46.9	RLQ	—	[13]
PKS 0438 − 436	2.85	1.5 ± 0.1	$5.8^{+2.6}_{-1.4}$ 72^{+44}_{-22} (qf)	1.47	47.0	RLQ	< 240	[14]
PKS 0528 + 134	2.07	2.64 ± 0.07	420 ± 90 (qf)	23.0	47.1	RLQ	119 ± 58	[11]
PKS 0537 − 286	3.11	1.4 ± 0.1	$2.8^{+1.0}_{-0.7}$ $16.7^{+19.3}_{-14.6}$ (qf)	1.95	47.3	RLQ	< 139	[14]
S5 0836 + 71^c	2.17	∼ 1.5	∼ 3.3 ± 0.7	2.78	47.5	RLQ	< 110	[14]
1101 − 264	2.15	$2.19^{+0.58}_{-0.47}$	$19.5^{+19.5}_{-15.5}$	5.68	45.8	RQQ	690 ± 560	[4]
1255 + 3536	2.04	$1.59^{+0.12}_{-0.11}$	< 6.6	1.22	46.3	RQQ	—	[4]
1352 − 2242	2.00	$1.66^{+0.25}_{-0.23}$	< 17.9	5.88	46.0	RQQ	—	[4]
1422 + 231	3.62	1.68 ± 0.14	< 151 (qf)	2.9	47.0	RLQ	< 263	[11]
RXJ 1430.3 + 4203	4.72	1.29 ± 0.05	< 5.3	1.4	47.1	RLQ	< 100	[21]
Q 1508 + 571	4.30	1.43 ± 0.08	< 477 (qf)	1.34	47.2	RLQ	< 156	[11]
PKS 1614 + 051	3.21	1.43 ± 0.14	< 250 (qf)	5.0	46.6	RLQ	< 132	[11]
Q 1745 + 624	3.89	1.68 ± 0.25	21^{+15}_{-14} 61^{+108}_{-59} (qf)	3.4	47.0	RLQ	< 180	[22]
PKS 2126 − 158	3.28	∼ 1.6 ± 0.1	$10.4^{+3.1}_{-2.3}$ 120^{+80}_{-50} (qf)	4.85	48.0	RLQ	< 107	[14]
PKS 2149 − 306	2.35	1.54 ± 0.05	$8.3^{+1.8}_{-2.2}$	1.91	47.8	RLQ	< 85	[14]
PKS 2351 − 154^d	2.67	1.92(fixed)	$12.7^{+3.7}_{-2.9}$ 200^{+90}_{-60} (qf)	2.18	—	RLQ	—	[23]

Notes: ^{a}(qf) absorption in the QSO frame as opposed to the observer's frame (all others).

^{b}Assumes a narrow Gaussian line at 6.4 keV in the QSO rest frame.

^{c}Absorption changes by $\Delta N_H \sim 8 \times 10^{20}$ in 0.8 yr.

^{d}Absorption changes on timescale of < 0.41 yr in the QSO frame.

easily swamp the underlying absorption, regardless of z. For RLQs, where the soft component is not as prominent, column densities of a few $\times 10^{22}$ cm^{-2} would flatten the observed spectrum at $\sim 1 - 2$ keV in the quasar frame. Such flattening could go undetected in low-z quasars in the following ways. For *Rosat*, the flattening falls too near the upper energy range of the detector. For *ASCA*, which covers a larger bandpass, flux from the soft component may average out the flattening so that it is not detected. On the other hand, for $z = 1 - 2$ the contaminating soft component is shifted out of the *ASCA/Rosat* band and the absorption cutoff is shifted to around $\sim 0.3 - 0.6$ keV in the observer's frame, into a region where it is more easily detectable. Higher quality spectra at low X-ray energies are needed to address this question.

It is interesting that QSOs do not show the same evidence for spectral flattening at energies > 10 keV [4,14] that is common in Seyfert galaxies [24] and is the signature of Compton reflection. Fe Kα emission is also weaker and/or much less common in quasars and QSOs than in Seyfert galaxies and nearby radio-loud objects like 3C 120 [25]. The equivalent width of the Fe Kα line decreases with increasing luminosity from log $L_X = 41.7 - 47.2$ [26]. This trend continues to $z = 2$ [27] with RLQs at high z rarely showing Fe Kα emission [14,11].

The lack of reflection and Fe Kα emission in QSOs suggests that they possess a different structure in their accretion flow compared to lower-luminosity galaxies. The difference may relate to the high luminosities and high degree of ionization in the their inner regions. A lack of Fe Kα emission may signify complete ionization of iron atoms in the accretion disk or that the inner parts of the disk have been completely blown away. High ionization would also allow the Compton reflection to suffer much less absorption in the disk. In this case, reflection is still present but it would have a shape almost indistinguishable from the continuum source. For RLQs, the X-ray emission associated with the jet may simply dominate the spectrum rendering spectral features from the accretion flow undetectable.

INTRINSIC X-RAY EMISSION MECHANISMS

The source of the continuum emission in AGN is thought to be ultimately linked to an accreting supermassive black hole. Within the vicinity of the black hole, X-rays can be produced via Compton upscattering of soft photons off either electron-positron pairs [28], a population of hot electrons [29], or bulk motion in the accretion flow [30,31]. The lack of annihilation lines in Seyfert spectra causes some problems for pair models [32] and so this discussion focuses on the latter two mechanisms for X-ray production. The emitting region is envisioned as lying somewhere just above the accretion disk, often represented as a smooth extended corona above the disk or small-scale flaring regions on the disk surface.

Thermal Comptonization or bulk-motion Comptonization predict that the distinct observational signature of a black hole is an emergent power law with a spectral turnover between $50 - 500$ keV. If both processes are occurring then the changes/differences in photon index can be explained as a trade-off between the two. If conditions are such that bulk motion Comptonization dominates, then for an accretion disk that orbits a Schwarzschild black hole with an inner radius extending to the event horizon, Γ approaches 2.5 for mass accretion rates $m_\odot >> 1$, where $m_\odot = M_\odot/(M_e)_\odot$ and $(M_e)_\odot$ is the Eddington rate. If thermal Comptonization dominates, the spectra are harder, with $\Gamma \geq 1.5$. This model has been successfully used to explain the spectra of Galactic black-hole candidates, which show $\Gamma = 1.3 - 1.9$ in their low state and $\Gamma = 2.5$ in their high state [33].

For RQQs, where we think accretion mechanisms dominate the X-ray spectrum, those with $\Gamma \geq 2$ may be cases where we see X-rays from bulk motion Comptonization without modifications by ionized absorbers or reflection, i.e., the raw emergent spectrum from the accreting black hole. Within the context of the accretion model, the hard X-ray component in RQQs can result from thermal Comptonization.

Different mechanisms dominate the spectra of RLQs. RLQs show a significant correlation between their $0.1 - 2.4$ X-ray and total radio luminosity at 5 GHz [34] and Γ decreases with increasing radio loudness [9] in the sense that flat spectrum radio quasars (FSRQs), which are core-dominated, have flatter X-ray spectra than their steep spectrum radio quasar (SSRQ) counterparts [16,35]. This implies the presence of two emission mechanisms and has led to the two-component beaming model, which explains the difference in observed properties as caused by the relative orientation of the source axis to the line of sight [36]. In FSRQs, the highly beamed component points toward us and we see mostly upscattering of low-energy photons off relativistic electrons in the jet, while in SSRQs we see mostly the isotropic emission.

UNIFICATION AND EVOLUTION

Unified schemes explain the diversity of AGN classes as resulting from inclination plus obscuration effects, e.g., Seyfert type 2 (narrow-line) galaxies are edge-on Seyfert 1 (broad-line) galaxies. These models succeed when applied to specific subsamples of AGN, but a global unification scheme has proved elusive. For radio-loud objects, two schemes seem to be required, one for low luminosity objects and one for high luminosity objects [37]. For RQQs, broad absorption line (BAL) QSOs can be unified with nonBAL QSOs through axis orientation, but there are many arguments against RQQs being edge-on RLQs, including differences in their radio morphologies and underlying galaxy hosts [38]. What we are left with is a fragmented "unification" scheme.

It may be more promising to examine evolution scenarios, i.e., are low luminosity Seyferts connected to distant and more powerful AGN? Seyfert galaxies have composite X-ray spectra that consist of an underlying power law with $\Gamma \sim 1.9$, a Compton reflection tail above ~ 10 keV, Fe Kα emission [24], and significant photoelectric absorption from neutral and/or ionized gas [39]. High luminosity AGN lack significant evidence of X-ray reprocessing and many lack significant absorption. An evolutionary scenario that accounts for the difference requires QSOs to have much less cold material in their cores or for the material to be very highly ionized. As the source becomes less luminous with time, the ionization decreases and/or more cold material collects in the galaxy core.

QSOs also differ from Seyferts by (apparently) having spectral components from two physically distinct emission mechanisms. One component is associated with a relativistic jet while the other is most likely related to the accretion mechanism. Since processes associated with accretion are also thought to produce the continuum emission in Seyfert galaxies, an evolutionary connection between RQQs and Seyferts is plausible. The soft X-ray spectra of RQQs are significantly steeper than Seyfert galaxies, but are also more consistent with the predictions of bulk-motion Comptonization. This would suggest that bulk motion Comptonization is the more important physical mechanism in RQQs while thermal Comptonization of soft photons by a hot corona is the more important physical mechanism in Seyfert galaxies.

THE TOTAL ENERGY OUTPUT OF THE UNIVERSE

Until recently, little attention has been given to sources of energy in the universe that are not directly visible at optical-UV wavelengths. It now seems probable that most AGN are heavily absorbed, and that their central engines are primarily visible via hard X-rays. The energy density of the X-ray background peaks at ~ 30 keV. Less than 15% of this total energy density can be accounted for by the *Rosat* AGN, which dominate the soft X-ray background. If AGN comprise the hard X-ray background, then most must have huge absorbing columns ($N_H \sim 10^{22} - 10^{25}$ cm-2) [40]. The importance of the hard (> 10 keV) X-ray band for studying the total energy output for these objects cannot be overemphasized. A significant fraction of the energy in the Universe may reside in absorbed AGN [41] and the total accretion energy released by these AGN may be comparable to the energy generated by nuclear burning by the total stellar population. If most of the accretion in the Universe is highly obscured, then the emitted power per galaxy based on optical, UV, or soft X-ray quasar luminosity functions will be underestimated.

SUMMARY AND FUTURE PROSPECTS

This article summarizes our current knowledge of the X-ray spectral properties of QSOs. X-ray observations from z = 0.1 to ~ 3 keV indicate two distinct X-ray components in their spectra. One component is soft with $\Gamma \sim 2.0 - 2.5$ and the other is hard with $\Gamma \sim 1.5 - 1.9$. In the soft X-ray band the spectra flatten with increasing z while in the hard X-ray band the spectra show little change with z. RLQs have flatter X-ray spectra than RQQs with the exception of high energies at high z, where both have similar spectral shapes. Accretion mechanisms such as thermal Comptonization or bulk-motion Comptonization are the most likely source of the continuum in RQQs while jet/beaming mechanisms dominate RLQs. Significant absorption is observed in RLQs at high z but the physical properties of the absorbing material are uncertain. More sensitive data will place stringent limits on the absorption cutoffs and properties of the absorbing gas, the spectral breaks between the soft and hard X-ray continuum components, and the signatures of X-ray reprocessing. Considering an evolutionary scenario, RQQs and Seyfert galaxies are possibly connected, with a different accretion mechanism dominating in each.

Results from X-ray experiments such as *ASCA* for low-z AGN suggest that much of the accretion in the Universe is highly obscured. So far, X-ray observations at intermediate to high z have shed little light on this question. Surveys by *XMM*, ABRIXAS, and *Chandra* (limited to $E < 10$ keV) will probe columns up to a few times 10^{23} cm^{-2}; while future missions such as *Constellation* $-$ *X* (Valinia, this volume) will probe columns up to 10^{25} cm^{-2} for fluxes as low as $\sim 1 \times 10^{-14}$ ergs cm^{-2} s^{-1} in reasonable exposure times.

REFERENCES

1. Nandra, K., George, I.M., Mushotzky, R.F., Turner, T.J. and Yaqoob, T. 1997, ApJ, 477, 602
2. Tanaka et al. 1995, Nature, 375, 659
3. Avni, Y. and Tananbaum, H. 1986, ApJ, 305, 83
4. Vignali, C., Comastri, A., Cappi, M., Palumbo, G.G.C., and Matsuoka, M. 1998 (astro-ph/9809076)
5. Zamorani, G. et al. 1981, ApJ, 245, 357
6. Schartel, N., Walter, R., Fink, N.H. and Trumper, J. 1996, AA, 307, 33
7. Laor, A., Fiore, F., Elvis, M., Wilkes, B. and McDowell, J. 1997, ApJ, 477, 93
8. Yuan, W., Brinkmann, W., Siebert, J. and Voges, W. 1998, AA, 330, 108
9. Wilkes, B.J. and Elvis, M. 1987, ApJ, 323, 243
10. Comastri, A., Setti, G., Zamorani, G., Elvis, M., Giommi, P., Wilkes, B. and McDowell, J. 1992, ApJ, 384, 62
11. Reeves, J.N., Turner, M.J.L., Ohashi, T. and Kii, T. 1997, MNRAS, 292, 468
12. Williams, O.R. et al. 1992, ApJ, 389, 157
13. Elvis, M., Fiore, F., Wilkes, B., McDowell, J. and Bechtold, J. 1994, ApJ, 422, 60
14. Cappi, M., Matsuoka, M., Comastri, A., Brinkmann, W., Elvis, M., Palumbo, G.G.C. and Vignali, C. 1997, ApJ, 478, 492
15. Siebert, J., Matsuoka, M., Brinkmann, W., Cappi, M., Mihara, T. and Takahashi, T. 1996, AA, 307, 8
16. Brinkmann, W., Yuan, W. and Siebert, J. 1997, AA, 319, 413
17. Fiore, F., Elvis, M., Giommi, P. and Padovani, P. 1998, ApJ, 492, 79
18. George, I.M., Nandra, K., Laor, A., Turner, T.J., Fiore, F., Netzer, H. and Mushotzky, R.F. 1997, ApJ, 491, 508
19. Elvis, M., Fiore, F., Giommi, P. and Padovani, P. 1998, ApJ, 492, 91
20. Serlemitsos, P., Yaqoob, T., Ricker, G., Woo, J., Kunieda, H. Terashima, Y. and Iwasawa, K. 1994, PASJ, 46, 43

21. Fabian, A.C., Iwasawa, K., Celotti, A., Brandt, W.N., McMahon, R.G. and Hook, I.M. 1998, MNRAS, 295, 25

22. Kubo, H. etal. 1997, MNRAS, 287, 328

23. Schartel, N., Komossa, S., Brinkmann, W., Fink, H.H., Trumper, J. and Wamsteker, W. 1997, AA, 320, 421

24. Nandra, K. and Pounds, K.A. 1994, MNRAS, 268, 405

25. Grandi, P., Sambruna, R.M., Maraschi, L., Matt, G., Urry, C.M. and Mushotzky, R.F. 1997, ApJ, 487, 636

26. Iwasawa, K. and Taniguchi, Y. 1993, ApJ, 413, L15

27. Nandra, K., George, I.M., Mushotzky, R.F., Turner, T.J. and Yaqoob, T 1997, ApJ, 488, L91

28. Zdziarski, A.A., Ghisellini, G., George, I.M., Svensson, R., Fabian, A.C. and Done C. 1990, ApJ, 363, L1

29. Titarchuk, L. and Lyubarskij, Y. 1995, ApJ, 450, 876

30. Titarchuk, L., Mastichiadis, A. and Kylafis, N. 1996, AASupl., 120, 171

31. Titarchuk, L., Mastichiadis, A. and Kylafis, N. 1997, ApJ, 487, 834

32. Haardt, F. 1997, in MEMORIE S.A.It, 68, 73

33. Ebisawa, K, Titarchuk, L. and Chakrabarti, S. 1996, PASJ, 48, 59

34. Baker, J.C., Hunstead, R.W. and Brinkmann, W. 1995, MNRAS, 277, 553

35. Worrall, D.M., Giommi, P., Tananbaum, H. and Zamorani, G. 1987, ApJ, 313, 596

36. Browne, I.W.A. and Murphy, D.W. 1987, MNRAS, 226, 601

37. Browne, I.W.A. and Jackson, N. 1992, in "Physics of Active Galactic Nuclei," eds. W.J. Duschl and S.J. Wagner (New York: Springer-Verlag), p. 618

38. Barthel, P.D. 1992, in "Physics of Active Galactic Nuclei," eds. W.J. Duschl and S.J. Wagner (New York: Springer-Verlag), p. 618

39. Reynolds, C.S. 1997, MNRAS, 286, 513

40. Madau, P., Ghisellini, G. and Fabian, A.C. 1994, MNRAS, 270, L17

41. Fabian, A.C., Barcons, X., Almaini, O. and Iwasawa, K. 1998, MNRAS, 297, L11

The X–ray Background: Echo of Black Hole Formation?

Günther Hasinger[*]

Astrophysikalisches Institut, An der Sternwarte 16, 14482 Potsdam, Germany

Abstract.
Deep X–ray surveys using ROSAT, ASCA and BeppoSAX have resolved a significant fraction of the cosmic X–ray background (XRB) into discrete sources and optical identifications are demonstrating that the XRB is largely due to accretion onto massive black holes, integrated over cosmic time. The deep soft X–ray surveys have detected a larger surface density of AGN than in any other waveband and find significant evolution in the space density of high–luminosity AGN contrary to the pure luminosity evolution which was the standard assumption so far. Considerable uncertainties still exist for the evolution of low-luminosity AGN. These findings are consistent with the notion that most larger galaxies contain black holes many of which have been active in the past. Of particular interest in the context of this conference is the space density of high–redshift AGN, which is consistent with a constant value in the range $2 < z < 5$. X–ray surveys are therefore prone to yield valuable insight into the formation and accretion history of black holes in the early universe.

However, the characteristic hard spectrum of the XRB can only be explained if most AGN spectra are heavily absorbed. Thus as much as 80–90% of the light produced by accretion may be absorbed by gas and dust clouds, which according to recent models could reside in nuclear starburst regions that feed the AGN. This scenario has important consequences for the current attempts to understand black hole and galaxy formation and evolution: The absorbed AGN will suffer severe extinction and therefore, unlike classical QSOs, will not be prominent at optical wavelengths. If most of the accretion power is being absorbed by gas and dust, it will have to be reradiated in the FIR range and be redshifted into the sub–mm band. AGN could therefore contribute a substantial fraction to the recently discovered cosmic FIR/sub–mm background which has already partly been resolved by deep SCUBA surveys. The AGN light therefore needs to be taken into account when studying the star formation history in the early universe.

INTRODUCTION

The X–ray background has been a matter of intense observational and theoretical studies in the last years. Deep pencil beam surveys in the soft (0.5–2 keV) and hard (2–10 keV) X–ray band using the ROSAT, ASCA and BeppoSAX observatories

CP470, After the Dark Ages: When Galaxies were Young (the Universe at 2 < z < 5),
edited by Stephen S. Holt and Eric P. Smith

have resolved significant fractions of the background radiation into faint discrete sources and, at the faintest fluxes, into source fluctuations. Wide-angle shallower surveys mainly with ROSAT in the soft band provided large samples of brighter X–ray sources and give a lever arm between the distant and the local universe. Optical identifications are yielding a substantial amount of information about the statistical properties of the source populations and their evolution with cosmic time, the main contributor being active galactic nuclei. Finally, population synthesis models together with detailed X–ray spectroscopy of bright, nearby X–ray sources have been able to provide a satisfactory description of the overall XRB spectrum and other observational constraints.

In this review I give a summary of our current understanding of the XRB with emphasis on the results of the ROSAT ultradeep HRI survey (section 2) and the ASCA and BeppoSAX surveys in context with population synthesis models (section 3). I discuss the X–ray luminosity function of AGN and its cosmological evolution in section 4 in relation to the formation of black holes and stars in the early universe. A Hubble constant of $H_0 = 50$ km/s/Mpc and a deceleration parameter $q_0 = 0.5$ are assumed throughout the paper.

THE ROSAT ULTRADEEP HRI SURVEY

The ROSAT Deep Survey (RDS) project has started as a collaboration between R. Burg (STScI), R. Giacconi (ESO), M. Schmidt (Caltech), J. Trümper (MPE), G. Zamorani (Bologna) and myself. We have obtained a pencil beam survey in the Lockman Hole with the deepest X–ray observation ever performed. Images of 200 ksec observation time with the ROSAT PSPC define the *Deep PSPC Survey* [1], and exposures totalling 1.2 Msec in a smaller area are the basis for the *Ultradeep HRI Survey* [2]. The ultradeep ROSAT HRI survey now reaches a surface density of ~ 1000 sources deg^{-2} at a flux of 10^{-15} erg cm^{-2} s^{-1} and 70–80% of the soft X–ray background has been resolved into discrete sources. The fluctuation analysis of the PSPC survey resolves about 85–100% [1].

The Lockman Hole is also target for other deep multifrequency surveys. In the optical band, a mosaic of UH 88" CCD images (B, R), Keck R–band CCD images, Palomar 5m drift scans (V, I) and UH 8K images (V, I) have been obtained, which form the basis of our spectroscopic follow-up identifications [3]. Deep VLA radio mosaic observations [4] and, recently, deep ASCA [5] and BeppoSAX [6] hard X–ray images have been acquired in the 0.3 deg^2 survey region. The field will soon be covered by AXAF and XMM hard X–ray surveys in GT and PV-time, respectively. The Lockman Hole was also covered by a deep and medium-deep 7μ and 15μ mid-IR survey with ISOCAM [7] and a 90μ far-IR survey with ISOPHOT [8] and is targeted by ongoing SCUBA observations. Finally, the Lockman Hole is one of the CADIS fields [9] and a deep K–band survey with the Omega–Prime camera at Calar Alto has been started.

Optical counterparts of the weakest X–ray sources are very faint ($R > 24$)

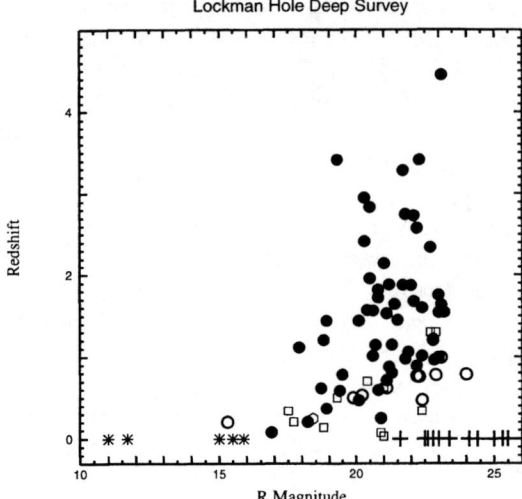

FIGURE 1. Correlation between optical R magnitude and redshift for all objects in the Lockman Hole survey. Filled circles are spectroscopically identified broad-line AGN (ID classes a-c [3]). Open circles are narrow-line AGN (ID classes d and e [3]). The open hexagon is one galaxy. Open squares are clusters of galaxies. Asterisks are stars. Plus signs are the as yet unidentified sources.

and require good, unconfused X–ray positions and high–quality optical spectra. For the Lockman Hole the optical follow–up spectroscopy could largely be done with long–slit and multi–slit spectroscopy at the Keck telescope. A catalogue of spectroscopic optical counterparts has been published for a complete sample of 50 ROSAT PSPC sources in the Lockman Hole with 0.5–2 keV fluxes brighter than 5.5×10^{-15} erg cm^{-2} s^{-1} [3]. The large majority ($> 80\%$) of the X–ray sources in the deep PSPC survey turned out to be AGN. Most of those are QSOs and Sy1 galaxies with at least one broad emission line in their optical spectra. A non-negligible fraction ($\sim 16\%$), however, shows only narrow emission lines in their spectra, which are interpreted as type 2 AGN because of the presence of high excitation [NeV] emission lines and/or high X–ray luminosity ($L_X > 10^{43}$ erg s^{-1}). The optical identification of the ultradeep HRI survey is discussed in [10]. Currently 83 out of 94 X–ray sources brighter than 1.2×10^{-15} erg cm^{-2} s^{-1} are spectroscopically identified. Among the identified objects is the highest redshift X–ray selected QSO at z=4.45 [11]. Figure 1 shows a correlation between optical magnitude and redshift for all X–ray sources. For the purpose of this conference, which is focussed on the redshift range $2 < z < 5$, it is important to note, that about one 20% of our AGN have $z > 2$ and about 40% $z > 1.5$. This sample provides the highest surface density of AGN observed so far.

ASCA AND BEPPOSAX SURVEYS

The X–ray background has a significantly harder spectrum than that of the sources resolved in the soft band. This led to the assumption that a large fraction of the background flux is due to obscured AGN, as originally proposed by Setti and Woltjer [12]. A model following the unified AGN schemes, assuming an appropriate mixture of absorbed and unabsorbed AGN spectra folded with cosmological AGN evolution models, is quite successfully explaining the shape of the background spectrum over the whole X–ray band as well as a number of other observational constraints [13]. This model predicts that in the hard X–ray band most of the contribution to the XRB comes from significantly absorbed objects, which are almost absent in the soft band, even at the faintest ROSAT limit. As a consequence, a significant test for this model would be the comparison of its predictions with the results of optical identifications of a complete sample of X–ray sources selected at low fluxes in the hard X–ray band.

Because of the technological challenge for X–ray imaging above 2 keV, hard X–ray surveys are just becoming available now. The deepest ASCA surveys [14,15] resolve source counts down to 2–10 keV fluxes of 5×10^{-14} erg cm^{-2} s^{-1}. At surface densities of typically 100 sources deg^{-2} these surveys are heavily confused due to ASCA's limited angular resolution. An analysis of the spatial fluctuations in deep ASCA images [16] probes the 2–10 keV X–ray source counts down to a flux limit of 2×10^{-14} erg cm^{-2} s^{-1}, resolving about 35% of the extragalactic 2–10 keV X–ray background.

A new High Energy Large Area Survey (HELLAS) has been started with BeppoSAX in the 5–10 keV band, which is particularly well suited for this instrument because of the relatively large throughput at high energies and a significantly sharper point spread function compared ASCA. A surface density of ~ 20 sources deg^{-2} is reported [17] at a 5–10 keV flux limit of 5×10^{-14} erg cm^{-2} s^{-1}, indicating that 30–40% of the background in this energy band has already been resolved.

The hard X–ray log(N)–log(S) data so far are in good agreement with the predictions of the Comastri *et al.* model [18]. Following these models, a large fraction of X–ray sources at fainter fluxes should be substantially absorbed and therefore their optical counterparts are expected to have optical spectra typical of Seyfert 2 galaxies. Programs to optically identify the sources from hard surveys have already started and indeed find mainly AGN counterparts [19,20,17], but the large positional uncertainty together with the relatively faint optical counterparts complicates the process.

Deep hard surveys with ASCA and BeppoSAX have also been taken in the Lockman Hole, where due to the existence of the ROSAT HRI data a cross–identification between the soft and hard X–ray data is readily available. Details of the ASCA and BeppoSAX surveys will be presented elsewhere [5,6]. A somewhat surprising result is that almost all hard X–ray sources in the ASCA and BeppoSAX images of the Lockman Hole have soft X–ray counterparts, apparently inconsistent with the simple XRB population synthesis models, which would predict a substantial fraction of hard sources not detectable in the soft band. We may see here effects that have been neglected in the population synthesis models, like e.g. leaky absorbers, unobscured soft spectral components in heavily obscured AGN or a dependence of the obscuration distribution on luminosity and/or redshift. Upcoming deep surveys with the Chandra observatory (AXAF) and XMM with very high sensitivity and good positional accuracy are expected to yield a solid statistical basis to disentangle these effects.

AGN COSMOLOGICAL EVOLUTION

Information about the cosmological evolution of the AGN population is a crucial input into the background synthesis models, but can not be obtained without taking into account the AGN absorption distribution. However, the AGN X–ray luminosity function in the 0.5–2 keV band has so far mainly been derived ignoring the effects of absorption. First global simultaneous fits of the XLF, X–ray background spectrum and absorption distribution have just been performed [21,22], but are still quite uncertain because of the large number of parameters and the possible hidden correlations involved.

Pioneering studies to determine the soft XLF of active galactic nuclei [23,24] were consistent with pure luminosity evolution models, similar to that found previously in the optical range. In the meantime optical identifications of a large number of ROSAT surveys at various flux limits and solid angle coverage have been completed, so that a new AGN soft X–ray luminosity function could be determined [25–27,21]. Contrary to the Boyle et al. findings, the new XLF is not consistent with pure luminosity evolution. For the first time we see evidence for strong cosmological evolution of the space density of low-luminosity AGN (e.g. Seyfert galaxy) XLF out to a redshift 1–2, incompatible with pure luminosity evolution. Pure density evolution proportional to $\sim (1+z)^5$ provided a reasonable fit to the ROSAT data [25], but overpredicts the total X–ray background, when extrapolated to lower luminosities. Therefore more complicated evolution models have to be taken into account. The latest treatments [21,26,27] agree that luminosity–dependent density evolution (LDDE) models, where the rate of density evolution is a function of luminosity, can match all constraints. This behaviour is similar to the most recently determined optical QSO evolution [28].

Figure 2 compares the luminosity function for local and high–redshift AGN to that of local normal galaxies. The low-redshift AGN XLF connects smoothly to the galaxy XLF at X–ray luminosities of $L_X \approx 10^{42}$ erg s^{-1}. Around this luminosity there is some ambiguity about the relative contribution between the nuclear AGN light and diffuse galactic X–ray emission processes [31]. Measurements of the high–redshift AGN XLF are shown for the two highest redshift shells ($1.6 < z < 2.3$ and $2.3 < z < 4.5$) from the data of [26]. Two luminosity–dependent density evolution models are shown, which fit all observational constrains well: the LDDE1 model

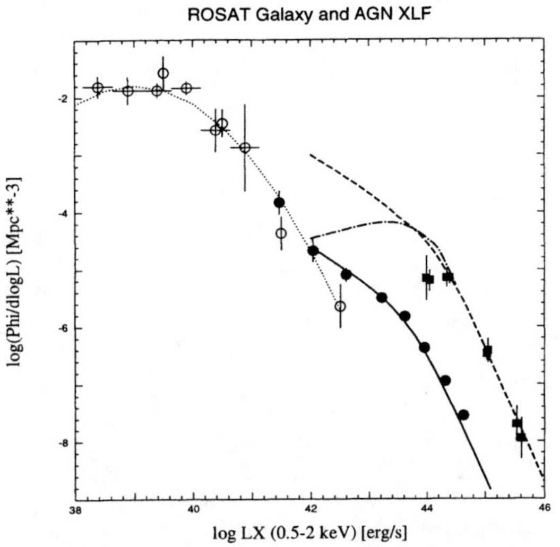

ROSAT Galaxy and AGN XLF

FIGURE 2. X–ray (0.5–2 keV) Luminosity functions for AGN and galaxies from ROSAT surveys. The local galaxy luminosity function (open circles) has been derived from the ROSAT Bright Survey [29,25] and from a volume-limited sample of local galaxies [30]. A log-normal distribution has been fit to the galaxy XLF (dotted curve). The AGN luminosity function (filled circles) is shown only for nearby objects ($z < 0.2$) and very distant objects ($z > 1.6$). Two different luminosity-dependent density evolution models have been fit to the data [27], one which is close to a pure density evolution model (LDDE2, dashed line) and one where evolution slows down substantially for low luminosities (LDDE1, dash–dotted line). Both models are consistent with all available constraints; their predictions for the density of low–luminosity AGN, however, diverge by more than an order of magnitude.

from [26] (dash–dotted line), which is similar to the LDDE model of [21] has a rapid slow–down of the density evolution below X–ray luminosities of 10^{44} erg s^{-1} and produces $\sim 60\%$ of the extragalactic 0.5-2 keV background. The LDDE2 model [27] is not very much different from a pure density evolution model and produces $\sim 90\%$ of the soft background. The constraints for the XLF of faint, high–redshift Seyfert galaxies, which can produce a significant fraction of the soft X–ray background and, depending on absorption properties, an even larger fraction of the hard X–ray background, are therefore still quite uncertain (the range is a factor of ~ 25 at $logL_X = 42$). The LDDE2 model predicts high–redshift AGN space densities which are in a similar range to that of normal galaxies just above the break of the luminosity function and to that of high–redshift galaxies selected as U–dropouts [32]. On the contrary, the LDDE1 model predicts a dearth of high–redshift Seyfert galaxies. A choice between these two possibilities will soon be possible with the even deeper X–ray surveys to be performed with the Chandra and XMM observatories.

The X–ray data can also give important new information on the AGN evolution

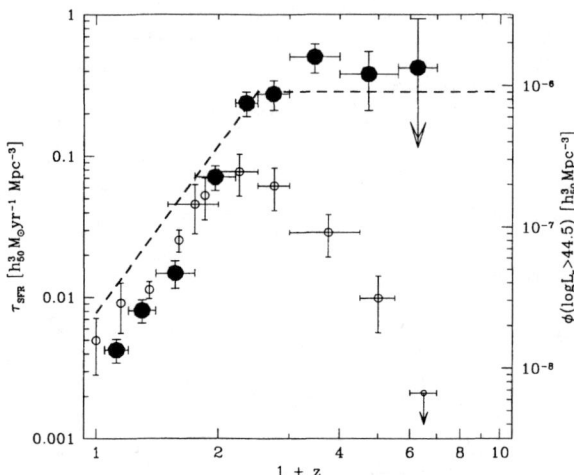

FIGURE 3. Cosmic star formation history τ_{SFR} (left Y-axis) compared to the space density ϕ of luminous X–ray selected QSOs (right Y-axis). Filled circles give the comoving number density of ROSAT QSOs with $log\ L_X > 44.5$ erg s^{-1} (from [26]). Open circles give the optical/UV measurements of the star formation rate compiled in [35]. The dashed line indicates the simplest star formation history model by Blain et al., which explains the whole FIR/sub–mm background light by dusty star formation. Note the similarity between this model and the QSO space density.

at very high redshifts and therefore on the epoch of black hole formation and the accretion history in the early universe. It is well known from optical samples that the strong evolution of the space density of high luminosity QSOs slows down beyond a redshift of ~ 1.5 and that the space density decreases significantly beyond $z \approx 2.7$ (see e.g. [33]). Different selection techniques have to be used below and above a redshift of ~ 2.2 leading to possible systematic uncertainties in the optical data. Radio–selected QSOs, however, confirm the decline at high redshift and indicate that it is apparently not due to an increase in dust obscuration [34]. The ROSAT sample of QSOs allows now for the first time to determine the space density of X–ray selected AGN in the whole range of $0 < z < 5$ with one technique. Figure 3 (right scale) shows the space density of AGN with X–ray luminosity $logL_X > 44.5$ erg s^{-1} as a function of redshift. The X–ray data does not show a significant decrease of the space density of high–redshift, high–luminosity X–ray selected AGN and appear to be marginally inconsistent with the optical and radio determinations [26]. However, the X–ray surveys still suffer from small sample sizes at high redshift (see Fig. 1), so that significantly larger solid angles have to be covered to a similar depth and optical completeness as the ultradeep HRI survey in order to get a clear picture of the AGN density at high redshift. The array of planned Chandra and XMM surveys in other fields than the Lockman Hole will be of great help in this respect.

The star formation history in the universe out to redshifts of four has been studied in the last few years by optical and NIR observations using ground–based telescopes and deep photometric surveys with the Hubble Space Telescope (see e.g. [35] for a recent review). The open circles in Fig. 3 show the compilation of the most recent observational determinations of the optical/UV star forming rate (SFR; left scale) by Blain et al., which suggest that star formation peaked at a redshift around 1–2. These data points, however, have to be regarded as lower limits of the true SFR because much of the light emitted in star bursts can be significantly obscured. Recently the far–infrared/sub–mm extragalactic background light (FIB), the equivalent of the X–ray background at very long wavelengths, has been discovered [36,37]. Deep SCUBA surveys have detected a population of optically faint galaxies, luminous in the sub–mm band, which could produce a significant fraction of the FIB signal [38–40]. Source counts of dusty galaxies and AGN in the sub–mm band are strongly weighted towards high redshift because of the large negative K–correction of the very steep dust spectra [41]. If all of the FIB should be due to star forming processes, a large population of strongly obscured star bursting galaxies would be missing from the optical/UV surveys at high redshifts. The dashed line in figure 3 sketches one of the SFR models by Blain et al. that is able to produce all of the FIB by early star formation. These models still have some drawbacks, however, because this massive early star formation would likely overproduce the heavy elements and consume a large fraction of all baryons in the universe into stars [35].

It is interesting to note that the star forming history required to fit the FIB light has a cosmic history which is very similar to the dependence of the AGN space density on redshift (see Fig. 3). Could it be that active galactic nuclei contribute significantly to the faint sub–mm source population? The X–ray background population synthesis models have recently been used by Almaini et al. [42] to predict the AGN contribution to the sub–mm background and source counts. Depending on the assumptions about cosmology and in particular on the AGN space density at high redshifts (see above) they predict that a substantial fraction of the sub–mm source counts at the current SCUBA flux limit could be associated with active galactic nuclei. Interestingly, the first optical identifications of SCUBA sources already indicate a significant AGN contribution [43].

Another, largely independent line of arguments leads to the conclusion that accretion processes may produce an important contribution to the extragalactic background light. Dynamical studies [44] come to the conclusion that massive dark objects, most likely dormant black holes, are ubiquitous in nearby galaxies. There is a correlation between the black hole mass and the bulge mass of a galaxy: $M_{BH} \approx 6 \times 10^{-3} M_{Bulge}$. Since gravitational energy release through standard accretion of matter onto a black hole is producing radiation about 100 times more efficiently than the thermonuclear fusion processes in stars, the total amount of light produced by accretion in the universe should be of the same order of magnitude as that produced by stars. A more detailed treatment following this argument comes to the conclusion that the AGN contribution should be about 1/5 of the stellar light in the universe [45].

Regardless of whether the FIR light of AGN is from dust heated by stellar processes or by accretion onto the massive black hole, these studies indicate that a large contribution to light emission in the early universe could come from sources associated with AGN, which are most easily pin-pointed by sensitive X–ray observations. Future sensitive joint sub–mm/X–ray surveys will therefore be a very powerful tool to disentangle the different processes dominating the universe in the redshift range $2 < z < 5$.

This work would not have been possible without the help of the whole ROSAT Deep Survey team and associated optical observers. I thank in particular the core team members R. Giacconi, M. Schmidt, J. Trümper and G. Zamorani, as well as R. Burg, J. Gunn, G. Luppino, J. MacKenty, M. McCaughrean, T. Stanke, D. Schneider, who have provided optical or NIR data for this study. I acknowledge very helpful discussions with I. Lehmann and T. Miyaji. This work has been supported in part by the DLR (former DARA GmbH) under grant 50 OR 9403 5 (G.H).

REFERENCES

1. Hasinger G. et al., 1993, A&A 275, 1
2. Hasinger G. et al., 1998, A&A, 329, 482
3. Schmidt M. et al. 1998, A&A, 329, 495
4. de Ruiter H. et al., 1997, A&A 319, 7
5. Ishisaki Y. et al., 1999, (in prep.)
6. Giommi P. 1999, proceedings of Taormina workshop, (in press)
7. Elbaz D., Aussel H., Baker A.C., 1998, Proceedings of the NGST workshop, June, 98, p. 47 (astro-ph/9807209)
8. Kawara K. et al., 1998, A&A 336, L9
9. Thommes E. et al., 1998, MNRAS 293, L6
10. Hasinger G. et al., 1999, in Highlights in X–ray astronomy in honour of Joachim Trümper's 65th birthday (astro-ph/9901103)
11. Schneider D., et al., 1998, AJ 115, 1230
12. Setti G. and Woltjer L. 1989, A&A 224, L21
13. Comastri A., Setti G., Zamorani G. and Hasinger G. 1995, A&A, 296, 1
14. Ogasaka Y. et al. 1998, AN 319, 43
15. Georgantopoulos I. et al., 1997, MNRAS 291, 203
16. Gendreau K.C., Barcons X. and Fabian A.C., MNRAS in press (astro-ph/9711083)
17. Fiore F. et al., 1999, First XMM workshop, astro-ph/9811149
18. Comastri A., et al., 1999, Adv. Space Res., in press
19. Boyle B.J. et al., 1998, MNRAS 296, 1
20. Akiyama M. et al., 1998, First XMM workshop (astro-ph/9811012)
21. Schmidt M. et al. 1999, in Highlights in X–ray astronomy in honour of Joachim Trümper's 65th birthday
22. Miyaji T. et al., 1999a, Adv. Space Res. (in press)

23. Della Ceca R. et al., 1992, ApJ 389, 491
24. Boyle B.J. et al., 1994, MNRAS 260, 49
25. Hasinger G., 1998, AN 319 ,37
26. Miyaji T., Hasinger G., Schmidt M., 1998, in Highlights in X–ray astronomy in honour of Joachim Trümper's 65th birthday (astro-ph/9809398)
27. Miyaji T. et al., 1999b, A&A (in prep.)
28. Wisotzki L., 1998, AN 319, 257
29. Hasinger et al., 1997, AN 318 ,329
30. Schmidt K.-H., Boller T., Voges W., MPE report 263, 395
31. Lehmann I. et al., 1998, in Highlights in X–ray astronomy in honour of Joachim Trümper's 65th birthday (astro-ph/9810214)
32. Pozzetti L. et al., 1998, MNRAS 298, 1133
33. Schmidt M., Schneider D.P., Gunn J.E., 1995, AJ 110 68
34. Shaver P.A. et al., 1998 (astro-ph/9801211)
35. Blain A.W. et al., 1998, MNRAS in press (astro-ph/9806062)
36. Puget J.-L. et al., 1996, A&A 308, L5
37. Fixsen D.J. et al., 1998, ApJ (in press, astro-ph/9803021)
38. Smail I., Ivison R.J., Blain A.W., 1997, ApJ 490, L5
39. Hughes D.H. et al., 1998, Nat 394, 241
40. Barger A.J. et al., 1998, Nat 394, 248
41. Blain A.W. & Longair M.S., 1993, MNRAS 264, 509
42. Almaini O. et al., 1999, MNRAS (submitted)
43. Smail I. et al. this volume (astro-ph/9810281)
44. Magorrian J. et al., 1998, AJ 115, 2285
45. Fabian A.C., Iwasawa K., 1999, MNRAS (in press)

Morphology and Dynamics of High Z Radio Galaxies and Quasars

K. C. Chambers

Institute for Astronomy, University of Hawaii, 2680 Woodlawn Drive, Honolulu, HI 96822

Abstract. The continuum morphologies of high redshift radio galaxies and quasars can be modeled as enormous bipolar reflection nebulae from shells of dust swept up by bipolar outflows. If the observed shape of a particular object is fit with an analytic function, then the velocity of the shell is specified by the equations of motion. The predicted kinematics can be compared with the observed emission line velocity field, and the resulting fit is excellent. The implications for massive galaxies at high redshift include the requirement of an initial epoch of star formation that creates dust distributed throughout a very large, diffuse, nearly virialized halo.

PHYSICAL MODEL OF THE ALIGNMENT EFFECT

High redshift radio galaxies have peculiar spatially extended optical and infrared morphologies that are generally aligned along the axis of their powerful radio sources [1,2]. This phenomena is called the "alignment effect" [3]. Dozens of distinct mechanisms for this phenomena have been proposed in the literature, but there has been no general consensus on the nature of the emission mechanisms. Among the various proposals for an optical alignment effect is the idea that the surrounding medium scatters a narrow anisotropic beam of light (e.g. a blazar) from an AGN and this beam appears as a visible pencil of light, like a searchlight beam scattered by fog [4–7]. Tadhunter et al. [4] argued that if the scattering medium had an optical depth $\tau \sim 0.1$, then the scattered intensity would be consistent with the brightest known blazars. This idea is inextricably mixed with proposals for the unification of radio sources [8]. In the unified scheme, radio galaxies are the unbeamed parent population of QSRs: a radio galaxy observed within 45° of the axis would have the appearance of a QSR.

Optical polarization has been detected in a number of HZRGs [7,9,10]. Spectropolarimetry has found specific QSR features in the polarized component of the spectra of HZRGs [11–13]. Furthermore, broad band optical and infrared polarimetry show the polarization properties are consistent with the characteristics of scattering by silicate-graphite dust grains rather than electron scattering [14]. The

CP470, After the Dark Ages: When Galaxies were Young (the Universe at 2 < z < 5),
edited by Stephen S. Holt and Eric P. Smith

discovery of polarization in HZRGS in the infrared [15] is crucial because it shows that the light redward of the 4000 Å break can be dominated by dust scattering. This is an important step forward in understanding the infrared alignments seen in HZRGs [16–18].

Manzini & di Serego Alighieri [19] proposed a specific model for dust scattering in high redshift radio galaxies where a diffuse spherical halo of dust was illuminated by a 45 degree bi-conical beam of quasar light, rather than a narrow blazar beam. Either model has a fundamental problem if the scattering medium is dust. They assume a unified scenario where radio galaxies are the parent population of the radio quasars, but if this assumption is true, these models would predict the radio quasars would be reddened when observed near the axis. This is not observed.

A solution to the problem can be found if the dust is distributed in an expanding bipolar shell of dust with an evacuated interior [20]. If the outward opening shell is illuminated by the active nucleus, then we can recover a self-consistent unification scheme. Although this is an oversimplification, I postulate that the morphologies and polarization properties are due to giant bi-polar dust nebulae [21]. For the high redshift radio galaxies, this hypothesis can account for an extraordinarily wide range of phenomena including: the alignment effect, the various features of quasar nebulosity, the presence of quasar spectral features in the spectropolarimetry of of high redshift radio galaxies, and the unification of quasars and radio galaxies without requiring reddening of quasars observed on axis.

In order to investigate this proposal further, I have modeled the morphology and dynamics of expanding bipolar dust shells illuminated from a central source [20]. By assuming an analytical form for an axisymmetric density distribution $\rho = Q(\theta)r^{-2}$, and bipolar wind force $P(\theta)$, the shape of a swept-up shell can be determined by quadrature [22,20]. If $P(\theta)$ and $Q(\theta)$ and the orientation i are chosen such that the resulting shell matches the observed morphology, then the model predicts the dynamics of the shell. The predicted velocity field can be compared with long slit spectroscopy.

An example of this kind of model is shown in Figure 1. The model has an inclination of 20 degrees from the plane of the sky and dust with a Henyey-Greenstein phase function. The functions $P(\theta)$ and $Q(\theta)$ were chosen to fit the morphology of the high redshift radio galaxy 3C265 [23]. The model spectrum has four components, two from the isotropic insitu photoionized emission line gas from the front and back surfaces of the shell, and two from the nuclear line emission scattered by dust grains swept up in the expanding shells. The model gives a remarkably good fit to the complex data set [24].

This model is the first physically self-consistent explanation of both the morphology and dynamics of high redshift radio galaxies [20]. Any alternative model for the complex emission line spectrum, e.g. entrainment in the jet, triggered star formation, or mergers, would be unlikely to reproduce these features seen in the velocity field data without introducing excessively ad-hoc components. Alternative scattering mechanisms for the polarization such as electron scattering or inverse Compton are similarly excluded. A prediction of the model is that the two scattered emission

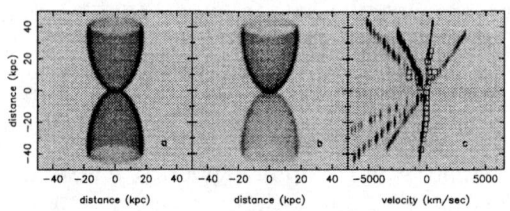

FIGURE 1. Dynamical model of the extended emission in high redshift radio galaxies as an expanding bipolar dust shell which scatters light from a quasar core (Chambers 1999). The functions $P(\theta)$ and $Q(\theta)$ were chosen to fit the morphology of the high redshift radio galaxy 3C265. A simulated narrow band image is constructed from the isotropic line emission (a), and a simulated broad band image is constructed from the scattered continuum (b). The model long slit spectrum has four components, two from the isotropic insitu photoionized emission line gas in the front and back shells, and two from the nuclear line emission scattered by dust grains swept up in the expanding shells. The predicted long slit spectrum (c) is compared with the observed $[OIII]$ velocity field of Tadhunter (1991) shown in boxes. The lumpiness of the model spectrum is an artifact of the course grid required by computing resources; the average surface brightness is representative. A falsifiable prediction of the model is that the scattered spectral components will be partially polarized and show the high ionization line ratios of the BLR whereas the isotropic components will not be polarized and will have low ionization line ratios.

line components will be partially polarized, whereas the two isotropic emission line components will not be polarized. Furthermore the scattered components should contain the high ionization lines of the broad line region. The continuum emission should show the same partial polarization as the scattered components (which will not be the same as the total line emission since some fraction is insitu and some fraction is isotropic).

IMPLICATIONS FOR GALAXY FORMATION

The strong evolutionary "turn on" of the alignment effect at $z \sim 1$ together with the ubiquity of the phenomena up to at least $z \sim 5$ has not been addressed in standard cosmogonies. This aspect of the alignment effect is particularly noteworthy given the success of the dynamical model discussed above. The model is largely dependent on the assumption of dust distributed through out a very large halo with an ambient density distribution roughly proportional to $1/r^2$. This implies a large, nearly viralized halo diffused with dust from a previous star formation episode. In particular, the halo cannot be very "lumpy" as one would expect from the merging of subgalactic units, or the dynamics and morphologies would be far more disorganized than they are. The presence of aligned structures out to redshifts $z > 4$ indicates that large halos were well organized at the time of the first major epoch of starformation [20].

REFERENCES

1. Chambers, K. C., Miley, G. K., van Breugel, W. J. M. 1987, Nature, 329, 604
2. McCarthy, P., van Breugel, W., Spinrad, H., Djorgovski, S. 1987, ApJ, 321, L29
3. Chambers, K. C., Miley, G. K., 1990, The Evolution of the Universe of Galaxies, ed. R. G. Kron (San Francisco: Astronomical Society of the Pacific) pp 373-388
4. Tadhunter, C., Fosbury, R., di Serego Alighieri, S. 1989, BL Lac Objects, ed. L. Maraschi, (Berlin: Springer-Verlag), p 79
5. Fosbury, R.A.E., 1989, ESO Workshop on Extra Nuclear Activity in Galaxies, ed. E.J.A. Meurs & R.A.E. Fosbury (Garching bei Munchen: ESO) p. 169
6. Fabian, A. C., 1989, MNRAS, 238, 41p
7. di Serego Alighieri, S., Fosbury, R. A. E., Quinn, P. J., Tadhunter, C.N., 1989 Nature, 341, 307
8. Barthel, P. D. 1989, ApJ 336, 606
9. Scarrott, S.M., Rolph, C.D., Tadhunter, C.N., 1990, MNRAS, 243,5p
10. Januzzi, B.T., Elston, R., 1991, ApJ 366, L69
11. Cimatti, A., van Breugel, W., Antonucci, R., Spinrad, H., 1996, ApJ 465, 145
12. Dey, A., Cimatti, A., van Breugel, W., Antonucci, R., 1996, ApJ 465, 157
13. di Serego Alighieri, S., Cimatti, A., Fosbury, R.A.E., Perez-Fournon, I., 1996, MN-RAS 279, 57p
14. Knopp, G. P. & Chambers, K. C. 1997, ApJ, 487, 644
15. Knopp, G. P. & Chambers, K. C. 1996 BAAS, 28, 1430
16. Chambers, K.C., Miley, G.K., Joyce, R.R., 1988, ApJ, 329, L75
17. Eisenhardt, P., Chokshi, A., 1989, ApJ, 351, L9
18. Chambers, K. C., Miley, G. K., van Breugel, W., M. Bremer, J. Huang, & N. Trentham, 1996, ApJ Supp, 106, 247
19. Manzini, A., di Serego Alighieri, S. 1996, A&A, 311,79
20. Chambers, K. C., 1999, ApJ, submitted
21. Chambers, K. C., 1996, Star Formation, Near and Far, ed. S. Holt & G. Mundy (New York: American Institute for Physics) pp 547-550
22. Chambers, K. C., 1990, Ph.D. thesis, Johns Hopkins University (Ann Arbor: University Microfilms International)
23. Longair, M. S., Best, P.N., Rottgering, H.J.A., 1995, MNRAS, 275, L47
24. Tadhunter, C.N., 1991, MNRAS 251, 46p

Strong Compact and Extended Radio Sources at High Redshifts

W. K. Rose

Department of Astronomy, University of Maryland, College Park, MD 20742

Abstract. Previously we have discussed models for strong compact and extended radio sources. In this paper we describe how the nature of these models is expected to depend on redshift and make a preliminary attempt to connect theory with observations. In particular we predict that the number of strong compact radio sources should decline more rapidly with redshift than the number of quasars. We also give physical explanations for the steepening of the average spectral index of radio sources with redshift as well as their smaller dimensions.

INTRODUCTION

Super-Eddington accretion onto massive black holes is the most plausible physical model for quasars and other very luminous AGNs. The most luminous of these objects radiate at 10^{47} - 10^{48} ergs^{-1}. Such high rates of emission correspond approximately to the Eddington luminosities of 10^9 - 10^{10} M$\odot$ black holes. The estimated mass of the black hole at the center of M87 (Virgo A) is $\simeq$ 2 x 10^9 M$\odot$ (Ford et al. 1994) [1]. Although gamma ray sources whose luminosities would be 10^{49} ergs^{-1} if their radiation were isotropic have been discovered (Dermer and Gehrels 1995) [2] it is likely that their radiation is anisotropic and therefore their power outputs do not exceed those given above. Strong compact extragalactic radio sources have radio luminosities of $\simeq$ 10^{42} - 10^{44} ergs^{-1} (Bridle and Perley 1984) [3] and therefore radio luminosities tend to be at least a factor of 10^3 less than total luminosities. The well known galactic object SS 433, which emits two counter streaming jets with outflow velocities of .26c, is often called a stellar mass prototype for quasars and other AGNs. Mass estimates of SS 433 indicate that it is $\simeq$ 10 M$\odot$ (Zwitter and Calvani 1989) [4] and therefore it is a black hole rather than a neutron star because neutron stars have masses $\lesssim$ 2 M$\odot$. Rose (1995) [5] has given a model for radio and X-ray emission from SS 433. The luminosity and radio luminosity of SS 433 are $\sim$ 4 x 10^{39} ergs^{-1} and 3 x 10^{34} ergs^{-1} respectively and therefore if its mass were increased by a factor of 10^8 its luminosity and radio luminosity would

CP470, After the Dark Ages: When Galaxies were Young (the Universe at 2 < z < 5),
edited by Stephen S. Holt and Eric P. Smith

be similar to those of the brightest quasars and other AGNs (i.e. 10^{47} - 10^{48} ergs^{-1} and $\sim 3 \times 10^{42}$ ergs^{-1} respectively).

Low and medium strength radio sources are undoubtedly accreting mass at highly sub-Eddington rates. Blandford and Znajek (1997) [6] introduced a process for the extraction of energy and angular momentum from rotating black holes in which magnetic fields thread the ergosphere and accelerate particles. A recent discussion of the Blandford-Znajek process with additional references is given by Ghosh and Abramowicz (1997) [7]. Radio loud quasars have radio luminosities that are typically 10^2 times greater than those of radio quiet quasars, which are approximately 20 times more numerous (Peterson 1997) [8]. If SS 433 were scaled to quasar luminosities then it would probably be classified as radio quiet because its luminosity is 10^5 times greater than its radio luminosity. Our point of view is that the relativistic electrons that emit synchrotron radiation are produced because interacting particle beams generate plasma turbulence by collisionless processes. The amount of energy available for electron acceleration is approximately equal to the electron kinetic energy and therefore radio luminosity is never more than $\simeq 10^{-3}$ times the luminosity of the source. If SS 433 were a rapidly rotating black hole then its radio luminosity might be a factor of about 10^2 times greater. Models for type II supernova remnants such as Cas A give the energy of relativistic electrons equal to 10^{48} erg whereas the total kinetic energy is $\simeq 10^3$ times greater (i.e. $\simeq 10^{51}$ erg).

It is well known that radio galaxies are often elliptical galaxies. Wilson and Colbert (1995) [9] have argued that radio loud quasars and AGNs have higher angular momenta than other AGNs because they were formed by mergers of galaxies with central black holes of comparable mass. It has been argued separately that elliptical galaxies are often formed by mergers. The average space density of galaxies is somewhat higher near elliptical galaxies than spiral galaxies. The rarity of radio loud objects is, therefore, a consequence of the necessity of the merger of two high mass black holes. The space density of quasars drops off rapidly for $Z \gtrsim 3$ (Hewitt and Burbidge 1993 [10], Warren, Hewett and Osmer 1994 [11], Peterson 1997) [8]. The above discussion indicates that the luminosities of the brightest objects depend on black hole mass whereas the corresponding radio luminosities depend on both black hole mass and angular momentum. It follows that our discussion indicates that the number of bright compact radio sources should decline with redshift more rapidly than the number of quasars.

In two previous papers (Rose 1987, 1989) [12,13] we have discussed the formation of large scale magnetic fields in extragalactic jets and presented solutions for these almost force free magnetic field configurations. A principal conclusion of these calculations is that large scale magnetic fields with strengths comparable to those of radio galaxies (i.e. $\simeq 10^{-4}$ - 10^{-5} gauss) can be generated by the interaction of mildly relativistic or relativistic jets with the 2.7 K cosmic background radiation. The intensity of the cosmic background radiation increases rapidly with redshift and therefore stronger magnetic fields are also predicted. Such fields in extended radio sources at larger redshifts provide a physical explanation for observed steeper radio spectra (McCarthy 1993) [14].

Extragalactic jets interact with ambient gas. This interaction is predominantly a collisionless interaction because of high jet velocities. We have previously discussed this collisionless interaction (Rose 1995 [5], Rose et al. 1984, 1987) [15,16]. A high speed jet excites resonant plasma waves initially via the two stream instability. Subsequent nonlinear development of resonant plasma waves, nonresonant plasma waves and ion acoustic waves is believed to lead to the generation of Langmuir solitons which accelerate electrons initially on the high velocity tail of a Maxwell-Boltzmann distribution to relativistic velocities. Therefore, the relativistic electrons responsible for observed synchrotron radiation from extended radio sources are produced by the collisionless interaction of jets with ambient gas. This implies that the dimensions of radio galaxies are determined by the amount of ambient gas. Because large amounts of ambient gas are likely to be present at higher redshifts it follows that the dimensions of radio galaxies should decrease with redshift as is observed (McCarthy 1993) [14].

REFERENCES

1. Ford, H. C., Harms, R. J. and Tsvetanov, Z. I. et al. 1994, Ap.J., 435, L27.
2. Dermer, C. D. and Gehrels, N. 1995, Ap.J., 447, 103.
3. Bridle, A. H. and Perley, R. A. 1984, Ann. Rev. Astron. Astrophys., 22, 319.
4. Zwitter, T. and Calvani, M. 1989, MNRAS, 236, 581.
5. Rose, W. K. 1995, MNRAS, 276, 1191.
6. Blandford, R. D. and Znajek, R. L. 1977, MNRAS, 179, 433.
7. Ghosh, P. and Abramowicz, M. A. 1997, MNRAS, 292, 887.
8. Peterson, B.M. 1997, Active Galactic Nuclei (Cambridge University Press: Cambridge).
9. Wilson, A.S. and Colbert, E. J. M. 1995, Ap.J., 438, 62.
10. Hewitt, A. and Burbidge, G. 1993, Ap.J.S., 87, 451.
11. Warren, S. J., Hewett, P. C., and Osmer, P. S. 1994, Ap.J., 421, 412.
12. Rose, W. K. 1987, Ap.J., 313, 146.
13. Rose, W. K. 1989, Ap.J., 337, 91.
14. McCarthy, P.J. 1993, Ann. Rev. Astron. Astrophys., 31, 639.
15. Rose, W. K., Guillory, J., Beall, J. H. and Kainer, S. 1984, Ap.J., 280, 550.
16. Rose, W. K., Beall, J. H., Guillory, J. and Kainer, S. 1987, Ap.J., 314, 95.

The Full Re-Ionization of Helium

James W. Wadsley, Craig J. Hogan, and Scott F. Anderson

University of Washington
Department of Astronomy
Box 351580
Seattle, WA 98195-1580

Abstract. Observations of resolved HeII Lyman alpha absorption in spectra of two QSO's suggest that the epoch of helium ionization occurred at $z \approx 3$. Proximity zones in the spectra of the quasars ($z = 3.18, 3.285$) at 304 Å resemble Stromgren spheres, suggesting that the intergalactic medium is only singly ionized in helium. We present models of the proximity effect which include the full physics of the ionization, heating and cooling and an accurately simulated inhomogeneous gas distribution. In these models the underdense intergalactic medium is heated to at least 10,000-20,000 K after cooling to as low as a few 1000K due to cosmological expansion, with higher temperatures achieved farther away from the quasar due to absorption-hardened ionizing spectra. The quasars turn on for a few $\times 10^7$ years with a fairly steady flux output at 228 Å comparable to the 304 Å flux output directly observed with HST. The recoveries in the spectra occur naturally due to voids in the IGM and may provide a fairly model-independent probe of the baryon density.

In the last few years it has become possible to observe details of absorption by singly ionized helium. The observations combine new information about the history of quasars, intergalactic gas, and structure formation. These phenomena can be disentangled with detailed quantitative models of the situation which we briefly describe here. Theoretical treatments of some of the effects modeled here were given by Zheng and Davidsen (1995), Croft et al. (1997), Miralda-Escudé et al. (1996), Zhang et al. (1998) and Fardal, Giroux and Shull (1998).

Early observations of the helium II Lyman alpha absorption spectral region included the quasars Q0302-003 (z=3.285, Jakobsen et al. 1994), HS 1700+64 (z=2.72, Davidsen et al. 1996) and PKS 1935-692 (z=3.18, Tytler & Jakobsen 1996). Higher resolution (GHRS) observations of Q0302-003 (Hogan, Anderson & Rugers 1997) and HE 2347-4342 (z=2.885, Reimers et al. 1997) revealed structure in the absorption which could be reliably correlated with HI absorption. The most recent published observations of PKS 1935-692 with STIS (Anderson et al. 1998) yield particularly good zero level estimates important for estimating the op-

CP470, After the Dark Ages: When Galaxies were Young (the Universe at 2 < z < 5),
edited by Stephen S. Holt and Eric P. Smith

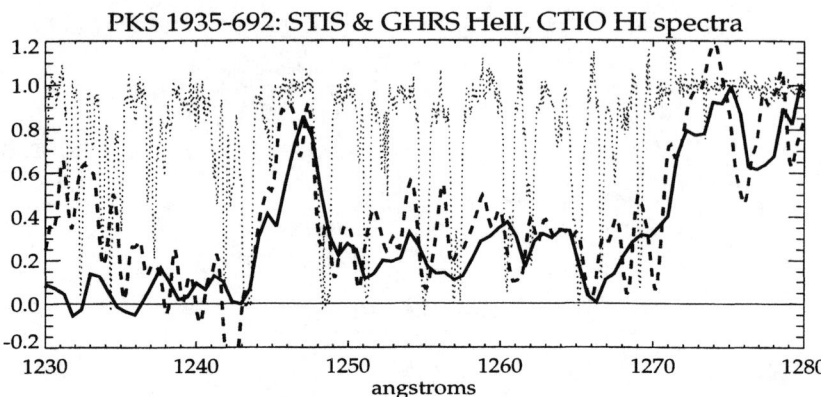

FIGURE 1. Observations of the HeII forest of PKS 1935-692 by Anderson et al. (1998) overlaid with a redshift-matched HI spectrum (dotted). The shelf structure in the GHRS (dashed)/STIS (solid) HeII spectra is best explained as the second-ionization of helium in the zone near the quasar. The flux recovers due to a void in the IGM at ≈ 1247Å. A similar pattern is found in the two other higher-z HeII quasars.

tical depth τ. Taken together, these data now appear to be showing the cosmic ionization of helium by quasars.

All of the objects show absorption with $\tau \gtrsim 1$ at redshifts lower than the quasar. For the higher redshift QSO's Q0302-033 and PKS 1935-692 (shown in Figure 1) there is a clear shelf of $\tau \gtrsim 1.3$ in a wavelength region of order 20 Å in observed wavelength blueward of the quasar emission line redshift, dropping to a level consistent with zero flux or $\tau \gtrsim 3-4$ beyond that. Anderson et al. conclude from these observations that gas initially containing helium as mostly HeII is being double-ionized in a region around the quasars. The lack of a strong emission line for HeII Lyman alpha suggests that ionizing flux is escaping so that the 228 Å flux may be similar to a simple power-law extension of the observed 304 Å rest frame flux. Hogan et al. used this to estimate the time required for quasars to create the double ionized helium region to be 20 Myr for a 20 Å shelf (dependent on the Hubble parameter, spectral hardness, cosmology, baryon density and the shelf size).

The features present in HeII Lyman alpha spectra are reflected in the HI Lyman alpha forest for these quasars. Attempts to model the HeII absorption with line systems detected in HI suggest that low column HI absorbers, difficult to differentiate from noise in HI spectra, provide a substantial contribution to the HeII absorption. Typically, in the shelf region, the ratio of HeII to HI ions is of order 20 or more, rising to at least 100 farther away (The cross-section for HeII Lyman alpha absorption is 1/4 that of HI). A dominant feature in PKS 1935-692 and HE 2347-4342 is a void in the HeII absorption near the apparent edge of the HeIII bubble with corresponding voids in the HI spectra.

To interpret the rich datasets we are constructing models which include a realistic inhomogeneous distribution of gas as well as the relevant gas and radiation physics.

We measured density and temperature along lines of sight through a SPH/N-body cosmological simulation (Wadsley & Bond 1998, CDM, $\Omega_b = 0.05$, $h = 0.5$) to use for modelling the radiative transfer of 54.4eV radiation from a newly turned on quasar. Very small systems produce significant HeII absorption features, prohibiting using large, poor resolution simulations. We generated the long line of sight by bouncing a ray inside a 5 Mpc comoving diameter typical, mean density simulation. There is thus no independent long wavelength structure in the spectra.

The quasar flux used was the power-law extension to 228 Å of the observed 304 Å rest frame flux of PKS 1935-692. There is only significant continuum absorption by HeII when it is the dominant form of helium. We track the radiation above 54.4eV (the ionization energy of HeII to HeIII) in 100 frequency bins. This is important because the ionization cross-section for falls off strongly with frequency as ν^{-3}. The radiation field thus becomes harder as it is absorbed moving away from the quasar.

The gas density was fixed at each point and hydrodynamic motions ignored, appropriate because of the rapid onset of ionization compared to hydrodynamical timescales. Non-equilibrium energy and ionization equations are evolved with all the heating, cooling and ionization processes required for a zero metallicity intergalactic medium: ionization heating, cosmological expansion, compton, bremsstrahlung, line cooling, radiative recombination, photoionization and collisional ionization. Shocks are a possible heating source but the time scales are sufficiently short that heating associated with bulk ionization is dominant.

We treated several lines of sight from the simulation and varied the flux history and baryon density. For a given baryon density, the key parameters are the integrated total luminosity from the quasar, determining the bubble size, and the flux level for the last few times 10^6 years (the recombination time) before the observation is taken, determining τ.

A typical model spectrum is shown in Figure 2. The basic features of the observed proximity shelves are straightforward to reproduce. To get substantial recovery ($\tau << 1$) in the spectrum a combination of a high flux and an empty void is required. Voids on the edge of the proximity shelf occur with sufficient frequency in the simulations that the ubiquity in the observations is not surprising. The 10,000-20,000K heating in the medium is greater away from the quasar (lower panel in Figure 2). The recombination rate goes as $\sim T^{-0.7}$ and thus distant voids are made emptier. This heating and ionization extends beyond the visible shelf by around 30% beyond which even harder photons get absorbed.

The time-averaged quasar fluxes could be substantially different from those observed. Quasars are known to vary by a factor of two over a period of years and the response times are order of 10^5 years. Observational bias favours selection of quasars currently at the bright end of their intrinsic variability. There is a trade off so that a higher flux can be used with a corresponding increase in the baryon density. Shelves resembling the real data can be constructed by suitably adjusting the lifetime and recent flux. If quasars are rare density peaks, the mean density nearer quasars is higher which will increase τ near the quasar and improve the model fit. However the density in void regions is seldom less than ~ 0.1 times the

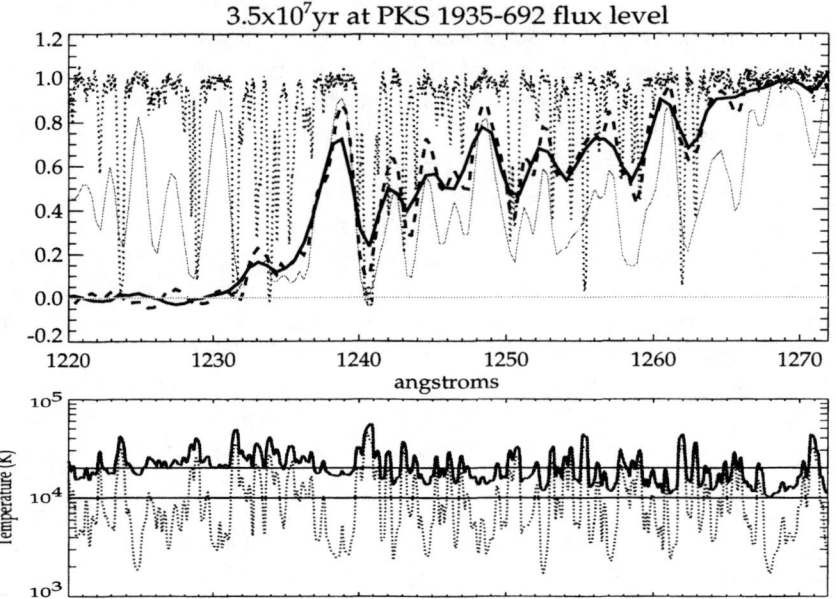

FIGURE 2. Simple model with PKS 1935-692 flux level at 228 Å and 35 Myr lifetime. Simulated spectra: GHRS (dashed), STIS (solid), HI (dotted) and HI optical depth times 25 (thin solid). The shelf and recovery resemble those found in the real data, but the Gunn-Peterson edge is not as pronounced; this can be fixed by allowing for quasar variability. The temperature before and after turn-on is shown in the lower panel; note the rise in temperature away from the quasar, especially the order-of-magnitude increase in the voids. The heated zone extends well beyond the edge of the detectable HeIII bubble.

cosmological mean and thus we do not have the freedom to increase the universal baryon fraction without raising the ionizing flux to compensate, lest the optical depth in the voids become too high. The voids might therefore offer a relatively model-insensitive constraint on mean baryon density.

REFERENCES

1. Anderson, S.F., Hogan, C.J., Williams, B.F. & Carswell, R.F. 1998, accepted AJ
2. Croft, R. A. C., Weinberg, D. H., Katz, N. & Hernquist, L. 1997, ApJ 488, 532
3. Davidsen, A.F., Kriss, G.A. & Zheng, W. 1996, Nature, 380, 47
4. Fardal, M. A., Giroux, M. L., & Shull, J. M. 1998, AJ in press
5. Hogan, C.J., Anderson, S.F. & Rugers, M.H. 1997, AJ, 113, 87
6. Jackobsen, P. *et al.* 1994, Nature, 370, 35
7. Miralda-Escudé, J., Cen, R., Ostriker, J.P. & Rauch, M. 1996, ApJ, 471, 582
8. Reimers, D., Köhler, S., Wisotzki, L., Groote, D., Rodriguez-Pascual, P. Wamsteker, W. 1997, A&A, 327, 890
9. Tytler, D. & Jackobsen, P. 1996, (unpublished)
10. Wadsley, J.W. & Bond, J.R. 1998, in preparation

11. Zhang, Y., Meiksin, A., Anninos, P., & Norman, M. 1998, ApJ 495, 63.
12. Zheng, W. & Davidsen, A. 1995, ApJ 440, L53

Evolution of the 1.4 GHz Radio Luminosity Function

Ian Waddington

Department of Physics & Astronomy, Arizona State University, Tempe AZ 85287–1504, USA, and The Institute for Astronomy, University of Edinburgh, Royal Observatory, Blackford Hill, Edinburgh EH9 3HJ, UK.

Abstract. The results of an optical and infrared investigation of a complete sub-sample of the Leiden-Berkeley Deep Survey are presented. Optical counterparts have been identified for 69 of the 73 sources in the two Hercules fields, and redshifts obtained for 49 of them. Photometric redshifts are computed from the $griK$ data for the remaining 21 sources. This complete sample is compared with the radio luminosity functions (RLFs) of Dunlop and Peacock (1990) [1]. The RLF models successfully trace the evolution of the radio sources with redshift, but there is some disagreement between the luminosity-dependence of the models and the data. The observed RLF for the lower luminosity population ($\log_{10} P < 26$) shows evidence for a cut-off at lower redshifts ($z \sim 0.5$–1.5) than for the more powerful objects.

INTRODUCTION

The purpose of the Leiden–Berkeley Deep Survey (hereafter "the LBDS") was to gain a better understanding of the nature of faint radio galaxies and quasars, and to determine their cosmological evolution. Several high latitude fields in the selected areas SA28, SA57, SA68 and an area in Hercules had been selected for the purpose of faint galaxy and quasar photometry, and a collection of good multi-color prime focus photographic plates had been acquired. Nine of these fields were then surveyed with the Westerbork Synthesis Radio Telescope at 21 cm (1.412 GHz), reaching a 5-σ limiting flux density of 1 mJy [2].

Following this selection of the radio sample, 171 of the radio sources (53%) were identified on the photographic plates, whilst for the Hercules fields there were 47 out of 73 sources identified [3,4]. Presented here are the results of an extensive optical/infrared investigation of the two Hercules fields, with the aim of completing the identification and redshift content of this sub-sample [5]. A cosmology of $H_0 = 50$ km s^{-1} Mpc^{-1}, $\Omega_0 = 1$ and $\Lambda = 0$ is assumed throughout.

CP470, After the Dark Ages: When Galaxies were Young (the Universe at 2 < z < 5),
edited by Stephen S. Holt and Eric P. Smith

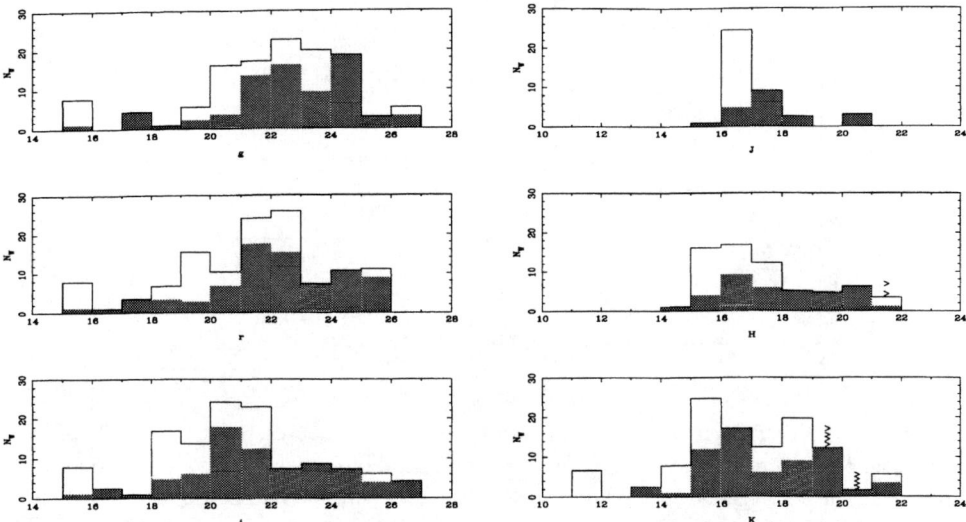

FIGURE 1. Magnitude distributions for the LBDS Hercules sample. Shaded histograms show the sources with $S_{1.4} \geq 2$ mJy. Arrows denote 3-σ upper limits at H and K.

THE DATA

The Hercules field was observed on the 200 inch Hale telescope at Palomar Observatory between 1984 and 1988. Multiple observations were made through Gunn g, r and i filters over the six runs. After processing and stacking of the multiple-epoch images, optical counterparts for 22 of the sources were found, leaving only four sources unidentified to $r \simeq 26$. Near-infrared observations have been made of the entire subsample at K, yielding 60/73 detections down to $K \simeq 19$–21. Half of the sources have been observed in H and approximately one-third in J. Observations of the brighter sources were made by Thuan et al. (1984) [6] and by Neugebauer et al. and Katgert et al. (priv. comm.). K-band observations of the sample were completed by the present authors at UKIRT.

Figure 1 presents the optical and infrared magnitude distributions. For those sources without CCD observations, photographic magnitudes from Kron et al. (1985) [4] have been transformed to the Gunn system [7]. It is seen that the distribution turns over at $r \sim 22$, a consequence of evolution in the redshift and/or luminosity distributions of the radio sources. The the r-band magnitude distribution is essentially unchanged from this milli-Jansky survey down to micro-Jansky surveys, a thousand times fainter in radio flux [8].

Prior to the start of the current work, only 16 of the 73 sources in the LBDS Hercules fields had redshifts published in the literature. Another 16 sources had unpublished redshifts. The author and collaborators have successfully observed a further 17 sources during the past few years, using both the 4.2 m William Herschel

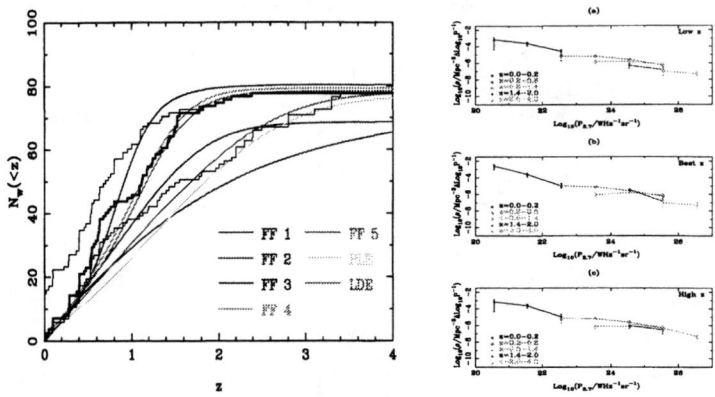

FIGURE 2. [Left] The cumulative redshift distribution of all sources in the 2-mJy Hercules sample. The bold histogram is computed from the best-fit photometric redshift distribution, the lighter histograms correspond to the lower and upper limits to the photometric redshifts. Lines are the model RLFs of [1]. [Right] The observed radio luminosity function for the 2-mJy Hercules sample, for each of the three photometric redshift distributions.

Telescope [5] and the 10 m W. M. Keck Telescope [9–11]. This brings the total number of redshifts to 49 out of 73 sources (67%).

Photometric redshifts were calculated for the remaining one-third of the sample. Using the spectral population synthesis models of Jimenez et al. (1998) [12], synthetic $griJHK$ magnitudes were computed and fitted to the observed magnitudes, giving the most-probable redshift and a measure of its uncertainty. Comparison of the estimated and the true redshifts for those sources with spectroscopic observations, showed that the average difference was ~ 0.1 in z.

THE 1.4 GHZ RADIO LUMINOSITY FUNCTION AND THE REDSHIFT CUT-OFF

Dunlop and Peacock (1990) [1] used a sample of radio sources brighter than 0.1 Jy at 2.7 GHz to investigate the radio luminosity function. They concluded that the comoving density of both flat- and steep-spectrum sources suffers a cut-off at redshifts $z \simeq 2$–4. This conclusion was drawn from the behavior of both free-form and simple parametric models (PLE/LDE), and the model-independent, banded V/V_{max} test. However, the results were crucially dependent upon the accuracy of their redshift estimates in the Parkes Selected Regions (PSR).

With a flux limit $\sim 100\times$ fainter than the PSR, the LBDS is well-suited to test the reliability of those RLF models and the redshift cut-off, via its potential to detect powerful radio galaxies at very high redshifts. In figure 2 [left] the cumulative redshift distribution of the LBDS Hercules sample (only sources with $S_{1.4} \geq 2$ mJy) is compared with the predictions of [1]. It is seen that two of the free-form models (FF-4 and FF-5) provide a reasonable fit to the data over all redshifts. The "bump" in the best-fit histogram at $0.4 \lesssim z \lesssim 1$ is due to two spikes in the redshift distribution, that may be the result of possible large-scale structures (sheets) along the line of sight.

The observed 1.4 GHz luminosity function presented in figure 2 [right] was also compared with the models. It was found that the two models which fit the cumulative counts (FF-4 and FF-5) do not predict the observed *luminosity* dependence of the data nearly as well as the overall redshift dependence. The observed RLF shows some indication that it turns over at $z \simeq 0.5$–1.5, and that the redshift of this cut-off is a function of the radio luminosity. However, the small number of sources makes it difficult to separate the redshift and luminosity dependence of the RLF sufficiently to be certain of this trend.

The full results of this project are presented in [5], and in forthcoming papers by the author and collaborators.

Acknowledgments: Many people have contributed data and knowledge to this project. In particular, I thank James Dunlop, Rogier Windhorst and John Peacock for their assistance. The financial support of the PPARC is acknowledged.

REFERENCES

1. Dunlop, J. S., Peacock, J. A. 1990, MNRAS, 247, 19
2. Windhorst, R. A., van Heerde, G. M., Katgert, P. 1984, Astron. Astrophys. Suppl., 58, 1
3. Windhorst, R. A., Kron, R. G., Koo, D. C. 1984, Astron. Astrophys. Suppl., 58, 39
4. Kron, R. G., Koo, D. C., Windhorst, R. A. 1985, Astron. Astrophys., 146, 38
5. Waddington, I. 1998, PhD Thesis, University of Edinburgh
6. Thuan, T. X., Windhorst, R. A., Puschell, J. J., Isaacman, R. B., Owen, F. N. 1984, ApJ, 285, 515
7. Windhorst, R. A., et al. 1991, ApJ, 380, 362
8. Windhorst, R. A., Waddington, I. 1998, in "The Birth of Galaxies: Proceedings of the Xth Rencontres de Blois", B. Guiderdoni et al. (eds), in press
9. Dunlop, J.S., Peacock, J. A., Spinrad, H., Dey, A., Jimenez, R., Stern, D., Windhorst, R. A. 1996, Nature, 381, 581
10. Spinrad, H., Dey, A., Stern, D., Dunlop, J. S., Peacock, J. A., Jimenez, R., Windhorst, R. A. 1997, ApJ, 484, 581
11. Dey, A. 1997, "The Hubble Space Telescope and the High Redshift Universe", N. Tanvir et al. (eds), World Scientific, p. 373
12. Jimenez, R., Dunlop, J. S., Peacock, J. A., Padoan, P., MacDonald, J., Jørgensen, U. G. 1998, MNRAS, submitted

Spectroscopy of Quasar Candidates from SDSS Commissioning Data

Xiaohui Fan[1], Michael A. Strauss[1], James Annis[4], James E. Gunn[1], Gregory S. Hennessy[2], Zeljko Ivezic[1], Gillian R. Knapp[1], Robert H. Lupton[1], Jeffrey A. Munn[3], Heidi J. Newberg[4], Donald P. Schneider[5], and Brian Yanny[4] for the SDSS Collaboration

(1) Princeton University Observatory, Princeton, NJ 08544
(2) U.S. Naval Observatory, 3450 Massachusetts Ave., NW, Washington, DC 20392
(3) US Naval Observatory, Flagstaff Station, PO Box 1149, Flagstaff, AZ 86002
(4) Fermi National Accelerator Laboratory, PO Box 500, Batavia, IL 60510
(5) Astronomy & Astrophysics, 525 Davey Lab, University Park, PA 16802

Abstract. The Sloan Digital Sky Survey has obtained images in five broad-band colors for several hundred square degrees. We present color-color diagrams for stellar objects, and demonstrate that quasars are easily distinguished from stars by their distinctive colors. Follow-up spectroscopy in less than ten nights of telescope time has yielded 22 new quasars, 9 of them at $z > 3.65$, and one with $z = 4.75$, the second highest-redshift quasar yet known. Roughly 80% of the high-redshift quasar candidates selected by color indeed turn out to be high-redshift quasars.

The Sloan Digital Sky Survey (SDSS; [1,2]) will use a dedicated 2.5m telescope at Apache Point Observatory in Southeast New Mexico to obtain CCD images to $\sim 23^m$ in five bands (u', g', r', i', z'; [3]) over 10,000 square degrees of high Galactic latitude sky. The imaging camera ([4]) contains 30 2048 × 2048 and 24 2048 × 400 CCDs in its focal plane, and takes data at a rate of 20 square degrees an hour in drift-scan mode in all five colors; the data rate is roughly 1 Gbyte per square degree. Specialized software has been written to carry out astrometric and photometric calibration of the data, and to find and measure the properties of all objects detected in the images. The brightest 10^6 galaxies and 1.5×10^5 quasar candidates will be followed up spectroscopically on the same telescope, using a pair of double spectrographs fed by a total of 640 fibers.

The SDSS obtained first light in imaging mode in May 1998, and is now undergoing intensive commissioning. We report here on the distribution of stellar objects in color-color space, the selection of quasar candidates, and follow-up spectroscopy

CP470, After the Dark Ages: When Galaxies were Young (the Universe at 2 < z < 5), edited by Stephen S. Holt and Eric P. Smith

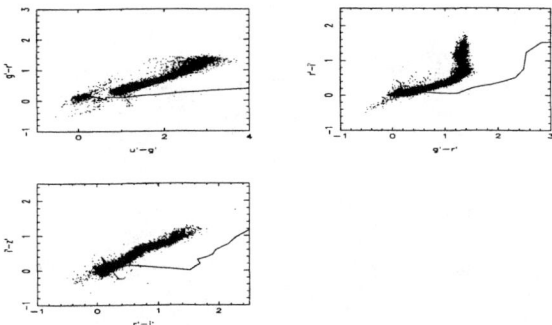

FIGURE 1. Simulated distribution of stellar objects in projections of SDSS color space over 10 square degrees towards the North Galactic Pole, which includes quasars, normal stars, white dwarfs and compact emission line galaxies to $r' = 20$. The solid line is the mean locus of quasars as a function of color.

with the Apache Point 3.5m telescope.

The SDSS will use the colors and morphology of objects to identify quasar candidates from the photometric data: objects with stellar appearance and colors that lie well outside the stellar locus in color space will be flagged for spectroscopic investigation.

The distinction between quasars and stars in color-color space is illustrated in Figure 1, which shows model distributions of stars and quasars in a series of three SDSS color-color diagrams, from the simulations of ref. [5]. These simulations put in realistic SED's for stars, quasars, and compact emission-line galaxies, and attempt to model the stellar populations and spatial distributions of stars for the North Galactic Pole. The mean locus of quasars as a function of redshift is shown as the solid line; for $z < 2.5$, quasars are very blue in $u' - g'$, and can be distinguished quite easily from stars (and hot white dwarfs as well, which tend to be bluer in $g' - r'$; see the discussion in ref. [5]). At higher redshifts, the Lyman forest, and eventually, Lyman-limit systems, move through the SDSS filters, causing the colors to become redder. Note that at most redshifts, the quasar locus is well-separated from the stellar locus; the pernicious exception is quasars at $z \approx 2.8$, which have very similar broad-band colors to an F star. The reddest bands will permit identification of quasars with redshifts higher than six.

Figure 2 shows the color-color diagram of stellar objects with $r^* < 20$ from 20 square degrees of SDSS imaging commissioning data taken in September 1998[1]. Notice the qualitative similarity to the simulations in Figure 1, and the narrowness of the distribution: this is a tribute both to the quality of the data, and the pipeline used to reduce it. As the SDSS spectrographs have not been commissioned as of this

---

[1] The asterisk * indicates that the final SDSS photometric system has not yet been defined; this is preliminary photometry, accurate to perhaps 0.05 mag.

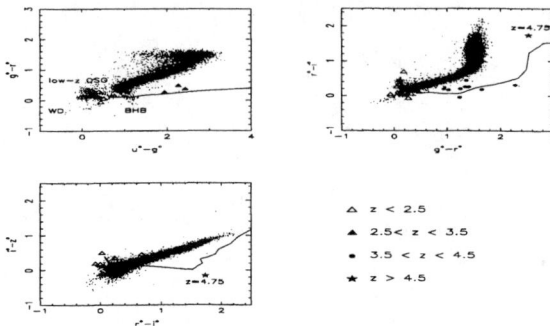

FIGURE 2. Observed color-color diagrams of 20 square degrees from the SDSS test data ($r^* < 20$). The positions of 22 newly discovered quasars (selected from 130 square degrees) are indicated. Already known quasars are not indicated in this figure.

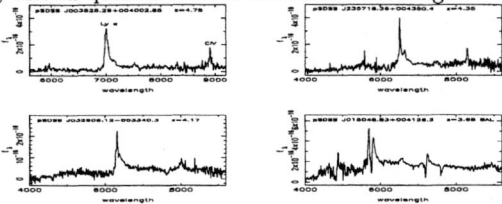

FIGURE 3. Spectra of 3 new SDSS quasars with $z > 4$, plus one with a broad absorption-line spectrum, obtained with the 3.5m ARC telescope and Double Imaging Spectrograph.

writing, we are using the Double Imaging Spectrograph on the Apache Point 3.5m telescope to carry out spectroscopy of promising high-redshift quasar candidates. Superposed on Figure 2 are the places in color-color space where the 22 new quasars we have identified thus far lie, based on roughly 130 square degrees of imaging data.

These quasars do not by any means constitute a complete sample. In the last two nights of spectroscopic data, we have concentrated on those objects which appeared from their broad-band colors to be high-redshift candidates. Out of 11 candidates, 9 are indeed quasars at $z > 3.65$ (the two high-redshift quasars previously known in the survey area also stood out cleanly in the color-color diagrams, and would have been selected as well). All are brighter than $i^* = 20$. This success rate far surpasses the typical 10% found in the literature for high-redshift quasar surveys [6–8], although again, we do not have a complete sample to make this quantitative.

Figure 3 shows our spectra of the three highest-redshift quasars we have found thus far, plus one which shows strong associated absorption. The one at $z = 4.75$ is the second-highest redshift quasar known (the current redshift holder is $z = 4.90$; see ref. [9]). These spectra are of quite low resolution, roughly 7Å pixel^{-1}, while the SDSS spectrographs will deliver 1-1.5Å pixel^{-1} over a similar wavelength coverage.

These objects were selected from roughly 1% of the sky that the SDSS will image. We therefore expect that there are enormously more high-redshift quasars to be discovered as part of the SDSS.

The Sloan Digital Sky Survey (SDSS) is a joint project of the University of Chicago, Fermilab, the Institute for Advanced Study, the Japan Participation Group, The Johns Hopkins University, Princeton University, the United States Naval Observatory, and the University of Washington. Apache Point Observatory, site of the SDSS, is operated by the Astrophysical Research Consortium. Funding for the project has been provided by the Alfred P. Sloan Foundation, the SDSS member institutions, the National Science Foundation, NASA, and the U.S. Department of Energy. XF and MAS acknowledge additional support from Research Corporation, NSF grant AST96-16901, the Princeton University Research Board, and an Advisory Council Scholarship. DPS acknowledges support from NSF grant AST95-09919.

REFERENCES

1. Gunn, J. E., & Weinberg, D. H. 1995, in *Wide-Field Spectroscopy and the Distant Universe*, ed. Maddox and Aragón-Salamanca (Singapore: World Scientific), 3
2. SDSS Collaboration, 1996, http://www.astro.princeton.edu/BBOOK/.
3. Fukugita, M., Ichikawa, T., Gunn, J.E., Doi, M., Shimasaku, K., & Schneider, D.P. 1996, AJ, 111, 1748
4. Gunn, J.E., Carr, M.A., Rockosi, C.M., Sekiguchi, M. *et al.* 1998, AJ, in press
5. Fan, X. 1998, AJ, submitted
6. Schneider, D. P., Schmidt, M., & Gunn, J.E. 1994, AJ, 107, 1245
7. Hall, P.B., Osmer, P.S., Green, R.F., Porter, A.C., & Warren, S.J. 1996, AJ, 462, 614
8. Kennefick, J.D. *et al.* 1995, AJ, 110, 78
9. Schneider, D. P., Schmidt, M., & Gunn, J.E. 1991, AJ, 102, 837

Star Formation and Dust Evolution in High Redshift Radio Galaxies

David S. De Young

Kitt Peak National Observatory/National Optical Astronomy Observatories, P.O. Box 26732, Tucson, AZ 85719

Abstract.
The alignment of optical emission along the axis of radio emission in high redshift radio galaxies is now commonly observed. In many cases this aligned component can be shown to be significantly polarized, a result that has led to models for the emission that employ scattering of light emitted from the AGN. Both electron scattering and dust scattering have been proposed, but one aspect of dust scattering that has not been explored previously is the response of the dust grains to the passage of the strong shock associated with the radio source. The survival of dust grains in such an environment after the passage of a high speed shock associated with the radio jet is calculated for a wide range of parameters. It is found that for most configurations the grains are destroyed as a scattering population by sputtering processes in a time much less than the minimum radio source lifetime of $\sim$ 10 million years. Thus polarization due to scattering by an $in-situ$ population of grains is somewhat problematic. Alternate methods for providing the needed grain population, either by grain replenishment via outward convection from a dust-rich galactic interior or by local production from a population of stars formed by passage of the radio jet are described.

CP470, After the Dark Ages: When Galaxies were Young (the Universe at 2 < z < 5),
edited by Stephen S. Holt and Eric P. Smith

Massive X-ray binaries and the X-ray Background

Priyamvada Natarajan

CITA, McLennan Labs., 60 St. George Street, Toronto M5S 1A7, Canada

Abstract. In this calculation we estimate the contribution of massive X-ray binaries to the hard X-ray background (XRB). Traditionally, due to their short life-times, MXRBs were thought not to be significant contributors to the XRB. However, given the recent progress in the observational determination of the global star-formation rate as a function of redshift probed both via emitted blue-light and that re-radiated in the sub-mm, assuming that the formation rate of MXRBs tracks the global star formation rate, their integrated contribution to the XRB can be computed and is shown to be non-negligible. While the origin of the bulk of the hard XRB can be attributed to obscured AGNs, we show that MXRBs can account for 10-30%.

INTRODUCTION

Obscured AGN are now considered to be the most likely explanation for the origin of the X-ray background. In a recent ultra-deep ROSAT survey (Hasinger et al. 1998), for example, with a 1 mega-second ROSAT HRI observation and follow-up spectroscopy with Keck, Schmidt et al. (1998) have identified X-ray sources deeper than ever before. All the sources found were AGN, with a mixture of obscured and unobscured types. In ASCA surveys in a harder band than ROSAT (2-10 keV) with deep pointed observations various groups have been able to resolve about 50% of the 2-10keV XRB into discrete sources. Identifications reveal them to be yet again, a mixture of ordinary and obscured (i.e. narrow-line) AGN (Sakano et al. 1998; Almaini 1998). We probably have to concede that obscured AGN can almost certainly explain the bulk of this background radiation (Hasinger 1997; Fabian & Iwasawa 1998), but emphasize that stellar processes could still be important, albeit at no more than the 10-30% level. A lot of these heavily obscured AGN which are emerging show very clear evidence for the existence of starburst activity in addition to the AGN, that is, optical line ratios which suggest copious star formation activity, meanwhile in the infra-red and hard X-rays they look like AGN (see also Heckman et al. 1997; Boyle et al. 1998). A nuclear starburst could itself kick up gas and dust to obscure the central active nucleus, rather than the standard Unified Scheme

CP470, After the Dark Ages: When Galaxies were Young (the Universe at 2 < z < 5),
edited by Stephen S. Holt and Eric P. Smith
© 1999 The American Institute of Physics 1-56396-855-X/99/$15.00

(e.g. Fabian et al. 1998), therefore star formation activity could indeed cause the obscuration.

THE X-RAY BACKGROUND

The energy density of the X-ray sky is dominated by a diffuse component which is extra-galactic in origin. While some fraction of the XRB must originate from known sources, no known source contributes more than 50% (for a comprehensive review see Fabian & Barcons 1992). Due to the resolution limits of current X-ray detectors this component need not be truly diffuse and could in principle be resolvable into individual sources. The isotropy of the XRB at energies $E > 3$ keV suggests an extra-galactic origin. There is a dipole component detected that is somewhat aligned with the CMB dipole (see Lahav et al. 1993) that arises due to the Compton-Getting effect from the motion of our Galaxy. This provides further evidence for the extra-galactic origin of the XRB. The contribution to the XRB can be decomposed into 2 energy bands: (i) $kT > 3$ keV (hard) and (ii) $kT < 3$ keV (soft). In the soft band less than 10% of the contribution is from extra-galactic sources, between 0.1 - 0.5 keV most of the background is Galactic in origin and is due to the local hot bubble at a temperature $T \sim 10^6$ K. The bulk of the energy in the XRB resides in the 20 - 40 keV range. Inter-stellar absorption by dust in our Galaxy obscures extra-galactic components. In the region from 3 - 300 keV, the background is best measured by the HEAO-1 A2 experiment.

The contribution from interstellar absorption and the contribution from our Galaxy are negligible in the 0.5 - 3 keV range, but extrapolated to higher energies, the fit to this component is softer but steep spectrum sources like AGN, clusters point to an excess here. The XRB intensity at 1 keV is roughly two times the value estimated by extrapolating the power-laws that are valid at higher energies. In summary, the fits to the XRB spectrum have 3 components: (i) a steep power law below 3 keV, (ii) an exponential of 40 keV and (iii) a broken power-law at higher energies. Most of the energy density is contained within 20 - 40 keV, most data exist for the 3 - 10 keV range which contains about 20% of the total energy. Once the contribution of our Galaxy has been removed the XRB shows large-scale anisotropy. The highest flux contributors are clusters and AGN. The X-ray evolution of clusters is such that the most luminous ones are at $z = 0$ and hence they do not contribute significantly to the total intensity.

The observed spectral shape and the number density of known AGN cannot account for the 3 - 300 keV XRB. Proposed solutions to the above spectral paradox problem are mechanisms for hardening the typical AGN spectrum (i) the role of reflection-dominated spectra, wherein slabs of cold gas in front of the central AGN hardens the spectrum as required, but there are problems with this model beyond the obvious fact that photo-electric absorption causes electrons to be reflected to lower energies and (ii) a population of galaxies undergoing advection-dominated accretion that produce intrinsically hard spectra (Di Matteo et al. 1998) or appealing to the existence of a large population of as yet undetected AGN. From the fluctuation analysis, the number density of the hitherto unknown classes of sources required to explain the XRB has to exceed that of AGNs but their individual luminosities must be less than that of a typical AGN, which strengthens the case for these low luminosity AGN accreting in the advective mode.

There are 2 primary extra-galactic sources of hard X-rays besides AGN that could account for the origin of the XRB: (i) Compton scattering in high-redshift radio sources - inverse Compton scattering of CMB photons by relativistic electrons (Felten & Morrison 1966) would be efficient at high redshift. To generate X-rays in the 1-10 keV energy range, electrons with Lorentz factors of a few thousand are required, therefore, radio sources with electrons in this energy range would be the optimal candidates, and (ii) young, star-forming galaxies, specially if high-z galaxies pass through a phase when their star-formation rates are significantly higher than the present day rate (Bookbinder et al. 1980; Griffiths & Padovani 1992; Treyer, Mouchet, Blanchard & Silk 1992) which is definitely the case for the sources detected in the sub-mm by SCUBA (inferred SFRs are of the order of a few hundred solar masses per year) as well for galaxies detected at $z \sim 3$ by the Lyman-break technique (Steidel et al. 1996).

CONTRIBUTION OF MASSIVE X-RAY BINARIES

Here, we compute the contribution of star-burst galaxies powered by the X-ray emission from massive binaries to the hard XRB. MXRBs are produced copiously in star-bursts and are composed of a compact object (either a neutron star or a black hole) accreting matter from a close O - B (Be) companion star. Hard X-ray emission E $\sim$ 20 keV results from the mass transfer process. About 50 MXRBs have been detected in the Milky Way and they have thermal spectra with $T > 15$ keV and $L \sim 10^{38} ergs^{-1}$. They are very short-lived, $t_{MXRB} \sim 10^{5-6} yr$, much shorter than LMXBs which last upto a Gyr. In low metallicity environments the X-ray emission from O stars is enhanced, therefore the contribution from star-bursts at high redshift (and low metallicity) might be more significant.

Spectra and contribution to the XRB

We argue here that for MXRBs in high-redshift star-burst galaxies, L_X/L_{opt} can be much larger than for local galaxies. We start with the ansatz that the total MXRB density is proportional to the total star-formation rate density as a function of redshift (Madau 1997), $\rho_{MXRB}(z) \propto \rho_{SFR}(z)$. The above proportionality involves several unknown factors: f_1 the fraction of energy emitted in the hard X-ray band given a fiducial spectral energy distribution (SED) for MXRBs, f_2 the fraction of mass that ends up in massive stars for every solar mass of stars formed with a specified IMF and finally, f_3 the fraction of massive stars that end up in MXRBs.

These uncertain factors are estimated as follows: (i) f_1: a typical SED for an MXRB can be fit by,

$$SED(E) \propto E^a \ E < E_c; \ SED(E) \propto E^b exp\left(\frac{E_c - E}{E_f}\right) \ E > E_c \qquad (1)$$

where $E_c = 19.28$ keV, $a = -0.217$ and $b = -2.67$ normalized at $E = 1$ keV such that $SED(E) = 1 \, kev^{-1}$. With the above SED, we find that approximately 60% of the energy is in the hard-band at $E > 19.28$ keV; (ii) f_2 - the fraction of mass that ends up in massive stars, for a Salpeter IMF roughly 0.035 $M_\odot$ ends up in massive stars for every solar mass of material processed; (iii) f_3 - in order to compute the fraction of massive stars that further end up in the appropriate binary, we use the results of stellar population synthesis models of Portegies Zwart & Verbunt (1996) that are calibrated to observations. In these models, 40-70% of massive stars end up as MXRBs. Employing the above estimates, we find that roughly 10-30% of the hard XRB can be accounted for by MXRBs (Natarajan 1998).

CONCLUSIONS

It has been shown convincingly that obscured AGN can probably account for the bulk of the total intensity (50-70%) of the XRB in the hard band. Since, many of the detected obscured AGN show evidence for star-burst activity, here we argue that X-ray emission from binaries in young star-forming galaxies could be important. Assuming that the MXRB rate is proportional to the star-formation rate, we estimate that MXRBs could contribute at the 10-30% level to the hard XRB.

REFERENCES

1. Almaini, O., 1998, Astronomische Nachrichten, vol. 319, no. 1, p. 55.
2. Bookbinder et al. 1980, ApJ, 237, 647.
3. Boyle, B., et al., 1998, MNRAS, 297, L53.
4. Fabian, A., Barcons, X., Almaini, O., & Iwasawa, K., 1998, MNRAS 297, L11.
5. Fabian, A., & Barcons, X., 1992, ARA&A, 30, 429.
6. Fabian, A., & Iwasawa, K., 1998, preprint.
7. Griffiths, R., & Padovani, P., 1990, ApJ, 360, 48.
8. Hasinger, G., et al., 1998, A&A, 329, 482.
9. Heckman, T., et al., 1997, ApJ 482, 114.
10. Lahav, O., et al., 1993, Nature, 364, 693.
11. Madau, P., 1997, IAU Symp. 186, 188.
12. Natarajan, P., 1998, ApJ Lett, submitted.
13. Portegies Zwart, S., & Verbunt, F., 1996, A&A, 309, 179.
14. Sakano, M., et al., 1998, ApJ, 505, 129.
15. Schmidt, M., et al., 1998, A&A, 329, 495.

16. Steidel, C., 1996, ApJ Lett, 462, 17.
17. Treyer, M., Mouchet, M., Blanchard, A., & Silk, J., 1992, A&A, 264, 11.

A Search for Emission-line Features in the X-ray Spectra of Quasars with $1 < z < 5$

Tahir Yaqoob and Saima Zobair

NASA/GSFC, Laboratory for High Energy Astrophysics, Greenbelt, MD 20771, USA.

Abstract. Following the detection of a puzzling emission-line feature in the X-ray spectrum of the $z = 0.654$ high-luminosity quasar PKS 0637−752 (centered at 1.6 keV in the quasar frame) we examined the spectra of all *ASCA* quasars (public as of 1998 July 7) with $1 < z < 5$ to search for any new, unexpected emission-line features. Out of 35 sources, we found only one source (PKS 2149−302, $z = 2.345$) with a significant emission-line feature, centered at ~ 17 keV in the quasar frame. The most likely explanation is blueshifted Fe-K emission (the EW is $\sim 300 \pm 200$ eV). Curiously, if the feature in PKS 0637−752 is due to blueshifted Oxygen VII emission the Doppler factor in both quasars is similar ($\sim 2.7 - 2.8$), implying high velocities of order $\sim 0.75c$.

INTRODUCTION

Recently, a puzzling emission-line feature was discovered in an *ASCA* X-ray spectrum of the $z = 0.654$, high luminosity ($L[2 - 10 \text{ keV}] \sim 10^{46}$ erg s^{-1}) radio-loud quasar PKS 0637−752 (Yaqoob *et al.* 1998). Modeled with a Gaussian, the emission line is centered at an energy 1.60 ± 0.07 keV, and has an equivalent width (EW) 59^{+38}_{-34} eV if it is narrow (both in the quasar frame). If the line-width is allowed to float we get a Gaussian width of $0.07^{+0.24}_{-0.07}$ keV, the same center energy and a slightly larger EW. The line was detected with a high statistical significance and it was demonstrated in Yaqoob *et al.* (1998) that the emission-line feature in PKS 0637−752 is highly unlikely to be due to an instrumental artifact.

The line is puzzling for two principal reasons. One is that even with conservative assumptions about systematic errors in the SIS energy scale, there is no obvious identification for the line. Secondly, even if we attribute the line to redshifted Si emission from infalling matter, an anomalous Si abundance relative to O, Ne, S and Fe is required because no other emission lines are detected. In particular, a strong Fe-K emission line would be expected but instead a 90% confidence upper limit on the EW of 80 eV was obtained (see Yaqoob *et al.* 1998).

CP470, After the Dark Ages: When Galaxies were Young (the Universe at 2 < z < 5),
edited by Stephen S. Holt and Eric P. Smith

It is important to try and understand the origin of the peculiar X-ray emission line in PKS 0637−752, especially in the context of X-ray spectroscopy of active galaxies and quasars (hereafter both referred to as AGNs) in general. The picture emerging from many X-ray studies over the last couple of decades of AGNs is that emission features in the X-ray spectra present in low-luminosity objects (Seyfert 1 galaxies) become scarce in AGNs with 2–10 keV intrinsic luminosity exceeding $\sim 10^{45}$ erg s^{-1} (synonymously, for our purpose, quasars). These trends are discussed at length in Nandra et al. (1997a) and Reeves et al. (1997; and references therein). The transition to a featureless X-ray power-law continuum (except for possible line-of-sight absorption at low energies) in the high luminosity AGNs, especially radio loud sources, is not fully understood but may be related to the complete ionization of matter responsible for emission-line features and/or beaming of the X-ray continuum swamping out emission-line features. Fiore et al. (1998) recently reported a feature at ~ 0.95 keV in the rest-frame of PG 1244+026 which could either be described as a broad emission line or an absorption edge. Also, Turner et al. (1998) reported a similar feature at ~ 1 keV in the rest-frame of Ton S 180. However, both of these are low-luminosity objects and the reported features are not similar to the one in PKS 0637−752. Thus, PKS 0637−752 is again unusual because it is anomalous as a high-luminosity quasar since no X-ray emission-line features are expected at all.

It is natural to wonder whether other quasars have unexpected emission-line features which may have previously gone undetected. This is quite plausible because with X-ray data in which features cannot be picked out 'by eye' (especially with over-sampled spectra), people generally only look for spectral features at expected energies.

In this paper we present the results of a systematic search in all quasars observed by *ASCA* in the redshift range $1 < z < 5$ for which the data were public as of 1998 July 7. All quasar spectra were fitted with a single power-law plus absorber model using data from all four *ASCA* instruments simultaneously. The absorber column density was placed at $z = 0$ and allowed to float. Next a narrow Gaussian emission line was added to the model, whose center energy was fixed at 100 values between 0.6 and 8 keV in linear steps (above 8 keV the signal-to-noise is too low and the background-subtraction uncertainty too great). This resulted in a set of $\Delta \chi^2$ values against energy for each object and would indicate the possible presence of any spectral line-like features.

RESULTS

Details of the data analysis follow closely the procedures described in Yaqoob et al. (1998). Our search resulted in only one source (out of a total of 35) with statistically significant line-like residuals, PKS 2149−306 ($z = 2.345$). Fig. 1 shows the ratio of SIS0 and SIS1 data to the best-fitting power-law plus absorber model. The feature is detected in both SIS0 and SIS1. When modeled with a narrow

Gaussian (σ fixed at 0.01 keV) we obtain a reduction in χ^2 of 10.3. For two free line parameters (intensity and center energy), this formally corresponds to a significance of $> 99\%$. Table 1 shows the complete spectral fitting results and other details of the observation. We note that instrumental artifacts are an unlikely explanation since the spectral feature is not detected in the other 34 sources in the sample, some of which are brighter than PKS 2149−306. The PKS 2149−306 observation occurred fairly early on in the *ASCA* mission (see Table 1) for which the SIS is best calibrated. The local background is over an order of magnitude below the on-source signal so cannot be a contributing factor. We also repeated the spectral fitting using released blank-sky background (from a different region of sky and a different time) and confirmed that the emission-line feature was unaffected. Note that the source is further off-axis in GIS2 and GIS3, so the signal-to-noise is less and the energy resolution much poorer than the SIS so the line feature is not detected in the GIS.

DISCUSSION AND CONCLUSIONS

The most likely explanation for the detection of an emission line centered at ~ 17 keV in the rest frame of the radio-loud quasar PKS 2149−306 ($z = 2.345$) is blueshifted Fe-K emission. We also note that the quasar-frame equivalent width of the line-feature is typical of the equivalent width of the Fe-K line in Seyfert

TABLE 1. Spectral Fitting Results

Observation Date	26–27 Oct 1994
Count Rates (ct/s) [a]	0.17–0.30
Exposure times (ks) [a]	19.4–19.9
N_H /Galactic (10^{20} cm^{-2})	2.12
N_H (10^{20} cm^{-2})	$7.5^{+2.4}_{-2.3}$
Γ	$1.556^{+0.053}_{-0.052}$
E_{line} [b] (keV)	$17.00^{+0.52}_{-0.49}$
I_{line} [b]	$4.5^{+3.1}_{-3.1}$
EW [b] (eV)	298^{+202}_{-205}
χ^2	574.1
egrees of freedom	575
$\Delta\chi^{2a}$	10.3
Flux [d] (0.5–2 keV)	3.6
Flux [d] (2–10 keV)	10.1
Luminosity [b] (2–10 keV)	5.8

[a] Difference in χ^2 resulting from adding a narrow Gaussian to a simple power law plus absorber model.

[b] *Intrinsic* luminosity (2–10 keV in the quasar frame) in units of 10^{47} ergs s^{-1} ($H_0 = 50$ km s Mpc^{-1}, and $q_0 = 0$)

[a] Range of values amongst the four *ASCA* instruments

[b] Line intensity in units of 10^{-5} photons s^{-1} cm^{-2}

[d] *Observed* flux in units of 10^{-12} ergs cm^{-2} s^{-1}

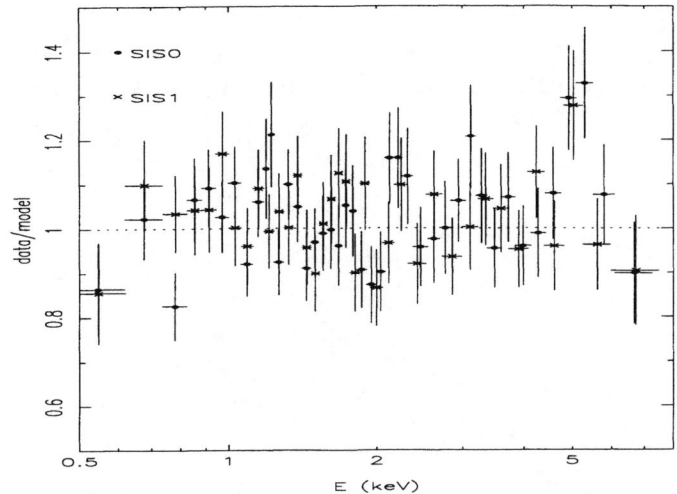

FIGURE 1. The ratio of data to best-fitting power-law plus absorber model for PKS 2149−302.

galaxies (e.g. see Nandra *et al.* 1997b). However, even this explanation has some difficulty in that it implies enormous velocities of $\sim 0.75c$. On the other hand, there is no physical reason against such high velocities (especially since radio-loud quasars are thought to be beamed). Yet another curious and remarkable fact is that if the strange emission-line feature discovered at 1.6 keV in the rest frame of PKS 0637−752 is due to blueshifted Oxygen VII (intrinsic energy 0.57 keV), then the blueshifting factor is similar to the blueshifting factor in PKS 2149−306 if the emission line in PKS 2149−306 is indeed due to Fe-K at an intrinsic energy ~ 6.4 keV (i.e. $1.6/0.57 = 2.8 \sim 17/6.4 = 2.7$). In PKS 0637−752 the Fe-K line would be blueshifted out of the *ASCA* bandpass (to 11.2 keV) and therefore rendered undetectable. In PKS 2149−306 the OVII line energy would be lower than the low-end of the *ASCA* bandpass.

There may be some new physical quasar phenomenon underlying these results and there are many questions to be answered. However, it is important that the results are first confirmed with further, higher signal-to-noise X-ray observations (including the sources which were too weak for detailed spectral analysis), with *ASCA* and/or future X-ray astronomy missions.

The authors would like to thank Ian George, Jane Turner, Paul Nandra and Andy Ptak for their contributions to this work.

REFERENCES

1. Fiore, F., *et al.* 1998, MNRAS, 298, 103
2. Nandra, K., *et al.* 1997a, ApJ, 488, L91

3. Nandra, K., *et al.* 1997b, ApJ, 477, 602.

4. Reeves, J.N., Turner, M.J.L., Ohashi, T., & Kii, T. 1997, MNRAS, 292, 468

5. Turner, T.J., George, I.M., Nandra, K. 1998, ApJ, in press, astro-ph/9806393

6. Yaqoob, T., *et al.* 1998, ApJ, 505, L87

8. Star Formation History

Faint Galaxies, Extragalactic Background Light, and the Reionization of the Universe

Piero Madau

Space Telescope Science Institute, Baltimore, MD 21218

Abstract.
I review recent observational and theoretical progress in our understanding of the cosmic evolution of luminous sources. Largely due to a combination of deep *HST* imaging, Keck spectroscopy, and *COBE* far-IR background measurements, new constraints have emerged on the emission history of the galaxy population as a whole. Barring large systematic effects, the global ultraviolet, optical, near- and far-IR photometric properties of galaxies as a function of cosmic time cannot be reproduced by a simple stellar evolution model defined by a constant (comoving) star-formation density and a universal (Salpeter) IMF, and require instead a substantial increase in the stellar birthrate with lookback time. While the bulk of the stars present today appears to have formed relatively recently, the existence of a decline in the star-formation density above $z \approx 2$ remains uncertain. The history of the transition from the cosmic 'dark age' to a ionized universe populated with luminous sources can constrain the star formation activity at high redshifts. If stellar sources are responsible for photoionizing the intergalactic medium at $z \approx 5$, the rate of star formation at this epoch must be comparable or greater than the one inferred from optical observations of galaxies at $z \approx 3$. A population of dusty, Type II AGNs at $z \lesssim 2$ could make a significant contribution to the FIR background if the accretion efficiency is $\sim 10\%$.

INTRODUCTION

There is little doubt that the last few years have been exciting times in galaxy formation and evolution studies. The remarkable progress in our understanding of faint galaxy data made possible by the combination of HST deep imaging [6] and ground-based spectroscopy [35], [14], [56] has permitted to shed new light on the evolution of the stellar birthrate in the universe, to identify the epoch $1 \lesssim z \lesssim 2$ where most of the optical extragalactic background light was produced, and to set important constraints on galaxy evolution scenarios [7], [57], [3], [23]. The explosion in the quantity of information available on the high-redshift universe at

CP470, After the Dark Ages: When Galaxies were Young (the Universe at 2 < z < 5),
edited by Stephen S. Holt and Eric P. Smith

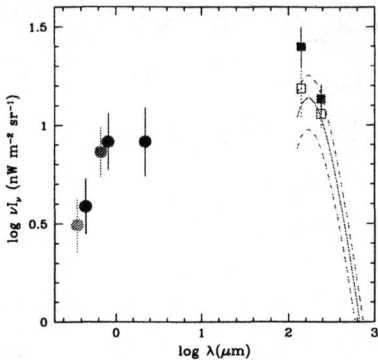

FIGURE 1. Spectrum of the extragalactic background light as derived from a compilation of ground-based and space-based galaxy counts in the U, B, V, I, and K-bands (*filled dots*), together with the FIRAS 125–5000 μm (*solid and dashed lines*) and DIRBE 140 and 240 μm (*filled squares*) detections. The *empty squares* show the DIRBE points after correction for WIM dust emission [32].

optical wavelengths has been complemented by the detection of the far-IR/sub-mm background by DIRBE and FIRAS [28], [5]. The IR data have revealed the optically 'hidden' side of galaxy formation, and shown that a significant fraction of the energy released by stellar nucleosynthesis is re-emitted as thermal radiation by dust. The underlying goal of all these efforts is to understand the growth of cosmic structures and the mechanisms that shaped the Hubble sequence, and ultimately to map the transition from the cosmic 'dark age' to a ionized universe populated with luminous sources. While one of the important questions recently emerged is the nature (starbursts or AGNs?) and redshift distribution of the ultraluminous sub-mm sources discovered by *SCUBA* [7], [5], [36], of perhaps equal interest is the possible existence of a large population of faint galaxies still undetected at high-z, as the color-selected ground-based and *Hubble Deep Field* (HDF) samples include only the brightest and bluest star-forming objects. In hierarchical clustering cosmogonies, high-z dwarfs and/or mini-quasars (i.e. an early generation of stars and accreting black holes in dark matter halos with circular velocities $v_c \sim 50$ km s^{-1}) may actually be one of the main source of UV photons and heavy elements at early epochs [44], [25], [26].

In this talk I will focus on some of the open issues and controversies surrounding our present understanding of the history of the conversion of cold gas into stars within galaxies, and of the evolution with cosmic time of luminous sources in the universe. An Einstein-deSitter (EdS) universe ($\Omega_M = 1$, $\Omega_\Lambda = 0$) with $h = H_0/100$ km s^{-1} Mpc$^{-1} = 0.5$ will be adopted in the following.

OPTICAL/FIR BACKGROUND

The extragalactic background light (EBL) is an indicator of the total luminosity of the universe. It provides unique information on the evolution of cosmic structures at all epochs, as the cumulative emission from galactic systems and AGNs is expected to be recorded in this background. Figure 1 shows the optical EBL from known galaxies together with the recent *COBE* results. The value derived by integrating the galaxy counts [49] down to very faint magnitude levels [because of the flattening at faint magnitudes of the $N(m)$ differential counts most of the contribution to the optical EBL comes from relatively bright galaxies] implies a lower limit to the EBL intensity in the 0.3–2.2 μm interval of $I_{\rm opt} \approx 12$ nW m^{-2} sr^{-1}.[1] When combined with the FIRAS and DIRBE measurements ($I_{\rm FIR} \approx 16$ nW m^{-2} sr^{-1} in the 125–5000 μm range), this gives an observed EBL intensity in excess of 28 nW m^{-2} sr^{-1}. The correction factor needed to account for the residual emission in the 2.2 to 125 μm region is probably $\lesssim 2$ [4]. We shall see below how a population of dusty AGNs could make a significant contribution to the FIR background. In the rest of this talk I will adopt a conservative reference value for the total EBL intensity associated with star formation activity over the entire history of the universe of $I_{\rm EBL} = 40\, I_{40}$ nW m^{-2} sr^{-1}.

COSMIC STAR FORMATION

It has become familiar to interpret recent observations of high-redshift sources via the comoving volume-averaged history of star formation. This is the mean over cosmic time of the stochastic, possibly short-lived star formation episodes of individual galaxies, and follows a relatively simple dependence on redshift. Its latest version, uncorrected for dust extinction, is plotted in Figure 2 (*left*). The measurements are based upon the rest-frame UV luminosity function (at 1500 and 2800 Å), assumed to be from young stellar populations [1]. The prescription for a 'correct' de-reddening of these values has been the subject of an ongoing debate. Dust may play a role in obscuring the UV continuum of Canada-France Reshift Survey (CFRS, $0.3 < z < 1$) and Lyman-break ($z \approx 3$) galaxies, as their colors are too red to be fitted with an evolving stellar population and a Salpeter initial mass function (IMF) [7]. The fiducial model of [7] had an upward correction factor of 1.4 at 2800 Å, and 2.1 at 1500 Å. Much larger corrections have been argued for by [50] ($\times 10$ at $z = 1$), [43] ($\times 15$ at $z = 3$), and [51] ($\times 16$ at $z > 2$). As noted already by [1] and [7], a consequence of such large extinction values is the possible overproduction of metals and red light at low redshifts. Most recently, the evidence for more moderate extinction corrections has included measurements of star-formation rates (SFR) from Balmer lines by [59] ($\times 2$ at $z = 0.2$), [21]

---

[1] Note that the direct detection of the optical EBL at 3000, 5500, and 8000 Å derived from *HST* data by [3] implies values that are about a factor of two higher than the integrated light from galaxy counts.

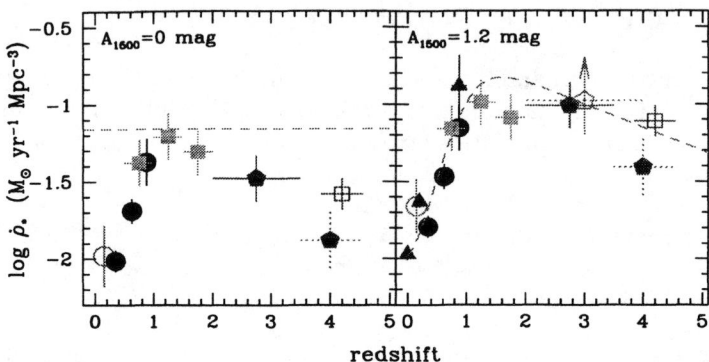

FIGURE 2. *Left*: Mean comoving density of star formation as a function of cosmic time. The data points with error bars have been inferred from the UV-continuum luminosity densities of [35] (*filled dots*), [10] (*filled squares*), [40] (*filled pentagons*), [59] (*empty dot*), and [57] (*empty square*). The *dotted line* shows the fiducial rate, $\langle \dot{\rho}_* \rangle = 0.054 \ M_\odot \ yr^{-1} \ Mpc^{-3}$, required to generate the observed EBL. *Right*: dust corrected values ($A_{1500} = 1.2$ mag, SMC-type dust in a foreground screen). The Hα determinations of [18], [58], and [20] (*filled triangles*), together with the SCUBA lower limit [29] (*empty pentagon*) have been added for comparison.

($\times 3.1 \pm 0.4$ at $z = 1$), and [48] ($\times 2 - 6$ at $z = 3$). *ISO* follow-up of CFRS fields [25] has shown that the star-formation density derived by FIR fluxes ($\times 2.3 \pm 0.7$ at $0 \leq z \leq 1$) is about 3.5 times lower than in [50]. Figure 2 (*right*) depicts an extinction-corrected (with $A_{1500} = 1.2$ mag, 0.4 mag higher than in [7]) version of the same plot. The best-fit cosmic star formation history (shown by the *dashed-line*) with such a universal correction produces a total EBL of 37 nW m^{-2} sr^{-1}. About 65% of this is radiated in the UV+optical+near-IR between 0.1 and 5 μm; the total amount of starlight that is absorbed by dust and reprocessed in the far-IR is 13 nW m^{-2} sr^{-1}. Because of the uncertainties associated with the incompleteness of the data sets, photometric redshift technique, dust reddening, and UV-to-SFR conversion, these numbers are only meant to be indicative. On the other hand, this very simple model is not in obvious disagreement with any of the observations, and is able, in particular, to provide a reasonable estimate of the galaxy optical and near-IR luminosity density.

STELLAR BARYON BUDGET

With the help of some simple stellar population synthesis tools it is possible at this stage to make an estimate of the stellar mass density that produced the integrated light observed today. The total *bolometric* luminosity of a simple stellar population (a single generation of coeval stars) having mass M can be well approximated by a power-law with time for all ages $t \gtrsim 100$ Myr,

$$L(t) = 1.3 \, L_\odot \frac{M}{M_\odot} \left(\frac{t}{1\,\text{Gyr}}\right)^{-0.8} \tag{1}$$

(cf. [8]), where we have assumed solar metallicity and a Salpeter IMF truncated at 0.1 and 125 $M_\odot$. In a stellar system with arbitrary star-formation rate per unit cosmological volume, $\dot{\rho}_*$, the comoving bolometric emissivity at time t is given by the convolution integral

$$\rho_{\text{bol}}(t) = \int_0^t L(\tau)\dot{\rho}_*(t - \tau)d\tau. \tag{2}$$

The total background light observed at Earth $(t = t_H)$ is

$$I_{\text{EBL}} = \frac{c}{4\pi} \int_0^{t_H} \frac{\rho_{\text{bol}}(t)}{1 + z}dt, \tag{3}$$

where the factor $(1 + z)$ at the denominator is lost to cosmic expansion when converting from observed to radiated (comoving) luminosity density. From the above equations it is easy to derive in a EdS cosmology

$$I_{\text{EBL}} = 740 \text{ nW m}^{-2}\,\text{sr}^{-1} \langle \frac{\dot{\rho}_*}{M_\odot \text{ yr}^{-1}\,\text{Mpc}^{-3}} \rangle \left(\frac{t_H}{13\,\text{Gyr}}\right)^{1.87}. \tag{4}$$

The observations shown in Figure 1 therefore imply a "fiducial" mean star formation density of $\langle \dot{\rho}_* \rangle = 0.054 \, I_{40} \, M_\odot \text{ yr}^{-1}\,\text{Mpc}^{-3}$. In the instantaneous recycling approximation, the total stellar mass density observed today is

$$\rho_*(t_H) = (1 - R) \int_0^{t_H} \dot{\rho}_*(t)dt \approx 5 \times 10^8 \, I_{40} \; M_\odot \, \text{Mpc}^{-3} \tag{5}$$

(corresponding to $\Omega_* = 0.007 \, I_{40}$), where R is the mass fraction of a generation of stars that is returned to the interstellar medium, $R \approx 0.3$ for a Salpeter IMF. The optical/FIR background therefore requires that about 10% of the nucleosynthetic baryons today [7] are in the forms of stars and their remnants. The predicted stellar mass-to-blue light ratio is $\langle M/L_B \rangle \approx 5$. These values are quite sensitive to the lower-mass cutoff of the IMF, as very-low mass stars can contribute significantly to the mass but not to the integrated light of the whole stellar population. A lower cutoff of 0.5 $M_\odot$ instead of the 0.1 $M_\odot$ adopted would decrease the mass-to-light ratio (and Ω_*) by a factor of 1.9 for a Salpeter function.

TWO SIMPLE MODELS

Based on the agreement between the $z \approx 3$ and $z \approx 4$ luminosity functions at the bright end, it has been recently argued [58] that the decline in the luminosity density of faint HDF Lyman-break galaxies observed in the same redshift interval

303

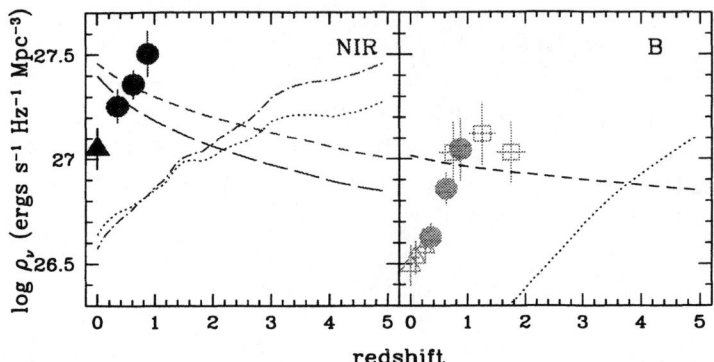

FIGURE 3. *Left*: Synthetic evolution of the near-IR luminosity density at rest-frame wavelengths of 1.0 (*long-dashed line*) and 2.2 μm (*short-dashed line*). The model assumes a constant star-formation rate of $\dot{\rho}_* = 0.054$ M$_\odot$ yr^{-1} Mpc^{-3} (Salpeter IMF). The *dotted* (2.2 μm) and *dash-dotted* (1.0 μm) curves show the emissivity of a simple stellar population with formation redshift $z_{\rm on} = 5$, and total mass equal to the mass observed in spheroids today [17]. The data points are taken from [34] (*filled dots*) and [19] (*filled triangle*). *Right*: Same but in the B-band. The data points are taken from [34] (*filled dots*), [14] (*empty triangles*), and [10] (*empty squares*).

[1] may not be real, but simply due to sample variance in the HDF. When extinction corrections are applied, the emissivity per unit comoving volume due to star formation may then remain essentially flat for all redshift $z \gtrsim 1$ (see Fig. 2). While this has obvious implications for hierarchical models of structure formation, the epoch of first light, and the reionization of the intergalactic medium (IGM), it is also interesting to speculate on the possibility of a constant star-formation density at *all* epochs $0 \le z \le 5$, as recently advocated by [47]. Figure 3 shows the time evolution of the blue and near-IR rest-frame luminosity density of a stellar population characterized by a Salpeter IMF, solar metallicity, and a (constant) star-formation rate of $\dot{\rho}_* = 0.054$ M$_\odot$ yr^{-1} Mpc^{-3} (needed to produce the observed EBL). The predicted evolution appears to be a poor match to the observations: it overpredicts the local B and K-band luminosity densities, and underpredicts the 1 μm emissivity at $z \approx 1$ from the CFRS survey. [2]

At the other extreme, we know from stellar population studies that about half of the present-day stars are contained in spheroidal systems, i.e. elliptical galaxies and spiral galaxy bulges, and that these stars formed early and rapidly [4]. The expected rest-frame blue and near-IR emissivity of a simple stellar population with formation redshift $z_{\rm on} = 5$ and total mass density equal to the mass in spheroids

---

[2] The near-IR light is dominated by near-solar mass evolved stars, the progenitors of which make up the bulk of a galaxy's stellar mass, and is more sensitive to the past star-formation history than the blue light.

observed today (see below) is shown in Figure 3. *HST*-NICMOS deep observations may be able to test similar scenarios for the formation of elliptical galaxies at early times.

TYPE II AGNS

Recent dynamical evidence indicates that supermassive black holes reside at the center of most nearby galaxies. The available data (about 30 objects) show a strong correlation (but with a large scatter) between bulge and black hole mass [42], with $M_{bh} = 0.006 \, M_{bulge}$ as a best-fit. The total mass density in spheroids today is $\Omega_{bulge} = 0.0036^{+0.0024}_{-0.0017}$ [18], implying a mean mass density of dead quasars

$$\rho_{bh} = 1.5^{+1.0}_{-0.7} \times 10^6 \, M_\odot \, Mpc^{-3}. \tag{6}$$

Noting that the observed energy density from all quasars is equal to the emitted energy divided by the average quasar redshift [63], the total contribution to the EBL from accretion onto black holes is

$$I_{bh} = \frac{c^3}{4\pi} \frac{\eta \rho_{bh}}{\langle 1 + z \rangle} \approx 18 \text{ nW m}^{-2} \text{ sr}^{-1} \eta_{0.1} \langle 1 + z \rangle^{-1}, \tag{7}$$

where $\eta_{0.1}$ is the efficiency for transforming accreted rest-mass energy into radiation (in units of 10%). A population of AGNs at (say) $z \sim 1.5$ could then make a significant contribution to the FIR background if dust-obscured accretion onto supermassive black holes is an efficient process [24], [15]. It is interesting to note in this context that a population of AGNs with strong intrinsic absorption (Type II quasars) is actually invoked in many current models for the X-ray background [37], [9].

SOURCES OF IONIZING RADIATION

The application of the Gunn-Peterson constraint on the amount of smoothly distributed neutral material along the line of sight to distant objects requires the hydrogen component of the diffuse IGM to have been highly ionized by $z \approx 5$ [53], and the helium component by $z \approx 2.5$ [12]. From QSO absorption studies we also know that neutral hydrogen at early epochs accounts for only a small fraction, $\sim 10\%$, of the nucleosynthetic baryons [33]. It thus appears that substantial sources of ultraviolet photons were present at $z \gtrsim 5$, perhaps low-luminosity quasars [26] or a first generation of stars in virialized dark matter halos with $T_{vir} \sim 10^4 - 10^5$ K [46], [25], [44]. Early star formation provides a possible explanation for the widespread existence of heavy elements in the IGM [11], while reionization by QSOs may produce a detectable signal in the radio extragalactic background at meter wavelengths [40]. Establishing the character of cosmological ionizing sources is an

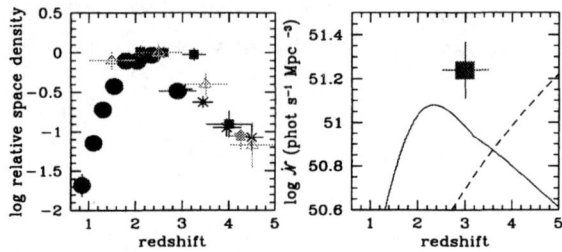

FIGURE 4. *Left:* comoving space density of bright QSOs as a function of redshift. The data points with error bars are taken from [26] *(filled dots)*, [60] *(filled squares)*, [51] *(crosses)*, and [30] *(filled pentagon)*. The *empty triangles* show the space density of the Parkes flat-spectrum radio-loud quasars with $P > 7.2 \times 10^{26}$ W Hz^{-1} sr^{-1} [27]. *Right:* comoving emission rate of hydrogen Lyman-continuum photons *(solid line)* from QSOs, compared with the minimum rate *(dashed line)* which is needed to fully ionize a fast recombining (with ionized gas clumping factor $C = 30$) EdS universe with $\Omega_b h^2 = 0.02$. Models based on photoionization by quasar sources appear to fall short at $z \sim 5$. The data point shows the estimated contribution of star-forming galaxies at $z \approx 3$, assuming that the fraction of Lyman continuum photons which escapes the galaxy H I layers into the intergalactic medium is $f_{\rm esc} = 0.5$ (see [35] for details).

efficient way to constrain competing models for structure formation in the universe, and to study the collapse and cooling of small mass objects at early epochs.

What keeps the universe ionized at $z = 5$? The problem can be simplified by noting that the *breakthrough epoch* (when all radiation sources can see each other in the Lyman continuum) occurs much later in the universe than the *overlap epoch* (when individual ionized zones become simply connected and every point in space is exposed to ionizing radiation). This implies that at high redshifts the ionization equilibrium is actually determined by the *instantaneous* UV production rate [39]. The fact that the IGM is rather clumpy and still optically thick at overlapping, coupled to recent observations of a rapid decline in the space density of radio-loud quasars and of a large population of star-forming galaxies at $z \gtrsim 3$, has some interesting implications for rival ionization scenarios and for the star formation activity at $< 3 < z < 5$.

The existence of a decline in the space density of bright quasars at redshifts beyond ~ 3 was first suggested by [45], and has been since then the subject of a long-standing debate. In recent years, several optical surveys have consistently provided new evidence for a turnover in the QSO counts [27], [61], [52], [31]. The interpretation of the drop-off observed in optically selected samples is equivocal, however, because of the possible bias introduced by dust obscuration arising from intervening systems. Radio emission, on the other hand, is unaffected by dust, and it has recently been shown [54] that the space density of radio-loud quasars also decreases strongly for $z > 3$. This argues that the turnover is indeed real and that dust along the line of sight has a minimal effect on optically-selected QSOs (Figure 4, *left*). The QSO emission rate (corrected for incompleteness) of hydrogen ionizing photons per unit comoving volume is shown in Figure 4 (*right*) [39].

Galaxies with ongoing star-formation are another obvious source of Lyman continuum photons. Since the rest-frame UV continuum at 1500 Å (redshifted into the visible band for a source at $z \approx 3$) is dominated by the same short-lived, massive stars which are responsible for the emission of photons shortward of the Lyman edge, the needed conversion factor, about one ionizing photon every 10 photons at 1500 Å, is fairly insensitive to the assumed IMF and is independent of the galaxy history for $t \gg 10^7$ yr. Figure 4 (*right*) shows the estimated Lyman-continuum luminosity density of galaxies at $z \approx 3$. [3] The data point assumes a value of $f_{esc} = 0.5$ for the unknown fraction of ionizing photons which escapes the galaxy H I layers into the intergalactic medium. A substantial population of dwarf galaxies below the detection threshold, i.e. having star-formation rates $< 0.3 \, M_\odot \, \text{yr}^{-1}$, and with a space density in excess of that predicted by extrapolating to faint magnitudes the best-fit Schechter function, may be expected to form at early times in hierarchical clustering models, and has been recently proposed by [44] and [39] as a possible candidate for photoionizing the IGM at these epochs. One should note that, while highly reddened galaxies at high redshifts would be missed by the dropout color technique (which isolates sources that have blue colors in the optical and a sharp drop in the rest-frame UV), it seems unlikely that very dusty objects (with $f_{esc} \ll 1$) would contribute in any significant manner to the ionizing metagalactic flux.

REIONIZATION OF THE IGM

When an isolated point source of ionizing radiation turns on, the ionized volume initially grows in size at a rate fixed by the emission of UV photons, and an ionization front separating the H II and H I regions propagates into the neutral gas. Most photons travel freely in the ionized bubble, and are absorbed in a transition layer. The evolution of an expanding H II region is governed by the equation

$$\frac{dV_I}{dt} - 3HV_I = \frac{\dot{N}_{\text{ion}}}{\bar{n}_{\text{H}}} - \frac{V_I}{\bar{t}_{\text{rec}}}, \tag{8}$$

where V_I is the proper volume of the ionized zone, $\dot{N}_{\text{ion}}$ is the number of ionizing photons emitted by the central source per unit time, $\bar{n}_{\text{H}}$ is the mean hydrogen density of the expanding IGM, H is the Hubble constant, and $\bar{t}_{\text{rec}}$ is the hydrogen mean recombination timescale,

$$\bar{t}_{\text{rec}} = [(1 + 2\chi)\bar{n}_{\text{H}}\alpha_B \, C]^{-1} = 0.3 \, \text{Gyr} \left(\frac{\Omega_b h^2}{0.02}\right)^{-1} \left(\frac{1 + z}{4}\right)^{-3} C_{30}^{-1}. \tag{9}$$

---

[3] At all ages $\gtrsim 0.1$ Gyr one has $L(1500)/L(912) \approx 6$ for a Salpeter mass function and constant SFR [6]. This number neglects any correction for intrinsic H I absorption.

One should point out that the use of a volume-averaged clumping factor, C, in the recombination timescale is only justified when the size of the H II region is large compared to the scale of the clumping, so that the effect of many clumps (filaments) within the ionized volume can be averaged over (see Figure 5). Across the I-front the degree of ionization changes sharply on a distance of the order of the mean free path of an ionizing photon. When $\bar{t}_{rec} \ll t$, the growth of the H II region is slowed down by recombinations in the highly inhomogeneous medium, and its evolution can be decoupled from the expansion of the universe. Just like in the static case, the ionized bubble will fill its time-varying Strömgren sphere after a few recombination timescales,

$$V_I = \frac{\dot{N}_{ion}\bar{t}_{rec}}{\bar{n}_H}(1 - e^{-t/\bar{t}_{rec}}).$$

(10)

In analogy with the individual H II region case, it can be shown that hydrogen component in a highly inhomogeneous universe is completely reionized when the number of photons emitted above 1 ryd in one recombination time equals the mean number of hydrogen atoms [39]. At any given epoch there is a critical value for the photon emission rate per unit cosmological comoving volume,

$$\dot{\mathcal{N}}_{ion}(z) = \frac{\bar{n}_H(0)}{\bar{t}_{rec}(z)} = (10^{51.2} \text{ s}^{-1} \text{ Mpc}^{-3}) C_{30} \left(\frac{1+z}{6}\right)^3 \left(\frac{\Omega_b h^2}{0.02}\right)^2,$$

(11)

independently of the (unknown) previous emission history of the universe: only rates above this value will provide enough UV photons to ionize the IGM by that epoch. One can then compare our estimate of $\dot{\mathcal{N}}_{ion}$ to the estimated contribution from QSOs and star-forming galaxies. The uncertainty on this critical rate is difficult to estimate, as it depends on the clumpiness of the IGM (scaled in the expression above to the value inferred at $z = 5$ from numerical simulations [22]) and the nucleosynthesis constrained baryon density. The evolution of the critical rate as a function of redshift is plotted in Figure 4 (*right*). While $\dot{\mathcal{N}}_{ion}$ is comparable to the quasar contribution at $z \gtrsim 3$, there is some indication of a deficit of Lyman continuum photons at $z = 5$. For bright, massive galaxies to produce enough UV radiation at $z = 5$, their space density would have to be comparable to the one observed at $z \approx 3$, with most ionizing photons being able to escape freely from the regions of star formation into the IGM. This scenario may be in conflict with direct observations of local starbursts below the Lyman limit showing that at most a few percent of the stellar ionizing radiation produced by these luminous sources actually escapes into the IGM [34].[4]

It is interesting to convert the derived value of $\dot{\mathcal{N}}_{ion}$ into a "minimum" SFR per unit (comoving) volume, $\dot{\rho}_*$ (hereafter we assume $\Omega_b h^2 = 0.02$ and $C = 30$):

---

[4] Note that, at $z = 3$, Lyman-break galaxies would radiate more ionizing photons than QSOs for $f_{esc} \gtrsim 30\%$.

FIGURE 5. Propagation of an ionization front in a 128^3 cosmological density field produced by a mini-quasar with $\dot{N} = 5 \times 10^{53}$ s^{-1}. The box length is 2.4 comoving Mpc. The quasar is turned on at the densest cell, which is found in a virialized halo of total mass 1.3×10^{11} $M_\odot$. The solid contours give the position of the I–front at 0.15, 0.25, 0.38, and 0.57 Myr after the quasar has switched on at $z = 7$. The underlying greyscale image indicates the initial H I density field. (From [1].)

$$\dot{\rho}_*(z) = \dot{\mathcal{N}}_{\rm ion}(z) \times 10^{-53.1} f_{\rm esc}^{-1} \approx 0.013 f_{\rm esc}^{-1} \left(\frac{1+z}{6}\right)^3 \; M_\odot \, {\rm yr}^{-1} \, {\rm Mpc}^{-3}. \qquad (12)$$

The star formation density given in the equation above is comparable with the value directly "observed" (i.e., uncorrected for dust reddening) at $z \approx 3$ [7]. The conversion factor assumes a Salpeter IMF with solar metallicity, and has been computed using a population synthesis code [6]. It can be understood by noting that, for each 1 $M_\odot$ of stars formed, 8% goes into massive stars with $M > 20 M_\odot$ that dominate the Lyman continuum luminosity of a stellar population. At the end of the C-burning phase, roughly half of the initial mass is converted into helium and carbon, with a mass fraction released as radiation of 0.007. About 25% of the energy radiated away goes into ionizing photons of mean energy 20 eV. For each 1 $M_\odot$ of stars formed every year, we then expect

$$\frac{0.08 \times 0.5 \times 0.007 \times 0.25 \times M_\odot c^2}{20 \, {\rm eV}} \frac{1}{1 \, {\rm yr}} \sim 10^{53} \, {\rm phot \, s}^{-1} \qquad (13)$$

to be emitted shortward of 1 ryd.

REFERENCES

1. Abel, T., Norman, M. L., & Madau, P. 1998, preprint (astro-ph/9812151)
2. Barger, A. J., et al. 1998, Nature, 394, 248

3. Baugh, C. M., et al. 1998, ApJ, 498, 504

4. Bernardi, M., et al. 1998, ApJ, 508, L143

5. Bernstein, R. A. 1998, preprint

6. Bruzual, A. C., & Charlot, S. 1999, in preparation

7. Burles, S., & Tytler, D. 1998, preprint (astro-ph/9803071)

8. Buzzoni, A. 1995, ApJS, 98, 69

9. Comastri, A., et al. 1995, A&A, 296, 1

10. Connolly, A. J., et al. 1997, ApJ, 486, L11

11. Cowie, L. L., et al. 1995, AJ, 109, 1522

12. Davidsen, A. F., Kriss, G. A., & Zheng, W. 1996, Nature, 380, 47

13. Dwek, E., et al. 1998, ApJ, 508, 106

14. Ellis, R. S., et al. 1996, MNRAS, 280, 235

15. Fabian, A. C., & Iwasawa, K. 1999, MNRAS, in press

16. Fixsen, D. J., et al. 1998, ApJ, 508, 123

17. Flores, H., et al. 1998, ApJ, in press (astro-ph/9811202)

18. Fukugita, M., Hogan, C. J., & Peebles, P. J. E. 1998, ApJ, 503, 518

19. Gallego, J., et al. 1995, ApJ, 455, L1

20. Gardner, J. P., et al. B. E. 1997, ApJ, 480, L99

21. Glazebrook, K., et al. 1998, MNRAS, submitted (astro-ph/9808276)

22. Gnedin, N. Y., & Ostriker, J. P. 1997, ApJ, 486, 581

23. Guiderdoni, B., et al. 1997, Nature, 390, 257

24. Haehnelt, M. G., Natarajan, P., & Rees, M. J. 1998, MNRAS, 300, 817

25. Haiman, Z., & Loeb, A. 1997, ApJ, 483, 21

26. Haiman, Z., & Loeb, A. 1998, ApJ, 503, 505

27. Hartwick, F. D. A., & Schade, D. 1990, ARA&A, 28, 437

28. Hauser, M. G., et al. 1998, ApJ, 508, 25

29. Hook, I. M., Shaver, P. A., & McMahon, R. G. 1998, in The Young Universe: Galaxy Formation and Evolution at Intermediate and High Redshift, ed. S. D'Odorico, A. Fontana, & E. Giallongo (San Francisco: ASP), p. 17

30. Hughes, D., et al. 1998, Nature, 398, 241

31. Kennefick, J. D., Djorgovski, S. G., & de Carvalho, R. R. 1995, AJ, 110, 2553

32. Lagache, G., et al. 1999, preprint (astro-ph/9901059)

33. Lanzetta, K. M., Wolfe, A. M., & Turnshek, D. A. 1995, ApJ, 440, 435

34. Leitherer, C., et al. 1995, ApJ, 454, L19

35. Lilly, S. J., et al. 1996, ApJ, 460, L1

36. Lilly, S. J., et al. 1998, astro-ph/9807261

37. Madau, P., Ghisellini, G., & Fabian, A. C. 1994, MNRAS, 270, L17

38. Madau, P., et al. 1996, MNRAS, 283, 1388

39. Madau, P., Haardt, F., & Rees, M. J. 1998, ApJ, in press

40. Madau, P., Meiksin, A., & Rees, M. J. 1997, ApJ, 475, 429

41. Madau, P., Pozzetti, L., & Dickinson, M. E. 1998, ApJ, 498, 106

42. Magorrian, G., et al. 1998, AJ, 115, 2285

43. Meurer, G. R., et al. 1997, AJ, 114, 54

44. Miralda-Escudé, J., & Rees, M. J. 1998, ApJ, 497, 21

45. Osmer, P. S. 1982, ApJ, 253, 280

46. Ostriker, J. P., & Gnedin, N. Y. 1996, ApJ, 472, L63
47. Pascarelle, S. M., Lanzetta, K. M., & Fernandez-Soto, A. 1998, ApJ, 508, L1
48. Pettini, M., et al. 1998, ApJ, 508, 539
49. Pozzetti, L., et al. 1998, MNRAS, 298, 1133
50. Rowan-Robinson, M., et al. 1997, MNRAS, 289, 490
51. Sawicki, M., & Yee, H. K. C. 1998, AJ, 115, 1329
52. Schmidt, M., Schneider, D. P., & Gunn, J. E. 1995, AJ, 110, 68
53. Schneider, D. P., Schmidt, M., & Gunn, J. E. 1991, AJ, 101, 2004
54. Shaver, P. A., et al. 1996, Nature, 384, 439
55. Songaila, A. 1997, ApJ, 490, L1
56. Steidel, C. C., et al. 1996, ApJ, 462, L17
57. Steidel, C. C., et al. 1998, ApJ, 492, 428
58. Steidel, C. C., et al. 1998, ApJ, submitted (astro-ph/9811399)
59. Tresse, L., & Maddox, S. J. 1998, ApJ, 495, 691
60. Treyer, M. A., et al. 1998, MNRAS, 300, 303
61. Warren, S. J., Hewett, P. C., & Osmer, P. S. 1994, ApJ, 421, 412
62. Williams, R. E., et al. 1996, AJ, 112, 1335
63. Soltan, A. 1982, MNRAS, 200, 115

Deep sub-mm Surveys With SCUBA

Ian Smail,[1] Rob Ivison,[2] Andrew Blain[3] & Jean-Paul Kneib[4]

[1] *Department of Physics, University of Durham, South Road, Durham DH1 3LE*
[2] *Dept. of Physics & Astronomy, University College London, London WC1E 6BT*
[3] *Cavendish Laboratory, Madingley Road, Cambridge CB3 0HE*
[4] *Observatoire de Toulouse, 14 Avenue E. Belin, 31400 Toulouse, France*

Abstract. We review published deep surveys in the submillimeter (sub-mm) regime from the new Sub-millimetre Common User Bolometer Array (SCUBA, [1]) on the 15-m James Clerk Maxwell Telescope (JCMT), Mauna Kea, Hawaii. Summarising the number counts of faint sub-mm sources determined from the different surveys we show that the deepest counts from our completed SCUBA Lens Survey, down to 0.5 mJy at $850\,\mu$m, fully account for the far-infrared background (FIRB) detected by *COBE*. We conclude that a population of distant, dust-enshrouded ultraluminous infrared galaxies dominate the FIRB emission around 1 mm. We go on to discuss the nature of this population, starting with the identification of their optical counterparts, where we highlight the important role of deep VLA radio observations in this process. Taking advantage of the extensive archival *Hubble Space Telescope* (*HST*) observations of our fields, we then investigate the morphological nature of the sub-mm galaxy population and show that a large fraction exhibit disturbed or interacting morphologies. By employing existing broadband photometry, we derive crude redshift limits for a complete sample of faint sub-mm galaxies indicating that the majority lie at $z < 5$, with at most 20% at higher redshifts. We compare these limits to the initial spectroscopic results from various sub-mm samples. Finally we discuss the nature of the sub-mm population, its relationship to other classes of high-redshift galaxies and its future role in our understanding of the formation of massive galaxies.

Introduction

The extragalactic background light is the repository for all emission from the distant Universe and thus contains unique information about the star-formation history of the Universe. The far-infrared component of this (the FIRB) was detected by *COBE* (e.g. [2]) at a level comparable to that seen in the optical background [3], suggesting that a large proportion of the stars seen in the local Universe were formed in dust-obscured galaxies at high redshifts [4]. Such strongly star-forming, dusty, distant galaxies would be luminous sub-mm sources, due to the re-radiation in the rest-frame sub-mm of the UV/optical starlight absorbed by the dust. Thus deep sub-mm observations, at $\lambda \gtrsim 100\,\mu$m, would be a fruitful avenue to pursue in the

CP470, After the Dark Ages: When Galaxies were Young (the Universe at 2 < z < 5),
edited by Stephen S. Holt and Eric P. Smith

search for these forming galaxies. The strong negative K-corrections provided by the thermal dust spectrum of galaxies also means these systems are easily observable out to high redshift [5]. A *doppelgänger* for Arp 220, with an star-formation rate (SFR) of $\gtrsim 100\,M_\odot$ yr^{-1}, would have a 850-μm flux density of $\gtrsim 3\,$mJy[1] out to $z \sim 10$ ($\gtrsim 0.3\,$mJy for $q_o = 0.05$ [5]). The recently commissioned SCUBA camera on JCMT can achieve this flux limit across a 5 sq. arcmin field in one night.

Faint Galaxy Counts in the Sub-mm

At the time of writing, 850-μm counts of faint extragalactic sources have been published by four groups [6,7,5,9]. The first indications of the surface density of mJy 850-μm galaxies was given by Smail, Ivison & Blain [6], who took advantage of gravitational amplification by massive cluster lenses to increase the sensitivity of their SCUBA maps, and derived a source count of $(2.5 \pm 1.4) \times 10^3\,$deg^{-2} down to a flux density limit of 4 mJy on the basis of 6 detections. This surface density has been broadly confirmed by a number of subsequent studies of blank fields, which spurn lens amplification, preferring simple brute-force integration to obtain the necessary sensitivity.[2] These studies include maps of the Canada-France Redshift Survey fields (CFRS, [9]), the Lockman Hole and Hawaii Survey Fields [5] and the *Hubble Deep Field* [7], and detect 11, 2 and 5 sources above their respective flux limits. The latter two surveys reach the blank-field confusion limit of the JCMT [10] in their deepest integrations. The sub-mm source densities derived from the different surveys are plotted in Fig. 1 to show the broad level of agreement reached.

The latest results from the analysis of the completed SCUBA Lens Survey [12] are also shown in Fig. 1. The complete sample comprises a total of 17 galaxies detected at $3\,\sigma$ significance or above, and 10 detected above $4\,\sigma$, in the fields of seven massive and well-studied cluster lenses at $z = 0.19$–0.41. The total surveyed area is 0.01 degree^2, with a sensitivity of better than 2 mJy in the image planes. The analysis of these catalogues [13] makes use of well-constrained lens models for all the clusters (e.g. [14]) to accurately correct the observed source fluxes for lens amplification. For the median source amplification, $\sim 2.5\times$, our survey covers an area of the source plane equivalent to roughly three times the SCUBA *HDF* map at a comparable sensitivity and with a factor of two finer beam size. At higher amplifications, the survey covers a smaller region, but at a correspondingly higher sensitivity (e.g. ~ 1 sq. arcmin at $\sigma_{850} \sim 0.1\,$mJy) and resolution. The uncertainties associated with our lensing analysis are included in the final error quoted on the derived counts. *The total uncertainty in the lensing correction is at most comparable to the typical error in the absolute SCUBA calibration.* The magnification produced by the massive cluster lenses allows us to constrain the source counts down to 0.5 mJy [13], four times fainter than the deepest blank-field

---

[1] We assume $q_o = 0.5$ and $h_{100} = 0.5$ throughout.
[2] Several other groups are pursuing surveys of lensing clusters using SCUBA, including those headed by Scott Chapman and Paul van der Werf, first results from these should appear soon.

counts published. Moreover, these observations are less affected by confusion noise, due to the expanded view of the background source plane provided by the cluster lenses.

At 450 μm, which SCUBA provides simultaneously with the 850-μm maps, only a few sources have been detected in any of the published surveys. This is due to a combination of the lower atmospheric transmission at 450 μm in normal conditions on Mauna Kea, the lower efficiency of the JCMT dish surface at 450 μm and the relatively high redshifts of the bulk of the sub-mm population [7].

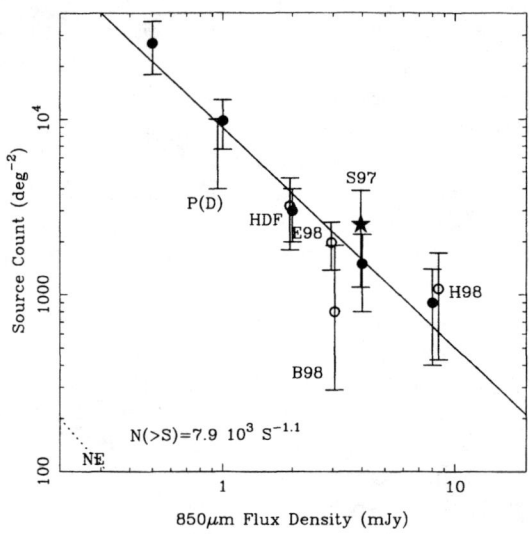

FIGURE 1. A comparison of 850-μm galaxy counts from different groups (S97 [6]; *HDF* and *P(D)* [7]; B98 [8]; E98 [9]; H98 [11]). The filled circles come from the analysis of the completed SCUBA Lens Survey [13] and are corrected for lens amplification. The solid line shows the fit to all of the observations with the form $N(> S) = 7.9 \times 10^3 S^{-1.1}$, which is an adequate description of the counts at flux densities of 0.5–10 mJy. The dotted line (labelled NE) indicates the count expected on the basis of a non-evolving local *IRAS* luminosity function.

The cumulative 850-μm counts of Smail, Ivison & Blain (1997) accounted for roughly 30% of the FIRB detected by *COBE* (e.g. [2,15]). The counts from the HDF brighter than the confusion limit at 2 mJy account for close to 50% of the FIRB, while the deepest counts from the lens fields [13] indicate that the bulk of the FIRB is resolved at 0.5 mJy. This suggests that not only must the counts converge around 0.5 mJy, but that the FIRB is dominated by emission from the most luminous sources. This is a remarkable achievement given that the FIRB was detected only three years ago and SCUBA has been operating for a little over a year.

Having resolved the background we can now study the nature of the populations contributing to the FIRB and so determine at what epoch the background was emitted. Here again our survey has the advantage of lens amplification, this time in the optical and near-IR where the identification and spectroscopic follow-up are undertaken. Typically the counterparts of our sub-mm sources will appear ~ 1 magnitude brighter than the equivalent galaxy in a blank field.

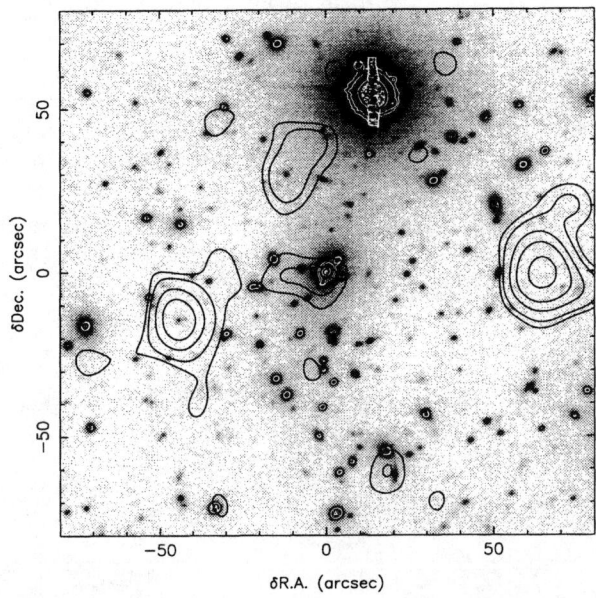

FIGURE 2. The SCUBA 850-μm map of the rich cluster A 1835 ($z = 0.25$), this is overlayed on a deep I-band image of the cluster taken with the Palomar 5.1-m Hale. Two bright sub-mm sources are visible either side of the central cluster galaxy, both are coincident with > 100-μJy radio sources in our deep VLA map [17]. The eastern source is identified with an $I \sim 21$ interacting galaxy at $z = 2.6$ [16,17], while the western source has no obvious counterpart at the position of the sub-mm/radio peak. Weak sub-mm emission is also detected from vigorous star formation in the central galaxy of this cooling-flow cluster [18].

Identifications and Morphologies of SCUBA Galaxies

The sub-mm fluxes of the sources detected in all the published surveys are in the range $S_{850} \sim 0.5$–10 mJy, equivalent to luminosities of $\log_{10} L_{\mathrm{FIR}} \sim 12$–$13$ if they lie at $z \gtrsim 1$, and so they class as ultraluminous infrared galaxies (ULIRGs). Smail et al. (1998, [12]) presented optical identifications obtained from deep *HST* and ground-based images for galaxies selected from the SCUBA Lens survey (e.g. Fig. 2). Down to a limit of $I \sim 25$, counterparts are identified for 14 of the 16 sources in the 3-σ sample and for 9/10 sources in the 4-σ catalog that lie within the optical fields. This rate of optical identification, 80–90%, down to $I \sim 25$ is similar to that achieved by the other sub-mm surveys [7,9]. The bulk of these sources are resolved in the optical images indicating that they are galaxies

We are undertaking near-IR imaging of all our fields to search for any extremely red counterparts which could have been missed in the optical identifications. A link has been suggested between the population of extremely red objects (EROs) and the sub-mm galaxies due to the recent sub-mm detection of one of the most well-studied EROs, HR10 [19]. We have so far identified only one possible ERO counterpart to a sub-mm source in our survey with the bulk of the sub-mm galaxies showing optical–near-IR colors more typical of the general field, $(I - K) \sim 2$–4.

We are using ultra-deep 1.4-GHz VLA maps to confirm the reliability of the identifications [20] and find radio counterparts brighter than $\sim 50\,\mu\mathrm{Jy}$ (equivalent to intrinsic flux densities of $\gtrsim 20\,\mu\mathrm{Jy}$) for over 60% of the sub-mm sources. Again this success rate is similar to that found for sub-mm/radio comparisons in the *HDF* [21]. The radio fluxes of the SCUBA sources are broadly in line with those expected if the sub-mm emission is powered by starbursts and the sources follow the locally determined L_{FIR}–L_{5GHz} correlation for galaxies. Similar analysis of the other surveys await deep radio observations and confirmation of the astrometric accuracy of the SCUBA maps.

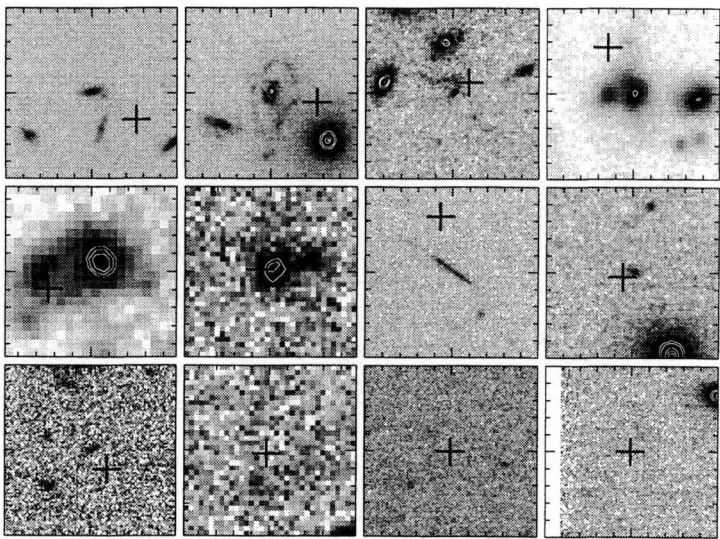

FIGURE 3. The 12 optically-faint sub-mm galaxies from the SCUBA Lens Survey sample for which we have high-resolution R- or I-band imaging [12]. Each panel is 10×10 arcsec corresponding to $\gtrsim 80\,\mathrm{kpc}$ at $z > 1$ and they are ordered from the upper-left on the basis of their morphologies: 6 disturbed/interacting, 4 compact/featureless (including a strongly-distorted arclet) and 2 blank fields. The images are centred on the most likely optical candidate where one is known. The centroids of the sub-mm sources are indicated by crosses. Note that these images span a range in exposure times and resolutions [12].

The morphologies of those galaxies for which we have high-resolution optical imaging fall into three broad categories: faint disturbed galaxies and interactions;

faint galaxies too compact to classify reliably; and dusty, star-forming galaxies at intermediate redshifts [12,22]. We show in Fig. 3 the R- or I-band images of the optically faint sub-mm galaxies, these illustrate the dominance of disturbed morphologies in this sample. About 70% of the faint galaxies are disturbed or interacting, suggesting that in the distant Universe, as in the local one, interactions remain an important mechanism for triggering starbursts and for the formation of ULIRGs [12]. The faint, compact galaxies may represent a later evolutionary stage of these mergers, or more centrally concentrated starbursts. It is likely that some of these will also host active galactic nuclei.

The Redshift Distribution of Faint Sub-mm Galaxies

An analysis of the optical colors of our sub-mm sample to search for the signature of the Lyman break in bluer passbands allows us to estimate a crude redshift distribution from their identification in deep B- and V-band images [12]. This indicates that $\gtrsim 75\%$ of the optically-identified galaxies have $z \lesssim 5.5$ whilst $\gtrsim 50\%$ lie at $z \lesssim 4.5$ on the basis of B-band identifications alone. Photometric redshifts have been used in the HDF [7] and CFRS [22] to place limits on the redshift distributions of sub-mm galaxies in these samples, the estimated redshifts span $z \sim 1$–4. Although the reliability of the photometric redshifts obtained for such dusty systems has yet to be tested, these results are consistent with the limits above. We conclude that the luminous sub-mm population is broadly coeval with the more modestly star-forming galaxies selected by UV/optical surveys of the distant Universe (e.g. [23]). However, the individual SCUBA galaxies have SFRs which are typically an order of magnitude higher than those of the optically selected galaxies, as well as being apparently more dust (and hence metal?) rich.

A further attraction of using lenses in our survey was the possibility of deriving redshift estimates for galaxies too faint for spectroscopic identification. Using our detailed mass models, redshifts can be determined for any background galaxy whose distortion can be measured from our HST imaging, using the relationship between source redshift and apparent shear for the lens model [24], a technique whose accuracy has been recently confirmed [25]. In this way we have estimated a redshift of $z = 1.6 \pm 0.2$ for an arclet detected in our survey with an intrinsic apparent magnitude of $I = 25.3$.

Preliminary results from spectroscopic surveys of the different sub-mm samples are beginning to appear, although none are yet complete. First indications from the existing CFRS redshift survey [22], for which the majority of galaxies have $z < 0.8$, unsurprisingly finds a low redshift for the identified sources, $z \sim 0.1$–0.7. Spectroscopic observations of the SCUBA Lens Survey sample with Keck, CFHT and WHT have identified a number of distant galaxies in the range $z \sim 1$–3 [16] as well as several galaxies at $z \sim 0.2$–0.4 in the foreground cluster lenses [16,18]. These sources are removed from our count analysis, but their detection does confirm that SCUBA has the ability to routinely detect star-forming galaxies at $z < 1$ [22]. The more distant systems include a $z = 2.8$ dusty AGN/starburst

[26] and a $z = 2.6$ starburst [16,17]. These spectroscopic observations support the photometrically-derived redshift limits for the bulk of the population as well as giving more information about the dominant emission processes in individual galaxies. In particular, the spectra provide some indication of the relative fractions of AGN and starbursts in the sub-mm population. Nevertheless, the spectroscopic surveys currently have nothing to say about the 10–20% of SCUBA sources that have no obvious optical counterpart. Owing to the large negative K-correction in the sub-mm, these sources may be at $z > 5$, and thus are the most interesting sources to followup.

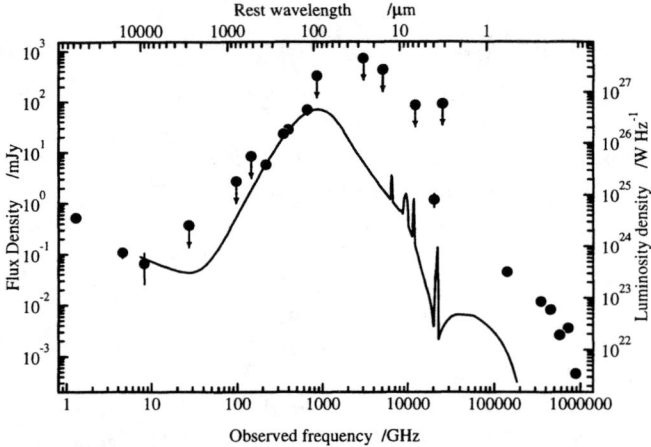

FIGURE 4. The spectral energy distribution (SED) of SMM J02399−0136 at $z = 2.8$, adapted from [26]. The intrinsic FIR luminosity of this source is 10^{13} L$_\odot$, after correction for lens amplification using our lens model, making it one of the most luminous galaxies known. The strong peak around $100 \, \mu$m in the restframe, probed by the SCUBA observations at $450 \, \mu$m–2 mm, is emission from dust at 40–50 K in the galaxy. The radio emission from this galaxy is at a level expected from the high SFR, as shown by the agreement with the rescaled composite SED of luminous *IRAS* galaxies plotted here [27].

The Physical Properties of Sub-mm Galaxies

Further insights into the nature of the sub-mm-selected galaxies come from detailed investigations of the physical properties of this population (e.g. SFR, T_{dust}, M_{dust}, M_{gas}, etc.). These studies are not easy though, in particular turning the observed sub-mm flux into a SFR or dust mass requires a number of uncertain steps, even in the absence of competing contributions from an AGN and a starburst within a single galaxy. The basis for all of these analyses are multi-wavelength observations, especially in the sub-mm. Fortunately the 450/850-μm arrays and 1.3- and 2.0-mm single-channel bolometers on SCUBA can provide the necessary sub-mm data to constrain quantities such as the dust temperature in high-redshift galaxies.

Radio observations are useful not only to provide more accurate positions, but also to rule out non-thermal contributions to the sub-mm flux.

A well-sampled SED is shown in Fig. 4 for the brightest source in the SCUBA Lens Survey, SMM J02399−0136 at $z = 2.8$ [26]. The amplification-corrected FIR luminosity of this galaxy is $10^{13} L_\odot$, its dust temperature is $T_{dust} = 40$–50 K and the dust mass is around $M_{dust} \sim 6 \times 10^8 M_\odot$. The estimated SFR for this galaxy is $\gtrsim 2000 M_\odot$ yr^{-1}, although the optical spectrum indicates that it hosts a Seyfert-2 nucleus, suggesting that some fraction of the L_{FIR} is probably attributed to the AGN [26]. SMM J02399−0136 has recently been detected in CO using the Owens Valley Millimeter Array [25], revealing the presence of $\sim 2 \times 10^{11} M_\odot$ of molecular gas, a dynamically important component of this massive galaxy and an enormous reservoir of fuel to power the star-formation. At this time, SMM J02399−0136 is the only sub-mm-selected galaxy for which such detailed observations are available.

The Nature of the Faint Sub-mm Population

We are beginning to build up a picture of the population of distant, luminous sub-mm galaxies which dominate the FIRB at wavelengths around 1 mm. The fact that the sub-mm population detected by SCUBA can account for all of the *COBE* background indicates that a substantial fraction (upto half) of the stars in the local Universe could be formed in these systems. The properties of this population are similar to those of ULIRGs in the local Universe, with the important distinction that they contribute a sub-mm luminosity density at early epochs that is more than an order of magnitude greater than the corresponding galaxies today. Detailed observations of this population can be used to trace the amount of high-redshift star-formation activity that is obscured from view in the optical by dust, and so is missing from existing inventories of star-formation activity in the distant Universe [6,7]. In this way a more complete and robust history of star formation for the Universe can be constructed.

As with local ULIRGs, there is uncertainty over the exact contributions from AGN and starbursts to this luminosity density. Insight may come through searches for hard X-ray emission from the sub-mm galaxies using *AXAF*, as these should detect all but the most heavily enshrouded AGN at $z > 1$ [29]. These searches will also provide an estimate of the total contribution from the dust-obscured AGN to the X-ray background [29,30]. At the current time we can only state that $\gtrsim 20\%$ of the sub-mm population shows obvious spectral signatures of an AGN — although this does not mean that the AGN dominates the emission in the sub-mm. Assuming that the bulk of the sub-mm emission arises from starbursts, we find that the total amount of energy emitted by dusty galaxies is about four times greater than that inferred from rest-frame UV observations, and that a larger fraction of this energy is emitted at high redshifts [4]. The simplest explanation for these results is that a large population of luminous strongly obscured galaxies at redshifts of $z \lesssim 5$ is missing from optical surveys of the distant Universe.

Finally, we come to the question of what class of object the sub-mm galaxies will evolve into by the present day? The similarities of these sources to local ULIRGs, which are expected to evolve into elliptical galaxies [1], along with their apparently high SFRs, which if sustained over $\sim 1\,\text{Gyr}$ would form an entire L^* galaxy at high redshift, suggest that the sub-mm population may be young, massive elliptical galaxies [6,9]. A question then arises concerning the relationship between the sub-mm population and the other class of high-redshift source which has been identified with forming ellipticals: the Lyman-break galaxies [32]. These objects appear to have typically lower SFRs and less dust than the sub-mm galaxies. The two populations could be naturally linked if the dust content of young galaxies is coupled to their masses or luminosities [33], such that the more massive galaxies are dustier. This behaviour would parallel the metallicity relation for elliptical galaxies, which show increasing metal enrichment with increasing galaxy mass, arising from the retention of processed gas by the potentials of the more massive galaxies. We would therefore identify the most massive, young ellipticals with the sub-mm galaxies and the less massive ellipticals and bulges with the Lyman-break objects. Thus detailed observations of the sub-mm population should give much needed observational information for models of the formation and evolution of massive galaxies [4]. In particular we look forward to further CO detections of sub-mm galaxies with both the existing and next-generation millimeter arrays to study the kinematics of these systems and hence determine their masses.

Acknowledgements

We thank Amy Barger and Len Cowie for the use of results from our on-going collaborative program of spectroscopy with the Keck telescopes and Omar Almaini, Steve Eales, Katherine Gunn, Dave Hughes, Andy Lawrence, Malcolm Longair, Chris Mihos, Ian Robson and Michael Rowan-Robinson for useful conversations. IRS thanks the organisers for support while at the conference and acknowledges a Royal Society Fellowship.

REFERENCES

1. Holland, W. S., et al., 1999, MNRAS, in press. (astro-ph/9809122)
2. Puget, J.-L., Abergel, A., Bernard, J.-P., Boulanger, F., et al., 1996, A&A, 308, L5
3. Bernstein, R. A., 1997, PhD Thesis, California Institute of Technology.
4. Blain, A. W., Smail, I., Ivison, R. J., Kneib, J.-P., 1999a, MNRAS, in press. (astro-ph/9806062).
5. Hughes, D. H., Dunlop, J. S., 1998, in *Highly Redshifted Radio Lines*, eds., C. Carilli, et al., PASP (astro-ph/9802260).
6. Smail, I., Ivison, R. J., Blain, A. W., 1997, ApJL, 490, L5.
7. Hughes, D. H., et al., 1998, Nature, 394, 241
8. Barger, A. J., et al., 1998, Nature, 394, 248.

9. Eales, S., et al., 1998, ApJL, in press. (astro-ph/9808040)
10. Blain, A. W., Ivison, R. J., Smail, I., 1998, MNRAS, 296, L29.
11. Holland, W. S., et al., 1998, Nature, 392, 788.
12. Smail, I., Ivison, R. J., Blain, A. W., Kneib, J.-P., 1998, ApJL, 507, L21.
13. Blain, A. W., Kneib, J.-P., Ivison, R. J., Smail, I., 1999b, ApJL, submitted.
14. Kneib, J.-P., Mellier, Y., Fort, B., Mathez, G., 1993, A&A, 273, 367.
15. Fixsen, D. J., Dwek, E., Mather, J. C., Bennett, C. L., Shafer, R. A., 1998, ApJ, submitted
16. Barger, A. J., et al., 1999, in prep.
17. Ivison, R. J., et al., 1999b, in prep.
18. Edge, A. C., Ivison, R. J., Smail, I., Blain, A. W., Kneib, J.-P., 1999, MNRAS, submitted.
19. Dey, A., Graham, J. R., Ivison, R. J., Smail, I., Wright, G. S., 1998, ApJ, in press.
20. Ivison, R. J., Owen, F. N., et al., 1999a, in prep.
21. Richards, E. A., et al., 1998, Nature, submitted.
22. Lilly, S. J., et al., 1998, preprint. (astro-ph/9807261)
23. Madau, P., et al., 1996, MNRAS, 283, 1388.
24. Kneib, J.-P., Ellis, R. S., Smail, I., Couch, W. J., Sharples, R. M., 1996, ApJ, 471, 643.
25. Ebbels, T. M. D., Ellis, R. S., Kneib, J.-P., Le Borgne, J.-F., Pelló, R., Smail, I., Sanahuja, B., 1998, MNRAS, 295, 79.
26. Ivison, R. J., Smail, I., Le Borgne, J.-F., Blain, A. W., Kneib, J.-P., Bézecourt, J., Kerr, T. H., Davies, J. K., 1998, MNRAS, 298, 583.
27. Guiderdoni B., Hivon E., Bouchet F.R., Maffei B., 1998, MNRAS 295, 877
28. Frayer, D. T., Ivison, R. J., Scoville, N. Z., Yun, M., Evans, A. S., Smail, I., Blain, A. W., Kneib, J.-P., 1998, ApJL, 506, L7.
29. Gunn, K. F., Shanks, T., 1999, MNRAS, submitted.
30. Almaini, O., Lawrence, A., Boyle, B. J., 1999, MNRAS, submitted.
31. Mihos, J. C., Hernquist, L., 1996, ApJ, 464, 641
32. Steidel, C. C., et al., 1996, 462, L17
33. Dickinson, M. E., 1998, in *The Hubble Deep Field*, eds. Livio, M., et al.

Starburst Galaxies: Implications at High-Redshift

Timothy M. Heckman [1]

[1] *Department of Physics & Astronomy, Johns Hopkins University, Baltimore, MD 21218, USA*

Abstract. Starbursts are an important component of the present-day universe, being the site of $\sim$ 25% of the high-mass star- formation. They also serve as local analogs of the processes that were important in the origin and early evolution of galaxies and in the heating and chemical enrichment of the inter-galactic medium. In this contribution I review starbursts from this cosmological perspective, stressing observations at ultra-violet, infrared, and X-ray wavelengths. These data show that: 1) Local starbursts are quite similar in their UV photometric and structural properties to the UV-selected 'Lyman-break' galaxies at high-redshift 2) Dust dramatically affects our view of starbursts 3) More massive galaxies host more metal-rich starbursts, which are in turn more heavily extincted by dust 4) More luminous starbursts are more heavily extincted by dust 5) The strong UV interstellar absorption lines directly trace outflows of metal-rich gas, and 6) X-ray observations show that these 'superwinds' are strongly mass-loaded flows that carry out gas at a rate comparable to the rate of star formation at inferred velocities that are close to the escape velocity from a massive galaxy. These results suggest that the Lyman-break galaxies are typically significantly reddened and extincted by dust (average factor of 3 to 10 in the UV), may have moderately high metallicities (0.1 to 1 times solar?), are probably building galaxies with stellar surface-mass-densities similar to present-day bulges and ellipticals, and may be suffering substantial losses of metal-enriched gas that can 'pollute' the inter-galactic medium. The sub-mm sources detected at high-z are most likely the high-luminosity tail of the Lyman-break population, and probably represent the most metal-rich (dustiest) starbursts occurring in the most massive halos.

INTRODUCTION

Starbursts are short-lived episodes of intense star-formation that usually occur in the circum-nuclear (kpc-scale) regions of galaxies, and dominate the integrated emission from the 'host' galaxy (cf. Leitherer et al 1991). Starbursts are major components of the local universe (cf. Gallego et al 1995; Soifer et al 1987), and are the sites of $\sim$ 20 - 30% of the total (high-mass) star-formation in the local universe (Heckman 1997), and out to redshifts of $z \sim 1$ (Flores et al 1998).

CP470, After the Dark Ages: When Galaxies were Young (the Universe at 2 < z < 5),
edited by Stephen S. Holt and Eric P. Smith

The cosmological relevance of starbursts has been dramatically underscored by the spectacular discovery of populations of high-redshift ($z > 2$) star-forming field galaxies selected by their rest-frame UV continuum emission (cf. Steidel et al 1996; Lowenthal et al 1997), rest-frame far-IR emission (e.g. Hughes et al 1998; Lilly et al 1998), Lyα emission (Hu, Cowie, & McMahon 1998), and Hα emission (Teplitz, Malkan, & McLean 1998; Mannucci et al 1998).The co-moving space density and luminosity of these galaxies implies that they almost certainly represent precursors of typical present-day galaxies and are responsible for the production of a significant fraction of the stars and heavy elements in the present-day universe (e.g. Madau et al 1996; Calzetti & Heckman 1998).

In this contribution I will discuss the relevance of local starbursts for understanding the high-redshift universe. I will take a 'panchromatic' perspective, but will primarily discuss the vacuum-UV properties of local starbursts. Only in this spectral regime can we clearly observe the direct spectroscopic signatures of the hot stars that provide most of the bolometric luminosity of starbursts (e.g. Leitherer, Robert, & Heckman 1995). Moreover, the vacuum-UV contains a wealth of spectral features, including the resonance transitions of most cosmically-abundant ionic species (cf. Morton 1991). These give UV spectroscopy a unique capability for diagnosing the (hot) stellar population and the physical and dynamical state of gas in starbursts. Also, optical observations of galaxies at high-redshifts sample the vacuum-UV portion of their rest-frame spectrum. Thus, a thorough understanding of how to exploit the diagnostic power of the rest-frame UV spectral properties of local starbursts will provide powerful tools for studying star-formation and galaxy-evolution in the early universe. I will also highlight X-ray observations of starburst-driven superwinds. Only such data allow us to directly probe the hot, tenuous fluid that contains the bulk of the energy, mass, and metals that these outflows may carry into the IGM.

THE VACUUM-UV PROPERTIES OF LOCAL STARBURSTS

Lessons Learned

My collaborators and I have recently completed analyses of vacuum-UV HST images (Meurer et al 1997 - hereafter M97; Meurer, Heckman, & Calzetti 1998 - hereafter MHC) and IUE spectra (Heckman et al 1998 - hereafter H98) of moderately large samples of local starbursts spanning broad ranges in metallicity, luminosity, and host galaxy mass. I'd like to first summarize our primary conclusions, and then briefly discuss their relevance to the high-z universe:

First, dust has a profound effect on the emergent UV spectrum.

Previous papers have established that various independent indicators of dust extinction in starbursts correlate strongly with one another. Calzetti et al (1994) show that the spectral slope in the vacuum-UV continuum (as parameterized by β, where $F_\lambda \propto \lambda^\beta$) correlates strongly with the nebular extinction measured in the optical using the Balmer decrement. M97 and MHC show that β also correlates well with the ratio of far-IR to vacuum-UV flux: the greater the fraction of the UV that is absorbed by dust and re-radiated in the far-IR, the redder the vacuum-UV continuum. The interpretation of these correlations with β in terms of the effects of dust are particularly plausible because the *intrinsic* value for β in a starburst is a robust quantity. Figures 31 and 32 in Leitherer & Heckman (1995) show that β should have a value between about -2.0 and -2.6 for the range of ages, initial mass functions, and metallicities appropriate for starbursts.

As shown by H98, the amount of dust-extinction and reddening is well- correlated with metallicity. In low metallicity starbursts (<10% solar) a significant fraction of the intrinsic vacuum-UV actually escapes the starburst ($L_{IR}/L_{UV} \sim$ unity), and the vacuum-UV colors are consistent with the intrinsic (unreddened) colors expected for a starburst population ($\beta \sim$ -2.3). In contrast, in high metallicity starbursts (> solar) 90% to 99% of the energy emerges in the far-IR ($L_{IR}/L_{UV} = 10$ to 100) and the vacuum-UV colors are very red ($\beta \sim 0$). Storchi-Bergmann, Calzetti, & Kinney (1994) had previously noted the correlation between metallicity and UV color.

These correlations have a straightforward interpretation: the vacuum-UV radiation escaping from starbursts suffers an increasing amount of reddening and extinction as the dust-to-gas ratio in the starburst ISM increases with metallicity.

Interestingly, H98 also find that the amount of vacuum-UV extinction in starbursts correlates strongly with the bolometric luminosity of the starburst: only starbursts with $L_{bol} <$ few $\times 10^9$ $L_\odot$ have colors expected for a lightly reddened starburst and have vacuum-UV luminosities that rival their far-IR luminosities. Starbursts that lie at or above the 'knee' in the local starburst far-IR luminosity function ($L_{bol} >$ few $\times 10^{10}$ $L_\odot$ - cf. Soifer et al 1987) have red UV continua ($\beta \sim$ -1 to +0.4) and are dominated by far-IR emission ($L_{IR} \sim 10$ to 100 L_{UV}). This is quite consistent with the recent comparison of the vacuum-UV and far-IR galaxy luminosity functions at low-redshift by Buat & Burgarella (1998): a census of star-formation would be dominated by UV-emission at the low-luminosity end and by far-IR emission at the high-luminosity end.

H98 also find that the amount of vacuum-UV extinction in starbursts correlates well with the absolute blue magnitude and the rotation speed of the galaxy 'hosting' the starburst: starbursts in more massive galaxies are more dust-shrouded. This almost certainly arises from the well-known mass-metallicity relation for star-forming galaxies (e.g. Zaritzky et al 1994), which also applies to starbursts (H98).

Second, local UV-selected starbursts are very similar in global structure to their larger and more luminous counterparts at high-z.

M97 found that local starbursts and high-redshift Lyman-break galaxies have similar rest-frame UV surface brightnesses. Using the empirical correlation between UV color and extinction for local starbursts, they argued that the relatively red UV colors of the Lyman-break galaxies imply that they suffer substantial amounts of extinction. If their proposed extinction-corrections are applied, the high-z galaxies can have very large bolometric luminosities (ranging up to nearly 10^{13} $L_\odot$ for H_0 = 75 km s^{-1} Mpc^{-1} and q_0 = 0.1). Interestingly, the bolometric surface- brightnesses of the extinction-corrected high-z galaxies are very similar to the values seen in local starbursts: typically $\sim 10^{10}$ to 10^{11} $L_\odot$ kpc^{-2}. The high-redshift Lyman-break galaxies thus appear to be 'scaled-up' (larger and more luminous) versions of the local UV-selected starbursts. The physics behind this 'characteristic' surface-brightness is unclear (cf. M97; Lehnert & Heckman 1996a).

Third, the metallicity of the starburst strongly affects the UV spectrum.

Apart from the effects of dust, a starburst's metallicity is the single most important parameter in determining its vacuum-UV properties. The properties of the vacuum-UV absorption-lines are strongly dependent on metallicity. Figure 1 shows that both the high-ionization (e.g. CIVλ1550 and SiIVλ1400) and low-ionization (e.g. CIIλ1335, OIλ1302, and SiII$\lambda\lambda$1260,1304) resonance absorption-lines are significantly stronger in starbursts with high metallicity.

The metallicity-dependence of the high-ionization lines (noted previously by Storchi-Bergmann, Calzetti, & Kinney 1994) is not surprising, given the likely strong contribution to these lines from stellar winds. Theoretically, we expect that since stellar winds are radiatively driven, the strengths of the vacuum-UV stellar wind lines will be metallicity-dependent. This is confirmed by available HST and HUT spectra of LMC and especially SMC stars (Walborn et al 1995; Puls et al 1996).

Figure 1 also shows a metallicity-dependence for the strengths of the UV absorption-lines that are of stellar-photospheric rather than interstellar origin (we know they are not interstellar lines because they correspond to transitions out of highly excited states). Such lines are generally rather weak in starburst spectra and/or blended with strong interstellar features. They include CIIIλ1175, SiIIIλ1417, CIII$\lambda\lambda$1426,1428, SVλ1502, SiIIIλ1892, and FeIII$\lambda\lambda$1925,1960.

The weak but statistically-significant correlation between metallicity and the strength of the low-ionization resonance lines (which are primarily formed in the interstellar medium of the starburst) is also unsurprising. Analyses of HST spectra (cf. Pettini & Lipman 1995; Heckman & Leitherer 1997) show that the strong interstellar lines are saturated (highly optically-thick). In this case, the equivalent width of the absorption-line (W) is only weakly dependent on the ionic column density (N_{ion}): $W \propto b[ln(N_{ion}/b)]^{1/2}$, where b is the normal Doppler line-broadening parameter. Over the range that H98 sample well, the starburst metallicity increases by a factor of almost 40 (from 0.08 to 3 solar), while the equivalent widths of the strong interstellar lines only increase by an average factor of about 2 to 3. This is consistent with the strong interstellar lines being quite optically-thick.

325

Fourth, the properties of the strong interstellar absorption- lines reflect the hydrodynamical consequences of the starburst, and do not straightforwardly probe the gravitational potential of the galaxy.

As noted above, analyses of HST UV spectra of starbursts imply that the interstellar absorption-lines lines are optically- thick. Their strength is therefore determined to first-order by the velocity dispersion in the starburst (see above). Thus, these lines offer a unique probe of the kinematics of the gas in starbursts. The enormous strengths of the starburst interstellar lines (equivalent widths of 3 to 6 Å in metal-rich starbursts) require very large velocity dispersions in the absorbing gas (few hundred km s^{-1}).

The most direct evidence for a non-gravitational component of the gas dynamics comes from analyses of HST and HUT spectra, which show that the interstellar lines are often blueshifted by one-to-several-hundred km s^{-1} with respect to the systemic velocity of the galaxy (Heckman & Leitherer 1997; Gonzalez-Delgado et al 1998a,b; Kunth et al 1998). This demonstrates directly that the absorbing gas is flowing outward from the starburst, probably helping to 'feed' the superwinds whose emission is readily observed in the optical, X-ray, and radio regime (as discussed below).

Implications at High-Redshift

The results summarized above have a variety of interesting implications for the interpretation of the rest-frame-UV properties of galaxies at high-redshift.

Powerful starbursts in the present universe are generally hosted by massive ($\sim$ L$_*$), metal-rich galaxies, and emit almost all their light in the far-infrared, not in the ultraviolet. Thus, an ultraviolet census of the local universe would not only underestimate the true star-formation-rate, it would *systematically under-represent the most powerful, most metal-rich starbursts occurring in the most massive galaxies.* Estimates of star-formation rates at high-redshift based on rest-frame vacuum-UV sample selection may suffer the same bias, and thus might under- represent young/forming massive elliptical galaxies. It is tempting to speculate (based on the empirical properties of local starbursts) that the submm-selected sources at high-z may represent just such objects.

In any case, using the strong correlation between the vacuum-UV color of local starbursts (β) and the ratio of far-IR to vacuum-UV light emitted by local starbursts, MHC estimate that the net extinction suffered by Lyman-break galaxies at z $\sim$ 3 ('U dropouts') is about a factor of $\sim$ 9 at 1600Å and the implied global rate of star-formation per co-moving volume element of the Lyman-break population is then comparable to rough estimates for the SCUBA source population (Hughes et al 1998; Lilly et al 1998). If the MHC extinction-corrections are applied, the most luminous Lyman- break galaxies have very large bolometric luminosities

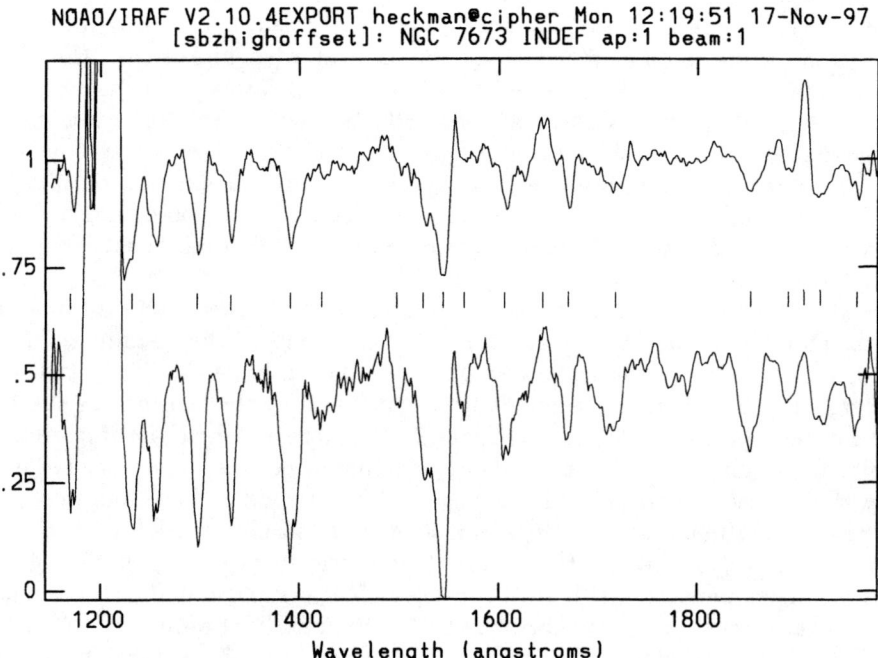

FIGURE 1. IUE spectra of local starbursts with low-metallicity (top) and high-metallicity (bottom). Each spectrum is a weighted average of the spectra of about 20 starbursts. The mean metallicities are 0.16 solar (top) and 1.2 solar (bottom). A number of features are indicated by tick marks and have the following identifications (from left-to-right): CIIIλ1175 (P), NVλ1240 (W), SiIIλ1260 (I), OIλ1302 plus SiIIλ1304 (I), CIIλ1335 (I), SiIVλ1400 (W;I), SiIIIλ1417 plus CIIIλ1427 (P), SVλ1502 (P), SiIIλ1526 (I), CIVλ1550 (W;I), FeIIλ1608(I), HeIIλ1640 emission (W), AlIIλ1671 (I), NIVλ1720 (W), AlIIIλ1859 (I;W), SiIIIλ1892 (P), CIII]λ1909 (nebular emission-line), FeIIIλ1925 (P), FeIIIλ1960 (P). Here, I, P, and W denote lines that are primarily of interstellar, stellar photospheric, or stellar wind origin. The strong emission feature near 1200 Å is geocoronal Lyα.

327

$(\sim few \times 10^{12} L_\odot$ for $H_0 = 75$ km s^{-1} Mpc^{-1} and $q_0 = 0.1$). They would there-fore overlap the population of SCUBA sources in terms of bolometric luminosity, and MHC show that at these luminosities, they have similar space densities as the SCUBA population (albeit with considerable uncertainty - cf. Lilly et al 1998).

The strong correlation found by H98 between vacuum-UV color (β) and metal-licity in local starbursts - if applied naively to high-z Lyman-break galaxies - would suggest a broad range in metallicity from substantially subsolar to solar or higher and a median value of perhaps 0.3 solar. This is somewhat higher than the mean metallicity in the damped Lyα systems (the major repository of HI gas at these redshifts), but this may be due to selection effects: the UV-selected galaxies are the most actively star- forming regions of galaxies, while the damped Lyα systems tend to sample the outer, less-chemically-enriched parts of galaxies or perhaps proto-galactic fragments (e.g. Pettini et al 1997), and will be biased against lines-of-sight through gas with both high column density and high metallicity (e.g. Pei & Fall 1995).

It would also be interesting to use the correlations between absorption- line strengths and metallicity in local starbursts to 'guesstimate' the metallicity of the high-z galaxies. One prediction based on the local starbursts (H98) is that the high-z galaxies should show a strong correlation between the strength of the UV absorption-lines (stellar and interstellar) and β (the more metal-rich local starbursts are both redder and stronger-lined). For the stellar wind lines, this correlation is exactly the opposite of what would be expected if the redder UV colors were due to either an aging burst or an IMF with a steeper-than-Salpeter high-mass slope.

As noted above (see M97), the typical bolometric surface brightnesses of the Lyman-break galaxies are $\sim 10^{10}$ to 10^{11} L$_\odot$ kpc^{-2}. It is intriguing that the implied average surface- mass-density of the stars within the half-light radius ($\sim 10^8$ to 10^9 M$_\odot$ kpc^{-2}) is quite similar to the values in present-day massive spheroids. Are we witnessing the formation of elliptical and bulges?

Finally, based on local starbursts - it seems likely that the gas kinematics that are measured in the high-z galaxies using the interstellar absorption-lines are telling us a great deal about the hydrodynamical consequences of high-mass star-formation for the interstellar medium, but only indirectly probe the gravitational potential or mass of the galaxy. Even the widths of the nebular emission-lines in local starbursts are not always reliable tracers of the galaxy potential well (cf. Lehnert & Heckman 1996a). This means that it will be tricky to determine masses for the high-z galaxies without measuring real rotation- curves via spatially-resolved spectroscopy.

On the brighter side, if the kinematics of the interstellar absorption-lines can be generically shown to arise in outflowing metal-enriched gas, we can then directly study high-redshift star- forming galaxies caught in the act of 'polluting' the IGM with metals in the early universe.

In fact, there is now rather direct observational evidence that this is the case at high-redshift, based on the systematic blueshift (redshift) of the UV inter-stellar absorption-lines (Lyα emission-line) with respect to the nebular Balmer and [OIII]λ5007 emission lines (Franx et al 1997; Pettini et al 1998; Lowenthal et al 1997). This is precisely the phenomenon exhibited by local starbursts (see Gonzalez-Delgado et al 1998a; Kunth et al 1998). The implied outflow rates are substantial: $\dot{M} \sim 50 \ (r_*/3 \ \text{kpc}) \ N_{21} \ (\Omega/4\pi) \ (v_{out}/300 \ \text{km s}^{-1}) \ M_{\odot} \ \text{yr}^{-1}$, where r_* is the starburst radius, N_{21} is the column density of the outflow in units of 10^{21} cm^{-2}, Ω is the outflow solid angle, and v_{out} is its velocity. The outflow rate is then comparable to the star-formation rate. *If* the outflowing gas escapes into the IGM, such flows could easily bring an IGM with $\Omega_{IGM} \sim 0.01$ h^{-2} up to a mean metallicity of $> 10^{-2}$ solar by a redshift of 2.5 (cf. Madau & Shull 1996). I will return to this issue in the next section.

PROPERTIES AND IMPLICATIONS OF STARBURST-DRIVEN SUPERWINDS

Over the last few years, observations have provided convincing evidence of the existence (and even the ubiquity) of 'superwinds' - galactic-scale outflows of gas driven by the collective effect of multiple supernovae and stellar winds in a starburst (cf. Heckman, Lehnert, & Armus 1993; Lehnert & Heckman 1996b; Bland-Hawthorn 1995). X-ray data have proved particularly crucial since they are the only direct probe of the hot gas that contains most of the energy and metals in the flow.

Soft X-ray emission (hot gas) is a generic feature of the halos of the nearest starburst galaxies (Dahlem, Weaver, & Heckman 1998 and references therein). The estimated thermal energy content of this gas represents a significant fraction of the time-integrated mechanical energy supplied by the starburst, while the soft X-ray luminosity is only a few % of the supernova heating rate. These results suggest that little of the mechanical energy supplied by supernovae and stellar winds in starbursts is radiated away. Thus, in principle superwinds may efficiently transport much of the mechanical energy supplied by high-mass stars into the IGM, making such outflows an important heating source for the IGM (e.g. Ponmon, Cannon, & Navarro 1998).

The temperature of the observed outflowing gas in starbursts is considerably cooler (few to ten million degrees) than would be expected for pure thermalized supernova+stellar wind ejecta (10^8 K). Given these temperatures, will this gas escape the galaxy gravitational potential, or will the gas remain bound and perhaps cool and return as a galactic fountain? Following Wang (1995), the 'escape temperature' for hot gas in a galaxy potential with an escape velocity v_{esc} is given by $T_{esc} \sim 4 \times 10^6 \ (v_{esc}/600 \ \text{km s}^{-1})^2$ K (where I have roughly chosen parameters appropriate to a typical spiral galaxy like our own).

The observed temperatures of the X-ray emitting gas in superwinds are generically observed to be one to ten million K, *independent of the galaxy rotation speed*, while T_{esc} should scale as the square of the rotation speed (*modulo* the extent of the dark matter halo). Thus, it appears that the X-ray emitting gas can easily escape from dwarf galaxies undergoing starbursts (Della-Ceca et al 1996) but its fate is unclear in the most massive starburst galaxies (Dahlem, Weaver, & Heckman 1998;

Wang 1995). As Larson & Dinerstein (1975) and many others have proposed, this selective loss of metal-enriched gas from the shallowest galaxy potential wells may be the physical mechanism that underlies the strong correlation between the metal abundance of the stellar population and the escape velocity in elliptical galaxies (e.g. Carollo & Danziger 1994). In this sense, superwinds should have a particularly devastating impact on dwarf galaxies (e.g. Dekel & Silk 1986; Marlowe et al 1995; Martin 1998; Della Ceca et al 1996).

If indeed superwinds carry substantial amounts of metal-enriched gas out of starbursts, we should see the cumulative effect of these flows in the form of a metal-enriched IGM and/or metal-enriched gaseous halos around galaxies. By now it is clear that typical MgII absorption-line systems at $z < 1$ seen in the spectra of distant QSOs arise in the metal-enriched halos of intervening galaxies (cf. Churchill, Steidel, & Vogt 1996). These galaxies appear to be normal systems (not starbursts), so it is unlikely that 'living, breathing' superwinds are implicated. Nevertheless, it is plausible that the metals in these galaxy halos may trace 'fossil' superwinds. That is, galactic halos may be polluted primarily by the episodic eruptions associated with powerful starbursts.

Can we say anything about the gas *outside* galaxies? The existence of an IGM in clusters of galaxies whose metal content exceeds that of all the stars in all the cluster's galaxies is one of the most remarkable phenomena in extragalactic astronomy. Recent ASCA X-ray spectra of this gas show solar or higher abundance ratios for the α-process elements like O, Ne, and Si relative to Fe. This implicates 'core-collapse' supernovae (the end product of high-mass stars) and - by inference - superwinds as the source of most of the metals in the cluster IGM (Gibson, Loewenstein & Mushotzky 1997; Renzini 1997).

If most of the metals in clusters of galaxies are floating around outside galaxies, could this be true more globally in the universe? We don't know the answer in the present-day universe, but the situation at high-redshift is very intruiging. Burles & Tytler (1996) use the detection of OVI absorption-lines in the spectra of background QSOs to estimate that the minimum metallicity of the entire gas-phase baryonic component of the universe at $z \sim 1$ is > 0.02 h solar for $\Omega_{B,gas} = 0.01$ h^{-2}. This is a lower limit, because it assumes all the Oxygen is the form of OVI and that the detected OVI absorption-lines are optically-thin. At $z > 2$, the presence of metals in the 'Lyα forest' (the 'cloudy' component of the early IGM) is certainly suggestive of the dispersal of chemically-enriched material by early superwinds (cf. Cowie et al 1995; Tytler et al 1995; Madau & Shull 1996). The Lyα forest material appears to have a metallicity of about 10^{-2} solar, but perhaps a super-solar ratio of Si/C that is suggestive of core-collapse supernovae as the source of metals (Cowie et al 1995; Giroux & Shull 1997; Boksenberg, Sargent, & Rauch 1998).

It is important to emphasize that metals in the form of the highly-ionized gas injected by superwinds could be largely invisible to QSO absorption-line spectroscopy. The X-ray gas in superwinds is so hot that the metals injected into the flow like O and C would be almost fully ionized and Si would be in the form of SiXII and higher. Since the recombination times in superwinds are estimated to exceed the adiabatic cooling times, it is likely that this gas remains highly ionized. The best bet for UV detection of hot inter-galactic metals would be to search for QSO absorption-lines due to coronal species like NeVIII $\lambda\lambda$770,780, MgX$\lambda\lambda$610,625, and SiXII $\lambda\lambda$499,521 (Verner, Tytler, & Barthel 1994). These species would all be abundant in gas at temperatures of one to several 10^6 K, and SiXII is even a significant ion at 10^7 K. All these doublets would be accessible for $z > 0.8$ if suitably clean (no intervening Lyman Limit systems) QSO sightlines can be found. X-ray spectroscopy could also in principle detect edges due to highly ionized metals. Future X-ray missions with good energy resolution and sensitivity like ASTRO-E and Constellation-X are promising in this regard.

The relatively large inferred masses of the X-ray emitting gas in local superwinds imply that we are primarily observing emission from ambient gas that has been 'mass-loaded' in some way into the superwind (cf. Suchkov et al 1996; Hartquist, Dyson, & Williams 1997; Strickland 1998). The large amount of mass-loading implies that the ratio of gas that is blown out of the starburst compared to the mass that is turned into stars ranges from of-order unity (if a normal complement of low-mass stars is formed in the starburst) to of-order ten (if only very high-mass stars are formed). To the extent that local starbursts are analogs to forming galaxies, this suggests that galaxy formation may have been an inefficient process with only a fraction of the initial complement of baryons being retained and converted into stars and the rest expelled into galactic halos or the IGM. This is consistent with measurements of the relative masses of gas and stars in galaxy clusters, and with more global comparisons of the ratio of Ω_B/Ω_* (cf. Fukugita, Hogan, & Peebles 1998).

CONCLUSIONS

Starbursts are defined as transient ($< 10^8$ years) episodes of intense star-formation that usually occur in the central-most kpc- scale regions of galaxies and dominate the integrated emission from the galaxy. They are a significant component of the present-day universe: they provide roughly 10% of the bolometric emissivity of the local universe and are the sites of $\sim$25% of the high- mass star-formation. Thus, they deserve to be understood in their own right.

They also offer unique 'laboratories' for testing our ideas about star formation, the evolution of high-mass stars, and the physics of the interstellar medium. They serve as local analogs of the processes that were important in the origin and early evolution of galaxies and in the heating and chemical enrichment of the inter-galactic medium (IGM).

In this contribution I have reviewed starbursts from this broad cosmological perspective, stressing several key lessons we have learned from starbursts:

1) Local UV-bright starbursts are quite similar in their UV photometric and structural properties to the UV-selected 'Lyman-break' galaxies at high-redshift.

Both have typical bolometric surface-brightnesses of $\sim 10^{10}$ to 10^{11} $L_\odot$ kpc^{-2}, and the Lyman-break galaxies appear to be scaled-up versions of the local starbursts. The implied stellar surface-mass densities are similar to those of present-day elliptical galaxies and galactic bulges.

2) Dust dramatically affects our view of starbursts both locally and at high-z. The strong empirical correlation between UV color and extinction defined by local UV-bright starbursts suggests that the Lyman-break galaxies are typically extincted by about two magnitudes in the UV.

3) More massive galaxies host more metal-rich starbursts, which are in turn more heavily extincted by dust. This probably reflects the well-known mass-metallicity relation for galaxies and the roughly linear dependence of the dust/gas ratio on metallicity. It implies that UV-selected samples will be biased towards low-metallicity galaxies.

4) More luminous starbursts are more heavily extincted by dust. This means that dust extinction changes the shape of the apparent UV luminosity function, and not just its normalization.

5) The sub-mm sources detected at high-z are most likely the high-luminosity tail of the Lyman-break population, and probably represent the most metal-rich (dustiest) starbursts occurring in the most massive halos. Using empirical methods for correcting the Lyman-break population for extinction, the most luminous Lyman-break galaxies overlap the SCUBA sub-mm population in luminosity and space-density.

6) The strong UV interstellar absorption lines directly trace outflows of metal-rich gas ('superwinds'). The broad blue-shifted lines imply that gas outflow rates may be comparable to the star-formation rate. It will be difficult to use these lines to straightforwardly probe the gravitational potential well of the host galaxy or halo.

7) X-ray observations of local starbursts suggest that these superwinds are strongly mass-loaded flows that carry out gas at a rate comparable to the rate of star formation at inferred velocities (of-order 10^3 km s^{-1}) that are *independent of the rotation speed of the host galaxy*, and are close to the escape velocity for a massive galaxy. The X-ray and UV evidence for outflows are in accord with the presence of substantial amounts of metals in the IGM.

Acknowledgements
I would like to thank my many collaborators who have worked with me on the data described in this paper, and especially Daniela Calzetti, Michael Dahlem, Roberto Della Ceca, Rosa Gonzalez-Delgado, Claus Leitherer, Gerhardt Meurer,

and Kim Weaver. This work was supported in part by the NASA LTSA program (grant NAG5-6400).

REFERENCES

1. Bland-Hawthorn, J. 1995, PASA, 12, 190
2. Boksenberg, A., Sargent, W., & Rauch, M. 1998, (astro-ph/9810502)
3. Buat, V., & Burgarella, D. 1998, A&A, 334, 772
4. Burles, S., & Tytler, D. 1996, ApJ, 460, 584
5. Calzetti, D., & Heckman, T. 1998, ApJ, in press (astro-ph/9811099)
6. Calzetti, D. Kinney, A., & Storchi-Bergmann, T. 1994, ApJ, 429, 582
7. Carollo, M., & Danziger, I.J. 1994, MNRAS, 270, 523
8. Churchill, C., Steidel, C., & Vogt, S. 1996, ApJ, 471, 164
9. Cowie, L., Songaila, A., Kim, T., & Hu, E. 1995, AJ, 109, 1522
10. Dahlem, M., Weaver, K., & Heckman, T. 1998, ApJS, 118, 401
11. Dekel, A., & Silk, J. 1986, ApJ, 303, 39
12. Della Ceca, R., Griffiths, R., Heckman, T., & MacKenty, J. 1996, ApJ, 469, 662
13. Flores, H., Hammer, F., Thuan, T., Cesarsky, C., Desert, F., Omont, A., Lilly, S., Eales, S., Crampton, D., & Le Fevre, O., 1998, ApJ, in press (astro-ph/9811202)
14. Franx, M., Illingworth, G., Kelson, D., van Dokkum, P., & Tran, K.-V. 1997, ApJL, 486, L75
15. Fukugita, M., Hogan, C., & Peebles, J. 1998, ApJ, 503, 518
16. Gallego, J., Zamorano, J., Aragon-Salamanca, A., & Rego, M. 1995, ApJL, 445, L1
17. Gibson, B., Loewenstein, M., & Mushotzky, R. 1997, MNRAS, 290, 623
18. Giroux, M., & Shull, S.M. 1997, AJ, 113, 1505
19. Gonzalez-Delgado, R., Leitherer, C., Heckman, T., Ferguson, H., & Lowenthal, J. 1998a, ApJ, 495, 698
20. Gonzalez-Delgado, R., Heckman, T., Leitherer, C., Meurer, G., Kinney, A., Koratkar, A., Krolik, J., & Wilson, A. 1998b, 505, 174
21. Hartquist, T., Dyson, J., & Williams, R. 1997, ApJ, 482, 182
22. Heckman, T. 1997, in "Cosmic Origins of Galaxies, Planets, and Life", ed. J.M Shull, C. Woodward, and H. Thronson, ASP
23. Heckman, T., & Leitherer, C. 1997, AJ, 114, 69
24. Heckman, T., Lehnert, M., & Armus, L. 1993, in "The Evolution of Galaxies and their Environments", Ed. M. Shull and H. Thronson, Kluwer, 455
25. Heckman, T., Robert, C., Leitherer, C., Garnett, D., and van der Rydt, F. 1998 (H98), ApJ, 503, 646
26. Hu, E., Cowie, L., & McMahon, R. 1998, ApJL, 502, L99
27. Hughes, D., Serjeant, S., Dunlop, J., Rowan-Robinson, M., Blain, A., Mann, R., Ivison, R., Peacock, J., Efstathiou, A., Gear, W., Oliver, S., Lawrence, A., Longair, M., Goldschmidt, P. & Jenness, T. 1998, Nature, 394, 241
28. Kunth, D., Mas-Hesse, J., Terlevich, E., Terlevich, R., Lequeux, J., and Fall, S.M. 1998, A&A, 334, 11
29. Larson, R.B., & Dinerstein, H.L. 1975, PASP, 87, 911

30. Lehnert, M., & Heckman, T. 1996b, ApJ, 462, 651

31. Lehnert, M., & Heckman, T. 1996a, ApJ, 472, 546

32. Leitherer, C., Walborn, N., Heckman, T., & Norman, C. 1991, "Massive Stars in Starburst Galaxies", Cambridge University Press

33. Leitherer, C., Robert, C., & Heckman, T. 1995, ApJS, 99, 173

34. Leitherer, C., & Heckman, T. 1995, ApJS, 96, 9

35. Lilly, S., Eales, S., Gear, W., Bond, R., Dunne, L., Hammer, F., Le Fevre, O., & Crampton, D. 1998, astro-ph/9807261

36. Lowenthal, J., Koo, D., Guzman, R., Gallego, J., Phillips, A., Faber, S., Vogt, N., & Illingworth, G. 1997, ApJ, 481, 673

37. Madau, P., & Shull, S.M. 1996, ApJ, 457, 551

38. Madau, P., Ferguson, H., Dickinson, M., Giavalisco, M., Steidel, C., & Fruchter, A. 1996, MNRAS, 283, 1388

39. Mannucci, F., Thompson, D., Beckwith, S., & Williger, G. 1998, ApJL, 501, L11

40. Marlowe, A., Heckman, T., Wyse, R., & Schommer, R. 1995, ApJ, 438, 563

41. Martin, C. L., 1998, ApJ, 506, 222

42. Meurer, G., Heckman, T., Leitherer, C., Lowenthal, J., & Lehnert, M. 1997, AJ, 114, 54

43. Meurer, G., Heckman, T., & Calzetti, D. 1998 (MHC), submitted to ApJ

44. Morton, D. 1991, ApJS, 77, 119

45. Pettini, M., & Lipman, K. 1995, A&A, 297, 63

46. Pettini, M., Smith, L., King, D., & Hunstead, R. 1997, ApJ, 486, 665

47. Pettini, M., Kellogg, M., Steidel, C., Dickinson, M., Adelberger, K., & Giavalisco, M. 1998, ApJ, 508, 539

48. Pei, Y.,& Fall, S.M. 1995, ApJ, 454, 69

49. Ponman, T., Cannon, D.,& Navarro, J. 1998, Nature, in press (astro-ph/9810359)

50. Puls, J. et al. 1996, A&A, 305, 171

51. Renzini, A. 1997, ApJ, 488, 35

52. Soifer, B.T., Sanders, D., Madore, B., Neugebauer, G., Lonsdale, C., Persson, S.E., & Rice, W. 1987, ApJ, 320, 238

53. Steidel, C., Giavalisco, M., Pettini, M., Dickinson, M., & Adelberger, K. 1996, ApJ, 462, L17

54. Storchi-Bergmann, T., Kinney, A.L. & Challis, P. 1995, ApJS, 98, 103

55. Strickland, D. 1998, Ph.D. Dissertation, University of Birmingham

56. Suchkov, A., Berman, V., Heckman, T., & Balsara, D. 1996, ApJ, 463, 528

57. Teplitz, H., Malkan, M., and McLean, I. 1998, ApJ, 506, 519

58. Tytler, D., et al. 1995, in QSO Absorption Lines, Ed. G. Meylan, Springer: Berlin, 289

59. Verner, D., Tytler, D., & Barthel, P. 1994, ApJ, 430, 186

60. Walborn, N., Lennon, D., Haser, S., Kudritzki, R., & Voels S. 1995, PASP, 107, 104

61. Wang, B. 1995, ApJ, 444, 590

62. Zaritzky, D., Kennicutt, R., & Huchra, J. 1994, ApJ, 420, 87

The Star Formation Rate Density of the Local Universe from the KPNO International Spectroscopic Survey

Caryl Gronwall*

*Astronomy Department, Wesleyan University, Middletown, CT 06457

Abstract. Understanding the nature and future evolution of star-forming galaxies found at 2 < z < 5 requires a thorough knowledge of the star formation process locally. The KPNO International Spectroscopic Survey (KISS) is a wide-field survey for extragalactic emission-line objects being carried out with the Burrell Schmidt at Kitt Peak. We have discovered approximately 1100 emission-line galaxies (ELGs) in a survey area of 68 sq. degrees (16.6 galaxies per square degree). These ELGs were identified via their $H\alpha$ flux and have been used to measure the star formation rate (SFR) density of the local universe. We find a SFR density approximately equal to that found by Gallego et al. (1995). We note, however, that our survey is incomplete for galaxies with $EW(H\alpha) < 25$ Å implying that the *total* local SFR density is higher than this value.

INTRODUCTION

The star formation rate (SFR) density of the local universe provides an important constraint on models of galaxy formation and evolution. In particular, measurements of the change in global SFR history with redshift is critically dependent on our knowledge of the local SFR. To date, this important number has only been measured by Gallego et al. [1], with a fairly small sample (~ 250) of galaxies at low redshifts ($z_{lim} = 0.045$). An independent check of this value is needed.

The KPNO International Spectroscopic Survey (KISS) is a wide-field survey for extragalactic emission-line objects being carried out with the Burrell Schmidt telescope at Kitt Peak. A full description of the survey can be found in [2]. The main difference between our survey and classical objective-prism searches for emission-line galaxies (ELGs), such as that of Gallego et al. [1], is our use of a CCD detector which allows us to probe to much fainter magnitudes ($B \sim 20$) and higher redshifts ($z_{lim} = 0.085$) than previous photographic surveys. As a result, our $H\alpha$ selected survey has detected 1126 ELG candidates in ~ 68 square degrees, for a surface

CP470, After the Dark Ages: When Galaxies were Young (the Universe at 2 < z < 5),
edited by Stephen S. Holt and Eric P. Smith

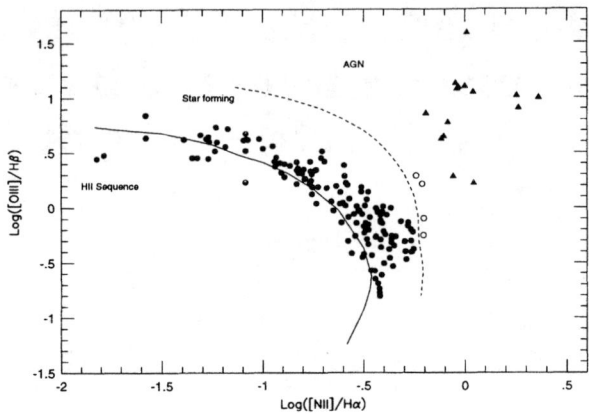

FIGURE 1. Line diagnostic diagram plotting the logarithm of [OIII]λ5007/Hβ against the logarithm of [NII]λ6583/Hα. Solid circles represent star-forming ELGs, open circles are on the border between star-forming galaxies and AGN and may be LINERS, and the triangles are Seyfert 2 galaxies. The solid line represents an HII model sequence at various metallicities from 0.1 Z$_\odot$ at the upper left to 2 Z$_\odot$ at the lower right.

density of 16.6 per square degree. In contrast, the surface density of Hα galaxies in the Gallego et al. survey is 0.56 per square degree.

FOLLOW-UP SPECTROSCOPY

In Spring 1998, we obtained follow-up spectroscopy for 223 ELG candidates selected via Hα emission using the WIYN 3.5 m telescope and the MDM 2.4 m telescope. These spectra provide: redshifts, Hα equivalent widths, and line fluxes for various important emission lines including Hβ, [OIII]λλ4959,5007, HeIλ5876, [OI]λ6300, Hα, [NII]λλ6548,6583, and [SII]λλ6717,6731. Measurements of the Balmer decrement allow us to directly measure the extinction (A_V) for each of these galaxies.

Over 90% of the galaxies observed were detected. 10% of the galaxies were active galactic nuclei (4 Seyfert 1's – not shown in Figure 1, 13 Seyfert 2's, and 4 potential LINER galaxies). A line diagnostic diagram plotting the logarithm of [OIII]λ5007/Hβ against the logarithm of [NII]λ6583/Hα is shown in Figure 1. The solid line represents the HII sequence with low metallicity, high-ionization ELGs in the upper left and high metallicity, low-ionization objects in the lower right part of the sequence. The star-forming ELGs follow the HII sequence allowing us to classify them from BCDs in the upper left to starburst nuclei in the lower right. We find many more high metallicity, low-ionization objects found than with tradition [OIII]λ5007-selected surveys. Hα-selected surveys better cover full range of ionization and metallicity. The mean A_V measured via the Balmer decrement is 1.3. A relationship between Balmer decrement and absolute B magnitude was used to correct the Hα flux measurements. The Hα flux and redshift measurements were used to calibrate flux and redshift measurements from the objective-prism spectra.

THE STAR FORMATION RATE OF THE LOCAL UNIVERSE

Because our redshifts and Hα fluxes for the majority of our sample were measured directly from the low-dispersion objective-prism spectra, our measurements are somewhat uncertain. Currently our Hα fluxes are calibrated using the follow-up spectra discussed above, and we have corrected for internal reddening using an empirical relation between M_B and the Balmer decrement also obtained from the follow-up spectra. We are planning to obtain higher-resolution follow-up spectroscopy for additional galaxies in hopes of creating a complete sample with full spectral information.

We have measured the Hα luminosity function of our sample using Hα luminosities derived from the measured line fluxes and redshifts, and the $1/V_{max}$ estimator. Note that our sample is *not* magnitude limited, but is instead limited by the line+continuum flux. We therefore determined the completeness of the survey using the V/V_{max} test as outlined by Salzer [3]. This procedure yields 808 galaxies in our "correctably complete" sample. Our Hα luminosity function is similar to that measured by Gallego et al. [1]; however, our sample is sensitive to substantially lower luminosities ($\sim$ 2 dex) than that of Gallego. The Schechter fit parameters of our LF are: $L^* = 10^{42.10 \pm 0.05}$, $\phi^* = 0.000666 \pm 0.0002$, and $\alpha = -1.29 \pm 0.05$.

The star formation rate (SFR) density of the local universe provides a critical constraint on our understanding of galaxy formation and evolution. The KISS sample allows us to determine this quantity out to a redshift of 0.085. If we integrate the Schechter luminosity function, we obtain a total Hα luminosity density of 1.06×10^{39} ergs s^{-1} Mpc^{-3}. Since this Hα luminosity reflects the number of ionizing photons from massive stars, we can use this number to compute the local SFR density. If we adopt the conversion factor of Madau et al. [4] and assume a Salpeter initial mass function, then

$$L(H\alpha) = 1.5 \times 10^{41} \frac{SFR}{M_\odot \text{yr}^{-1}} \text{ ergs s}^{-1}$$

which implies a local SFR density of 0.0071 $M_\odot$ yr^{-1} Mpc^{-3}. This value is consistent that found by Gallego et al. [1]. Our value for the local SFR density of the universe is plotted in Figure 2 along with other measurements over a range in redshift. We re-iterate that these results are *preliminary* and also represent a lower-limit to the

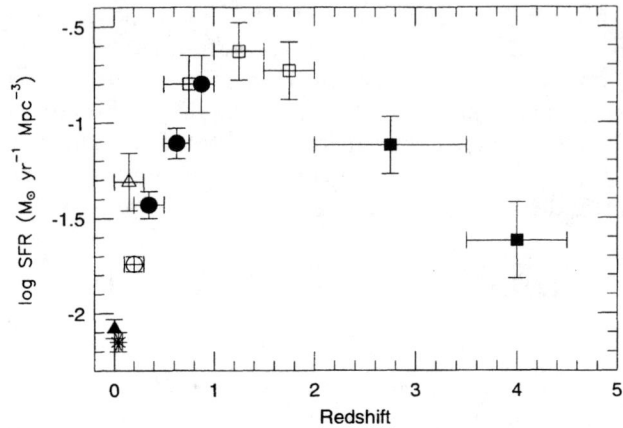

FIGURE 2. The observed redshift evolution of the star formation rate density of the universe. The asterisk represents our value, solid triangle is from [1], the open triangle is from [5], the open circle is from [6], the solid circles are from [7], the open squares are from [8], and the solid squares are from [9].

true SFR density of the local universe as our survey is incomplete for galaxies with EW($H\alpha$ < 25 Å).

ACKNOWLEDGMENTS

It is a pleasure to acknowledge my collaborators in the KISS project: John Salzer, Alexei Kniazev, Valentin Lipovetsky, Yuri Izotov, Todd Boroson, J. Ward Moody and Trinh Thuan. In addition, I would like to thank the many students who aided in processing the survey data: Laura Brenneman, Erin Condy, Mike Santos, Karen Kinemuchi, Julie Barker, Janice Lee, Kathy Rhode, and Kristin Kearns. Valuable software support was provided by Jose Herrero and Lisa Frattare.

REFERENCES

1. Gallego, J., Zamorano, J., Aragón-Salamanca, A. & Rego, M. 1995, ApJ 455 L1
2. Salzer, J. J. 1998, in "Dwarf Galaxies and Cosmology", eds. Thuan et al., in press
3. Salzer, J. J. 1989, ApJ 347 152
4. Madau, P., Pozzetti, L. & Dickinson M. 1998, ApJ 498 106
5. Treyer, M. A. et al. 1998, MNRAS 300, 303.
6. Tresse, L. & Maddox, S. J. 1998, ApJ 495 691

7. Lilly, S. J., LeFèvre, O., Hammer, F. & Crampton, D. 1996, ApJ 460 L1
8. Connolly, A. J., et al. 1997, ApJ 486, L11
9. Madau, P. et al. 1996, MNRAS 283, 1388

Chemical Constraints on the Star Formation History in High Redshift Galaxies

Henry A. Kobulnicky[*], Dennis Zaritsky[*], Robert C. Kennicutt[†], and
James L. Pizagno[†]

[*] Lick Observatory, UC Santa Cruz, Santa Cruz CA 95064
[†] Steward Observatory, University of Arizona, Tucson AZ, 85721

Abstract. It is now possible to study the chemical properties of distant galaxies
through emission line diagnostics which have long been used in local H II regions.
Initial results from a sample of compact, narrow emission line galaxies at $z = 0.2$ to
$z = 0.5$ show a range of metallicities from metal-poor like the SMC to super solar. The
oxygen abundances correlate strongly with optical luminosity just like local galaxies.
This suggests that the chemical production and retention in individual galaxies is determined predominantly by local characteristics, such as the integrated star formation
activity and the depth of the gravitational potential rather than the cosmic star formation history. The ratios of specific elements like nitrogen-to-oxygen can constrain
the evolutionary descendants of hi-z galaxies.

BACKGROUND AND MOTIVATION

The history of star formation within individual galaxies and on cosmic scales
is closely coupled to the production of heavy elements. The measurement of the
metallicity as a function of cosmic epoch is a consistency check on the star formation rates derived from imaging surveys. At the present time, absorption line
spectroscopy toward distant QSO background sources provides the only information
on the abundance of heavy elements are early times. Generally, 1) the metallicity
of damped Lyman Alpha absorbers decreases at higher redshifts [1,2], 2) there is
a considerable dispersion in DLA metal abundances at a given redshift [3], and 3)
the DLA systems apparently do not trace the bulk of star-forming material [4].

Now it is possible, with 8–10 m class telescopes, to measure the chemical properties of star-forming regions using nebular emission lines in much the same manner
as local H II regions [5–7,70]. This technique provides a complementary measure
of the chemical abundances in the immediate vicinity of star formation activity.

CP470, After the Dark Ages: When Galaxies were Young (the Universe at 2 < z < 5),
edited by Stephen S. Holt and Eric P. Smith

OBTAINING NEBULAR CHEMICAL ABUNDANCES FROM GLOBAL GALAXY SPECTRA

In cosmologically distant H II regions, only the brightest emission lines may be detectable, even using the collecting area and sensitivity of the largest telescope/instrument combinations envisioned today. A typical ground-based resolution element will encompass whole galaxies which have large internal variations in chemistry and temperature. Are global galaxy spectra in any way indicative of the physical properties of its ISM? Several effects may complicate the interpretation of global galaxy spectra, such as spatial variations in the ionization parameter, electron temperature, and metallicity within the galaxy. Figure 1 shows a comparison of the oxygen abundances derived from global galaxy spectra with the oxygen abundances derived from individual H II regions at 0.4 optical radii. The correlation is excellent, despite the presence of radial chemical gradients within the spiral galaxies. The correlation for small, metal-deficient galaxies is even better. These findings provide confidence that chemical properties do provide reliable information on the chemical content [9].

CONSTRAINING GALAXY EVOLUTION FROM HI-Z CHEMICAL MEASUREMENTS

We measured the H II–region oxygen and nitrogen abundances for 14 star-forming emission line galaxies (ELGs) at intermediate redshifts ($0.11 < z < 0.5$) using optical spectra obtained with the Keck II telescope and LRIS multi-object spectrograph [10]. The target galaxies exhibit a range of metallicities from slightly metal-poor like the LMC ($12+\log(O/H)\simeq8.4$) to super-solar ($12+\log(O/H)\simeq9.05$) where the solar value is $12+\log(O/H)\simeq8.89$ (Figure 3). Oxygen abundances of the sample correlate strongly with rest–frame blue luminosities. The metallicity–luminosity relation based on these 14 objects is formally indistinguishable from the one obeyed by galaxies in the local universe, although there is marginal evidence (1.1σ) that the sample is slightly more metal-deficient than local galaxies of the same luminosity. The observed galaxies exhibit smaller emission linewidths than local galaxies of similar metallicity, but proper corrections for inclination angle and other systematic effects are unknown. For 8 of the 14 objects we measure nitrogen-to-oxygen (N/O) ratios. Seven of the 8 systems show evidence for secondary nitrogen production, with $log(N/O) > -1.4$ similar to local spiral galaxies. These chemical properties are inconsistent with un-evolved objects undergoing a first burst of star formation. Comparison with local galaxies showing similar chemical properties suggests that these intermediate$-z$ objects contain substantial old stellar populations which were responsible for the bulk of the heavy element production presently seen in the ionized gas.

Four of the 14 galaxies exhibit small half-light radii and narrow emission line profiles (Compact Narrow Emission Line Galaxies—CNELGs; [11,12]) consistent

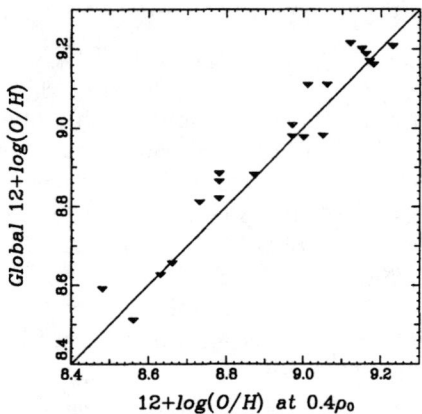

FIGURE 1. Oxygen abundances in 23 spiral galaxies with multiple H II regions and significant radial chemical gradients. The X-axis shows the characteristic oxygen abundance at a fiducial radius of 0.4 isophotal radii as tabulated by Zaritsky, Kennicutt, & Huchra (1994) from individual H II regions. The Y-axis show the oxygen abundance derived from the global nebular spectrum for each galaxy. Global spectra reproduce oxygen abundances which agree closely (0.1 dex) with the value at a fiducial radius of 0.4 isophotal radii (25 mag/arcsec in B). This shows that global nebular spectra provide robust measurements of the chemical properties of the ionized gas in star-forming galaxies even in the presence of ionization and chemical abundance variations.

with small dynamical masses despite their large optical luminosities and high levels of chemical enrichment. We find that the four CNELGs are indistinguishable from the 10 other emission line galaxies (ELGs) in the sample on the basis of their metallicity and luminosity alone. Because of their morphological similarity to H II and spheroidal galaxies, CNELGs have been proposed as the starburst progenitors of today's spheroidals. Our assessment of the stellar chemical abundances in nearby spheroidals reveals that the majority of the CNELGs are presently ~4 magnitudes brighter and ~0.5 dex more metal-rich than the bulk of the stars in well-known metal-poor dwarf spheroidals such as NGC 205 and NGC 185. Two of the four CNELGs exhibit oxygen abundances higher than the planetary nebula oxygen abundances in NGC 205, making an evolution between these two CNELGs and metal-poor dwarf spheroidals highly improbable. However, the data are consistent with the hypothesis that more luminous and metal-rich spheroidal galaxies like NGC 3605 may become the evolutionary endpoints of some CNELGs after 1 to 3 magnitudes of fading. We suggest that the $z = 0.1 - 0.4$ ELGs, and perhaps some of the CNELGs, are the precursors to today's spiral galaxies during an episode of vigorous bulge star formation ~5 Gyr ago.

342

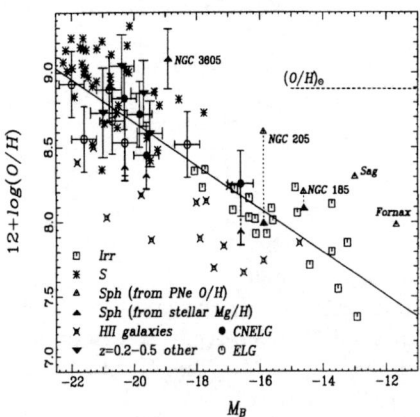

FIGURE 2. Oxygen abundance (12+log(O/H)) versus nitrogen-to-oxygen ratio (log(N/O)) for the target emission line objects and an assortment local galaxies. With the exception of object #2, all of the observed objects at $z > 0.2$ show relatively high oxygen abundances and evidence for secondary N production ($log(N/O) > -1.3$) from 4–8 solar mass stars. This degree of chemical enrichment indicates serveral prior episodes of star formation and makes them chemically most similar to local spiral galaxies.

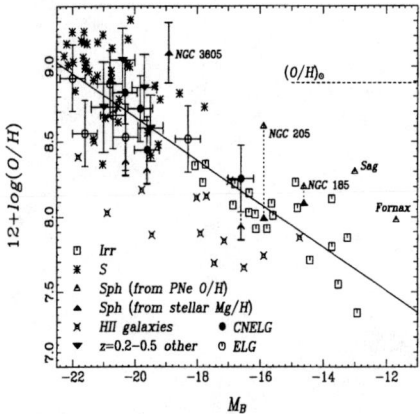

FIGURE 3. Oxygen abundance versus absolute blue magnitude for the sample. The solid line is a least squares fit to the local irregular and spiral galaxies. The data at $z = 0.4$ are consistent with the local relation between luminosity, L, and metallicity, Z, suggesting no evolution in the L-Z relation. Only H II galaxies seem to deviate significantly from the local relation. This uniformity of the L-Z relation at earlier epochs needs confirmation with a larger sample.

REFERENCES

1. Sargent, W. L. W., Boksenberg, A., & Steidel, C. C. 1988, ApJS, 68, 539
2. Steidel, C. C. 1990, ApJS, 74, 37
3. Pettini, M., Smith, L., King, D. L., & Hunstead, R. W. 1997, ApJ, 486, 680
4. Pettini, M. Ellison, S. L., Steidel, C. C., & Bowen, D. V. 1998, astro-ph 9802017
5. Aller, L. H. 1942, ApJ, 95, 52
6. Peimbert, M. 1975, ARA&A, 13, 113
7. Pagel, B. E. J. 1986, PASP, 98, 1009
8. Shields, G. A. 1990, ARA&A, 28, 525
9. Kobulnicky, H. A., Kennicutt, R. C., & Pizagno, J. L. 1999, ApJ, 514, in press
10. Kobulnicky, H. A. & Zaritsky, D. 1999, ApJ, 511, in press
11. Koo, D. C., Guzman, R., Faber, S. M., Illingworth, G. D. Bershady, M. A., Kron, R. G., & Takamiya, M. 1995, ApJ, 440, L49
12. Guzman, R., Koo, D. C., Faber, S. M., Illingworth, G. D., Takamiya, M., Kron, R. G., & Bershady, M. A. 1996, ApJ, 460, L5
13. Zaritsky, D., Kennicutt, R. C., & Huchra, J. P. 1994, ApJ, 420, 87

Implications of Faint Radio Sources for Star Formation History

Deborah B. Haarsma and R. Bruce Partridge

Haverford College, Haverford, PA 19041

Abstract. Faint radio sources provide important information about global star formation history. We make use of the correlation between radio and far infrared (FIR) flux to study the faint star forming galaxies producing the FIR and radio extragalactic background emission. Combining this with the typical radio spectral indices of these sources and their number counts, we estimate their mean redshift to be roughly between 1 and 2. In a second calculation, we use a simple model of the redshift distribution of faint radio sources, in combination with the number count information, to calculate the peak star formation rate density, which we can do without knowledge of the luminosity function.

INTRODUCTION

Faint radio sources provide important information about global star formation history. Sensitive observations of the HDF [1] and other radio fields [2] have found that sub-mJy radio sources are predominantly associated with star formation activity rather than AGN (see paper by Richards in these proceedings). Estimates of star formation based on radio observations have the advantage of being independent of dust content and extinction, as well as being more sensitive than far infrared (FIR) and sub-mm observations at redshifts of $z \sim 1$. Cram et al. [3] show the reliability of radio observations in calculating local star formation rates.

In this work, we make use of the tight correlation between radio and FIR luminosity for star forming galaxies [4] to compare the FIR and radio backgrounds and study the sources producing them. We also use a simple model of the redshift distribution of radio sources to determine the peak star formation rate.

FIR VS. RADIO BACKGROUNDS

This section follows our recent paper [5]. The far infrared background was recently detected with DIRBE [6], and is most likely the collective emission of star forming galaxies. We use the radio-FIR correlation for individual galaxies [7] to calculate the radio background associated with the FIR background. We then compare this to the observed radio background [8]. This allows us to draw several conclusions about the faint sources making up the FIR background:

1. The radio emission from these sources makes up about half of the observed extragalactic radio background.

2. Since (1.) is in agreement with other radio observations, the FIR-radio correlation appears to hold even for the very faint sources making up the FIR background. This confirms the assumption that the FIR background between about 140 and 240 μm is dominated by star-formation, not AGN activity.

3. By quantitatively comparing the radio and FIR backgrounds, we find the relationship shown in Figure 1, which allows us to estimate the mean redshift of the sources z given their radio spectral index α and the fraction A of the radio background they produce.

4. By extrapolating the counts of 3.6 cm radio sources to fainter flux densities, we estimate that most of the FIR background is produced by sources whose 3.6 cm flux density is greater than about 1 μJy. An RMS sensitivity of 1.5 μJy has already been reached in VLA observations [9]. The source counts indicate that the number density of 1 μJy sources is about 25/ arcmin^2, similar to some model predictions [10]. At this density, these sources will cause SIRTF to encounter confusion problems at 160 μm.

THE PEAK STAR FORMATION RATE

In this section, we assume that most faint radio sources lie in a shell at the redshift of peak star formation activity. This simplistic assumption is reasonable for faint radio sources due to a combination of evolutionary and cosmological geometrical effects [11]. Note also that some models show a significant peak in the star formation rate density as a function of redshift [12].

We can calculate the luminosity density $\rho(z)$ at the redshift of peak star formation z_p, without specific knowledge of the radio luminosity function and its evolution $\phi(L, z)$. The luminosity density is related to the luminosity function by

$$\rho(z) = \int L\phi(L, z)dL \tag{1}$$

The number of sources in a certain flux density interval is proportional to $\int \phi(L, z)dz$. Since we are assuming the sources lie in a small redshift range Δz_p at the peak z_p, we find the number density of sources to be

$$\frac{dN}{dS} \propto \Delta z_p \phi(L, z_p) \tag{2}$$

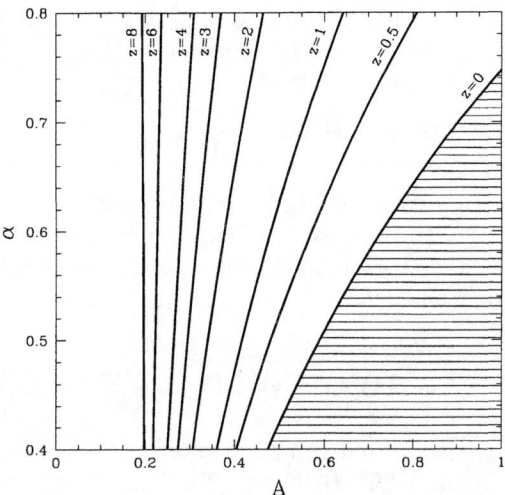

FIGURE 1. Relationship between the radio spectral index α, the ratio of star-formation flux to the total radio background A, and the typical redshift z for the sources making up the FIR background.

We then invert eq 2 and substitute into eq. 1 to find the luminosity density,

$$\rho(z_p) \propto \int \frac{S}{\Delta z_p} \frac{dN}{dS} dS. \tag{3}$$

Using an Einstein-De Sitter cosmology, and $\frac{dN}{dS} = -4.6S^{-2.3}$ at 3.6 cm for the counts of faint radio sources [9] with S down to $1\,\mu$Jy [5], the luminosity density becomes

$$\rho(z_p) = h\,3.1 \times 10^{18} \frac{(1+z_p)^{\alpha+\frac{5}{2}}}{\Delta z_p} \frac{\mathrm{W}}{\mathrm{HzMpc}^3} \tag{4}$$

at the redshift of peak star formation, where $H_0 = h100\,\mathrm{km\,s^{-1}Mpc^{-1}}$ and α is the typical radio spectral index of the sources.

The radio luminosity of a galaxy is proportional to its star formation rate [4], so the luminosity density $\rho(z_p)$ can be converted to the star formation rate density at the redshift of peak star formation,

$$\psi(z_p) = h\,0.012\,A \frac{(1+z_p)^{\alpha+\frac{5}{2}}}{\Delta z_p} \frac{\mathrm{M_\odot}}{\mathrm{yrMpc}^3}, \tag{5}$$

where A is the fraction of the radio background produced by star formation activity, assuming a Salpeter Initial Mass Function with masses $0.1M_\odot < M < 100M_\odot$.

Assuming, for instance, $h = 0.5$, $A = 0.5$, a radio spectral index of $\alpha = 0.4$, and star formation peaking at redshift $z_p = 1.5$ with width $\Delta z_p = 0.75$, we find that

$$\psi(1.5) = 0.057 \frac{M_\odot}{\text{yrMpc}^3} \tag{6}$$

This is somewhat lower than the estimates made by others for the peak star formation rate (*e.g.* [12,13]). To increase our value, a higher redshift of peak star formation (larger z_p), or a sharper peak (smaller Δz_p) would be needed. We are looking for ways to correct and refine this model further.

RADIO STAR FORMATION HISTORY

The optical identifications and redshifts will soon be known for nearly all of the micro-Jy radio sources in the HDF, HFF, the Medium Deep Survey, and the V15 field (Waddington & Windhorst, in preparation). In collaboration with I. Waddington, R. Windhorst, and E. Richards, we plan to use the redshift distribution of a complete radio sample, in combination with the radio luminosity function and its evolution, to calculate the star formation history as a function of redshift. Thus, we will make a radio version of the Madau et al. diagram [14], extending to redshifts of 2 or 3, without any need for assumptions about dust or extinction corrections.

We are happy to thank our collaborators, Jim Condon, Ed Fomalont, Ken Kellermann, Eric Richards, Ian Waddington, and Rogier Windhorst.

REFERENCES

1. Richards, E. A., Kellermann, K. I., Fomalont, E. B., Windhorst, R. A., & Partridge, R. B. 1998, AJ, 116, 1039
2. Windhorst, R. A., Fomalont, E. B., Kellermann, K. I., Partridge, R. B., Richards, E., Franklin, B. E., Pascarelle, S. M., & Griffiths, R. E. 1995, Nature, 375, 471
3. Cram, L., Hopkins, A., Mobasher, B., & Rowan-Robinson, M. 1998, ApJ, 507, 155
4. Condon, J. J. 1992, ARA&A, 30, 575
5. Haarsma, D. B., & Partridge, R. B. 1998, ApJ, 503, L5
6. Hauser, M. G., et al. 1998, ApJ, in press
7. Helou, G., Soifer, B. T., & Rowan-Robinson, M. 1985, ApJ, 298, L7
8. Bridle, A. H. 1967, MNRAS, 136, 219
9. Partridge, R. B., Richards, E. A., Fomalont, E. B., I.Kellermann, K., & Windhorst, R. A. 1997, ApJ, 483, 38
10. Guiderdoni, B., Hivon, E., Bouchet, F. R., & Maffei, B. 1998, MNRAS, 295, 877
11. Condon, J. J. 1989, ApJ, 338, 13
12. Dwek, E., et al. 1998, ApJ, in press
13. Rowan-Robinson, M., et al. 1997, MNRAS, 289, 490

14. Madau, P., Ferguson, H. C., Dickinson, M. E., Giavalisco, M., Steidel, C. C., & Fruchter, A. 1996, MNRAS, 283, 1388

A Radio Perspective on Star-Formation in Distant Galaxies

Eric A. Richards

National Radio Astronomy Observatory & University of Virginia
520 Edgemont Road
Charlottesville, VA 22903
email: erichard@nrao.edu

Abstract.
Determination of the epoch dependent star-formation rate of field galaxies is one of the principal goals of modern observational cosmology. Deep radio surveys, sensitive to starbursts out to $z \sim$ 1-2, may hold the key to understanding the evolution of the starburst phenomenon unhindered by the effects of dust. Using deep, high resolution radio observations of the Hubble Deep Field, we show that the μJy radio emission from field galaxies at $z \sim 0.4 - 1$ is primarily starburst in origin. In addition, we have discovered a population of optically faint, possibly obscured systems that are candidate high-z protogalaxies. At least one of these radio sources is identified with a sub-mm detection.

RADIO EMISSION AS A STAR FORMATION TRACER

The diffuse radio emission observed in local starbursts is believed to be a mixture of synchrotron radiation (excited by supernovae remnants and hence directly proportional to the number of supernovae producing stars) and thermal radiation (from HII regions and hence an indicator of the number of O and B stars in a galaxy). As the thermal and synchrotron radiation of a starburst dissipates on a physical time scale of $10^7 - 10^8$ years, the radio luminosity is a true measure of the instantaneous star-formation rate (SFR) in a galaxy, uncontaminated by older stellar populations. Since supernovae progenitors are dominated by $\sim$8 $M_\odot$ stars, synchrotron radiation has the additional advantage of being less sensitive to uncertainties in the initial mass function as opposed to UV and optical recombination line emission. However, the most obvious advantage of using the radio luminosity as a SFR tracer is its unsusceptibility to dust obscuration, as galaxies and the inter-galactic medium are transparent at centimeter wavelengths.

Comparison of the local radio luminosity function (LF) of star-forming galaxies [1] with those derived independently at FIR [2], Hα [3], and UV wavelengths [4]

CP470, After the Dark Ages: When Galaxies were Young (the Universe at 2 < z < 5),
edited by Stephen S. Holt and Eric P. Smith

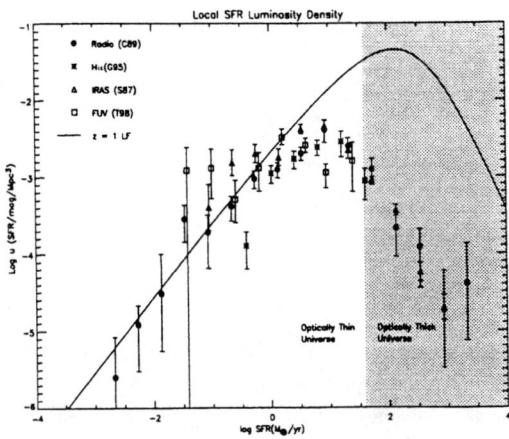

FIGURE 1. Shown is the contribution to the local star-formation luminosity density (u) per luminosity interval of star-forming galaxy. The radio, IRAS, Hα, and far-ultraviolet (FUV) luminosities have been converted to SFRs assuming a Salpeter IMF over 0.1-100 $M_\odot$ Notice that the four measures of SFRs all peak at approximately 10 $M_\odot$ yr^{-1} (see Cramm, ApJL, 506, 85 for a fuller discussion). The radio and IRAS points are in particularly good agreement, reflecting the tight FIR/radio correlation in star-forming galaxies. The shaded region represents what may be a dust curtain beyond which optical surveys are blind to star formation. If the SFR luminosity function evolves as L $\propto$ $(1 + z)^{3.5}$, then by $z = 1$, it will appear as the solid line. This analysis suggests that the bulk of global star-formation at high$-z$ is hidden from optical surveys.

shows surprising agreement. Figure 1 shows the four LFs in units of SFRs. This plot suggests that the bulk of local star formation is occurring in modest starbursts with SFR $\sim$ 10 $M_\odot$ yr^{-1}. However, past the peak in the LF, the Hα and UV estimates begin to drop below the radio/FIR rates, and beyond 50 $M_\odot$ yr^{-1} has entirely vanished. **This is direct evidence that optically selected surveys are incapable of detecting the most extreme and dust obscured starbursts.**

RADIO EMISSION FROM DISTANT GALAXIES

Our 1.4/8.5 GHz study of the Hubble Deep Field using the VLA and MERLIN has demonstrated that 70% of μJy sources are identified with morphologically peculiar, merging and/or interacting disk galaxies, many with independent evidence of star-formation (blue colors, infra-red excess, HII optical spectra) [5] [6]. Radio morphologies from the high resolution VLA/MERLIN observations of the HDF [7] indicate that 95% of μJy radio sources are resolved at 0.2″ resolution and suggests a median size of 2″, comparable to the optical extent of these $z \sim 0.4 - 1$ systems. These data exclude AGN as the dominant contributor to the radio luminosity in the majority of these systems. Thus the cosmological faint radio population is dominated by the distant analogs of local IRAS galaxies with suggested star-formation rates of 10-1000 $M_\odot$ yr^{-1}.

We have detected over 100 radio sources in complete samples within 4.5′ of

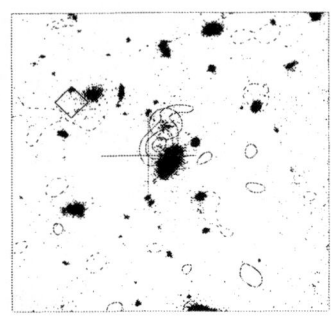

FIGURE 2. The greyscale shows a $20'' \times 20''$ HDF I-band image containing the SCUBA detection HDF850.1. The contours correspond to 1.4 GHz emission at the -2, 2, 4 and 6 σ level ($\sigma = 7.5$ μJy). The three sigma position error circle for HDF850.1 is shown after shifting to the VLA coordinate frame. The original position of HDF850.1 taken from Hughes et al. (1998) is denoted by the diamond. The ISO detection is marked with a cross with three sigma position errors (Aussel et al. 1998, A & A in press). The 0.1'' radio/optical registration clearly rules out association with the bright spiral. VLA3649+1221 may be the most obscured part of a larger galaxy 3-633.1 at $z = 1.72$ (located directly underneath the SCUBA error circle; Fernandez-Soto et al. 1998, AJ in press).

the HDF where deep optical imaging is available. Ninety percent of these radio sources are identified with galaxies of mean magnitude $I_{AB} \sim 22$. However, approximately 10% remain unidentified to $I_{AB} = 28$ in the HDF and $I_{AB} = 26$ in the HDF flanking fields. NICMOS imaging of one of these unidentified radio sources (VLA3642+1331) has revealed a H = 22.7 *disk* galaxy at unknown redshift (see I. Waddington elsewhere in these proceedings). Another radio source, VLA3649+1221, is identified with a $I_{AB} = 28$ object in the HDF. NICMOS imaging by Dickinson et al. (1998, private communication) shows that this object has a steeply rising optical/near infrared spectrum possibly suggesting a high redshift. We have subsequently shown that the brightest sub-mm/SCUBA source in the HDF (HDF850.1) (Hughes et al. 1998, Nature, 393, 241) is identified with VLA3649+1221. *It is our hypothesis that these systems are extreme, dust obscured starburst galaxies.* The surface density of these candidate high-z radio objects is comparable to the sub-mm population at ~ 0.1 arcmin^{-2}.

Of the five 850 μm sources in the HDF, two are solidly detected at radio wavelengths (Fig. 2 and 3), while two have less secure radio identifications [8]. Two of these radio/sub-mm sources are associated with $z \sim 0.5$ starbursts (HDF850.3 and HDF850.4). The other two detections (HDF850.1 and HDF850.2) must have redshifts less than 1.5 or be contaminated by AGN based on radio luminosity arguments. This radio analysis suggests the claim, based on low resolution sub-mm observations alone, that the optical surveys underestimate the $z > 2$ global star-formation rate are premature. On the other hand, the $z < 1$ star-formation history may have been underestimated if a significant fraction of the sub-mm population

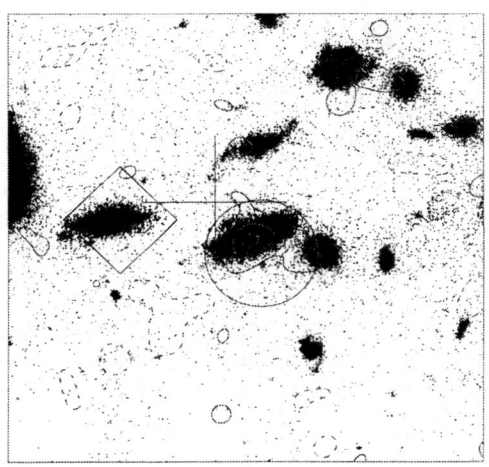

FIGURE 3. Radio 1.4 GHz contours drawn at the -2, 2, 4 and 6 σ level are overlaid on the HDF I-band image centered on the position of HDF850.4 ($20''$ on a side). A 15 μm detection from the complete catalog of Aussel et al. (1998) has been associated with this radio source and suggests likely starburst activity in the disk galaxy. The symbols are the same as for Figure 2. We estimate a SFR = 150 $M_\odot$ yr^{-1}.

lies at relatively low redshift, in agreement with the large number of $z = 0.4$-1 radio starbursts we are finding in the HDF. Analysis of the evolving radio star-forming luminosity function promises to shed light on the prevalence of dust enshrouded starbrust activity at high$-z$ (see Haarsma & Partridge, elsewhere in these proceedings).

It is a pleasure to thank my collaborators K. Kellermann, E. Fomalont, B. Partridge, R. Windhorst, T. Muxlow, I. Waddington, and D. Haarsma. Support for part of this work was provided by NASA through grant AR-6337.*-96A from the STSI, which is operated by AURA, Inc., under NASA contract NAS5-2655, and by NSF grant AST 93-20049.

REFERENCES

1. Condon, J. J. 1989, ApJ 338, 13
2. Soifer, B. T. et al. 1987, ApJ, 320, 238
3. Gallego, J., Zamorano, J., Aragon-Salamanca, A. & Rego, M. 1995,ApJL, 459, 43
4. Treyer, M., Ellis, R., Milliard, B., Donas, J. & Bridges, T. 1998, MNRAS, in press
5. Richards, E. A., et al. 1998, AJ, 116, 1039
6. Richards, E. A. 1998, in preparation
7. Muxlow, T. W. et al., in preparation
8. Richards, E. A. 1998, ApJL, submitted

A Tentative Detection of the Cosmic Infrared Background at 3.5 μm from *COBE*/DIRBE Observations

E. Dwek* and R. G. Arendt[†]

*Laboratory for Astronomy and Solar Physics, Code 685, NASA/Goddard Space Flight Center, Greenbelt, MD 20771. e-mail address: eli.dwek@gsfc.nasa.gov
[†]Raytheon STX, Code 685, NASA/Goddard Space Flight Center, Greenbelt, MD 20771

Abstract. Foreground emission and scattered light from interplanetary dust (IPD) particles and emission from Galactic stellar sources are the greatest obstacles for determining the cosmic infrared background (CIB) from diffuse sky measurements in the ~ 1 to 5 μm range. We describe a new method for deriving the CIB at near infrared wavelengths which reduces the uncertainties associated with the removal of the Galactic stellar emission component from the sky maps. The method produces positive residuals at 3.5 and 4.9 μm, of which only the 3.5 μm residual is nearly isotropic. We consider our result as a tentative detection of the CIB at this wavelength.

INTRODUCTION

Determination of the cosmic infrared background (CIB) from diffuse sky measurements is greatly hampered by the presence of foreground emission and scattered light from the interplanetary dust (IPD) cloud, and emission from discrete and unresolved stellar components in our Galaxy, and from dust in the interstellar medium (ISM). In a recent publication, Hauser et al. (1998; hereafter H98) presented the results of the search for the CIB in the 1.25 to 240 μm wavelength region that was conducted with the Diffuse Infrared Background Experiment (DIRBE) on the *Cosmic Background Explorer* (*COBE*) satellite. Careful subtraction of foreground emission from the IPD cloud (Kelsall et al. 1998) and from stellar and interstellar Galactic emission components (Arendt et al. 1998) revealed a residual emission component in the DIRBE skymaps that, after detailed analysis of the random and systematic uncertainties, was consistent with a positive signal at 100, 140, and 240 μm. Subsequent rigorous tests showed that only the 140 and 240 μm signals were isotropic, a strict requirement for their extragalactic origin. Only upper limits for the CIB intensity were given for $\lambda = 1.25 - 60$ μm, where the CIB detection

CP470, After the Dark Ages: When Galaxies were Young (the Universe at 2 < z < 5),
edited by Stephen S. Holt and Eric P. Smith
© 1999 The American Institute of Physics 1-56396-855-X/99/$15.00

was hindered by residual emission from the IPD cloud. In the 1.25 to 4.9 μm wavelength region, uncertainties in the subtraction of the Galactic stellar component contributed to the uncertainties as well. The upper limits on the CIB determined by H98 can be found in Table 2 of their paper.

Here we briefly summarize a new method for the subtraction of the Galactic stellar emission component. Instead of using a statistical model to characterize the Galactic stellar emission (Arendt et al. 1998), we create a spatial template of this emission from the DIRBE data itself. The method is described in more detail by Dwek & Arendt (1998; hereafter DA98). In this contribution we emphasize isotropy tests conducted on the residual emission to examine its possible extragalactic origin.

I Description of the Method

A Subtraction of the Galactic Stellar Emission

We use the DIRBE 1.25, 2.2, 3.5, and 4.9 μm all sky maps from which the emission from interplanetary dust (IPD) has been subtracted (Kelsall et al. 1998) as the starting point in the analysis. The intensity of these maps should, in principle, contain only the Galactic emission (starlight and ISM emission) and the CIB.

We then use the IPD−emission subtracted 2.2 μm skymap to create a template map for the Galactic stellar emission component at this wavelength. The 2.2 μm map is particularly suitable for our analysis, since the contribution from interstellar dust emission is negligible at this wavelength. Furthermore, ground−based galaxy counts in the K-band (e.g. Gardner 1996) provide a strict lower limit of 7.4 nW m^{-2} sr^{-1} to the CIB intensity at this wavelength which can be used as the nominal value for the CIB at 2.2 μm. The spatial template of the Galactic stellar emission component is derived by subtracting this uniform intensity from the 2.2 μm zodi−subtracted skymap. The final results of our analysis depend only weakly on the adopted 2.2 μm CIB intensity, and we explicitly state their dependence on this value.

The next step in the analysis consists of correlating the 2.2 μm stellar emission template with the zodi−subtracted skymaps at 1.25, 3.5, and 4.5 μm (see Figure 1 in DA98).

The slopes of the correlations represent the colors of stellar emission, and are in very good agreement with the colors of M and K giants shown in Figure 2 of Arendt et al. (1994). The intercepts of the correlations represent the residual emission. At 3.5 and 4.9 μm we had to subtract the small contribution of emission from interstellar dust.

The final step in the analysis consisted of testing the residual emission components for positivity and isotropy, the two criteria required to ascertain their extragalactic nature (H98). The analysis concentrated on high quality (HQ) regions of the sky, identified as such for their location at high Galactic latitudes (b)

and ecliptic latitudes (β), in which the contributions from the Galactic and zodiacal emission components are relatively small compared to other regions of the sky. Figure 1 in H98 depicts the location of these HQ regions.

The error analysis (see DA98 for further details) shows that our method significantly reduces the errors associated with the subtraction of Galactic starlight. However, only at 3.5 and 4.9 μm are the residuals positive, and as described in more detail below, only the 3.5 μm residual is nearly isotropic. A panel of three DIRBE 3.5 μm maps depicting the as observed sky, the sky after the removal of the interplanetary dust emission, and the final residual map after the subtraction of the Galactic starlight and ISM emission was presented by Cowen (1998).

B Isotropy Tests

Any extragalactic signal is expected to be isotropic, barring small fluctuations expected from the discrete distribution of galaxies. We therefore conducted several isotropy tests on the residual intensities in the 3.5 and 4.9 μm bands, which in increasing order of complexity consisted of: (a) comparison of mean intensities in the various HQ regions; (b) examination of large scale gradients; and (c) analysis of the two point correlation function in the HQ regions.

Table 1 in DA98 shows that the residual 3.5 μm emission is isotropic, as evident from the agreement between the mean intensities in the various HQ regions. The 4.9 μm residuals failed this simple isotropy test.

Examination of large scale gradients in the residual 3.5 μm map shows that they are significantly reduced compared to those found in the residual intensity map presented by H98. A comparison between the cuts in the residual intensity maps derived by the two methods is presented in Figure 1.

Although greatly reduced, the bottom panel of Fig. 1 still shows the presence of large scale gradients in the residual map. At low Galactic latitudes ($\beta \approx 60°$, $240°$ in Figure 1) the gradient is largely due to residual Galactic stellar emission which is affected by dust extinction. Dust extinction is negligible in the HQ regions. Additional gradients are due to the incomplete subtraction of the IPD emission.

The same gradient seen in the 3.5 μm image is significantly stronger in the 4.9 μm residual map, and is responsible for the larger dispersion in the mean intensities in the various HQ regions. We therefore consider the 4.9 μm result as only an upper limit.

We also performed a two point correlation analysis of the residuals, a strict isotropy test adopted by H98. In this test the two point correlation function of the residual emission is compared to that of a simulated flat background possessing random Gaussian uncertainties estimated from the DIRBE data (H98). Figure 2 shows the results of our analysis.

The top panel shows the two point correlation function of the residuals presented by H98 (Figure 3 in their paper). The middle panel shows the two point correlation function of the residuals using our method for subtracting the Galactic stellar emission component, but with the same uncertainties in the residual maps as those adopted by H98. These uncertainties are dominated by unremoved stellar emission, visible as small–scale structure in the residuals shown in Figure 1. With these uncertainties, the residual emission is isotropic. The bottom panel shows the two point correlation function of the residuals obtained by using the new method for foreground subtraction, and their reduced uncertainties. Note the large change in scale compared to the previous panels. Because of the smaller uncertainties, the residual emission does not pass the strict two point correlation isotropy test.

If we used only the mean intensities in the various HQ patches as the criterion for isotropy, the 3.5 μm residual would classify as a definite detection of the CIB. However, because of the persistence of large scale gradients (albeit largely reduced compared to the previous analysis), and the existence of correlations between the pixel intensities beyond what one would expect from a flat background, we only regard the derived residual intensity as a tentative detection of the CIB at 3.5 μm with an intensity given by:

$$\nu I_{res}(\text{nW m}^{-2} \text{ sr}^{-1}) = 9.9 + 0.312 \; \Delta_{CIB}(2.2 \; \mu m) \pm 2.9 \tag{1}$$

where the quoted errors represent 1 σ uncertainties, and $\Delta_{CIB}(2.2 \; \mu m) \equiv [I_{\text{CIB}}(2.2 \; \mu m) - 7.4]$ represents the difference in the actual value of the CIB at 2.2 μm (in nW m^{-2} sr^{-1}) and the nominal value adopted in our model. Our analysis also yields new upper limits (95% confidence limit) on the CIB at 1.25 and 4.9 μm of 68 and 36 nW m^{-2} sr^{-1}, respectively.

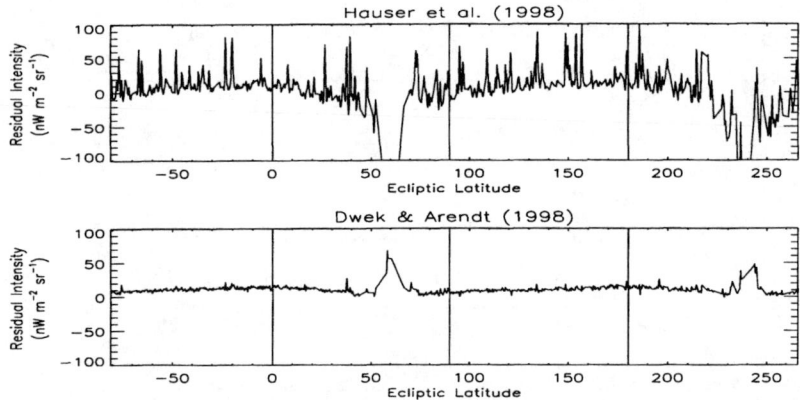

FIGURE 1. (Top) Intensities in the Hauser et al. (1998) 3.5 μm residual map along a great circle at ecliptic longitude 0°and 180°(the latter plotted as latitudes > 90°). All foreground models have been removed, and the locations of bright stars have been blanked. (Bottom) The same intensity slice after the removal of the improved model of the Galactic stellar emission (Dwek & Arendt 1998).

REFERENCES

1. Arendt, R. G. et al. 1994, ApJ, 425, L85
2. Arendt, R. G. et al. 1998, ApJ, 508, 74
3. Dwek, E. & Arendt, R. G. 1998, ApJ, 508, L9
4. Dwek, E. et al. 1998, ApJ, 508, 106
5. Fixsen, D. J., Dwek, E., Mather, J. C., Bennett, C. L., & Shafer, R. A. 1998, ApJ, 508, 123
6. Cowen, R. 1998, Science News, 154, 326
7. Gardner, J. P. 1996, in Unveiling the Cosmic Infrared Background, ed. E. Dwek (New York: AIP Press), p. 127
8. Hauser, M. G. et al. 1998, ApJ, 508, 25
9. Kelsall, T. J., et al. 1998, ApJ, 508, 44

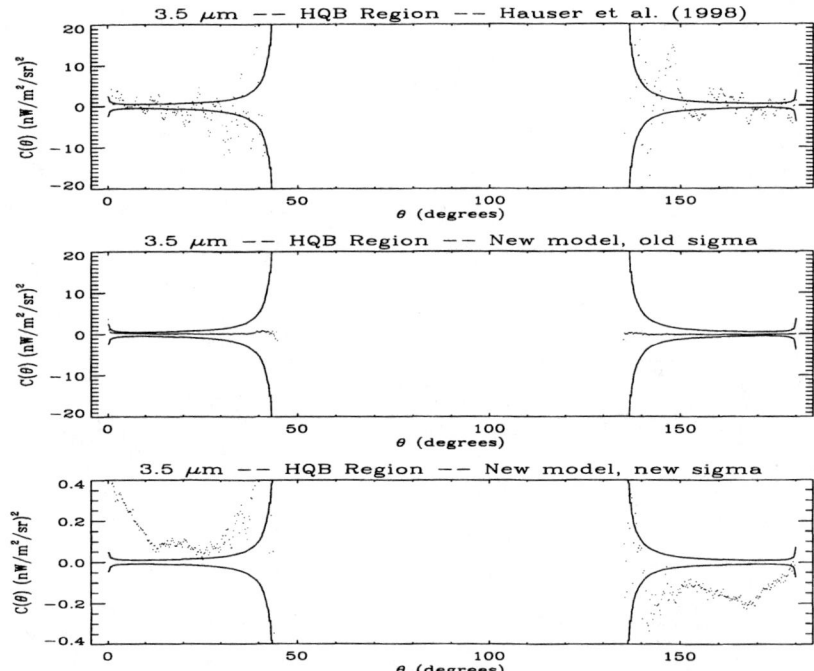

FIGURE 2. Two-point correlation functions of the 3.5 μm residual emission in the HQB region. (Top) The result after foreground subtractions as presented by Hauser et al. (1998). (Middle) The result after foreground subtractions using the improved model of the Galactic stellar emission (Dwek & Arendt 1998) shown with the same uncertainties as the Hauser et al. (1998) result. (Bottom) The result after foreground subtractions from Dwek & Arendt (1998), now shown with uncertainties that do not include a component allowing for random stellar variations. Note the large change in scale.

Calibrating UV Emissivity And Dust Absorption At $z \approx 3$

Gerhardt R. Meurer*, Timothy M. Heckman*, and Daniela Calzetti[†]

*The Johns Hopkins University, Baltimore, MD 21218
[†]Space Telescope Science Institute, Baltimore, MD 21218

Abstract. We detail a technique for estimating the UV extinction and luminosity of UV selected galaxies using UV quantities alone. The technique is based on a tight correlation between the ratios of far infrared (FIR) to UV flux ratios and UV color for a sample of local starbursts. A simple empirical fit to this correlation can be used to estimate UV extinction as a function of color. This method is applied to a sample of Lyman-break systems selected from the HDF and having $z \approx 3$. The resultant UV emissivity is at least nine times higher than the original Madau et al. [1] estimate. This technique can be readily applied to other rest-frame UV surveys.

INTRODUCTION

Most of the light from high mass stars is emitted in the ultraviolet (UV; $\lambda \approx 1100 - 3000$Å), making it an attractive passband for tracing cosmic star formation evolution. This utility is accentuated with increasing redshift as the rest-frame UV emission enters the optical where modern detectors have quantum efficiencies approaching unity. Unfortunately, star formation occurs in a dusty environment, and dust efficiently absorbs and scatters UV radiation. This must also be the case in the early universe since dust has been observed in objects with $z > 4$ (e.g. [2]).

The challenge of interpreting rest-frame UV emissivities is to devise an adequate prescription to account for dust absorption. Currently there is much debate in the literature on what the proper dust correction prescription is, resulting in different groups estimating $\lambda = 1600$Å dust absorption factors ranging from a factor of about 3 (e.g. [3]) to 20 [4] at $z \approx 3$. The amount of high-z dust absorption has a direct bearing on interpretting how galaxies evolve. Small dust corrections favor hierarchical models of galaxy formation, while large corrections favor monolithic collapse models [5].

Here we consider the UV luminosity density at $z \approx 3$ derived mainly from the U-dropouts in the Hubble Deep Field (HDF) [6]. Our technique [7] is based on the

CP470, After the Dark Ages: When Galaxies were Young (the Universe at 2 < z < 5),
edited by Stephen S. Holt and Eric P. Smith

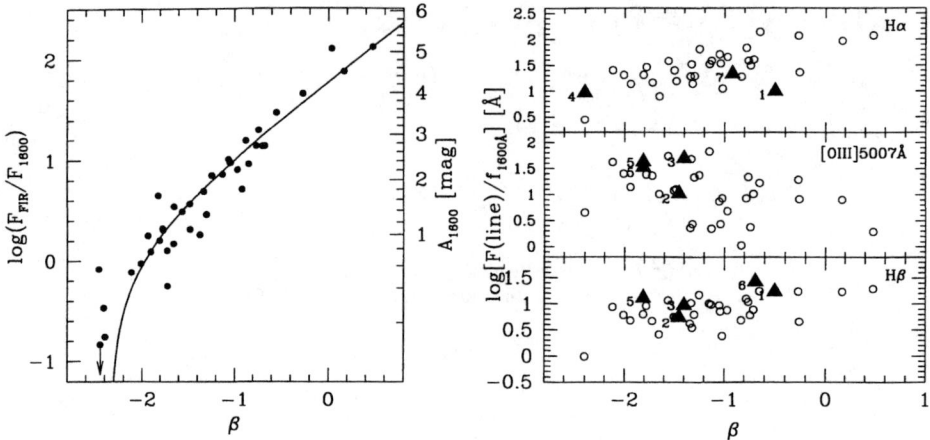

FIGURE 1 (left). FIR to UV flux ratio compared to ultraviolet spectral slope β for local UV selected starbursts.

FIGURE 2 (right). Ratio of emission line flux to UV flux density for local starbursts (circles), and HDF U-dropouts (numbered triangles).

strong similarity between local starburst galaxies and Lyman-break systems (e.g. [8]). Throughout this paper we adopt $H_0 = 50\,\mathrm{km\,s^{-1}\,Mpc^{-1}}$, $q_0 = 0.5$.

METHOD

In earlier works [9,10], we showed that for local UV selected starburst galaxies, the ratio of far infrared (FIR) to UV fluxes correlates with UV spectral slope β ($f_\lambda \propto \lambda^\beta$ - β is essentially an ultraviolet color). This is illustrated in Fig. 1. Since F_{FIR} is dust reprocessed UV flux, *this empirical correlation can be used to recover the intrinsic UV flux from UV quantities alone.* In addition, for starbursts, the y axis can be transformed directly into a UV absorption [7]. The fitted line is a simple linear fit to the transformed data of the form: $A_{1600} \propto \beta - \beta_0$.

We selected our HDF U-dropout sample from a corner cut out of the $U_{300} - B_{450}$ versus $V_{606} - I_{814}$ color-color diagram ($V_{606} - I_{814} < 0.5$; $U_{300} - B_{450} \geq 1.3$) and adopted the same magnitude limits as Madau et al. [1]. We select in $V_{606} - I_{814}$ instead of $B_{450} - I_{814}$ [1] because (1) V_{606} is less affected by the Lyman forest and edge than B_{450}, and (2) this selection yields fairly even cutoff in β, and hence in A_{1600}. Note that our selection recovers high-z galaxies in the "clipped corner" of the Madau et al. [1] selection area, and includes no known low-z interlopers.

We applied our absorption law fit to broad-band $V_{606} - I_{814}$ colors transformed into β. The transformation was derived from high-quality IUE spectra that were "redshifted" through the $z = 2$ to 4 range of U-dropouts. The transformation is

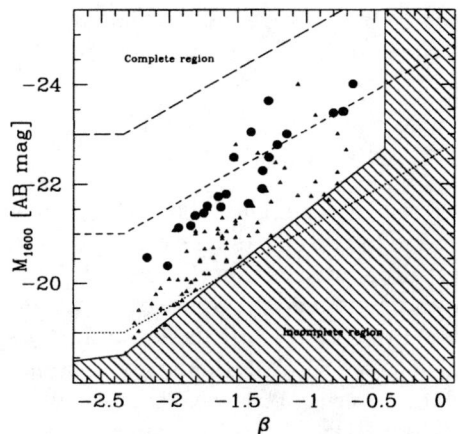

FIGURE 3. Color - absolute magnitude diagram of HDF U-dropouts.

linear in color with a quadratic z correction. The z correction is needed to account for the Lyman forest and Lyman-edge creeping into the V_{606} band at high-z.

Figure 2 shows a test of our technique. It compares the ratio of (rest frame) optical emission line flux to UV continuum flux density for local starbursts, and seven U-dropouts [3,11,12,5]. These ratios are not corrected for dust absorption. The overlap of the two samples indicates that U-dropouts are ionizing populations to the same degree as local starbursts. Hence their intrinsic UV spectrum should be similar. Pettini et al. [3] claim that U-dropouts probably suffer from little dust absorption since they tend to have fairly low $F_{H\beta}/f_{1600}$ values. However, this ratio can be misleading. In fact, $F(\text{line})/f_{1600}$ is not a good indicator of dust absorption: it does not correlate strongly with β, which we know to be a good indicator of dust absorption (Fig. 1). This is the case for both the local and U-dropout samples. The reason for this was first proposed by Fanelli et al. [14]: H II emission lines are seen through a larger column of dust than the general UV continuum thus cancelling the expected benefit in opacity of observing in the optical instead of the UV.

RESULTS

Figure 3 plots the absorption corrected absolute AB magnitude of the HDF U-dropouts versus β. The broken lines show $M_{1600\text{Å}}$ in the absence of absorption correction. The data show an apparent color - luminosity correlation. This is in part due to the selection limits, but the lack of very luminous blue galaxies is real. This implies that there is a mass - metallicity relationship at $z \approx 3$. It also shows that the most luminous galaxies tend to have the most dust absorption. A similar color - luminosity correlation is seen in local starbursts [15].

Summing the results for the HDF U-dropouts yields *lower limits* to the intrinsic UV emissivity, and hence the star formation rate density:

$$\rho_{1600,0} \gtrsim 1.5 \times 10^{27} \, \mathrm{erg \, s^{-1} \, Hz^{-1} Mpc^{-3}}$$

$$\rho_{\mathrm{SFR}} \gtrsim 0.19 \, \mathcal{M}_\odot \, \mathrm{yr^{-1} \, Mpc^{-3}}$$

We find that $\rho_{1600,0}$ is factor of 9.2 higher than ρ_{1600} first estimated by Madau et al. [1]. This difference is due to two effects: the dust absorption correction (factor of 5.5), and the improved U-dropout selection (factor of 1.7). These emissivities are still lower limits because we have made no completeness corrections, and because our $V_{606} - I_{814}$ selection is only sensitive to galaxies with $A_{1600} \lesssim 3.4$ mag.

Recently, Madau et al. [5] (see also Madau, this volume) have fit models to cosmological emissivity data covering rest-wavelengths from the FIR to the UV and redshifts out to ~ 4. Their HDF U-dropout sample now has a selection similar to ours. Our $\rho_{1600,0}$ estimate for the U-dropouts is a factor of 2.5 larger than their preferred model, which simulates the heirarchical collapse scenario and which includes a small amount of dust absorption. However it is only 30% larger than their "monolithic collapse" model. Hence, the initial phase of galaxy collapse was probably more rapid than predicted by heirarchical models and somewhat obscured from our view by at least modest amounts of dust.

REFERENCES

1. Madau, P., Ferguson, H.C., Dickinson, M.E., Giavalisco, M., Steidel, C.C., & Fruchter, A. 1996, MNRAS, 283, 1388
2. Guilloteau, S., Omont, A., McMahon, R.G., Cox, P., & Petitjean, P. 1997, A&A, 328, L1
3. Pettini, M., Kellogg, M., Steidel, C.C., Dickinson, M., Adelberger, K.L., & Giavalisco, M. 1998, ApJ, submitted
4. Sawicki, M., & Yee, H.K.C. 1997, AJ, 115, 1329
5. Madau, P., Pozzetti, L., & Dickinson, M. 1998, ApJ, 498, 106
6. Williams, R.E., et al. 1996, AJ, 112, 1335
7. Meurer, G.R., Heckman, T.M., & Calzetti, D. 1998, ApJ, submitted
8. Lowenthal, J.D., Koo, D.C., Guzmán, R., Gallego, J., Phillips, A.C., Faber, S.M., Vogt, N.P., Illingworth, G.D., & Gronwall, C. 1997, ApJ, 481, 673
9. Meurer, G.R., Heckman, T.M., Leitherer, C., Kinney, A., Robert, C., & Garnett D.R. 1995, AJ, 110, 2665
10. Meurer, G.R., Heckman, T.M., Lehnert, M.D., Leitherer, C., & Lowenthal, J. 1997, AJ, 114, 54
11. Wright & Pettini, M. 1998, private communication.
12. Bechtold, J., Yee, H.K.C., Elston, R., & Ellingson, E. 1997, ApJ, 477, L29
13. Bechtold, J., Elston, R., Yee, H.K.C., Ellingson, E., & Cutri, R.M. 1998, preprint (astro-ph/9802230)
14. Fanelli, M.N., O'Connell, R.W., & Thuan, T.X. 1988, ApJ, 334, 665

15. Heckman, T.M., Robert, C., Leitherer, C., Garnett, D.R., & van der Rydt, F., 1998, ApJ, in press

The NICMOS Parallel Program: Grism Survey Results and Emission-Line Candidates

Harry I. Teplitz[1,2], Jonathan P. Gardner[1,2], Eliot Malumuth[1,3], Sara Heap[1], & Patrick McCarthy[4], Lin Yan[4], Matthew Malkan[4]

[1] *Code 681, Goddard Space Flight Center, Greenbelt MD 20771*
[2] *NOAO Research Associate,*　　　[3] *Raytheon STX*
[4] *Carnegie Observatories, 813 Santa Barbara Street, Pasadena CA 91101*
[5] *UCLA Department of Astronomy & Physics, Los Angeles CA 90095*

Abstract. In its parallel pointings, NICMOS has given us a unique window on the evolving Universe. We have analyzed the deep parallel NICMOS grism observations, covering more than 60 square arcminutes and down to continuum $H_{1.6}=22$ in the deepest fields. With this slitless grism spectroscopy, NICMOS reveals 33 new line-emitting galaxies. In most cases, the single emission line is likely to be Hα at $0.7 < z < 1.9$, and their volume density is comparable to that of the brighter galaxies selected by the Lyman break method. Our first Keck LRIS spectrum of one of these galaxies confirms this identification. The inferred star formation rates for the emission-line galaxies are 2–480M$_\odot$/year. We present the details of the grism survey and our analysis so far.

INTRODUCTION

The NICMOS grism spectra provide a new probe of star-forming galaxies in the epoch just after the "Dark Ages". The spectra are sensitive to the primary tracer of star formation, Hα (Kennicutt 1983), at redshifts up to $z = 1.9$. Previous estimates of the high-z star formation density at the highest redshifts ($z > 2.5$) have relied on the UV continuum flux (Steidel et al. 1996) which can be highly affected by reddening; Pettini et al. (1998) suggest a factor of 3–6 extinction. Other estimates of star-formation density at $z > 2$ have been made from IR emission-line searches centered on QSO or absorber environments (e.g. Teplitz et al. 1998) but may be biased by the cluster environment. NICMOS spectra are a significant advance, since they are less affected by extinction and represent a field population of galaxies. The full details of this survey may be found in McCarthy et al. (1999).

CP470, After the Dark Ages: When Galaxies were Young (the Universe at 2 < z < 5),
edited by Stephen S. Holt and Eric P. Smith

OBSERVATIONS

Parallel observations with the G141 grism of the NICMOS Camera 3 were taken throughout 1998. The data obtained cover the 1.1 μm-1.9 μm spectral region, at an effective resolution of R=150 (worse for highly extended objects). The typical observing strategy consisted of two dither positions per field, offset in the spectral direction by 2″. Observations were taken in pairs of F160W images and G141 spectra.

Spectra were extracted by members of the team at GSFC in a process similar to that employed by the NICMOS-Look software (see discussion McCarthy et al. 1999). Spectral positions on the chip were determined by the known offset from the object location in the direct image. Extractions were performed on individual (unregistered) frames. Each spectrum was extracted in a weighted box, based on the extent of a point source spectrum (most emission-line objects are found to be compact). The wavelength solution was calculated by independent analysis of calibration observations of a planetary nebula. Flat-fielding was performed using a data-cube of narrow-band observations of a uniform source, and interpolation to the wavelengths of the extracted pixels. The flux calibration was performed using independent analysis of the white dwarf observations. Finally, individual spectra were registered in wavelength space. Line fluxes and equivalent widths were measured using gaussian fitting to the line and polynomial fitting of the continuum.

Emission line candidates were identified by eye. A number of false-positives had to be eliminated. Artifacts that appear similar to emission lines include the following: unrejected cosmic rays, persistent images, and the zeroth order image of other objects. The first two problems were identified by examining the data in each dither position; only real features will move with the spectra. Zeroth order spectra can usually be identified by the presence of an object in the direct image, or the presence of the first order spectrum in the G141 frame; emission lines in a small "danger zone" where neither check is possible must be regarded as more suspect.

RESULTS

NICMOS grism observations were obtained for 100 different fields, with total integration times of 2000–20,000 seconds. Some fields were discarded due to excessive contamination by cosmic rays, and others we exclude from the current analysis due to their high star counts. In the 85 remaining pointings (59 square arcminutes), continuum spectra were measured for objects with $H_{1.6}$=20–22. Average emission-line flux limits of 4×10^{-17}ergs/cm^2/s (3σ) were achieved. 33 emission-line objects were identified, in 22 different fields (6 fields had multiple candidate objects). No objects were found with more than one emission line.

We consider the most likely identification for all of the emission lines to be Hα, redshifted into the G141 spectra range. Hα is the strongest probable line, particularly for star-forming galaxies. Other rest-frame optical lines like Hβ and

[OIII]5007Å are typically substantially weaker than Hα. [OII] is the second strongest line, but would require $z \sim 3$ and is thus are less likely due to the relative brightness of the objects. Lyα at very high redshift is unlikely for the same reason. Lines at longer wavelength than Hα are unlikely for the opposite reason, as they would require too low a redshift, and in many cases would place multiple lines in the G141 spectral range.

We have obtained and optical spectrum of one of the brightest of our candidates. This object, denoted J0055+8518 A, has a continuum magnitude of $H_{1.6}$=19.7 and a strong emission line at 1.160 μm, with an observed line flux of 5.3×10^{-16} ergs/cm^2/s. Our optical spectroscopy with LRIS (Oke et al. 1995) on Keck II confirms the inferred redshift of this object of $z = 0.76$. Figure 1 shows the image and 1D and 2D spectra of this object.

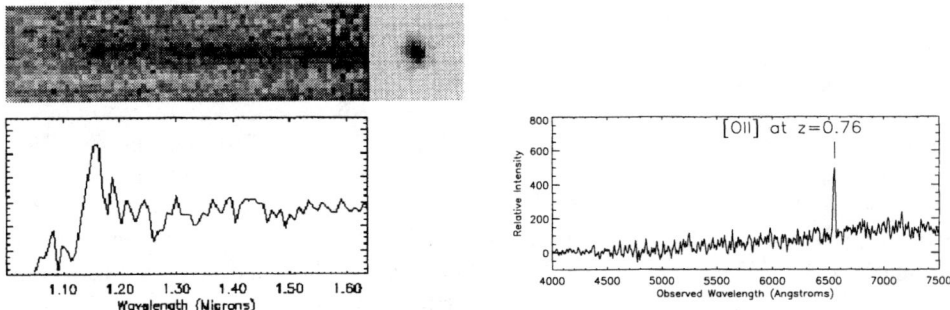

FIGURE 1. Emission-line object in the parallel field at 0055+8518. *top:* Image and 2D spectrum. *left:* R=180 Grism Spectrum *right:* LRIS spectrum confirming the redshift of an Hα-emitter in the G141 grism parallels.

DISCUSSION

Assuming that the observed emission lines are Hα, we can infer star formation rates (SFRs) for each object (see Kennicutt 1983). We find a range of 2 to 480M$_\odot$/yr ($H_0 = 50$ km/s/Mpc, $q_0 = 0.5$), assuming a $\sim 30\%$ correction for an unresolved [NII] contribution to the line flux. The average SFR of 34M$_\odot$/yr is consistent with other estimates for starbursts at $z \geq 1$ from Hα surveys (see for example Teplitz et al. 1999). It is also consistent with the estimates for Lyman Break Galaxies at $z > 3$ assuming a factor of 3–5 extinction correction to SFR inferred from the UV continuum (Pettini et al. 1998).

We can also estimate the density of Hα-emitting objects. Table 1 compares the density of such objects from several Hα surveys. While it is difficult to make direct comparisons, since surveys have uneven flux limits, some general trends do emerge. The density of objects in the field sample measured by NICMOS is broadly consistent with the other surveys not targeted to absorption system redshifts. The density enhancement seen in absorption fields by Teplitz et al. (1998) as compared to the present work is similar to the enhancement between the Mannucci and Thompson surveys (both taken with the same instrument and similar depths).

TABLE 1. Comparison of $H\alpha$ Surveys

redshift	density $10^{-4}\mathrm{Mpc}^{-3}$ (comoving)	3σ flux limit 10^{-16} ergs/cm^2/s	$<SFR>$ $M_\odot$/yr	No. of Galax.	reference	notes on field selection
0.7–1.9	2	0.6	50	30	this work	random
0.95–1.0	4.7	1.3	35	2	Teplitz 1999	QSO em.
1.5	< 120	~5.0	...	0	Teplitz 1999	QSO em.
2.0–2.7	< 70	10	...	0	Bunker 1999	DLA[c]
2.2–2.5	0.33	3.4	250[a]	1	Thompson 1996	QSO em.
2.28–2.29	< 120	0.9	...	0	Pahre 1995	QSO em.
2.3–2.4,0.89	9	4.8	70	18	Mannucci 1998	Abs. line
2.3–2.5	60	1.0	50	5	Teplitz 1998 [b]	Abs. line

[a]See Beckwith et al. 1998 for a discussion.
[b]We have counted objects with line fluxes $> 1 \times 10^{-16}$ ergs/cm^2/s.
[c]longslit spectroscopy.

REFERENCES

1. Kennicut, R. 1983, ApJ 272, 54
2. Steidel, C.C., Giavalisco, M., Pettini, M., Dickinson, M., & Adelberger, K.L., 1996, ApJ Letters 462, L17
3. Pettini, M., Kellogg, M., Steidel, C., Dickinson, M., Adelberger, K., Giavalisco, M., 1998, ApJ 508, 539
4. Teplitz, H., Malkan, M., & McLean, I.S., 1998, ApJ, 506
5. McCarthy, P.J., et al., 1999, submitted to ApJ
6. Oke, J. B., et al.,1995, P.A.S.P, 107, 375
7. Teplitz, H.I., McLean, I.S., & Malkan, M.A., 1999, submitted to ApJ
8. Bunker, A.J., Warren, S.J., Clements, D.L., Williger, G.M., & Hewett, P.C., 1999, submitted to MNRAS
9. Thompson, D., Djorgovski, S, & Beckwith,S.V.W, 1994, AJ, 107 , 1
10. Pahre, M.A. & Djorgovski, S.G., 1995, ApJL, 449, 1
11. Mannucci, F., Thompson, D., Beckwith, S.V.W., & Williger, G.M., 1998, ApJLetters, 501, L11

9. Gamma Ray Bursts

Recent Discoveries in Gamma Ray Burst Astronomy

Neil Gehrels

NASA/GSFC, Code 661, Greenbelt, MD 20771

Abstract. Gamma-ray bursts are among the most energetic explosions in the Universe, and yet their origin has been a mystery since they were discovered in 1973. Early instrumentation showed that the bursts have a huge variety of time profiles and are not associated with any bright or unusual steady sources at other wavelengths. The BATSE instrument on CGRO is more sensitive than any previous GRB detector and is providing a huge (> 2000) and uniform sample. From it we have learned that GRBs are isotropic on the sky and have an intensity distribution that flattens toward fainter bursts. Recently, tremendous progress has been made by measurements performed and enabled by the BeppoSAX mission. X-ray, optical and radio afterglow has been discovered for several bursts. From the accurate positions in these wavelength bands it has been possible to obtain redshifts for 3 bursts and to identify host galaxies for ~ 10 bursts. From this we now know that most, if not all, GRBs are cosmological in origin with typical distances of ~ 2 Gpc ($z \approx 1$).

INTRODUCTION

Gamma ray bursts (GRBs) are short flashes of gamma rays and x-rays, lasting from a few milliseconds to several minutes. They occur at random positions on the sky with a frequency of about 1000 yr^{-1} (full sky rate). The typical GRB spectral energy distribution is nonthermal, peaking in the range from 100 keV to 10 MeV with power-law tails extending into the GeV range (see Fig. 1). There is prompt keV x-ray emission during the burst, that in most cases, falls below the extrapolation of the 10-100 keV spectrum, but, in some cases, can exceed the extrapolation [1].

GRBs were discovered by the Vela satellites in the late 1960's and announced in 1973 [2]. The discovery was serendipitous, as the gamma-ray detectors onboard were flown to monitor the Nuclear Test Ban Treaty of 1963. From this series of missions, the rough temporal and spatial characteristics of GRBs were first learned. Many small GRB detectors were flown on American, Russian and European missions in the 1970's and 1980's, often onboard interplanetary spacecraft. By comparing the onset times of bursts at the different instrument locations, it was possible

CP470, After the Dark Ages: When Galaxies were Young (the Universe at 2 < z < 5),
edited by Stephen S. Holt and Eric P. Smith

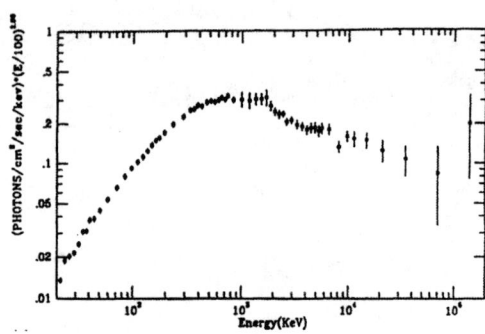

FIGURE 1. Spectral energy distribution of GRB910503 observed by the four instruments on CGRO. From Schaefer (1995).

TABLE 1.

Instrument	Energy Range	Field of View
Burst And Transient Source Experiment (BATSE)	0.03 - 1.9 MeV	full sky
Oriented Scintillation Spectrometer Experiment (OSSE)	0.05 - 10 MeV	4° x 11°
Imaging Compton Telescope (COMPTEL)	0.8 - 30 MeV	1 sr
Energetic Gamma Ray Experiment Telescope (EGRET)	20 MeV - 30 GeV	0.6 sr

in some cases to obtain positions to arcminute accuracy, compared to the $\sim 10°$ positions from the Vela's [3,4]. Even with these improved positions, searches for quiescent counterparts at other wavelengths were unsuccessful [5–7]. From the observations of the 1970's and 1980's a burst paradigm evolved in which GRBs were thought to be associated with local (< 300) pc neutron stars.

OBSERVATIONS BY CGRO/BATSE

The GRB field was greatly advanced by the launch of the Compton Gamma Ray Observatory (CGRO) in 1991 [8]. All four instruments onboard (Table 1) have contributed significantly, with BATSE studying large numbers of bursts (see this section), OSSE observing low-energy gamma-ray afterglow from bursts [9], COMPTEL imaging bursts and detecting ~ 10 MeV emission [10], and EGRET detecting 100 MeV - GeV emission and afterglow (see next section).

The BATSE instrument was designed to make comprehensive observations of GRBs [11]. Its 8 large detectors (1800 cm^2 each) monitor the entire visible sky (~ 7 sr from low-Earth orbit) for bursts down to a flux threshold of about 0.2 photons cm^{-2} s^{-1} (50-300 keV). The instrument detects about 300 GRBs yr^{-1} which, when corrected for viewing factor and deadtime, corresponds to a full-sky rate of ~ 1000 GRBs yr^{-1}.

A sample of BATSE light curves is shown in Figure 2. The variety of shapes, variability and duration is tremendous. A plot of durations (Figure 3) shows a bimodal distribution of short and long bursts [12,13]. Also, it is found that weaker bursts have longer durations on average than stronger bursts [14]. The magnitude of the effect may be consistent with cosmological time dilation.

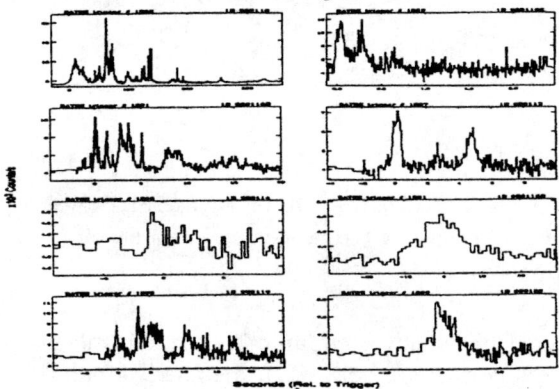

FIGURE 2. Sample of GRB time profiles from BATSE. From Fishman et al. (1994).

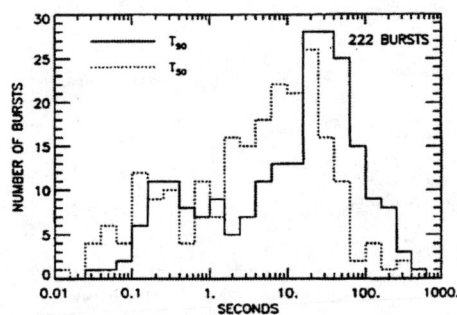

FIGURE 3. The duration distribution of 222 GRBs from BATSE for 2 different trigger criteria. From Fishman et al. (1994).

The two most important BATSE results concern the sky and brightness distributions of GRBs. The sky distribution for 2119 bursts detected by BATSE during the first 7 years of the CGRO mission is shown in Figure 4. The burst locations are completely consistent with an isotropic distribution. No evidence for global anisotropy (dipole, quadrupole) or local concentrations is seen [11,15,16]. Particularly important is the lack of a concentration along the galactic plane as would be expected for most galactic populations.

The BATSE brightness distribution of GRBs is show in Figure 5. The brightest bursts follow a power law with index -3/2 as would be expected for a population of equal-luminosity GRBs filling Euclidean space homogeneously. For fainter bursts,

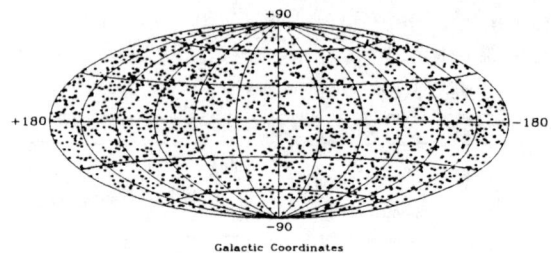

Galactic Coordinates

FIGURE 4. Sky distribution of 2119 GRBs observed by BATSE between April 1991 and April 1998. Plotted is an Atoff projection with the galactic center at the origin. Each dot is one GRB. From C. Meegan (priv. comm.).

BATSE finds a marked deficiency compared to a homogeneous population. The log N - log P shape is consistent both with confined population models such as a galactic halo (e.g. [17,18]) and cosmological models [19,20]. For cosmological models, the deficit of weak GRBs is due to redshift effects.

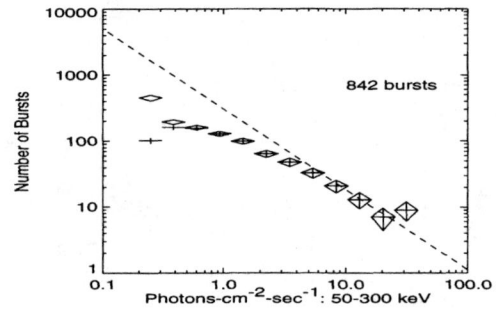

FIGURE 5. Peak flux distribution for GRB's on 1.024 second timescale observed by BATSE. Plotted is the integral number of bursts greater than a given peak flux. The crosses include a correction for BATSE trigger efficiency. From C. Meegan (priv. comm.)

HIGH ENERGY GAMMA RAY EMISSION

A key result from EGRET is that GRBs can have high energy > 100 MeV emission during the burst and extending for about an hour afterward [21–23]. The best example is shown in Figure 6 for GRB940217. The low-energy gamma-ray profile observed by Ulysses is plotted along with data points for individual EGRET

374

photons. The sparcity of points during the peak of the burst is due to the fact that EGRET was deadtime limited for these high rates. Remarkably, high-energy radiation continued for about 6000 seconds (interrupted by an Earth occultation). A very high-energy photon at 18 GeV was detected at 4700 seconds.

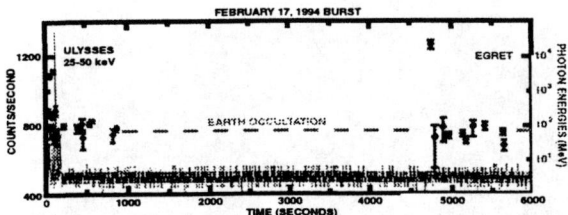

FIGURE 6. Light curves for GRB970228 from Ulysses and EGRET. The Ulysses data (left axis) peak at early times. EGRET data (right axis) are plotted as points, for individual photons. EGRET data continues for ~ 6000 seconds, interupted by an earth occultation. Adapted from Hurley et al. (1994).

To date, EGRET has detected high-energy gamma rays from 4 bursts [23]. All bright bursts that occurred in the instrument field of view were detected, so it is probable that bursts are accompanied by high-energy radiation. In only the one case mentioned above (GRB940217) was long-lasting afterglow detected. The afterglow is very constraining on models of the interaction of the GRB blastwave with its surrounding medium. Future observations are promised by the GLAST mission discussed below.

BEPPOSAX AFTERGLOW ERA

The past two years have seen a huge step forward in GRB research due to discoveries related to the BeppoSAX mission. BeppoSAX [24] is an Italian/Dutch mission with four instruments relevant to GRBs onboard: a Gamma-ray Burst Monitor (GBM; 60-600 keV, full sky), a Wide Field Camera (WFC; 2 - 28 keV; 20° x 20° FWHM field of view), and two Narrow Field Instrument (NFI) x-ray telescopes (0.1 - 10 keV; 1 arcminute field of view). The important feature of BeppoSAX for GRB studies is its combination of GRB detectors (GBM and WFC) and sensitive x-ray telescopes (NFIs). Although the spacecraft is not autonomous, it can be repointed by ground command in approximately 6-8 hours. The operational sequence is: (1) WFC detects transient, (2) confirmation by the GBM that it is a GRB, (3) spacecraft slewed to point NFIs at GRB, (4) NFI observations of GRB.

The afterglow discovery burst for BeppoSAX was GRB 970228 [25]. The burst was detected by the GBM and WFC at UT 02h 58m on 1997 February 28. Observations by the NFIs 8 hours later revealed a faint x-ray source (2.4×10^{-12} erg cm^{-2} s^{-1} in 2-10 keV range). Another observation with the same instruments at 3.5 days showed that the flux had decreased by a factor 20 (see Figure 7), giving strong indication that it was the counterpart of the GRB. Data from BeppoSAX, ROSAT and ASCA reveal that the x-ray source faded away as $t^{-1.3}$.

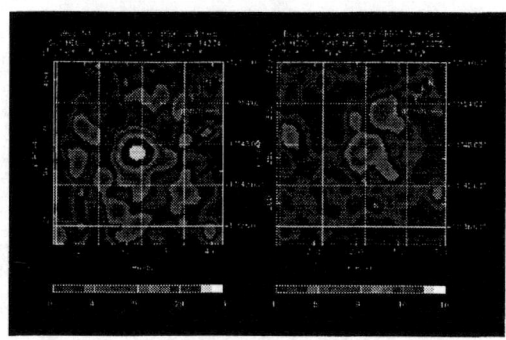

FIGURE 7. BeppoSAX observations of fading x-ray afterglow from GRB970228 at 8 hours and 3.5 days. From [25]. Reprinted by permission from Nature 387: 783 © 1997 Macmillan Magazines Ltd.

Based only on the WFC position, optical observations were made by the WHT telescope on La Palma at +20 hours and +8 days. A fading optical source was discovered in the field (and later confirmed to be in the NFI error box), decreasing from V=21.3 to V> 23 [26]. Later observations by HST [27,28] showed that the fading source was superimposed on an extended object of magnitude R=24.6 that is likely to be a distant galaxy.

The first GRB redshift measurement was made for GRB970508, for which a fading optical source had spectral absorption lines (see Figure 8) consistent with a redshift of 0.835 [29]. This was the first definitive proof of a cosmological origin for GRBs. GRB970508 was also the first burst with a radio counterpart. The VLA observations by Frail et al. [30] were important in that they showed marked variability for the first 60 days and then decreasing (Figure 9). The observers suggested that the variability was due to interstellar scintillations and used the disappearance of the variability to infer a relativistic expansion of the source [30].

A flurry of activity followed the discovery of afterglows. BeppoSAX was joined by RXTE (often trigger by BATSE) to provide prompt arcmin positions for several GRBs. A large number of space- and ground-based observatories were commandeered to track the afterglow, identify host galaxies and obtain redshift measurements. Table 2 summarizes the current situation (as of December 1998) where 13 GRB afterglows have now been followed in detail. Redshift measurements have been obtained for two more cases with a surprisingly large redshift of z=3.418 for

TABLE 2.

Burst GRB	Trigger Mission	Afterglow X-ray	Afterglow Optical	Afterglow Radio	Galaxy R(mag)	z
970111	SAX	x				
970228	SAX	x	x			
970402	SAX	x				
970508	SAX	x	x	x	25.8	0.835
970828	B/X*	x				
971214	SAX	x	x		25.5	3.418
971227	SAX	x			>22	
980326	SAX	?	x		25.3	
980329	SAX	x	x	x	25.4	
980425	SAX	?	supernova	x	14.3	0.008
980519	SAX	x	x	x	24.7	
980613	SAX	x	x		24.4	
980703	B/Xª	x	x	x	>22	0.966

ª B/X= BATSE/XTE

GRB971214 [31] and z=0.966 for GRB980703 [32]. There is even evidence in the continuum spectrum of GRB980329 for a redshift of ~ 5 [33]. The GRB energy release is more than 10^{53} ergs for the most distant of these bursts!

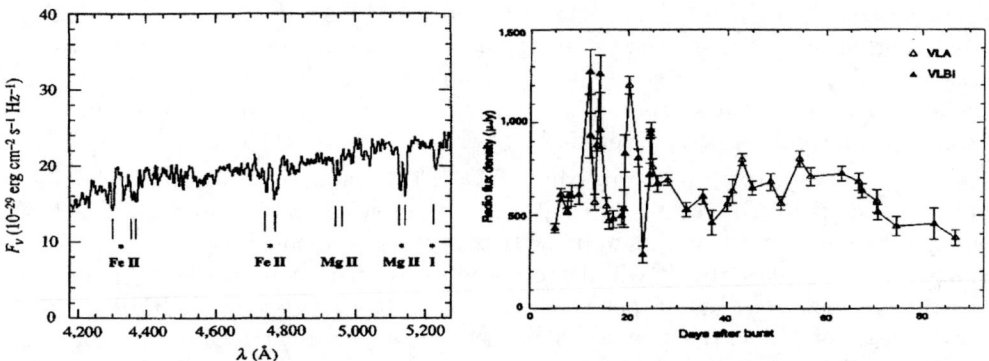

FIGURE 8. (*left*) Redshifted Fe lines in the Keck-II spectrum of the optical afterglow from GRB970508. From [29]. Reprinted by permission from Nature 387: 879 © 1997 Macmillan Magazines Ltd.
FIGURE 9. (*right*) VLA observations of the radio counterpart to GRB970508 at 4.86 GHz. From [30]. Reprinted by permission from Nature 389: 261 © 1997 Macmillan Magazines Ltd.

Of the bursts detected so far with x-ray afterglow, approximately 1/2 of them have been found to have optical afterglow. Some of the ones not detected where not observed soon enough with large telescopes, but in other cases good observations were made and nothing detected. An example of a firm non-detection is for GRB980828 for which no optical emission was seen to a limit of R=23.8 at 4 hours after the burst [34]. It appears that there are different kinds of events in terms of optical afterglow. A remarkable result is that for those events with optical afterglow detected and arcsecond positions determined, faint galaxies are seen in all cases at the position of the afterglow. These bursts at least, and possibly all bursts, are associated with host galaxies. From Table 2 it is seen that the range of brightnesses for the hosts is very narrow, with all clustered near R=25. Most are too faint for redshift measurements, but the brightnesses are consistent with galaxies in the z=1-3 redshift range.

An unusual event is GRB980425. In the optical search triggered by the GRB, a supernova was found spatially coincident with one of two x-ray sources in the WFC error circle [35]. Even more remarkably, the estimated core collapse time of the SN Ib/c is consistent with the time of the GRB. One factor against an association of the two events is that the x-ray source that is coincident with the supernova is steady, whereas the other x-ray source in the field was the one that faded. With no firm conclusion on the association, we can only wait to see if other supernovae make GRBs.

FUTURE OBSERVATIONS

The current suite of GRB instruments (BeppoSAX, BATSE, Ulysses, and RXTE) are aging but continue to operate and promise more years of service. In addition, new missions are being developed and proposed that offer great advances in GRB capabilities. They have factors of 10 improvement in angular resolution, follow-up response time and sensitivity compared to the current instruments.

HETE-II (see http://space.mit.edu/HETE/) is an approved mission that is being built for launch in 2000. It combines a wide-field gamma-ray instrument of arcminute burst location capability with a soft x-ray camera of ~ 10 arcsecond burst location capability. It will detect ~ 50 bursts per year with the gamma-ray instrument and ~ 16 bursts per year with the x-ray instrument. GRB locations will be telemetered to the ground within seconds of receipt, triggering fast optical response with ground-base telescopes.

GLAST (see http://glast.gsfc.nasa.gov/) is a mission in NASA's strategic plan for high energy (20 MeV - 300 GeV) gamma-ray observations to be launched in 2005. It will have excellent GRB capability and will follow up the EGRET discoveries of high-energy emission during and after bursts. It will have an order of magnitude better sensitivity than EGRET, better angular resolution and much less deadtime for high-rate events.

A dream of the GRB community is to have a dedicated NASA MIDEX mission for GRB studies. In the current (1998) MIDEX round, several missions were proposed that offer superb capabilities such as prompt arcsecond locations, hundreds of bursts detected, better sensitivity than BATSE, and onboard multiwavelength follow-up observations. Hopefully, one of these will be selected for flight.

REFERENCES

1. Strohmayer, T., et al. 1998, ApJ, 500, 873.
2. Klebesadel, R. W., et al. 1973, ApJ, 182, L85.
3. Atteia, J. L., et al. 1987, ApJS, 64, 305.
4. Hurley, K., et al. 1994, Nature, 372, 652.
5. Vrba, F. J., Hartmann, D. H., & Jennings, M. C. 1995, ApJ, 446, 115.
6. Boer, M., et al. 1993, A&A, 277, 5.
7. Schaefer, B. 1994, in Gamma Ray Bursts, 2nd Huntsville Workshop, eds. G. J. Fishman, J. J. Brainerd, and K. Hurley (New York: AIP) 382.
8. Gehrels, N. & Shrader, C. 1996, A&AS, 120, C1.
9. Matz, S., et al. 1998, ApJ, in preparation.
10. Kippen, R. M., et al. 1998, ApJ, 492, 246.
11. Meegan, C. et al. 1992, Nature, 355, 143
12. Kouveliotou, C., et al. 1993, ApJ, 413, L101.
13. Klebesadel, R. W. 1992, Proc. Los Alamos Workshop on GRBs, 1990 Taos, eds. C. Ho, R. Epstein, and E. Fenimore (Cambridge: Cambridge Univ. Press) 161.
14. Norris, J. P., et al. 1995, ApJ, 439, 542.
15. Fishman, G. J., et al. 1994, ApJS, 92, 229.
16. Briggs, M., et al. 1995, ApJ, 459, 40.
17. Brainerd, J. J., 1992, Nature, 355, 522.
18. Fabian, A. & Podsiadlowski, P. 1993, MNRAS, 263, 49.
19. Piran, T. 1992, ApJ, 389, L45.
20. Emslie, A. G. & Horack, J. M. 1994, ApJ, 435, 16.
21. Hurley, K. 1994, in Gamma Ray Bursts, 2nd Huntsville Workshop, eds. G.J. Fishman, J. J. Brainerd, and K. Hurley (New York: AIP) 687.
22. Sommer, M., et al. 1994, ApJ, 422, L63.
23. Dingus, B. L., Catelli, J. R., and Schneid, E. J. 1997, in 25th ICRC Durban, South Africa, Vol. 3, eds. M. S. Potgieter, B. C. Raubenheiner, and D. J. van der Walt (ICRC press) 29.
24. Boella, G., et al. 1997, A&AS, 122, 299.
25. Costa, E., et al. 1997, Nature, 387, 783.
26. van Paradijs, J., et al. 1997, Nature, 368, 686.
27. Sahu, K. C., et al. 1997, Nature, 387, 476.
28. Fruchter, A. S., et al. 1998, ApJ, astro-ph/9807265.
29. Metzger, M. R., et al. 1997, Nature, 387, 879.
30. Frail, D. A.,et al. 1997, Nature, 389, 261.
31. Kulkarni, S. R., et al. 1998, Nature, 393, 35.
32. Djorgovski, G., et al. 1998, GCN Cir. 139.
33. Fruchter, A. S., et al. 1998, ApJ, astro-ph/9810224.
34. Groot, P. J., et al. 1998, ApJ, 493, L27.
35. Galama, T. J., et al. 1998, Nature, 395, 670.
36. Schaefer, B. 1995 in Gamma Ray Bursts, 3rd Huntsville Symposium, eds. C. Kouveliotou, M. F. Briggs, G. J. Fishman (New York: AIP) 248.

Gamma-ray Bursts at High Redshift

Ralph A.M.J. Wijers

Dept. of Physics and Astronomy, SUNY at Stony Brook
`rwijers@astro.sunysb.edu`

Abstract. Gamma-ray bursts are much brighter than supernovae, and could therefore possibly probe the Universe to high redshift. The presently established GRB redshifts range from 0.83 to 5, and quite possibly even beyond that. Since most proposed mechanisms for GRB link them closely to deaths of massive stars, it is a reasonable ansatz to assume that their rate density in the past was proportional to the star formation rate. Work with Bloom, Bagla, and Natarajan (1997), as well as by other groups, does indeed show that the GRB flux distribution calculated from this assumption agrees well with the data. This means that GRB are bright lamps which illuminate an era of the Universe that may predate the quasar age somewhat, so they can become useful beacons illuminating the early Universe.

It also seems that GRB can become well calibrated lamps, like supernovae. The theory that attributes the afterglows to relativistic blast waves emitting synchrotron radiation has been quite successfully tested, and in the case of GRB 970508 so well measured that the physical parameters of the burst are known. If this could be repeated regularly, GRB can become as precise probes of cosmic geometry at redshifts 2–5 as ordinary supernovae are at redshifts $\lesssim 1$.

INTRODUCTION

After having resisted serious challenges to their resolution for the better part of three decades, they dropped their veil of secrecy with surprising speed: before the first afterglow detection in optical and X rays on 28 February 1997 (Costa et al. 1997, Van Paradijs et al. 1997) a small majority of researchers had been won over hesitantly to the cosmological distance camp. A week or so after the announcement of the redshift lower limit of 0.83 on the burst of 8 May 1997, few still contested that GRB are at high redshift. This revolution has shifted the field into high gear, mostly because afterglows have now been seen from radio to X-ray wavelengths, and in many bands their detection does not require the world's leading facility. So now gamma-ray bursts are an exciting game for the many, rather than an esoteric activity of those who only count photons that pack a real punch.

A systematic overview of the data and the observational developments are given elsewhere in this volume by Gehrels, so I shall present this paper from a theoretical stance and illustrate my points with an eclectic choice of data where appropriate. In a nutshell, my points are the following. The data show that some GRB are at high redshift, and they have nothing that distinguishes them from the bulk of GRB as a group. By the Principle of Insufficient Reason, the bulk of GRB are therefore at high redshift. The afterglows of GRB fit well to the expected behavior of relativiistic blast waves. At these distances they require an energy release approaching $M_\odot c^2$, and the rapid variability indicates that the region into which it is released is not much larger than the Sun's Schwarzschild radius; so a roundup of the usual suspects yields a cast of characters all linked to the death of massive stars. A plausible ansatz is then that GRB happen where massive stars are born, i.e. the GRB rate in Universe should trace the star formation rate, and indeed it does. The implied redshifts of the faintest GRB from fits by various groups indicate that we may have already seen GRB from $z = 4 - 8$ in the faint end of the BATSE catalog (but cannot say which ones they are specifically). This means GRB really are among the most distant probes of the Universe, so we can investigate how much useful cosmology can be done with them. Since the blast wave model has only a limited number of free parameters, achievable amounts of data on an afterglow can in fact constrain all of them. For GRB 970508 this has been done, and this means that a decent sample of good cases can provide a set of calibrated lamps, of the same quality for doing cosmology at high redshift as supernovae are at $z < 1$.

BLAST WAVES

When enough energy is deposited in a small volume, an explosion results in which usually the bulk of the explosion energy is carried away as kinetic energy of ejecta. If they expand into vacuum that is the end of the story, but if there is an ambient medium it will decelerate the ejecta, reconverting a large fraction of its kinetic energy back into heat. Depending on conditions in the hot material, it will radiate some fraction of the heat away and thus give rise to an observable object. Thus a supernova gives rise to first a bright few-week flash at the explosion, then a period of nearly invisible 'free expansion', and then a supernova remnant phase in which we can see the expanding shocked gas again.

When the expansion is initiallyrelativistic, i.e. when the explosion endows the particles with an energy much greater than their rest mass, the situation is not much different in principle. First there is free expansion, then there is the interaction phase during which the ambient medium is shocked and its radiation makes the source visible again. The differences are due to the relativistic nature of the expansion in the GRB case. In the rest frame of the shock the incoming ambient medium has energy $\Gamma M c^2$, whereas the initial mass only has its rest energy $M_0 c^2$. When the two are equal, i.e. when M_0/Γ has been swept up, the shock is significantly modified by the influence of the added swept-up mass. If no energy is radiated the total energy of the shocked mass is constant. To an external observer, the energy per unit mass of the shocked gas is not Γc^2 as it is in the shock frame, but $\Gamma^2 c^2$. The extra factor Γ comes from the fact that the (thermal) energy of shocked gas appears blueshifted to the external observer. At late times, when $M_0 \ll M$, constant energy therefore means that $M\Gamma^2$ is fixed, i.e. for a uniform ambient medium $r^3\Gamma^2$ is constant, so $\Gamma \propto r^{-3/2}$ for so-called adiabatic evolution. In the other extreme, radiative, limit all thermal energy behind the shock is immediately radiated isotropically in the rest frame of the shock (i.e. in the shock rest frame there is no radiation reaction force on the material). Then ΓM is constant, which implies $\Gamma \propto r^{-3}$. In these simple limits, the dynamics is then fully fixed. In the intermediate regime, dynamics and radiation have to be solved simultaneously.

To calculate the synchrotron emission from the blast wave, we need to know what fraction of the total post-shock energy density $U = 4\Gamma^2 n m_{\mathrm{p}} c^2$ is in a power-law distribution of relativistic electrons and what fraction is in magnetic fields. This is not doable ab initio, so we simple parameterize the post-shock conditions with the ratios of electron and field energy to total energy: $\epsilon_{\mathrm{e}} = U_{\mathrm{e}}/U$ and $\epsilon_B = U_B/U$. With some algebra, and assuming uniform conditions behind the shock, we can then write five observable quantities in terms of distance and local conditions (Wijers and Galama 1998): a spectral slope, three frequencies where the slope changes (self-absorption, peak, and cooling), one flux (usually at the peak). In addition, for some bursts the redshift can be measured. Against this, there are five parameters of the blast wave: total energy, ambient density, electron energy and power-law slope, and magnetic field energy. In principle, a complete theory of relativistic shocks would provide the last 3 in terms of the others, so the problem would be greatly over-determined. Lacking such a theory, solving for all parameters exactly uses up the measurable quantities. For only one burst, GRB 970508, has this succeeded (Galama and Wijers 1998) since it is the only one with a complete enough set of data. Its measured spectrum 12 days after the burst is shown in Fig. 1. Since the break frequencies move to long wavelengths rapidly, catching them for more bursts will require much more rapid follow-up, for a substantial sample of bursts, than has been possible thus far.

REDSHIFT DISTRIBUTION

Of course, the best way of deriving the redshift distribution of GRB is by measuring redshifts. This is not very practical now because, first, there are only a handful of GRB with measured redshifts, and second, this sample is clearly biased towards brighter GRB, so it may not answer the most intriguing question of what the highest redshifts are, since these are (presumably) among the dimmest bursts. Table 1 lists the events with known redshift. GRB 970508 and GRB 980329 (Fruchter 1998) are based on the afterglow light itself, the other two on the redshift of the galaxy they coincide with. It is clear that the luminosities of these bursts range widely.

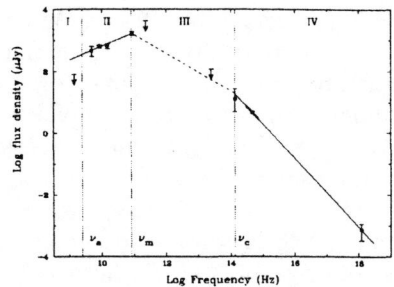

FIGURE 1. The X-ray to radio spectrum of GRB 970508 on May 21.0 UT (12.1 days after the event) from Galama et al. (1998). Indicated are the inferred values of the break frequencies ν_a, ν_m and ν_c for May 21.0 UT.

Another method of constraining the redshift distribution is to fit the observed flux distribution of GRB. It has the advantage of a large sample, and the ability to correct fairly accurately for detection threshold effects. The disadvantage is that the flux distribution is a convolution of the redshift distribution with other properties, such as luminosity (or luminosity function), spectral shape variation, detection threshold, etc. The point where theories of GRB may assist is in helping to decide what, for a given model, is more likely to be a constant quantity between bursts related to flux, fluence, etc. The quantity usually plotted is P, the energy flux at peak in fixed range of energy in the observer frame (generally not too wide, e.g. 50–300 keV for BATSE on CGRO). It has the advantage of being easily measurable and closely related to what the instrument directly see and trigger on, but certainly for a wide redshift range one has no reason to expect it to be constant between bursts. A theoretical quantity used frequently when trying to estimate theoretical distributions is L_0, the luminosity in the burst rest frame over a generally wider energy range (e.g. 30–2000 keV). It may be more likely to be constant (though, e.g., the fireball model does not in any way suggest or demand that it is) but requires an iterative procedure to determine L_0 for the bulk of bursts of which z is unknown, and requires a spectral model fit to extrapolate the spectrum to cover the

TABLE 1. Properties of gamma-ray bursts with known redshifts. The 1998 April 25 burst is considered (by me) to be of a different class.

GRB	z	$F_{peak,\gamma}$ $(\times 10^{-6}\, erg\, cm^{-2}\, s^{-1})$	$L^{iso}_{peak,\gamma}$ $(\times 10^{51}\, erg\, s^{-1})$
970508	0.83	0.17	0.34
971214	3.42	0.34	5.3
980329	5.	7.9	145.
980703	0.97	1.9	2.8
980425	0.0085		5.5×10^{-5}

whole (rest frame) range (Fenimore and Bloom 1995, Wijers et al. 1998). Another frequent choice is the fluence of the burst, based on the notion that a burst event represents a more or less constant energy reservoir being emptied. The history of its use is marred by the complicated selection effects involved, since the probability of detecting a burst is fairly unrelated to its fluence (unless its time profile is not resolved by the detector). Now that we have some idea of what the models may be like, this issue should be revisited. E.g., if the emission mechanism of the GRB itself is synchrotron radiation, like the afterglow, it should be possible to constrain the range of free fit parameters, e.g., by seeing whether the break that is usually seen in broken power-law GRB spectra is one of the standard synchrotron breaks.

Several groups have in recent times attempted to fit the GRB flux distribution with a model of sources with specified distribution properties in redshift and luminosity. Due to the many free functions involved, there is no unique answer to the problem (Horváth, Mészáros and Mészáros 1995). Specifically, the redshift one derives for the dimmest BATSE bursts ranges from below 1 to above 4. The earliest models assume no evolution of the GRB population with redshift, and standard candles, i.e. the luminosity function has zero width. The typical value then found for the redshifts of GRB with $P = 1\,\mathrm{cm}^{-2}\,\mathrm{s}^{-1}$ is 0.8–1 (Mao and Paczyński 1992, Wickramasinghe et al. 1993). At the other extreme, one can assume that the GRB population evolves very strongly. Specifically, there are good reasons to think that the GRB rate may trace the cosmic star formation rate (SFR) at high redshift (see below). The SFR is indeed a very strongly evolving function, and moreover it is measurable independent of GRB. As derived from the rest frame blue light density in the Universe (Madau 1996, and this volume, Lilly et al. 1996) it increases steeply from the present time back to a peak 20 times higher than present at about $z = 1.5$, then slowly declines again. But the behavior past the peak is uncertain even from blue light density studies; it is furthermore asserted by some studying obscured star formation in far IR that at high z the curve should be a factor few higher to account for those contributions to the SFR. Studies of this type, with standard-candle bursts and the Madau-Lilly SFR were first published by Totani (1997) and by Wijers et al. (1998). Totani found a maximum redshift of about 2.5–4 for the dimmest bursts using a simple power-law spectrum for the GRB to get K corrections. A more sophisticated treatment of the K corrections was done by Wijers et al. following the method of Fenimore and Bloom (1995) to account for the spectral shape variation in bursts. The result was a very high value of the redshifts of the dimmest bursts, $z_{\mathrm{dim}} = 6.2$, but we found out that there is an error of a factor $(1 + z)$ in our GRB rate density. Our corrected value for z_{dim} is about 4.5, closer to Totani's value.

Since it has become clear that GRB are not standard candles (Table 1) a few studies have been done to gauge the effect of a wide luminosity function on the derived redshifts. Krumholz, Thorsett, and Harrison (1998) took a wide, power-law distribution of luminosities to check whether a wide luminosity function could make the GRB counts consistent with a non-evolving population; they found that it could, providing one chooses the slope of the luminosity function equal to the faint end of the flux distribution, a result known from studies of halo GRB in a previous era (Wasserman 1992, Ulmer, Wijers, and Fenimore 1995). An important lesson is implicit in their work: by assuming a wide luminosity function, one may bring down the average redshift, but at the cost of making some very distant GRB detectable, so the fraction of GRB at very high z tends to be comparable, despite the smaller median redshift. Kommers et al. (1998) did the most extensive and thorough job thus far in this area. They extended the sample of BATSE bursts to fainter GRB by searching the data for untriggered bursts, and found that the flux distribution gets much less steep below the usual BATSE threshold. Their fits appear to indicate that non-evolving populations are not consistent with the data. However, they rightly note that there is enough uncertainty left in both the SFR and in the choice of a shape of luminosity function that one probably could find an acceptable fit without evolution. The circumstantial evidence, though, that GRB are related to stars, plus the fact that too many GRB have very high redshifts makes a strong case against such models.

In the fits of Kommers et al., the fraction of detectable bursts with $3 < z < 4$ is less than 5%, and even less than 0.7% for $z > 4$. Given that we have one each in those ranges among the 4 bursts with known redshift, this presents a puzzle. Perhaps the star formation rate is just different at high z than what we have been assuming, but it could also be that other effects play a role. As Kommers et al. emphasize, the number of bursts added to their sample by halving the detection threshold is relatively small, and the bursts detected are often well above threshold. So there may be a deficit of very faint bursts. Another way of hitting this point home is to consider Fig. 2. It seems to indicate no hint of the usual trend (e.g. in magnitude-limited quasar samples) that sources lie near the detection threshold, and more so as redshift increases; in fact, the highest-flux source by far is also the most distant one! It is quite tricky to derive intrinsic luminosity distributions from such plots (in quasars, one might naively infer that the 'typical' quasar luminosity increases with distance squared, since all members of the sample will have nearly the same flux regardless of redshift), and a lot hinges right now on GRB 980703. Still, I submit that we should consider luminosity evolution with some care. It is to my mind a serious candidate for resolving the apparent discrepancy found by Kommers et al. (1998) between the low probability of seeing $z > 3$ GRB according to their best-fit redshift distribution and the fact that half of the GRB with known redshift have $z > 3$.

FUTURE PROGRESS

With the large energies required and the fact that all GRB with arcsecond locations have been associated with galaxies (sometimes this requires a significant wait, say half a year, since the galaxy is so small and faint that it is not seen initially) there is growing evidence that GRB are associated somehow with young stars. Even with the phenomenal energies they can release, say as hypernovae

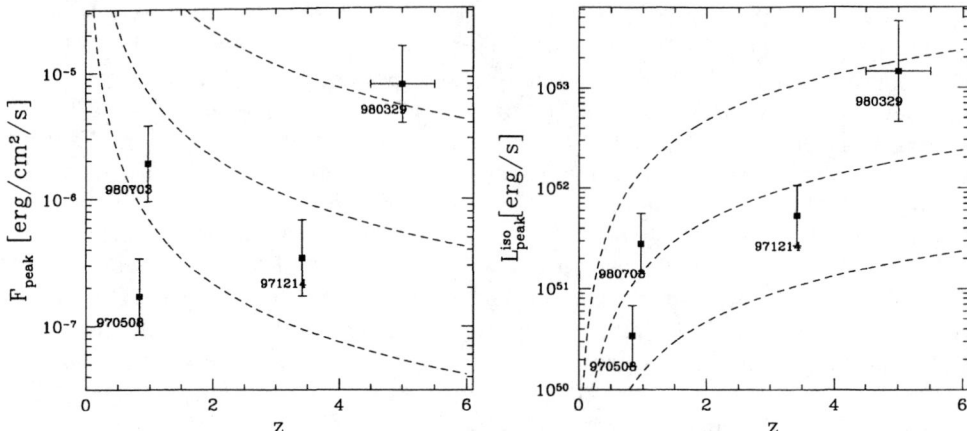

FIGURE 2. The fluxes and derived isotropic luminosities in gamma rays of GRB with known redshifts. (left) Fluxes; dashed curves are gamma-ray flux as function of redshift for a source with a power-law spectrum $F(E) \propto E^{-2}$ and fixed rest frame luminosity. (right) Isotropic luminosities; dashed curves are inferred bolometric rest frame luminosity as a function of redshift for a fixed gamma-ray flux.

(Paczýnski 1998), it is beginning to look as if gamma-ray bursts are not isotropic emitters, a perfectly logical consequence of an origin in the collapse of a rotating object (Mészaros, Rees, and Wijers 1998). The likely mechanism may be Poynting flux from near a spinning black hole (Blandford and Znajek 1977), since a Poynting flux has the advantage of naturally circumventing the baryon pollution problem (i.e., the requirement that very few of the many baryons present in the source may become part of the outflow, since otherwise there is not enough energy to get to the required Lorentz factor of many hundreds). If there is luminosity evolution, then it is not clear that the current model that attributes GRB in the gamma-ray phase to internal shocks in a relativistic wind would accommodate it. A stellar collapse followed by relativistic ejection is not really expected to change its properties significantly with redshift. Perhaps some of the gamma-ray emission can still be attributed to the forward external shock; in that case, the higher density of the ambient medium at high redshift could give brighter bursts (the peak synchrotron flux is proportional to the square root of the ambient density).

In the issue of high-redshift bursts, it is clear that a fruitful paradigm has emerged which encourages collaboration between fields, using cosmology to help pin down the redshifts and location of GRB, and gamma-ray bursts as lamps to illuminate the early Universe possibly beyond quasars. Also, with some luck GRB afterglows will remain as simple as they appear to be now, in which case they may become to the Universe at redshifts 2–5 what type Ia supernovae are to the 'nearby' (well, $z \lesssim 1$) Universe. But if indeed even luminosity evolution is important, it will be a while before all the unknowns are pinned down.

REFERENCES

1. Blandford, R. D. and Znajek, R. L.: 1977, MNRAS, 179, 433
2. Fenimore, E. E. and Bloom, J. S.: 1995, ApJ, 453, 25
3. Fruchter, A. S.: 1998, ApJ Lett., accepted
4. Horváth, I., Mészáros, P., and Mészáros, A.: 1996, ApJ, 470, 56
5. Kommers, J. M., Lewin, W. H. G., Kouveliotou, C., van Paradijs, J., Pendleton, G. N., Meegan, C. A., and Fishman, G. J.: 1998, ApJ, submitted
6. Krumholz, M. R., Thorsett, S. E., and Harrison, F. A.: 1998, ApJ Lett., in press
7. Lilly, S. J., Le Fèvre, O., Hammer, F., and Crampton, D.: 1996, ApJ, 460, L1
8. Madau, P.: 1996, in *Star Formation Near and Far*, AIP Conf. Proc., astro–ph/9612157, AIP, New York
9. Mao, S. and Paczyński, B.: 1992, ApJ, 388, L45
10. Mészáros, P., Rees, M. J., and Wijers, R. A. M. J.: 1998, Phys. Rep., in press (astro–ph/9808106)
11. Paczyński, B.: 1998, ApJ, 494, L45
12. Totani, T.: 1997, ApJ, 486, L71
13. Ulmer, A., Wijers, R. A. M. J., and Fenimore, E. E.: 1995, ApJ, 440, L9
14. Wasserman, I.: 1992, ApJ, 394, 565
15. Wickramasinghe, W. A. D. T., Nemiroff, R. J., Norris, J. P., Kouveliotou, C., Fishman, G. J., Meegan, C. A., Wilson, R. B., and Paciesas, W. S.: 1993, ApJ, 411, L55
16. Wijers, R. A. M. J., Bloom, J. S., Bagla, J. S., and Natarajan, P.: 1998, MNRAS, 294, L13
17. Wijers, R. A. M. J. and Galama, T. J.: 1998, ApJ, accepted

A Crisis in the Theory of Gamma Ray Bursts?

Howard D. Greyber

10123 Falls Road, Potomac, MD 20854, U.S.A.

Abstract. The estimated GRB energy of around 10^{55} ergs for GRB980329 implies that all stellar mass models are inadequate. An original "beam on target" model, based on gravitational collapse forming a galaxy in the presence of a uniform primordial magnetic field, and producing a powerful, highly relativistic, intense, storage ring, is proposed. The afterglow or "fireball" is produced by blowoff of target material as the target crosses the beam.

THE HISTORY

Astronomers frequently are shocked by (and resist fiercely) brand new ideas. Old ideas are so much more comfortable. Eddington criticized bitterly and acidly Chandra's mathematical result predicting the black hole. When first suggested by Baade & Zwicky, the idea of tiny neutron stars was ridiculed. During the Seventies over a hundred theoretical models for Gamma Ray Bursts (GRB) were proposed. During the Eighties the conventional, unanimous dogma that evolved was that surely, no doubt, gamma ray bursts (GRB) originated from neutron stars within our own Galactic disk only 100-200 pc away. This firm conviction was shattered in 1992 with the observations from the Compton Gamma Ray Observatory (CGRO) showing the isotropy of the sources in the sky. Then, for a few years, the obvious cosmological hypothesis for gamma ray bursts battled with ever larger galactic halo neutron star models.

A dramatic new epoch opened in February 1997 when the BeppoSax satellite observed the first X-ray afterglow from a GRB. Rapidly expanding hot plasma optical and radio afterglows from many GRBs were quickly observed, and soon large optical redshifts from either the optical transient (OT) or from the underlying host galaxy were measured. It is clear that the GRB afterglows are much more than 100 times brighter than supernovae. The cosmological hypothesis for GRBs now appears secure.

CP470, After the Dark Ages: When Galaxies were Young (the Universe at 2 < z < 5),
edited by Stephen S. Holt and Eric P. Smith

THE PRESENT

The expanding afterglow has been nicknamed "Fireball", and delineated in elaborate mathematical models, by Rees and Meszaros, Waxman, Wijers, Piran, Sari, Katz and others. Although for a few observations some details are in doubt, mostly, the afterglow concept of a rapidly expanding hot relativistic plasma appears correct. For a physical model of the GRB itself, an "explosion" of some type is commonly invoked, either in the form of binary mergers or else massive stellar collapse. Examples of the former are merging neutron stars (Rees and Meszaros), or a neutron star merging with a black hole. The latter are hypernovae type models associated with the names of Woolsey, Eichler and Pacyznski. Both models involve stellar size masses.

However, despite these suggested physical models, the basic, ultimate astrophysical source of GRBs is still unknown. In a very recent paper, Andrew S. Fruchter points out that there is little direct evidence that GRBs occur in regions of high extinction (1). He adduces strong evidence that GRB980329 may be at Z 5, and that with a standard cosmology, the fluence in the BATSE 50-300 KeV range implies an isotropic burst would emit 5 x 1054 ergs in this narrow gamma ray range alone. In the title of another paper, Hogg and Fruchter ask "Why are GRBs Found in Faint Galaxies?". Fruchter notes that, already, with only nine bursts well

identified with optical transients, this clearly implies "$\sim$ 20% of bursts are at $z > 3$ and $\sim$ 10% of bursts are at $z > 5$ " .

A recent Nature article by S. R. Kulkarni et al.(supported by two Letters in the same issue) identifies a host galaxy having a $z = 3.42$ for GRB971214, (2) . Kulkarni, Djorgovski et al calculated that the burst emitted 3 x 1053 ergs in gamma rays alone in a very small volume indicated by the rapid variability of GRBs. They say "GRB971214 was not a particularly faint event, and thus statistically we expect many fainter events to arise from larger redshifts". This implies even higher burst energies to come. Both energy estimates, GRB980329 and GRB971214 are clearly underestimates, omitting energy production in all other wavelengths outside of the very narrow range observed.

When one considers that the GRB980329 energy estimate by Fruchter equals the conversion of the mass of an entire 2.5 solar rest mass object into a narrow energy range of gamma rays, one senses a severe crisis in understanding the 30 year old puzzle of the origin of GRBs. Stellar mass models are simply not adequate, barring unrealistic, extremely strong, assumed gamma ray beaming. Based on research by Woolsey and others, Bohdan Paczynski has proposed a hypernova scenario (3). However, Paczynski admits his scenario is very schematic, and emphasizes that the assumed efficient magnetic energy transport from the massive stellar core to the envelope is a speculation.

An even stronger argument against stellar mass models is from Piran and Sari, based on the very long-delayed gamma ray photons observed in some GRBs (4). They conclude "The 'inner engine' must operate up to hundreds of seconds, and produce highly variable winds. This directly rules out all explosive models." E.

Fenimore et al. came to a similar conclusion. "Explosion" means energy is supplied by some interior process in the source.

A DIFFERENT PARADIGM

The only logical astrophysical source for bursts with energies of 10^{55} ergs and higher is the gravitational collapse energy of an entire galaxy. This is a huge reservoir of 10^{67} ergs or more. Of course we know that trillions of optically faint, young galaxies must be forming by collapsing under gravitation in, and after, "The Dark Ages" of the Early Universe. A totally different paradigm, that can produce the "fireball", or optical transient (OT), that is observed at the site of a GRB, and yet is not an "explosion", is Greyber's Beam On Target (BOT) model. It was mentioned in 1992, described in 1993 and described in more detail in 1994 (5,6,7)

BOT is simply the very common, ordinary method used on Earth to produce gamma rays in accelerators. This author's old Strong Magnetic Field model (SMF) posits that an enormous, intense, very relativistic current loop (termed Storage Ring) is generated during the original gravitational collapse of the pregalactic plasma cloud in the presence of an almost uniform primordial magnetic field as the galaxy or quasar forms. The concept was introduced in 1961 to explain the origin of spiral arm structure and answer Oort's questions regarding spiral galaxies. However the new galactic component has proved valuable in several other ways, producing a physical model for the AGN-Quasar central engine, which clearly creates the very long, narrow, often straight radio and optical jets that are observed (8,9,10,11).

One can see such a storage ring forming in Figure 1, from a brilliant, pioneering article by Leon Mestel and Peter Strittmatter (1967) on "The Magnetic Field of a Contracting Gas Cloud". In II. they analyzed the effect of Ohmic diffusion on the magnetic field distribution of a spherical, gravitationally bound, uniform density ionized gas cloud permeated by a uniform magnetic field, as it contracts (12). Their MHD calculation illustrated how the magnetic field topology changes by contraction under self gravitation. The cloud field detaches from the background field and a growing equatorial current loop is formed. As the loop shrinks and self-pinches, gravitational energy of collapse is converted into charged particle kinetic energy and thus into magnetic field energy.

In reality the current grows far faster than MHD indicates, since the well-known plasma physics Pinch Effect for an increasing current causes constant accretion of huge numbers of charged particles from the plasma to join the current stream, increasing the number of current carriers. A gigantic computer calculation with many millions of particles, like that pioneered by Prof. Oscar Bunemann of Stanford University, but with the laws of plasma physics added, is needed for a good simulation.

As magnetofluiddynamics dictates, the intense, coherent, highly relativistic current loop maintains a cavity around itself inside the massive, slender toroid of

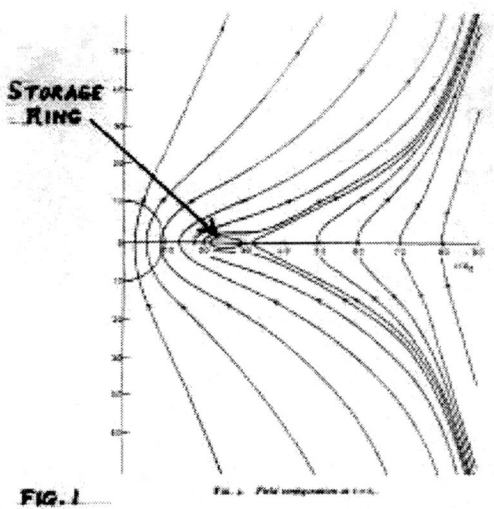

FIGURE 1. Figure from Mestel and Strittmatter (1967)

plasma which is bound to the loop by the Maxwell "frozen in" magnetic field condition.(13,14) The bursting force of this very strong unified magnetic field system is in equilibrium, balancing the strong gravitational force between the toroidal plasma and the central massive object.

To understand the storage ring, the clue is to realize that the only radiation loss from a completely coherent, constant intensity, undisturbed, highly relativistic current loop is the classical curvature radiation, which is extremely small when the loop radii are larger than 10 Astronomical Units. Using a result from a classic paper by Julian Schwinger (15), one finds a undisturbed one GeV loop current with a radius of 10 A.U. would last for times far, far longer than the age of the Universe (see Table in ref. 10). For loop radii in the many kiloparsec range, the effect of gravitational collapse on the storage ring could create particles in the TeV, PeV, and even the cosmic ray energy range, for a short time in the early stages of the gravitational contraction of the pregalactic plasma cloud. Notice that when gravitational collapse ceases, all interactions will weaken and dissipate the storage ring. A remnant storage ring in our Local Group galaxies is unlikely.

Stability of the storage ring has been discussed in earlier papers. Both thick and thin rotating accretion disk structures have been found to posses many serious instabilities. However, it is significant that O. M. Blaes, using hydrodynamics, analyzed constant specific angular momentum tori, and demonstrated that any instability becomes vibrational (i.e., stable) for extremely slender tori such as the SMF storage ring (16). Probably the same result will occur for the MHD case.

Once formed, the completely coherent relativistic current loop is *uniquely stable*. Due to the coherence, one part of the undisturbed steady loop current will not radiate in the magnetic field of another part. However, if a fluctuation or "bump" occurs somewhere along the loop, the particles in the "bump", now out of coherence, will suddenly radiate furiously in the extremely strong magnetic field, the energy in the fluctuation will dissipate rapidly, and the ring will return quickly to its undisturbed configuration. Equipartition of the particle energy with the magnetic field energy is irrelevant since equipartition occurs only when the physics demands it does.

The morphology and energetics of objects of galactic dimension are determined in SMF by the ratio of magnetic field energy to rotational energy in the particular object. This ratio is the highest for quasars, BL Lacs and blazars, decreasing steadily for giant radio galaxies, Seyferts, Markarians, is low for spirals and even lower for ordinary ellipticals. It is pertinent to recall that the Russian astrophysicist Ya. B. Zel'dovich emphasized, "A major challenge is to understand strong magnetic fields whose energies greatly exceed that of hydrodynamic motions" (17). SMF deals precisely with that topic. By that criterion, the local magnetic field can be a microgauss or less, as long as it is strong enough to control the slow motion of the extremely low density plasma.

BEAM ON TARGET MODEL (BOT)

In 1995, in an unpublished paper, Greyber showed that BOT met all the severe constraints on theoretical models imposed by the careful, detailed analysis by 21 C.G.R.O. authors, of the giant, intense GRB930131 burst.(18,19). Sommer et al point out that the requirement on photon collimation to avoid pair production implies a huge Lorentz factor if the source is at cosmological distances. Therefor a minimum of millions of times higher than the observed rate of sources (about 800 per year) is required if none repeat. In BOT the trillions of young faint galaxies forming under gravity in the Early Universe satisfies this requirement. Also, Sommer et al.ruled out any black body spectrum model with a temperature less than 1 GeV, which is satisfied by BOT. Sommer et al.require for GRB930131 that the high energy particle lifetime is at least 25 seconds (many hundreds of seconds in later bursts). Of course BOT satisfies this since the storage beam particles live for millions of years until a target removes them.

In BOT, when large "rocks" (perhaps a neutron star, white dwarf, asteroid or comet) race across the immensely powerful storage ring in a young forming galaxy, strongly beamed bursts of gamma rays are produced, (Figure 2). The timing fits the observations. A target crossing the minor diameter of the beam can take 30 seconds or more, while one crossing a short chord might take far less than one second. Since it is well known from space physics that currents in space are often made up of slender filaments, the millisecond spikes observed in many bursts probably reflect the storage beam structure in that galaxy (20). The afterglow or "fireball" is produced in BOT when the intense, extremely relativistic current beam evaporates or "blows off" the surface of the target. This simulates an explosion. As I wrote

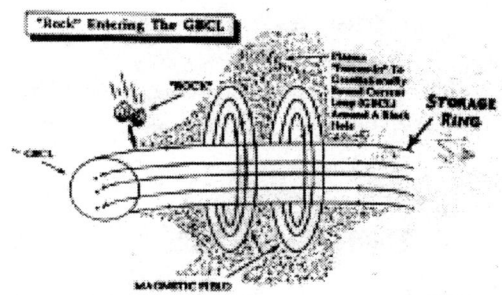

FIGURE 2. "Beam-on-target" model

in 1993, "one may now have a means of studying a population of objects even earlier in the Universe than the quasars" (6). Presumably colleagues at CERN or FermiLab could modify their large beam-target interaction computer programs to fit the BOT scenario and verify the predictions. This would be analogous to the mid-Fifties when my colleagues at Lawrence Livermore Laboratory modified our thermonuclear weapons computer program to create one of the first computer models of stellar thermonuclear burning for astrophysics.

In October 1996, four GRBs were observed within two days in the same few square degrees of the sky. The probability of this being accidental is very small. A scenario from BOT would be, like Comet Shoemaker-Levy 9 approaching Jupiter, if the target broke up into pieces as it approached the storage ring. BOT implies the existence of a very large-scale, almost uniform (over a pregalactic gas cloud), primordial magnetic field that has not yet been observed. An original model was proposed by Greyber that explains the origin of such a primordial magnetic field as well as the observed quasi-regular, cell-like structure of very thin spatially curved sheets of rich superclusters around huge voids. (21,22). This is in stark contrast to the popular galactic dynamo model for cosmical magnetism. Eugene Parker wrote, "In the extreme case, the magnetic activity is driven by some colossal energy source, such as gravitational collapse within the galaxy - -" (23). Clearly SMF is in accord with Professor Parker's statement, but for all AGN and quasars.

SMF obviously predicts that very young evolving galaxies that are also radio sources must have evidence of magnetic fields since the Storage Ring is formed right during the gravitational collapse forming the galaxy. Research by P.P. Kronberg confirms this prediction (24). A. M. Wolfe commented "the prevalent dynamo model leads one to expect that the fields do not form until much later after a galaxy has completed several full rotations".

Walter Elsasser, a geophysicist, once related to a friend Thomas Cowling's attempt in 1934 to make an astrophysical dynamo. "If that simple idea does not work", remarked Albert Einstein, "then dynamo theory will not work either". The result was Cowling's famous anti-dynamo theorem proving that axisymmetric fluid flows cannot generate magnetic fields.

Acknowledgements

It is a pleasure to acknowledge the late Donald H. Menzel for advice and warm encouragement, and Gart Westerhout for permission to use the U. S. Naval Observatory Library.

REFERENCES

1. Fruchter, A. S., preprint, astro-ph9810224, 15 Oct. 1998
2. Kulkarni, S. R., Djorgovski, S.G. et al, 1998, Nature 393, 35
3. Paczynski, B. , preprint, 1998
4. Piran, T. & Sari, R. preprint, 1997
5. Greyber, H. D., 1992, in "Testing the AGN Paradigm", A.I.P. Conference Series 254, edited by S. Holt, S.G. Neff and C. M. Urry, p.467
6. Greyber, H. D., 1993 in "Compton Gamma Ray Observatory", A.I.P. Conference Series 280, edited by M. Friedlander, N. Gehrels and D. Macomb, p.569
7. Greyber, H. D., 1994, in "Cosmical Magnetism- Contributed Papers" in honour of Prof. L. Mestel, FRS, Institute of Astronomy, Cambridge Univ., England, edited by D. Lynden-Bell, p.110
8. Greyber, H. D., 1961, in Transactions of the I.A.U. XIB, 332; Report of Commission 33
9. Greyber, H. D., 1967, in "Instabilite Gravitationelle et Formation des Etoiles, des Galaxies, de Leurs Structures Caracteristique", Memoirs of the Royal Society of Liege, XV, p.189 and p.197
10. Greyber, H. D., 1989, Comments On Astrophysics, 13, 201
11. Greyber, H. D., 1990, Annals New York Acad. of Sciences, 571, 239; 14th Texas Symposium .
12. Mestel, L. and Strittmatter, P., 1967, MNRAS, 137, 95
13. Sozou, C. & Loizou, G., 1966, J. Fluid Mech., 25, 761
14. Downs, A. M. et al, 1966, Q. J. Mech. Appl. Math. XIX (part1), 27
15. Schwinger, J., 1949, Phys. Rev. 75, 1912
16. Blaes, O. M., 1985, MNRAS, 212, 37
17. Zel'dovich, Ya. B., 1983, in "Magnetic Fields in Astrophysics", Gordon & Breach, p.8
18. Greyber, H. D., 1995, "On the Beam On Target Model for Gamma Ray Bursts", unpublished
19. Sommer M. et al, 1994, ApJ 422, L63-L66
20. Alfven, H., 1981, in "Cosmic Plasma", D. Reidel Publishers, Chapter II.
21. Greyber, H. D., 1995, in "Dark Matter", A.I.P. Conference Proceedings 336, edited by S. S. Holt and C. L. Bennett, p.509
22. Greyber, H. D., 1996, in "Clusters, Lensing and the Future of the Universe", A.S.P. Conference Series, Vol. 88, edited by Virginia Trimble and Andreas Reisenegger, p.298
23. Parker, E. N., 1979, in "Cosmical Magnetic Fields", Oxford Univerisity Press, p.816

24. Kronberg, P. P., 1987, in "Interstellar Magnetic Fields", eds, R. Beck and R. Grave, Springer-Verlag
25. Kronberg, P. P., 1994, "Extragalactic Magnetic Fields", in Reports on Progress in Physics, vol. 57, 325

GRB Redshift Distribution
is Consistent with GRB Origin
in Evolved Galactic Nuclei

V.I. Dokuchaev[1], Yu.N. Eroshenko[1], and L.M. Ozernoy[2,3]

[1] *Institute for Nuclear Research, Russian Academy of Sciences, Moscow, 117312, Russia*
[2] *George Mason University, Fairfax, VA 22030-4444, USA*
[3] *Goddard Space Flight Center, Code 685, Greenbelt, MD 20771, USA*

Abstract. Recently we have elaborated a new cosmological model of gamma-ray burst (GRB) origin [1], which employs the dynamical evolution of central dense stellar clusters in the galactic nuclei. Those clusters inevitably contain a large fraction of compact stellar remnants (CSRs), such as neutron stars (NSs) and stellar mass black holes (BHs), and close encounters between them result in radiative captures into short-living binaries, with subsequent merging of the components thereby producing GRBs, typically at large distances from the nucleus.

In the present paper, we calculate the redshift distribution of the rate of GRBs produced by close encounters of NSs in distant galactic nuclei. To this end, the following steps are undertaken: (i) we establish a connection between the parameters of the fast evolving central stellar clusters (*i.e.* those for which the time of dynamical evolution exceeds the age of the Universe) with masses of the forming central supermassive black holes (SMBHs) using a dynamical evolution model; (ii) we connect these masses with the inferred mass distributions of SMBHs in the galactic nuclei and the redshift distribution of quasars by assuming a certain 'Eddington luminosity phase' in their activity; (iii) we incorporate available observational data on the redshift distribution of quasars as well as a recently found correlation between the masses of galaxies and their central SMBHs. The resulting redshift distribution of the GRB rate, which accounts for both fast and slowly evolving galactic nuclei is consistent with that inferred from the BATSE data if the fraction of fast evolving galactic nuclei is in the range $\varepsilon \simeq 0.016 - 0.16$.

INTRODUCTION

In our previous work [1], we have considered the generation of GRBs due to the radiative collisions of CSRs (for simplicity, taken to be solely NSs) in the central stellar clusters of evolved galactic nuclei. Depending on its initial radius and mass, the characteristic dynamical evolution time of a cluster, t_e, can either

CP470, After the Dark Ages: When Galaxies were Young (the Universe at 2 < z < 5),
edited by Stephen S. Holt and Eric P. Smith
© 1999 The American Institute of Physics 1-56396-855-X/99/$15.00

exceed the age of the Universe, t_0, (we call such clusters as '*slowly evolving clusters*' hereinafter) or be less than t_0 ('*fast evolving clusters*'). The GRB rate from slowly evolving clusters does not depend on redshift of the host galactic nuclei. Following an approach developed in [2], the mean rate of radiative collisions of NSs in the fiducial galactic nucleus with mass of a central cluster M, radius R, and stellar velocity dispersion $v = (GM/2R)^{1/2}$ is given by [1]:

$$\dot{N}_c \simeq 9\sqrt{2}\left(\frac{v}{c}\right)^{17/7}\frac{c}{R} \simeq 5.8 \cdot 10^{-6}\left(\frac{M}{10^7\,M_\odot}\right)^{17/14}\left(\frac{R}{0.1\,\mathrm{pc}}\right)^{-31/14}\frac{\mathrm{events}}{\mathrm{yr\ galaxy}}\,, \quad (1)$$

where we use for normalization the likely values of M and R to fit the inferred rate of GRBs. The GRB rate from a unit comoving volume is $\dot{n}_0 = \langle \dot{N}_c \rangle n_g$, where the number density of galaxies $n_g \sim 10^{-2}\ \mathrm{Mpc}^{-3}$ and $\langle \ldots \rangle$ is an averaging throughout all types of slowly evolving galactic nuclei.

A galactic nucleus with the gravitationally dominating SMBH of mass $M_h > M$ might belong to slowly evolving galactic nuclei as well. The corresponding GRB rate from such a nucleus is given by [1]

$$\dot{N}_{c,h} \simeq \frac{9}{2\sqrt{2}}\left(\frac{Nm}{M_h}\right)^2\left(\frac{v}{c}\right)^{17/7}\frac{c}{R}\,, \quad (2)$$

which differs from that given by Eq. (1) by a factor $(M/M_h)^2 \ll 1$.

One more possible contribution to GRB rate from a slowly evolving system is due to the coalescence of tight NS binaries in the galactic discs (which is a commonly believed scheme of cosmological GRB origin). In this paper, we assume that there was no strong evolution with cosmological time of the number of tight NS binaries both in the galactic disks and in the slowly evolving nuclei so that all these contributions result in the redshift-independent GRB rate $\dot{n}_0 \simeq 10^{-8}\ \mathrm{yrs}^{-1}\ \mathrm{Mpc}^{-3}$.

Meanwhile the actual parameters of galactic nuclei vary in a wide range and some central clusters may, in fact, be fast evolving ($t_e < t_0$). The GRB rate from the fast evolving clusters depends on redshift. Below, we find the redshift distribution of GRB rate from both slow and fast evolving clusters by (i) modelling the dynamical evolution of a fast evolving cluster in the galactic nucleus, which leads to the production of a SMBH; (ii) connecting the parameters of a fast evolving cluster with the mass of its central SMBH, and (iii) using the observed distribution in luminosity and redshift for quasars and that in mass for SMBHs in the galactic nuclei.

FAST EVOLVING GALACTIC NUCLEI

We introduce a dimensionless time variable $y \equiv t_0/t_e = (1+z_e)^{3/2}$, where t_e is the time necessary for a SMBH formation, z_e is the redshift at that instant. Evidently, $y > 1$ for clusters collapsed to the present epoch $t = t_0$ (fast evolving clusters) and $y < 1$ for clusters which do not collapse by t_0 (slowly evolving clusters). As we show elsewhere [3], the dynamical evolution of a stellar cluster, in the framework of a homologous Fokker-Plank approximation [2], leads to the following relationship between the initial mass of the cluster and the mass of the central SMBH formed there:

$$M_f = 2 \cdot 10^6 y^{2.22} \left[\frac{M_i}{2.8 \cdot 10^8 M_\odot} \right]^{5.44} M_\odot. \tag{3}$$

There are serious observational indications [4], [5], [6] that a fraction of galaxies with a central SMBH might be as high as $\varepsilon \sim 0.1$ and that there is a correlation between the central SMBH mass, M_h, and the luminosity of the host galaxy bulge, L_s, which in a simplified form can be expressed as $M_h \simeq 10^{-2.5}(L_s/L_\odot)M_\odot$. This correlation, taken together with the Schechter luminosity function of galaxies [7], gives the mass distribution of SMBHs, $\phi_1(M_h)dM_h$ with $\phi_1(M_h) \propto \varepsilon$. Then we use the observed [8] distribution of quasars in absolute magnitude M_B and redshift z, $\phi_2(M_B, z)dM_B dz$, for $z \leq 3$. Finally, we assume the existence of an 'Eddington luminosity phase' in the evolution of quasars, when they shine with the Eddington luminosity (or with its fraction $\lambda \leq 1$) during a certain finite time interval, which duration only depends upon λ and the initial SMBH mass. In our calculations, the 'final' (i.e., by the end of the dynamical evolution) mass of a galactic nucleus is the initial mass of its new-born SMBH. The latter then experiences the 'Eddington luminosity phase', during which its mass grows up to the SMBH mass in the actually observed quasar. As a result, we find the mass distribution of *active* SMBHs, $P(M_f, y)dM_f dy$, in terms of the variables y and mass M_f at the instant of the SMBH formation (in the limit $M_f < 3 \cdot 10^6 M_\odot$). Thus, these active SMBHs arise due to the dynamical evolution of galactic nuclei with fast evolving central clusters.

DISTRIBUTION OF GRB RATE IN REDSHIFT

In fast evolving galactic nuclei, the resulting rate of GRB generation per unit of comoving volume is given as a function of redshift z by

$$\dot{n}_{ev}(z) = \int dM_f \int_1^{x(z)^3} P(M_f, y)\dot{N}_{ci}(M_i(M_f), y) \frac{dy x^{3\xi}}{(x^3 - y)^\xi}, \tag{4}$$

where $x = (1+z)^{1/2}$, $\xi \simeq 0.6$, $\dot{N}_{ci}$ is found from Eq. (1) with $M = M_i$ and $R = R_i$, and function $M_i(M_f)$ is the inverse function given by Eq. (3). Numerical integration over the variable M_f in Eq. (4) proceeds in the range $M_f \sim 10^5 - 3 \cdot 10^6 M_\odot$, which approximately correspond to the range of $M_i \sim 10^7 - 2 \cdot 10^8 M_\odot$.

The total rate of GRBs from *all* kinds of galactic nuclei, $\dot{n}(z)$, includes also the contribution $\dot{n}_0 \simeq 10^{-8}$ Mpc^{-3} yr^{-1} from all slowly evolving galactic nuclei and disks with $t_e > t_0$, i.e. $\dot{n}(z) = \dot{n}_0 + \dot{n}_{ev}(z)$. The total GRB rate as a function of z can be fitted by a one-parametric function: $\dot{n}(z) = \dot{n}_0(1+z)^\beta$, where the range

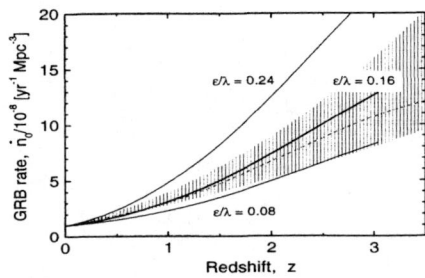

FIGURE 1. The GRB rate as a function of redshift. The hatched region confines the allowed GRB redshift distribution from the BATSE data [9]. Three solid lines show the GRB rate from fast evolving stellar clusters in distant galactic nuclei (a constant contribution from the slowly evolving systems is added) as given by our model for the three different values of ε/λ ratio, where ε is the fraction of galactic nuclei in which SMBHs are formed, and λ is the average quasar luminosity normalized to the Eddington luminosity. For comparison, the dashed line shows the rate of GRBs from NS coalescences in tight binaries produced in the distant star-forming galaxies [11].

of β, $1.5 < \beta < 2$, accounts for the statistical uncertainties [9]. This function is shown in Fig. 1. The consistency between the calculated and inferred GRB rates is achieved at $\varepsilon/\lambda \simeq 0.16$. Since quasars shine at $L \simeq (0.1 - 1) \, L_{Ed}$ (e.g., [10]), it implies that $\varepsilon \simeq 0.016 - 0.16$, which is consistent with available data [4]-[6].

REFERENCES

1. Dokuchaev, V.I., Eroshenko, Yu.N., & Ozernoy, L.M., 1998, ApJ, 502, 192
2. Quinlan, G.D. & Shapiro, S.L., 1987, ApJ, 321, 199
3. Dokuchaev, V.I., Eroshenko, Yu. N., & Ozernoy, L.M., 1998, in preparation
4. Van Der Marel, R.P., 1997, astro-ph/9712076
5. Magorrian, J. et al., 1997, submitted to AJ, astro-ph/9708072
6. Kormendy, J. & Richstone, D., 1995, ARA&A, 33, 581
7. Efstathiou, G., Ellis, R.S., & Peterson, B.A., 1988, MNRAS, 232, 431
8. Boyle, B.J., 1991, in "Proc. Texas/ESO–CERN Symposium on Relativistic Astrophysics, Cosmology, Fundamental Physics", ed. J. D. Barrow et al., p.14
9. Horack J.M. et al., 1995, ApJ, 447, 474
10. Wandel, A., 1998, astro-ph/9808171
11. Bagot, P. et al., 1998, astro-ph/9802094

10. Next Generation Capabilities

The Golden Age for Near–IR Astronomy

Eric P. Smith and John C. Mather

Laboratory for Astronomy & Solar Physics, NASA/GSFC, Greenbelt, MD 20771

Abstract.
The previous National Academy of Sciences decadal survey for astronomy, the "Bahcall report", heralded the 90's as the "decade of the infrared". The report outlined the many near–infrared (here defined as 1-30 μm) facilities that would become available. Their vision is now coming to fruition with the advent of large ground–based telescopes equipped with infrared instruments and upcoming space–based infrared (IR) missions. For astronomers, the first decade of the 21st century promises to be the one in which the floodgates open for the IR data from these facilities. In addition, there are many new capabilities for near–IR astronomy in various stages of planning for the future. We outline the current capabilities for near IR astronomy and project how these new observatories and instruments will impact the field. We also list the online sources of information for each of the facilities.

INTRODUCTION

Astronomical sources of near–IR emission will play a critical role in furthering our understanding of many important formative processes in the universe. Locally, proto-stellar and proto-planetary objects are hidden from direct optical investigation by dust. These regions are more accessible to observation by the infrared radiation that is not blocked by the dust. Cosmologically, the peak in the blackbody radiation from the first stars and galaxies to form after the universe cooled from the big bang is redshifted into the near–IR. Hence, the recent advances made in infrared astronomy are driven both by a recognition of the exciting discovery space yet available in this portion of the spectrum, and the development of better infrared detectors. Indeed, in the 1990s the rapid advancement of IR detector technology in the areas of format size for imaging arrays (increased by an order of magnitude), and quantum efficiencies (approaching that of CCDs) have made these devices as indispensable to astronomers as CCDs became in the 1980s. We will give examples of ground and space capabilities, then describe near term capabilities for both types of observatories and conclude with a discussion of the Next Generation Space Telescope and other future missions. To set the stage however we

CP470, After the Dark Ages: When Galaxies were Young (the Universe at 2 < z < 5),
edited by Stephen S. Holt and Eric P. Smith

show a plot (Figure 1) of the sensitivity of near-IR observatories widely available c. 1997 prior to installation of the NICMOS on HST. In discussing the new missions we show the rapid advances expected over the next few years will vastly improve astronomers reach faint IR flux levels. The facilities we discuss will come on–line or begin construction within the 1998-2008 time frame.

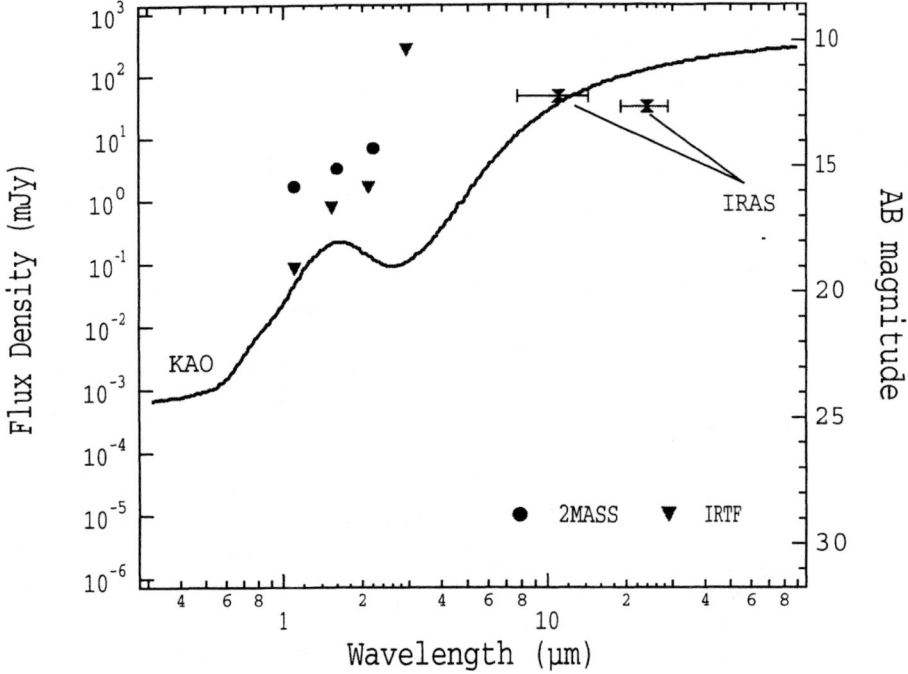

FIGURE 1. *Wide area coverage, general user facilities for near–IR astronomy: Sensitivities of 2MASS, IRTF, and KAO facilities*

GROUND–BASED FACILITIES

Even though ground based telescopes are severely limited by atmospheric transparency and emission, it is possible to achieve good success with adaptive optics in the near-IR, with Strehl ratios approaching unity around 2 μm. New, large facilities coming on–line presently are the Gemini, the Keck, and the VLT observatories. In addition, the 2MASS 2 Micron All Sky Survey will form an observational data base comparable in importance to the Palomar Sky Survey, and the SOFIA airborne telescope will offer frequent opportunity for new instrument deployment for specialized observations.

2MASS. The 2MASS telescopes are 1.3 m in diameter, deployed at Mt. Hopkins and Cerro Tololo, and utilize improved IR detectors to create an all-sky 2 μm map to levels $\sim 10^4$ times deeper than the existing survey [1]. With arrays operating at three wavelengths (J, H, K), and a pixel size of 2 arcsec, they scan 8.5 arcmin by 6 degrees wide strips at the rate of 1 arcsec per second, giving an effective exposure time of 7.8 sec. The sensitivity is in the mJy range, or equivalently $m_{AB} \sim 15$. They will achieve 90 per cent completeness for extended sources at $|b| > 30$ degrees, and 95 per cent coverage for $|b| > 10$ degrees. The survey has already found several brown dwarfs and led to the recognition of two new stellar spectral classes [2]. See http://pegasus.phast.umass.edu/ for additional information.

Gemini. The Gemini 8.1 m telescopes at Mauna Kea and Cerro Pachón are facility class observatories with many science goals. The North telescope will initially have three instruments: the Near Infrared Imager (NIRI), Mid-IR imager/spectrometer (Michelle), and Near Infrared Spectrograph (NIRS), while the South telescope will have the Mid-Infrared Imager (MIRI) and the NIRS-2. The NIRI, from the University of Hawaii, will be ready in March 1999, and will provide 1 to 5 μm imaging and grism spectroscopy. It will provide 0.02 arcsec pixels with a 20 arcsec field of view for adaptive optics work; 0.05 arcsec pixels with a 51 arcsec field for tip-tilt corrections only; and 0.12 arcsec pixels with a 123 arcsec field for wide field observations. The NIRS, from the NOAO, will provide spectroscopy with resolving powers of 2000 and 8000, and the NIRS-2 will be similar. The Michelle, from the Royal Observatory of Edinburgh, will be available in June 1999, with a 7-25 μm imager and long slit spectrometer, 0.10 arcsec pixels, and a 25 arcsec field of view, providing spectral resolutions from 10 to 20,000. The MIRI, from the University of Florida, will provide 0.09 arcsec pixels for 8 to 26 μm, with a 25 μm field of view, and will be diffraction limited at 8 μm. Future instruments will include a Small Field Optimized Near IR Imager (coronagraph) with 0.01-0.05 arcsec pixels, operating from 0.8 to 5 μm, and an infrared multiobject spectrometer. Visible-near IR sensitivities are measured in tens of nJy, $m_{AB} \sim 27 - 29$. See http://www.gemini.edu/.

Keck. The two famous Keck 10 meter telescopes I and II are general purpose facility class observatories. Their primary infrared instruments are,the Near Infrared Camera (NIRC), Long Wavelength Spectrograph (LWS), and Long Wavelength Infrared Camera (LWIRC). NIRC, from Caltech, covers the wavelength range from 1 to 5 μm with 256^2 InSb array detectors. In K band it achieves a sensitivity of 22.3 mag., 10 σ in 1 hour. The LWS, from the University of California at San Diego, covers the range from 3 to 25 μm with 128^2 Si:As detectors, and provides diffraction limited imaging at 10 and 20 μm, with spectral resolutions of 100 and 1400. The LWIRC, from the University of California, Berkeley, will be operational in late 98, and will cover the wavelength range from 7 to 13 μm with 0.055, 0.11, and 0.22 arcsec/pixel modes. Future instruments include the NIRC2, with 1024^2 InSb arrays and 0.02 arcsec pixels at 1.1 μm. The NIRSPEC, an echelle spectrograph from the University of California, Los Angeles, will be operational in October 1998. The most ambitious undertaking, an interferometer operating between the two telescopes, will provide an 85 m baseline and angular resolutions of 0.005 to 0.025 arcsec at wavelengths of 2 and 10 μm. See `http://www2.keck.hawaii.edu:3636/index.html`.

VLT. The Very Large Telescope (VLT) comprises four 8.2 m telescopes and several auxiliary 1.8 m telescopes, which can be operated separately or together in interferometric mode. Initial results show that the seeing at the VLT site, Paranal, is excellent, as good as 0.25 arcsec in the published photographs. This extraordinary system will provide instrumentation comparable to those described above for Gemini and Keck, with the advantage that they can be left attached to the telescopes since there are four telescopes. See `http://www.eso.org/vlt/` for the telescopes, and `http://www.eso.org/instruments/` for descriptions of the instrumentation.

SOFIA. The SOFIA (Stratospheric Observatory for Infrared Astronomy) is the logical next step after the Kuiper Airborne Observatory. It is a 2.5 m Cassegrain telescope, whose mirror is now being light-weighted at the REOSC facility near Paris. It will be mounted in a specially modified Boeing 747-SP aircraft, and will be operational in 2001 for a planned 20 year life. It is being built and operated by a partnership comprised of the University Space Research Association (USRA), United Airlines, Raytheon E-Systems, Sterling Software, the University of California, the SETI Institute, and the ASP (Astronomical Society of the Pacific) in the USA, and DLR, MAN, and Kayser-Threde in Germany. The telescope will provide an image with 80 per cent of the encircled energy within 1.5 arcsec. Planned facility instruments include the FORCAST (a mid-IR imager from 5 to 40 μm), the AIRES Echelle Spectrograph, and the EXES (an Echelon - Echelle Spectrograph covering 5-28 μm with a resolving power of 1500 to 100,000). See `http://sofia.arc.nasa.gov/` for details.

SPACE–BASED IR FACILITIES

In addition to the completed IRAS, COBE, and ISO infrared space missions, we have one currently operating near infrared instrument, the NICMOS, aboard the Hubble Space Telescope. The WIRE mission is ready for launch in early 1999, and the SIRTF is under construction. The Next Generation Space Telescope is under study.

NICMOS. The NICMOS (Near Infrared Camera Multi Object Spectrometer) from the University of Arizona is on the Hubble Space Telescope. It uses HgCdTe arrays cooled by solid nitrogen. The initially expected lifetime was not achieved, because of distortion of the cryostat by the expansion of the solid nitrogen, so a reverse turbo Brayton cycle cooler has been developed and was tested on the October 1998 Shuttle flight. This cooler will be installed on the next HST servicing mission (2000). The NICMOS is a general purpose instrument on the 2.4 m HST and operates from 0.8 to 2.5 μm with imaging, coronagraphic, and grism modes. It provides 0.04-0.2 arcsec pixels and about 22 nJy point source sensitivity (3σ, 5 orbit exposure time). With very deep exposures the sensitivity reaches the nJy level ($m_{AB} \sim 31$). See `http://www.stsci.edu/`.

WIRE. The Wide Field Infrared Explorer (WIRE) has a 30 cm Ritchey-Chrétien telescope, cooled by a two stage solid hydrogen cooler that is predicted to last for four months. It awaits its Pegasus XL vehicle for a March 1999 launch. It carries cameras operating at 9 - 15 and 21 - 27 μm with 128^2 Si:As BIB array detectors, 15.5 arcsec pixels, and a 33 arcmin square field of view. It is designed to answer three questions: What fraction of the luminosity of the Universe at a redshift of 0.5 and beyond is due to starburst galaxies? How fast and in what ways are starburst galaxies evolving? Are luminous protogalaxies common at redshifts less than 3? Simulations show that its sensitivity will be limited by confusion noise at around 1 mJy ($m_{AB} \sim 16 - 17$). See http://www.ipac.caltech.edu/wire/ for details.

SIRTF. The Space Infrared Telescope Facility, now planned for launch on a Delta 7920H rocket in December 2001, will have a lifetime of 2.5 years with a goal of 5 years. It will be in solar orbit so that the system can be radiatively cooled, and liquid helium will cool the 85 cm $f/12$ beryllium telescope to less than 5.5 K. It will be diffraction limited at 6.5 μm. Using new generation infrared array detectors, it will achieve sensitivity far better than was possible before. The Infrared Array Camera (IRAC) will cover the 3 - 9 μm range in four bands. The Multiband Imaging Photometer for SIRTF (MIPS) will provide up to 5 arcmin fields with mapping and high resolution imaging from the mid-infrared to the submm. The InfraRed Spectrograph (IRS) will provide low resolution (R<1000) spectroscopy from 5 - 40 μm. SIRTF observing time will be open to the community for at least 75 per cent of the time. There will be significant Legacy Science programs, large, coherent investigations of lasting importance. The data will be placed in a public archive immediately to promote follow-on observations with SIRTF and other telescopes. Sensitivities range from 0.6 to 10 μJy ($m_{AB} \sim 21 - 25$). Extensive information is available online at http://sirtf.jpl.nasa.gov/.

NEXT GENERATION SPACE TELESCOPE

Objectives

As capable and ground-breaking as the previous facilities are or will be there are still astronomical investigations beyond their reach. To address such investigations NASA has been studying a Next Generation Space Telescope. The NGST study was chartered in 1995 by NASA Headquarters, with scientific requirements from the "HST and Beyond" report of [4]. Early concepts had already been explored in the 1989 conference at the Space Telescope Science Institute [3]. The minimum set of engineering requirements was called the "core" and called for a 5 year mission with a 4 m aperture, optimized for the 1 - 5 μm region, radiatively cooled, with sensitivity limited by the zodiacal light brightness at 1 AU from the Sun, and diffraction limited at 2 μm wavelength. Headquarters also required that the NGST be built for a construction cost of \$500 M (FY 1996 dollars) and a life cycle cost (including 10 years of operations, and the launch) of less than \$900 M (same units). The report also recommended extension to both shorter and longer wavelengths, with a goal of 0.5 - 30 µm coverage, a larger aperture, and a 10 year life. The required orbit is in deep space, to allow radiative cooling, and an orbit around the Lagrange point L2, about 1.5×10^6 km away, seems the likely choice. The SPIE Conference Proceedings, 1998, vol. 3356, contains a wealth of details about NGST and other space telescopes and instruments. For current NGST details, see http://www.ngst.nasa.gov/.

The science goals were further developed by a volunteer science team in 1996, which produced a Design Reference Mission (DRM) document, a list of observations with the parameters needed to achieve the key scientific goals laid out in the "HST and Beyond" report. The DRM is useful in making decisions about engineering capabilities and scientific importance. Since the NGST will be a general user facility like the HST, with time allocated competitively, the DRM is only a guide to design, and not a commitment to a particular observing program. The scientific context may change radically in the decade between now and launch, but the capabilities will be sufficiently general that they should address the new problems that arise. The DRM study was updated by the participants in the "Science with NGST" conference in April 1997, and published in the report "Next Generation Space Telescope - Visiting a Time When Galaxies were Young," in July 1997 [5]. The conference "Science With NGST" [6] developed the case in more detail, and the 1998 ESA conference "The Next Generation Space Telescope" combined both science and technology [7]. In 1998, the DRM was modified and enhanced by the Ad Hoc Science Working Group (ASWG), which was competitively selected by NASA Headquarters, and augmented with specific expertise from the US, Canada, and ESA. The refined DRM contains 5 major elements, listed here with the proposed time allocations for them: 1) Cosmology and the Structure of the Universe (21 %), 2) The Origin and Evolution of Galaxies (33 %), 3) The History of the Milky Way and its Neighbors (15 %), 4) The Birth and Formation of Stars (16 %), and 5) The Origins and Evolution of Planetary Systems (15 %). Note that this is not equivalent to an evaluation of the relative importance, since some observations are brief because the sources are bright, while others are intrinsically long.

According to the calculations of the ASWG, the original 4 m goal of the "HST and Beyond" report is not sufficient to meet the scientific objectives. The team found that a 6-8 m aperture is required to do the whole DRM program in 2.5 years, leaving time for general observers and projects not yet conceived. There was no strong call for special capabilities such as real-time interaction, rapid response to targets of opportunity, moving target capability, or extraordinary timing or positioning capabilities.

Plans and Participants

The NGST study was begun with a nucleus of people at NASA Goddard Space Flight Center and the Space Telescope Science Institute in October 1995. The organizations involved now include the Jet Propulsion Laboratory, Marshall Space

Flight Center, Langley Research Center, Ames Research Center, and Lewis Research Center, as well as several large aerospace firms. The Department of Defense and the National Reconnaissance Organization have agreed to pursue joint technology development and demonstrations with the NGST study. NASA and the European Space Agency have signed a letter of intent to cooperate, aiming for a $200 M contribution to NGST in return for a 15 % guaranteed share of observing time. The Canadian Space Agency has submitted plans for a $50 M contribution to the mission as well. The three space agencies have agreed to a joint schedule of activities. The Phase A request for proposals for two mission concept studies will be issued by NASA in January 1999. In June 1999, the three agencies will meet to decide which instrumentation category will be provided by ESA. In 2001, the Phase B/C/D competition will be held, leading to a single prime contractor to continue through the construction phase, which will begin in 2003. Launch will be in 2007 or 2008. All of these plans are of course contingent on budget approvals and successful technology development. The competitions for instruments will be held in 2001.

Many studies are already under way. In 1996, NASA funded two mission concept studies, led by TRW and Lockheed Martin. In 1997, NASA selected TRW and Ball Aerospace for further mission concept and technology development studies. It also chose the University of Arizona and Composite Optics Inc. to produce test mirrors with glass face sheets in the 1.5 m size range, capable of cryogenic operation. In 1998 NASA selected Ball Aerospace to lead a beryllium mirror demonstration at the 0.5 m scale, and a winner has been chosen to polish and test a 0.5 m C-SiC mirror. The mirror is a carbon fiber-silicon carbide material made by carbonizing a felt material and then reacting it with liquid silicon.

NASA has also selected six instrument teams to study concepts and develop the needed technologies. ESA has selected one team to study instrument concepts for the near IR, one team to consider mission concepts and the instrument module, and is soliciting an additional visible instrument study. The CSA has selected 3 instrument studies and 4 industry studies, and plans to support a professional science historian for the NGST as well. NASA has also selected 5 contracts for actuator studies.

In 1998, NASA designated the Space Telescope Science Institute to carry out the flight operations. ESA would contribute to the operations, as they do for the HST. Early designation of the Institute makes a large scientific resource available to the Project, avoiding many of the management challenges of the HST project.

Mission Concepts

The mission concepts under serious consideration have most major characteristics in common. All have deployable 3 mirror anastigmat telescopes with an effective diameter of about 8 m, with three or more segments, and all would adjust the alignment of the segments with actuators after deployment. Each concept has a deployable sunshield with multiple layers, capable of lowering the telescope temperature to 20-50 K. The back surface of the sunshield radiates toward the telescope and provides the dominant source of stray light. The final architecture for NGST will be determined by competition among the prime contractors in 2001. TRW, Ball Aerospace, and Lockheed Martin have all expressed their intention to bid.

The sunshield itself must be carefully designed to avoid an excessive imbalanced solar radiation pressure torque. The TRW concept would use electrochromic panels to adjust the reflectivity of the surface, while other concepts depend on adjustable trim tabs. The torque is dependent on orientation relative to the Sun. Some kind of thruster is also required to maintain the spacecraft in its unstable equilibrium orbit around the L2 point. The main concern about thrusters is the potential for contamination.

Because the spacecraft and telescope are large and somewhat flexible, a fine guidance sensor in the main telescope focal plane will be used to provide feedback to a fast steering mirror located near the image of the primary mirror formed by the tertiary mirror. With a bandwidth of a few Hz, it can correct for slow changes of the pointing and allow a simplification of the coarse guidance system. A small deformable mirror at the pupil can be used to correct residual small errors of the primary mirror, and might enable diffraction limited imaging at $0.6\,\mu$m.

The data rate would be comparable to that of the HST. The NGST instrument concepts provide very large focal planes, but onboard processing can reduce the data rate by lossless compression, and by removal of cosmic ray glitches prior to telemetry. It is hoped that the spacecraft will be capable of autonomous operation and simple commanding, because the observations are not interrupted by the Earth as they are in low orbit, and the system will have routine daily contact with a ground station.

Instrument Concepts

The yardstick mission includes four instruments. Two cover the core wavelength range from 1 - 5 μm with a camera and multiobject spectrometer, and two extend the range from 5 to 28 μm with a camera and ordinary slit spectrometer. The ASWG has found that there is significant value to extending the wavelength range into the visible band, at least to $0.6\,\mu$m, the limit set by gold coatings on the mirrors and sensitivity of the detectors. No additional detectors would be required for this, since the InSb needed for the 1-5 μm range has good sensitivity down to $0.6\,\mu$m. These detectors function well at 30 K, which can be provided by radiative cooling. Extension past 5 μm requires a different type of detector, either a HgCdTe detector with a different band gap, or Si:As BIB detectors. The HgCdTe detectors can operate at a higher temperature than the Si:As, which requires an operating temperature of about 7 K meaning an active cooler or stored cryogen. Several types of active coolers are of interest, including a reverse turbo Brayton cycle device similar to the one planned for NICMOS on the next Hubble servicing mission, or a

410

sorption pumped cooler using hydrogen. Stirling cycle coolers would be of interest if their vibration levels could be reduced.

The near IR multiobject spectrometer could use a micromirror array or a microshutter array as a programmable entrance aperture for a dispersive spectrometer. It would be capable of taking up to 1000 simultaneous spectra, at programmable locations within a field of view of several arcmin. This would be comparable to the multifiber spectrometers that have been so successful in mapping the redshift structure of the universe from the ground. Neither microdevice has yet been used for astronomical spectroscopy, but laboratory demonstrations have been very promising. One micromirror array is sold by Texas Instruments for use in projection televisions, but is not the right design for cryogenic operation and does not have the ideal pixel size. The ESA studies will consider integral field spectroscopy, using image slicers, microlens arrays, and other approaches. One of the selected US study teams is also investigating an imaging Fourier spectrometer, which would provide simultaneous images and spectra over a large detector area. It has greatest advantage for low resolution spectroscopy, and has the virtue that there are no slit losses or order sorting filters required. One of the CSA funded studies is supporting technology development for this concept.

The sensitivity of the NGST is shown in Figure 2 along with all the missions and observatories discussed above. With a collecting area about ten times that of HST, it will be about 100 times faster for deep imaging, so the Hubble Deep Field could be observed in about an hour. With an aperture ten times that of SIRTF, it will be 10^4 times faster for deep imaging, and because its angular resolution is far better it is not expected to be limited by confusion noise. The Hubble Deep Field images are still approximately 99 per cent dark sky. In the 1-2 μm band, the zodiacal dust is about 20 times darker than the sky at Mauna Kea, even between the OH lines, so there will be a comparable speed advantage for deep imaging and spectroscopy. However, high resolution spectroscopy in this band is still competitive from the ground, taking advantage of the large number of telescopes becoming available, the dark regions between the OH lines, and the benefits of adaptive optics. At wavelengths greater than 2 μm, thermal emission from the telescope and atmosphere make ground based equipment up to 10^6 times slower than space observations with a cold telescope, even at wavelengths where the atmosphere transmits. The limiting sensitivity of the NGST for a very deep field, with weeks of exposure time, would be under 0.11 nJy, or 33.9 magnitudes on the AB scale.

OTHER FUTURE SPACE TELESCOPES

Far Infrared and Submillimeter Telescope, FIRST. The FIRST mission, now planned for launch with the Planck surveyor on an Ariane rocket around 2005, will carry a 3.5 m class telescope, radiatively cooled to around 70 K or lower. It will carry two instruments: 1.) Large format imaging array detectors for 85 - 900 μm, using Ge:Ga photodetectors and very cold (< 0.1 K) bolometers, with low (R=3) and

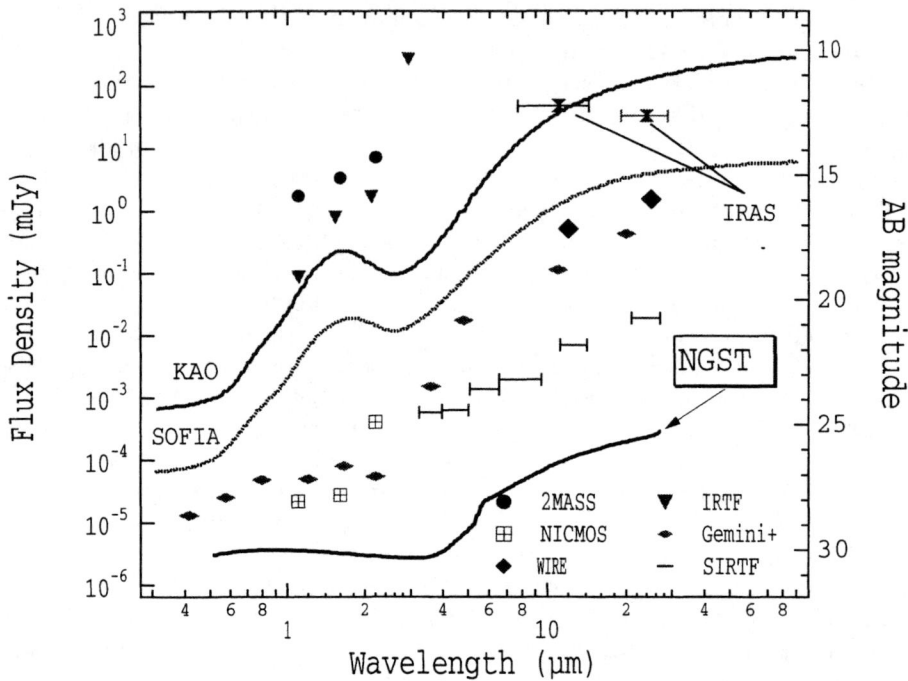

FIGURE 2. *IR sensitivity "phase space" in 2008: Sensitivities of 2MASS, IRTF, NICMOS, Gemini, WIRE, SIRTF, KAO, SOFIA, IRAS, and NGST. Ten years from now astronomers will have access to facilities that reach 4 to 6 orders of magnitude fainter in the near infrared than those available as recently as three year ago.*

medium (R=10,000) resolution spectroscopy, and 2.) Coherent receivers with cryogenic SIS mixers and acousto-optical spectrometers for high resolution spectroscopic surveys from 500-1200 GHz, and possibly up to 3000 GHz, depending on progress and demand. It will provide diffraction limited imaging, assuming the mirror contributed by the US does well. See `http://astro.estec.esa.nl/SA-general/Projects/First/first.html`.

ASTRO-F. The ASTRO-F (InfraRed Imaging Surveyor, or IRIS) mission will be in low (750 km) Earth sun-synchronous orbit by 2003 and will provide a 70 cm helium cooled telescope with a silicon carbide mirror for astronomy from 2 to 200 μm. It will have diffraction limited imaging at 10 μm, with pixels less than 1 arcmin even at 100-200 μm. Cameras will cover the near IR, 10 μm, and 20 μm bands with 10 arcmin square fields. Low resolution spectroscopy will be provided in the near and mid IR with grisms, and from 50 to 200 μm a Fourier transform spectrometer is being considered. Stirling cycle coolers will prolong the life of the liquid helium. See `http://koala.astro.isas.ac.jp/Astro-F/ index-e.html`.

SUMMARY AND CONCLUSIONS

The coming decade may well be the Decade of the Infrared, in which new capabilities come on line and spectacular surprises are found. It is an area in which new technology is opening up windows to completely unexplored research topics. The redshifts of the expanding universe bring early galaxies into the infrared, while obscuration by dust requires infrared to see into the denser regions of our own galaxy. Room temperature bodies like planets emit at infrared wavelengths where their chemical composition may be discerned, and there is a real possibility that such objects could be detected and measured. By the end of this decade we will have witnessed the universe light up with its first stars and galaxies. Furthermore, we will have deep surveys our local Galactic neighborhood at near–IR wavelengths that will point the way for future telescopes to study nearby planetary systems.

REFERENCES

1. Neugebauer, G. & Leighton, R. B. 1969, Calif. Inst. Technology, NASA, (XXFE)
2. Kirkpatrick, J. D. 1999, BAAS, 30, 1374
3. Bély, P.Y, Burrows, C.J., and Illingworth, G.D, eds. 1989, "The Next Generation Space Telescope (NGST)," Proc. Workshop, Space Telescope Science Institute. Avail. Space Telescope Science Institute, Baltimore, MD.
4. Dressler, A., et al. 1996, "HST and Beyond," Avail. Space Telescope Science Institute, Baltimore, MD. or http://www.ngst.nasa.gov/project/ bin/HST_Beyond.PDF
5. Stockman, H.S., ed., 1997, "The Next Generation Space Telescope: Visiting a Time When Galaxies were Young," Space Telescope Science Institute, Baltimore. http://oposite.stsci.edu/ngst/initial-study/
6. Smith, E.P., and Koratkar, A., eds., 1998, "Science with the Next Generation Space Telescope," ASP Conference Proceeding, Vol. 133
7. 34th Liège International Astrophysics Symposium, 1998, "The Next Generation Space Telescope,", ESA SP-429.

Far-Infrared and Submillimeter Observations of High Redshift Galaxies

David A. Neufeld

Department of Physics and Astronomy, The Johns Hopkins University, 3400 North Charles Street, Baltimore, MD 21218

Abstract.
   Observations at far-infrared and submillimeter wavelengths promise to revolution-ize the study of high redshift galaxies and AGN by providing a unique probe of the conditions within heavily extinguished regions of star formation and nuclear activity. Observational capabilities in this spectral region will expand greatly in the next decade as new observatories are developed both in space and on the ground. These facilities include the Space Infrared Telescope Facility (SIRTF), the far-infrared and submil-limeter telescope (FIRST) and the millimeter array (MMA). In the longer term, the requirements of high angular resolution (comparable to that of HST), full wavelength coverage, and high sensitivity (approaching the fundamental limit imposed by photon counting statistics) will motivate the development of far-IR and submillimeter space interferometry using cold telescopes and incoherent detector arrays.

INTRODUCTION

Key scientific questions about the Universe that have been raised at this meeting and elsewhere include

- What is the history of star formation in the Universe?

- What is the history of metallicity and dust content in the Universe?

- What is the origin of the extragalactic background observed by the DIRBE experiment on COBE?

- What are the relative contributions of stars and of active galactic nuclei to the luminosity of the Universe, and how do they vary with redshift?

In this paper, I will argue that observations in the far-infrared and submillimeter wavelength region (40 – 1000 μm, corresponding to rest wavelengths in the range 5 – 300 μm for galaxies at $z = 2-5$) offer a unique probe of the high redshift Universe

CP470, After the Dark Ages: When Galaxies were Young (the Universe at 2 < z < 5),
edited by Stephen S. Holt and Eric P. Smith

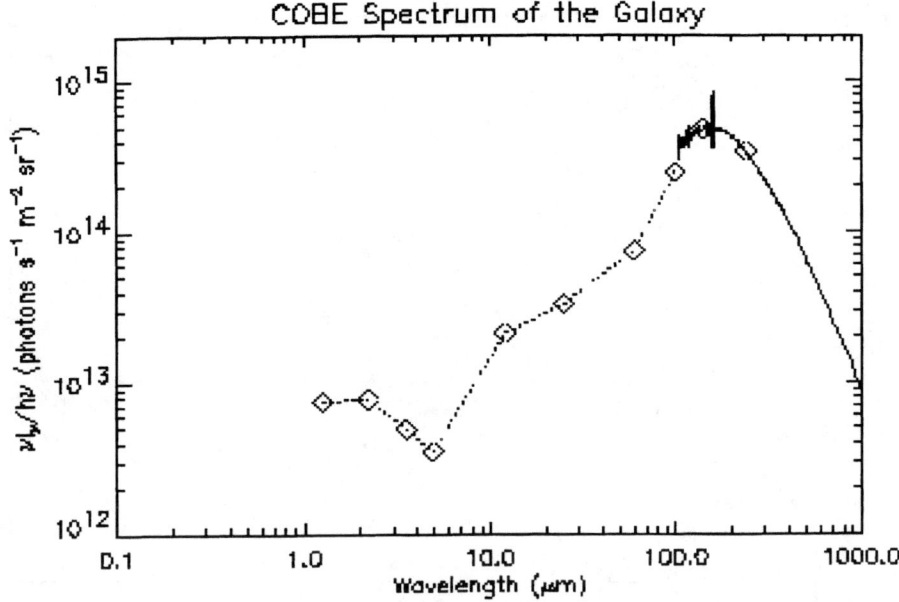

FIGURE 1. COBE spectrum of the Milky Way (from Mather et al. 1998)

that will address these questions. The happy coincidence of several astronomical facts make far-infrared and submillimeter observations particularly powerful.

First, galaxies are extremely luminous at far-infrared rest wavelengths. The spectrum of our own Milky Way galaxy, for example, shown in Figure 1, exhibits two distinct peaks, the first at around 1 μm resulting from the integrated emission from stars, and the second at around 100 μm resulting from interstellar dust emission. The representation given here (in which equal areas correspond to equal rates of photon emission) shows that most of the Galaxy's *photons* emerge in the far-infrared region. The Milky Way is entirely unremarkable in this regard: indeed, in many starburst galaxies even the *energy* output is dominated by far-infrared radiation. The strength of the far-infrared emission from galaxies simply reflects that fact that the average visual extinction is sufficient to allow a significant fraction of the starlight to be reprocessed by interstellar dust.

Second, the opacity of interstellar dust is a strongly decreasing function of wavelength, allowing embedded regions of star formation and nuclear activity that are invisible at optical and even near-infrared wavelengths to be detected in the far-infrared and submillimeter spectral regions. A second implication of the strong wavelength-dependence of the dust opacity is that the submillimeter region provides a unique cosmological window to the high redshift Universe, the background caused by dust emission from $z \sim 0$ (local galaxies) dropping rapidly with

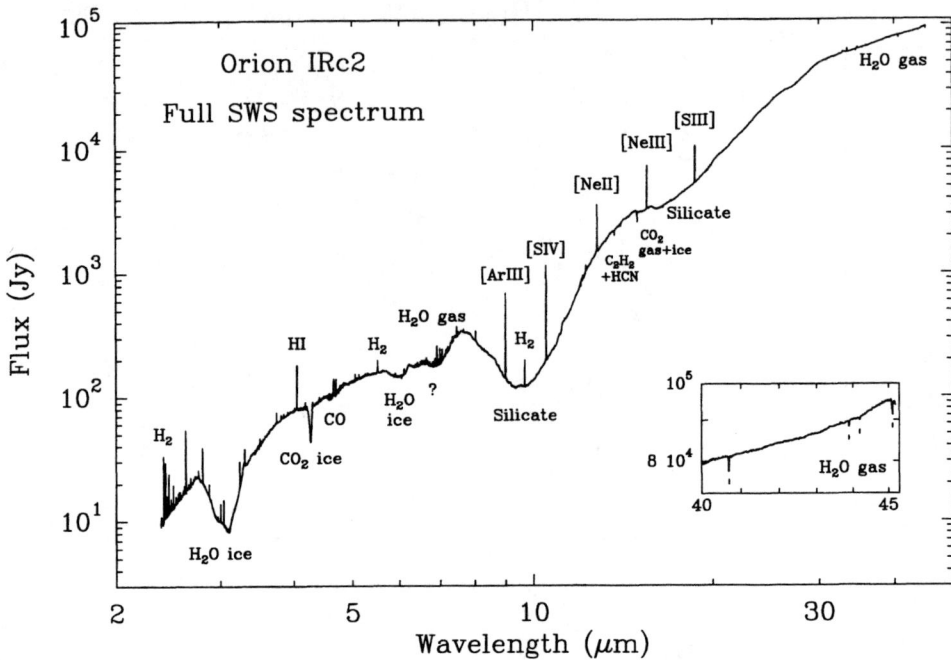

FIGURE 2. ISO Short Wavelength Spectrometer spectrum of Orion IRc2, from the paper of van Dishoeck et al. (1998).

increasing wavelength longward of $\sim 200\,\mu$m and the background from $z = 1500$ (the CMB) dropping rapidly with decreasing wavelength shortward of $\sim 800\,\mu$m.

Third, the 5 – 300 μm (rest) wavelength range is extremely rich in atomic and molecular diagnostics that can serve as powerful probes of the physics and chemistry of interstellar gas and dust. The remarkable richness of the mid- and far-infrared spectral region is demonstrated by the Infrared Space Observatory (ISO) spectrum of the Orion region (van Dishoeck et al. 1998) in Figure 2, which shows numerous rotational lines of H_2 and H_2O, fine structure emissions from a wide variety of atomic ions, as well as several broader features associated with interstellar dust. In the next section, I will discuss the potential importance of such spectral features in constraining the properties of high-redshift galaxies.

EMISSION MECHANISMS AT FAR-INFRARED AND SUBMILLIMETER WAVELENGTHS

Interstellar dust is the dominant source of far-infrared and submillimeter radiation in galaxies. Recent SCUBA observations of dust continuum radiation – reported, for example, at this meeting by Ian Smail – have already demonstrated the power of submillimeter observations to probe the Universe at high redshift. Some – although by no means all – of the identified SCUBA sources are galaxies at redshifts $z > 2$ (e.g. Ivison et al. 1998), and the $\sim 25\%$ of SCUBA sources for which no optical counterpart can be found may well be sources at very high redshift (or alternatively low-redshift galaxies or AGN that are very heavily extinguished). Although the field is currently in its infancy, observations of dust continuum radiation will ultimately allow the the effects of dust absorption to be corrected for quantitatively in models for the luminosity history of the Universe, and will elucidate the relative contribution of sources at different redshifts to the extragalactic background detected by the DIRBE experiment on COBE (Hauser et al. 1998).

Although it shows a continuum spectrum, emission from dust is not featureless. It exhibits several broad features of large equivalent width, most notably the silicate feature at 9.7 μm and several bands in the $3.3 - 11.3$ μm range that have been attributed to polycyclic aromatic hydrocarbons (Allamandola, Tielens & Barker 1985), and these features may allow redshifts to be estimated from far-infrared observations of very modest spectral resolving power.

Interstellar gas emits a rich spectrum of atomic and molecular line radiation in the $5 - 300$ μm range, which – although a negligible contribution to the overall far-infrared and submillimeter emission – dominates the cooling of the interstellar gas and provides valuable diagnostics of the physical and chemical conditions.

Fine-structure emissions from the low-ionization species C^+ and O dominate the cooling of neutral atomic gas clouds. The C^+ $^2P_{3/2} - ^2P_{1/2}$ line at 158 μm has an upper state energy (E_u/k) corresponding to only 92 K and is therefore readily excited in cold atomic clouds. In most galaxies, the C^+ 158 μm line is the strongest source of line emission and accounts for $0.2 - 1\%$ of the total far-infrared luminosity (Malhotra et al. 1997), this percentage representing the fraction of the absorbed radiant energy from stars that is deposited in the interstellar gas rather than the dust.[1] Spectroscopic observations of the C^+ 158 μm line along with the O 63μm and 145μm lines ($^3P_1 - ^3P_2$ and $^3P_0 - ^3P_1$ with $E_u/k = 227$ and 326 K respectively) from high-redshift galaxies will yield reliable redshifts and will allow the heating rate for the interstellar gas to be determined.

---

[1] Note, however, that the *relative* strength of the C^+ 158 μm line is considerably smaller in those galaxies that show the strongest far-infrared continuum emission. Malhotra et al. (1997) have argued that this effect likely arises because the larger ultraviolet fluxes incident upon cold clouds within such galaxies lead to larger positive charges on the interstellar dust grains and a resultant decrease in the efficiency of grain photoelectric emission that is the primary mechanism for heating the gas.

In dense regions of the interstellar medium that are well shielded from starlight, the gas is primarily molecular and its emission is dominated by rotational emissions from molecules. Gas temperatures in molecular regions range from $\sim 10\,\mathrm{K}$ in quiescent clouds to several hundred Kelvin in gas that has been heated by a nearby star or protostar, or even several thousand Kelvin in shocked regions. The radiative cooling of molecular clouds is an essential feature of the star formation process, because cloud collapse involves the conversion of gravitational potential energy to thermal energy and can proceed only if the latter is efficiently removed. Theoretical calculations (e.g. Neufeld, Lepp & Melnick 1995) predict that over a wide range of physical conditions the radiative cooling of molecular gas is dominated by rotational transitions of the molecules H_2, CO and H_2O in the 7 – 600 μm region. At low temperatures, submillimeter transitions of CO are the primary coolant, while at higher temperatures pure rotational lines of H_2 (e.g. the S(0), S(1), S(2), S(3), S(4) and S(5) lines at 28.3, 17.0, 12.3, 9.66, 8.03, 6.91 μm) and of H_2O (many lines in the 40 - 600 μm region) are expected to dominate the cooling. This prediction is corroborated by recent ISO observations of H_2 and H_2O emissions from nearby regions of star formation (e.g. van Dishoeck et al. 1998, see Figure 2; Harwit et al. 1998) and well as by extensive ground-based observations of CO carried out previously toward both nearby and high-redshift galaxies (e.g. Omont et al. 1996). Measurements of line ratios permit the density, temperature, and molecular abundances within the molecular gas to be constrained.

In addition to probing cold atomic and molecular gas, far-infrared and submillimeter observations of high redshift galaxies also promise to yield invaluable information about photoionized regions. Many galaxies are luminous sources of mid IR fine structure emissions from NeII (12.8 μm), OIII (52, 88 μm), NeIII (15.6, 36.0 μm), NeV (14.3, 24.2 μm) and several other ions that result from photoionization by radiation shortward of the Lyman limit. Such mid-IR lines provide unique information about the metallicity and gas density in ionized regions, as well the spectral shape of the ionizing radiation field (e.g. Voit 1992). These transitions show several important advantages over the optical wavelength lines traditionally used to study HII regions: they are not heavily extinguished by interstellar dust; their luminosities are only weakly dependent on temperature and therefore provide model-independent estimates of metallicity; and they provide line ratios that are useful diagnostics of density over a wide dynamic range (e.g. Spinoglio & Malkan 1992). The availability of rare gas elements (e.g. Ne, Ar) allows the metallicity to be determined without the complicating effects of interstellar depletion, and the availability of a wide range of ionization states (e.g. NeII and NeV) provides an excellent discriminant between regions that are ionized by hot stars and those that are ionized by a harder source of radiation such as an AGN. The power of that discriminant has been demonstrated by the recent ISO observations shown in Figure 3 (Moorwood et al. 1996). Here the otherwise similar spectra of the starburst galaxy M82 and the Circinus galaxy (which contains an active nucleus) are distinguished by the presence in Circinus of a variety of highly ionized species such as NeV and Ne VI that can only result from a very hard source of ionizing radiation.

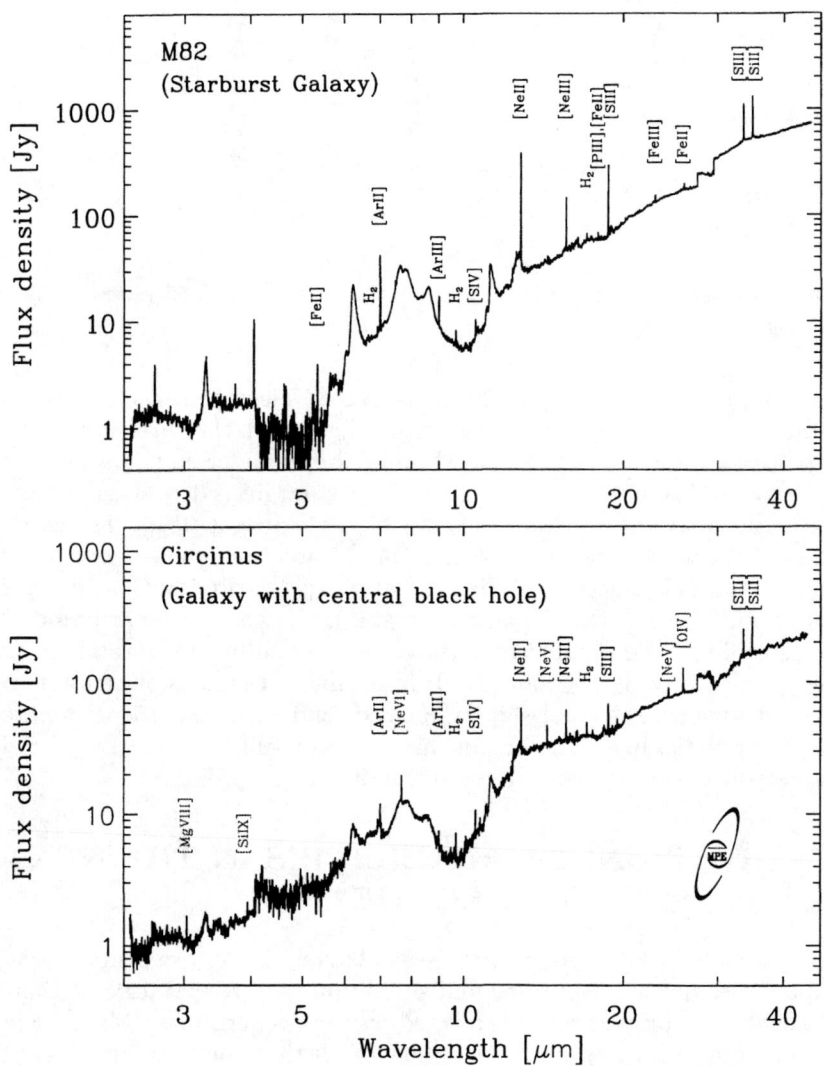

FIGURE 3. Comparison of ISO Short Wavelength Spectrometer spectra of M82 and Circinus, demonstrating the power of mid-IR fine structure lines as discriminants of the ionizing spectrum (from the ISO science gallery, credit: ESA/ISO, SWS, Moorwood; see also Moorwood et al. 1996)

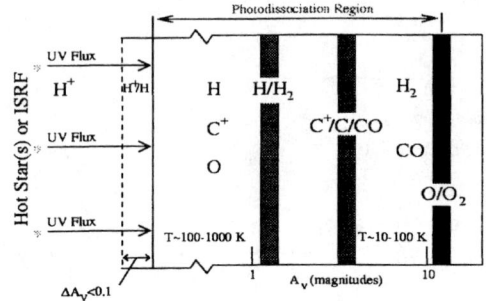

FIGURE 4. Interaction of starlight with the interstellar medium, from the review of Hollenbach & Tielens (1997).

Figure 4, from the review paper of Hollenbach & Tielens (1997), is a schematic representation of the interaction between starlight and the interstellar medium that summarizes the various far-infrared and submillimeter emission mechanisms described above. Most of the radiant energy from starlight is deposited at moderate visual extinctions, $A_V \sim 1$. Roughly 99% of the starlight heats the interstellar dust and is reprocessed as far-infrared continuum radiation, while very roughly 1% heats the gas and is reprocessed as line radiation (primarily the C^+ 158 μm line). At visual extinctions $A_V > 3$ (rightmost region), the gas is primarily molecular, and the gas cooling is dominated by infrared and submillimeter rotational lines of molecules, particularly H_2, CO, and H_2O. In unshielded regions of high ultraviolet flux (leftmost region), the gas is highly ionized, and the cooling is dominated by optical and ultraviolet line emission. In this zone, mid-infrared fine structure lines, although not the major coolant, are powerful diagnostic probes.

OBSERVATIONAL CAPABILITIES IN THE NEXT DECADE

The next decade (2001 – 2010) promises substantial improvements in observational capabilities in the far-infrared and submillimeter spectral regions, thanks to several new observatories that are expected to begin operations. My goal in this article is not to give a comprehensive review of all these new facilities but rather to discuss briefly selected observatories that will offer capabilities most directly relevant to the study of galaxies at high redshift.

The **Space Infrared Telescope Facility** (SIRTF)[2], scheduled for launch at the end of 2001, will deploy a liquid helium cooled 85 cm telescope capable of carrying out observations of extremely high sensitivity. The Multiband Imaging Photometer (MIPS) instrument on board SIRTF will offer diffraction-limited imaging using

---

[2] The SIRTF home page is at http://sirtf.jpl.nasa.gov

sensitive detector arrays at wavelengths of 24, 70 and 160 μm, as well as very low resolution ($\lambda/\Delta\lambda \sim 10$) spectroscopy in the 50 – 100 μm region. The principal limitation of SIRTF for the detection of galaxies at far-infrared wavelengths is the large size of the diffraction-limited beam: particularly at 160 μm, MIPS will be source confusion limited for observations of relatively short duration. The Infrared Spectrograph (IRS) instrument will be capable of moderate resolution spectroscopy ($\lambda/\Delta\lambda \sim 600$) over the 10 – 37 μm range, with a large spectral multiplex advantage that will allow full, high-quality spectra to be obtained very much more quickly than was possible with ISO. While the wavelength coverage of IRS does not quite reach the 40 – 1000 μm range that is the subject of this article, IRS deserves mention here because of its capability for detecting mid-infrared line emission from ions in HII regions.

The **Far Infrared and Submillimeter Telescope** (FIRST)[3] will be a space observatory with a much larger ($\sim$ 350 cm) primary mirror, but one that is not actively cooled. Current plans call for the launch of FIRST in 2007, with instrumentation capable of carrying out broad band photometry, imaging spectroscopy, and high-resolution heterodyne spectroscopy. The wavelength coverage will extend to much longer wavelengths than SIRTF, allowing a far wider range of atomic and molecular line emissions to be studied spectroscopically. Again, the relatively large diffraction limit at these wavelengths for any single dish instrument of reasonable size means that source confusion will be significant except for observations of short duration (e.g. Blain, Ivison & Smail 1998). Thus interferometers will be critical for the study of all but the most luminous galaxies at high redshift, and the most important impact of SIRTF and FIRST on studies of high redshift galaxies is likely to be in measuring spectra of low redshift galaxies that can be used as templates for understanding future interferometric observations.

The **Millimeter Array** (MMA)[4] will have an extremely powerful interferometric capability, providing spatial resolution as good as $\sim 0.01''$, wavelength coverage down to 350 μm, and high spectral resolution. MMA promises to allow large numbers of high redshift sources to be detected routinely and associated unambiguously with optical counterparts. It will make use of $\sim$ 36 antennae of diameter $\sim$ 10 m that can be deployed over baselines of several kilometers on a high plateau site in Chile. An observatory of more modest collecting area – the Smithsonian Astrophysical Observatory's Submillimeter Array (SMA) – will operate in a Northern Hemisphere site (Mauna Kea). The primary limitations of these ground-based facilities are those imposed by Earth's atmosphere, which only permits observations in a series of submillimeter windows all longward of 300 μm, and by the fundamental sensitivity limits set by heterodyne receivers and warm telescopes.

---

[3] The FIRST home page is at http://astro.estec.esa.nl/SA-general/Projects/First/first.html
[4] The MMA home page is at http://www.mma.nrao.edu

THE LONGER TERM: FAR-INFRARED AND SUBMILLIMETER INTERFEROMETRY FROM SPACE

The ideal instrument for the study of far-infrared and submillimeter emissions from high redshift galaxies would combine (1) full wavelength coverage; (2) HST-like spatial resolution; (3) sensitivity approaching the fundamental limit imposed by photon-counting statistics; (4) high spectral resolution ($\lambda/\Delta\lambda$ of at least 10^4). The first and third of these capabilities require a space observatory; the second requires interferometry; and the third requires a cooled telescope (barely warmer than the CMB) equipped with a new (not presently existing) generation of incoherent detectors rather than heterodyne receivers; and the fourth can be accomplished by means of a Fabry-Perot or Michelson interferometer.

In a recent white paper (Mather et al. 1998), we have presented a preliminary study of such an instrument – dubbed the Submillimeter Probe of the Evolution of Cosmic Structure, SPECS[5] – in which we envisaged a Michelson interferometer providing spatial and spectral interferometry with three, cold, free-flying elements of diameter ~ 3 m deployable over baselines ~ 1 km. Although such a facility may lie significantly beyond what could be built today, Mather et al. 1998 have emphasized the importance of developing key technologies over the next decade to make such an instrument feasible in the decade 2011 – 2020; those technologies include formation flying, active cooling of large mirrors, and the development of sensitive incoherent detector arrays. In particular, photon-counting incoherent detectors – which do not yet exist at these wavelengths but would likely be some type of superconductive device – would offer enormous sensitivity advantages for faint sources both over current bolometers and relative to the fundamental limit of a heterodyne receiver.[6]

I gratefully acknowledge the support of a grant from NASA's Long Term Space Astrophysics Research Program. I thank Ewine van Dishoeck and David Hollenbach for making available Figures 2 and 4. It is a pleasure also to acknowledge helpful discussions with Mark Voit, Harvey Moseley and John Mather.

REFERENCES

1. Allamandola. L., Tielens, A.G.G.M., & Barker, J.R. 1985, ApJ, 290, L25
2. Blain, A.W., Ivison, R.J., & Smail, I. 1998, MNRAS, 296, L29
3. Harwit, M., Neufeld, D.A., Melnick, G.J., & Kaufman, M. 1998, ApJ, 497, L105
4. Hauser, M.G., et al. 1998, ApJ, 508, 25
5. Hollenbach, D.J., & Tielens, A.G.G.M. 1997, ARA&A, 35, 179

---

[5] The SPECS home page is at http://www.gsfc.nasa.gov/astro/specs

[6] A perfect photon-counting detector is more sensitive than a perfect heterodyne receiver by a factor $\sim (\Delta\nu/R)^{1/2}$, where $\Delta\nu$ is the bandwidth and R is the photon arrival rate, a factor much larger than unity for faint extragalactic sources.

6. Ivison, R.J., Smail, I., Le Borgne, J.-F., Blain, A.W., Kneib, J.-P., Bezecourt, J., Kerr, T.H., & Davies, J.K. 1998, MNRAS, 298, 583

7. Malhotra, S., et al. 1997, ApJ, 491, L27

8. Mather, J.C., et al. 1998, astro-ph/9812454

9. Moorwood, A.F.M., Lutz, D., Oliva, E., Marconi, A., Netzer, H., Genzel, R., Sturm, E. & de Graauw, T. 1996, A&A, 315, L109

10. Neufeld, D.A., Lepp, S., & Melnick, G.J. 1995, ApJS, 100, 132

11. Omont, A., Petitjean, P., Guilloteau, S., McMahon, R.G., Solomon, P.M., & Pecontal, E. 1996, Nature, 382, 428

12. Spinoglio, L., & Malkan, M.A. 1992, ApJ, 399, 504

13. van Dishoeck, E.F., Wright, C., Cernicharo, J., Gonzalez-Alfonso, E., de Graauw, T., Helmich, F.P., Vandenbussche, 1998, ApJ, 502, L173

14. Voit, M. 1992, ApJ, 339, 495

Imaging Distant Dust and Gas: The Millimeter Array

Alwyn Wootten*, Min S. Yun[†]

*National Radio Astronomy Observatory (NRAO)[1], 520 Edgemont Road, Charlottesville, VA
22903
[†]NRAO P. O. Box 0, Socorro, NM 87801

Abstract. Although the richness of the millimeter/submillimeter sky has been known
for the last two decades, precision imaging of the sky has been impossible at all but
the longest wavelengths. The Millimeter Array (MMA) will consist of 36 precision 10m
antennas located above 5000m altitude in the Atacama Desert of Chile. It will apply
aperture synthesis techniques for precision imaging of the cool thermal sky, particularly
those objects which can now only be studied in the far infrared with coarse angular
resolution. The MMA will offer unprecedented sensitivity–it will image a galaxy with
the dust mass of the Milky Way at z=1 in one minute. The images the MMA will
produce of this and high z systems will be high resolution images: the MMA will
offer unprecedented angular resolution for imaging cool matter, providing images at a
few hundredths arcsecond resolution. The MMA provides extinction-free spectroscopy
of dusty high z systems. The MMA is currently under development; construction is
scheduled to begin in 2001 with operations scheduled to begin in 2008.

INTRODUCTION

The Millimeter Array (MMA) is a millimeter wavelength telescope funded for
design and development by the National Science Foundation during 1998-2001.
The MMA design specifications include:
- Thirty-six 10-meter antennas on the 16400 foot elevation Llano Chajnantor,
Chile
- Imaging instrument in all atmospheric windows between 10 mm and 350 μm
- Spatial resolution of 10 milliarcseconds, 10 times better than the VLA or the
Hubble Space Telescope
- Velocity resolution under 1 km/sec
- Array configurations from approximately 80 meters to 10 km

---

[1] The National Radio Astronomy Observatory is a facility of the National Science Foundation
operated under cooperative agreement by Associated Universities, Inc.

CP470, After the Dark Ages: When Galaxies were Young (the Universe at 2 < z < 5),
edited by Stephen S. Holt and Eric P. Smith

- Able to image sources arcminutes to degrees across at one arcsecond resolution
- By more than an order of magnitude the largest and most sensitive instrument in the world at millimeter and submillimeter wavelengths

During the 1980s, astronomers identified millimeter arrays as a goal for instrumental development. Caltech and Berkeley operated and upgraded their facilities, the Smithsonian Institution entered a design and development phase for their Submillimeter Array on Mauna Kea, while NRAO planned for a national facility, the Millimeter Array (MMA), submitting a proposal to NSF in 1990. In 1991, the Astronomy and Astrophysics Survey Committee of the National Research Council identified the Millimeter Array as one of the highest priority telescope programs to be undertaken in the 1990s. Funding for the Design and Development Phase of the MMA began 1 June 1998. This phase is expected to continue at NRAO, BIMA and OVRO through 1 June 2001. Construction of the MMA is expected to commence in FY2000 at Llano Chajnantor, Chile. This phase will culminate with operation at the end of 2007 or early 2008.

Other countries have also identified a large millimeter array as national priorities for astronomical instrumentation. In Europe ESO, IRAM, NfRA, MPIfR and OSO along with the UK have studied a Large Southern Array, or LSA, with much larger collecting area than the MMA proposed. An agreement between ESO, France, the UK and others is being pursued with a goal of synchronizing a development phase of the LSA with that of the MMA. In Japan, astronomers have planned a Large Millimeter and Submillimeter Array (LMSA) of intermediate collecting area. As a result, the possibility of merging the instruments into a Large Millimeter Array, or LMA, is being pursued and will be the focus of an URSI workshop in Toronto in August 1999.

THE SITE

The location of the 16400 foot Cerro Chajnantor site is just south of the road from the mining town of Calama in northern Chile as it extends toward the Paso de Jama into Chile. With only 10cm of rain a year, this is one of Earth's driest regions. Extensive site testing has shown it to be drier, with a more stable atmosphere, than Mauna Kea. Less than one millimeter of precipitable water lies above the site for half each year (day plus night, winter plus summer), resulting in an optical depth in the critical submillimeter windows of about one or less for much of the time. Nearby at 8000 foot elevation lies the charming ancient town of San Pedro de Atacama, whose water supply has drawn visitors since Inca times. A tourist town, it is a starting point for expeditions to nearby geysers, salt lakes and volcanos as well as the Inca ruins. It will become a center of MMA operations.

Operation at such a superb site as Llano Chajnantor, the large collecting area of the array (up to a factor of ten for the LMA), its planned large bandwidth (16 GHz as compared to 1 GHz on existing arrays), antennas and receivers optimized for operation on the excellent site, and the long baselines available on the site will provide sensitivity improvement measured in orders of magnitude over presently operating or planned millimeter or submillimeter instrumentation. Currently, the weakest sources in these bands which have ever been detected measure about one millijansky in flux. With the MMA, this sensitivity will be reached in less than one second. Since thermal science is the target of the array, good brightness temperature sensitivity is critical to the array. Even on the longest baselines, the array will offer sufficient sensitivity to, for example, image CO emission from the Milky Way Galaxy to beyond z=1. High sensitivity without sufficient spatial resolution can result in confusion by weak unresolved sources. The MMA brings to the submillimeter regime the precision imaging capability of the VLA, HST, Keck or Gemini, providing resolution of 0.05″or better. While confusion will limit the usable sensitivity of other submillimeter instruments, such as JCMT/SCUBA, CSO/SHARC, FIRST, SOFIA, Planck Surveyor or BOLOCAM, **the MMA will execute extremely deep galaxy surveys without reaching the confusion limit** (Blain, Ivison and Smail 1998) [1].

CONTINUUM OBSERVATIONS

The distinct nature of the submillimeter sky is nicely demonstrated by comparison of the Hubble Deep Field (HDF) in optical (and radio by Richards et al. 1998 [2]) and in submillimeter (SCUBA 850μm by Hughes et al. 1998 [3]). Rapidly increasing volume with redshift combined with strong K-correction ($S_\nu \propto \nu^{3-4}$) means that the brightest sources at submillimeter wavelengths are the distant (z>1) dusty galaxies. Higher star formation rates in the earlier epochs [4] adds to the dominance of high redshift objects in the submillimeter.

Comparison of the expected blackbody spectrum of the ultraluminous starburst galaxy Arp 220 as seen at high redshifts with the frequency coverage and the sensitivity of the MMA suggests that The Millimeter Array should be able to detect Arp 220-like dusty starburst galaxies out to a redshift of 10 or more. Galaxies like the present day Milky Way can also be detected out to z beyond 1.

Predicted detection rates for mm/submillimeter sources by the existing and future instruments at 5 sigma significance have been calculated by Blain 1998 [5]. The expected detection speed of the MMA is about 100 times faster than the current SCUBA and comparable to that of the BOLOCAM when installed in the future LMT (Olmi and Mauskopf 1998) [6], without suffering from confusion.

SPECTROSCOPIC OBSERVATIONS

Along with the dust continuum emission, various redshifted spectral lines such as all rotational transitions of CO and infrared fine structure lines such as [C II] and [O I] can be detected and studied to derive the molecular gas content and star forming activity. Secure detections of redshifted CO lines from dusty galaxies at cosmological distances already exist (e.g. z=2.3 FSC 10214+4724 [Brown & Van-den Bout 1992] [7], z=4.7 BR 1202-07 [Ohta et al. 1996 [8], Omont et al. 1996] [9]) and z=2.6 H1413+117(e.g. Barvainis et al 1997) [10], among others. The search for CO emission from luminous submillimeter galaxies identified by deep SCUBA surveys are also underway, and the detection of CO (3-2) emission in the z=2.8 sub-millimeter galaxy SMM02399-0136 was reported recently. **The high frequency resolution attained with the MMA, enabling dynamical studies of distant objects, is a capability unmatched by other observatories planned for the next decade, such as NGST, SIRTF.**

The Spectral Energy Distribution (SED) of the first luminous submillimeter galaxy detected by SCUBA, SMM02399-0136 at z=2.8 (Ivison et al. 1998) [11] is essentially identical to the z=2.3 hyperluminous infrared galaxy FSC 10214+4724 and has the same shape, dominated by the FIR/submillimeter peak, as in the prototype ultraluminous galaxy Arp 220. The apparent luminosity is about 100 times larger however, partly due to gravitational magnification by a foreground cluster.

Redshifted CO (3-2) emission from the z=2.8 submillimeter galaxy SMM02399-0136 is detected and mapped using the Caltech millimeter array by Frayer et al. (1998) [12], confirming the optical identification of the distant faint optical galaxy as the submillimeter continuum source. The CO spectrum obtained measures the redshift of the host galaxy to be z=2.808, coincident with the optical galaxy within the uncertainty of registering the optical HST image.

Although detection of the CO emission from SMM02399-0136 required about 40 hours of observation with OVRO, such a 6 mJy line should be detected to about 3σ in one minute with the MMA with a velocity resolution of 25 km s^{-1}; owing to the larger MMA bandwidth, offsets between molecular and optical redshifts will not compromise the ability to detect a galaxy in molecular line emission.

REFERENCES

1. Blain, A.W., Ivison, R.J., and Smail, I. 1998 MNRAS 296, L29
2. Richards et al. 1998, AJ, 116, 1039
3. Hughes, D. et al. 1998, Nature, 394, 241
4. Blain, A.W., Smail, I., Ivison, R.J., Kneib, J.-P. 1998, MNRAS, in press
5. Blain, A.W. 1998, to appear in Wide-field Surveys in Cosmology, (astro-ph/9806369)
6. Olmi, L. and Mauskopf, P. 1998 in *The Young Universe* ASP Conference Series, / Vol 146, D'Odorico, Fontana and Giallongo, eds San Francisco: ASP), p. 371.
7. Brown, R. and vanden Bout, P. 1992 ApJ 397, L19.
8. Ohta et al. 1996 Nature 382, 426.
9. Omont et al. 1996 Nature 382 428.
10. Barvainis et al. 1997 ApJ 484, 695.
11. Ivison et al. (1998) MNRAS 298 583.
12. Frayer et al 1998 Apj 506, L7.

Calibration of the ASTRO-E XRS Detector

M. D. Audley*,†, K. C. Gendreau*,†, R. L. Kelley†, K. A. Arnaud*,†, K. R. Boyce†, R. Fujimoto‡, F. S. Porter†, C. K. Stahle†, A. E. Szymkowiak†

*University of Maryland, College Park, MD 20742
†NASA/GSFC, Greenbelt, MD 20771
‡ISAS, 3-1-1 Yoshinodai, Sagamihara, Kanagawa 229-8510, Japan

Abstract. The US-Japanese ASTRO-E observatory, is scheduled to be launched in early 2000. ASTRO-E carries four X-ray CCD detectors and a hard X-ray detector. The CCDs are located at the focus of grazing incidence X-ray mirrors and will primarily provide imaging over 0.4–12 keV bandpass. ASTRO-E also carries the XRS microcalorimeter X-ray detector. A platinum X-ray mirror will focus X-rays onto a 32-element array of microcalorimeter pixels for high-throughput, high-resolution spectroscopy with limited spatial resolution. The mean measured energy resolution of the XRS flight model detector is about 12 eV at 6 keV for the nominal operating temperature of 65 mK. We present results from our calibration of the XRS flight model detector. We describe the methods used to determine the spectral redistribution of the detector and the overall detection efficiency.

THE X-RAY SPECTROMETER (XRS)

XRS is a new generation of X-Ray Spectrometer. It will measure the spectrum of celestial objects in the soft X-ray range (200 eV to 10 keV). XRS will, for the first time, provide both high resolution and high throughput in one instrument.

XRS is a 32-element array of X-ray quantum calorimeters [1] fabricated from anisotropically etched silicon. A platinum conical foil X-ray mirror will focus X-rays onto the microcalorimeter array for high resolution spectroscopy with limited spatial resolution. The mirror has an effective area of 300 cm^2 at 6 keV and 400 cm^2 at 1 keV.

The average measured energy resolution of the XRS instrument is about 12 eV at 6 keV for the nominal operating temperature of 65 mK or about 10 times better than typical solid-state detectors.

CP470, After the Dark Ages: When Galaxies were Young (the Universe at 2 < z < 5),
edited by Stephen S. Holt and Eric P. Smith

In contrast to a bolometer which measures an integrated radiant flux, a calorimeter measures the energy of individual X-ray photons by the temperature rise they cause when absorbed in a small sensing element. The concept of a calorimeter is similar to that of a bolometer. However, it is subject to more stringent design constraints due to the necessity of thermalizing individual photons rapidly.

When an X-ray photon hits one of XRS's HgTe absorbers its energy is converted into heat. In its simplest form ($\Delta T << T$) the resulting temperature rise is $\Delta T = E/C$ where E is the energy of the photon and C is the heat capacity of the calorimeter. This is sensed by an ion-implanted thermistor. Each calorimeter is supported by micromachined silicon legs that provide a thermal link to a heat bath. These legs conduct heat away from the calorimeter so that the temperature decays exponentially to the baseline value with a thermal time constant $\tau = C/G$ where G is the thermal conductance of the legs.

XRS is the prime instrument aboard the Japanese Astro-E spacecraft, to be launched early in the year 2000. It is a joint project of NASA and the Japanese space science agency ISAS. In addition to XRS, ASTRO-E carries four X-ray CCD detectors and a hard X-ray detector.

CALIBRATION HARDWARE

The XRS detector and cryostat are in a horizontal test dewar. There is a ^{55}Fe radioactive source in the test dewar that can be moved in and out of the detector's field of view.

There are two X-ray monochromators that can be connected directly to the test dewar. A double crystal monochromator (DCM) covers the energy range 2–10 keV with Ge and Si crystal pairs. The X-ray source has ten anodes of different materials on a rotatable stage. For low energy calibration (0.2–2 keV) we use a Surface Normal Reflection Monochromator (SNR [2]).

An electron impact source with a Mo target is also available. This can be operated either as a continuum source, or in a mode where an external motorized wheel brings different targets into the beam to produce fluorescent lines. X-rays from this source enter the test dewar through a Be window.

RESULTS

Figure 1 shows the Mn Kα spectrum from a ^{55}Fe radioactive source detected by a single pixel of the flight model detector. Fitting a Gaussian-smeared double Lorentzian model yields an energy resolution of 9.7 ± 0.2 rm eV FWHM.

Figure 2 contains spectra from monochromatic and continuum sources. The first illustrates spectral redistribution of incident photons into escape lines and an electron loss continuum. The second (continuum) spectrum shows the absorpti on edges in the HgTe absorbers.

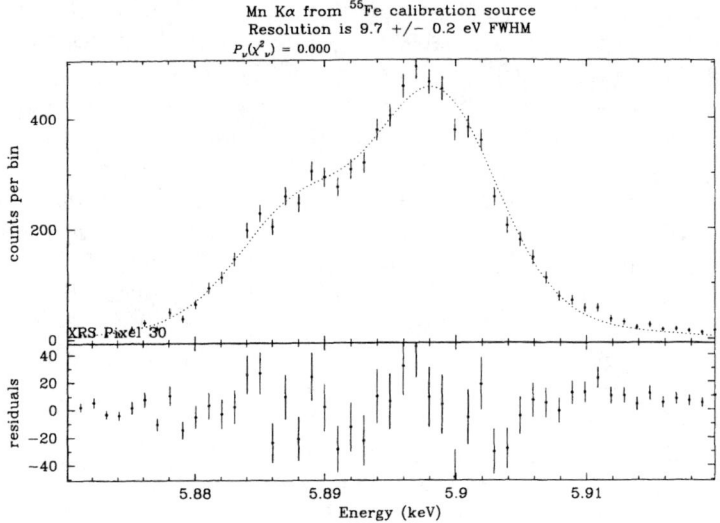

FIGURE 1. Mn Kα spectrum from flight model detector. The spectrum is fitted to two Lorentzians smeared by a Gaussian. The Kα_1 and Kα_2 lines, separated by 11 eV, are partially resolved. The fitted resolution is 9.7 ± 0.2 eV FWHM.

The low energy detection efficiency is determined by thermal/optical blocking filters located along the various thermal stages of the instrument. There are five aluminized polyimide filters. XRS's high spectral resolution makes us sensitive to fine structure in the X-ray absorption edges as shown in Figure 3 for the oxygen and aluminum edges around 0.5 and 1.5 keV.

PLANS

We intend to measure the X-ray transmission of the flight model filters at room temperature and to investigate the temperature dependence of EXAFS in the engineering model filters. We also plan to make a response model whose physical parameters are c onstrained by our calibration data.

ACKNOWLEDGMENTS

We would like to thank Susan Breon and the members of GSFC's Code 713, Dave Bloom and the GSFC I&T staff, Mark Bautz, Mike Hettrick, Greg Madejski, and Fred Finkbeiner.

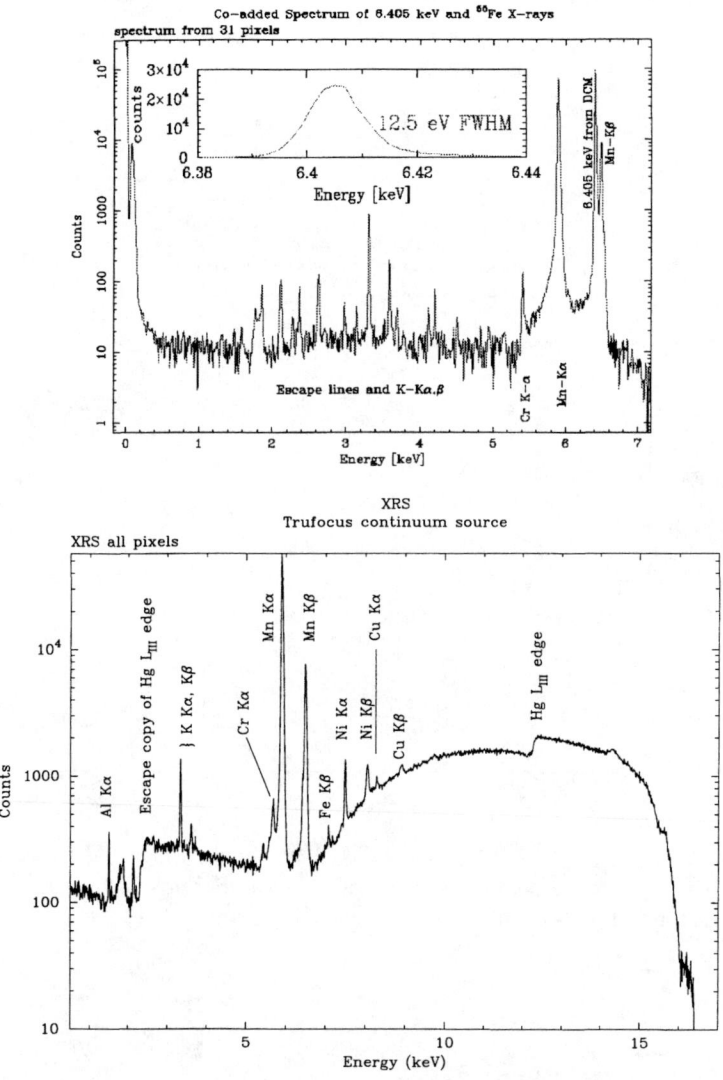

FIGURE 2. Left: Spectral redistribution of monochromatic 6.405 keV X-rays from DCM. Right: Continuum spectrum from electron impact source showing Hg L-edge and escape copy.

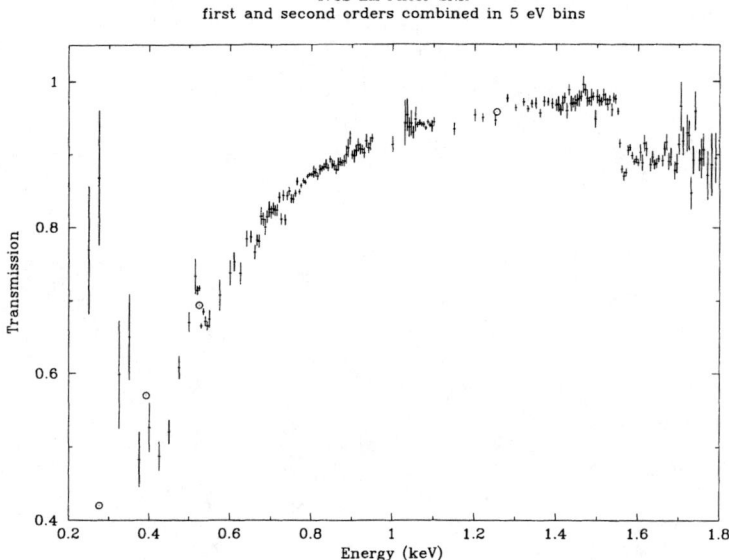

FIGURE 3. Engineering Model IVCS filter (1000 Å polyimide, 1000 Å Al) transmission measured with SNR. The circled points are independent measurements with a flow proportional counter. This is the transmission of only one of five of the filters in XRS.

REFERENCES

1. Moseley, S.H., Mather, J.C., and McCammon, D. 1984, J. Appl. Phys., 56, 1257
2. Hettrick, M. C. 1992, Appl. Opt., 31, 7174

Science with the Constellation-X Observatory

Azita Valinia[1,2], Nicholas White[1], Harvey Tananbaum[3], and the Constellation-X Team[4]

[1] *NASA's Goddard Space Flight Center, Code 662, Greenbelt, MD 20771*
[2] *Department of Astronomy, University of Maryland, College Park, MD 20742*
[3] *Harvard-Smithsonian Center for Astrophysics, 60 Garden St, Cambridge, MA 02138*
[4] *see http://constellation.gsfc.nasa.gov*

Abstract. The Constellation X-ray Mission is a high throughput X-ray facility emphasizing observations at high spectral resolution (E/ΔE $\sim$ 300–3000), and broad energy bandpass (0.25–40 keV). Constellation-X will provide a factor of nearly 100 increase in sensitivity over current high resolution X-ray spectroscopy missions. It is the X-ray astronomy equivalent of large ground-based optical telescopes such as the Keck Observatory and the ESO Very Large Telescope. When observations commence toward the end of next decade, Constellation-X will address many fundamental astrophysics questions such as: the formation and evolution of clusters of galaxies; constraining the baryon content of the Universe; determining the spin and mass of supermassive black holes in AGN; and probing strong gravity in the vicinity of black holes.

CONSTELLATION-X

The prime objective of Constellation-X mission is high resolution X-ray spectroscopy. It will cover the $0.25 - 40$ keV X-ray bandpass by utilizing two types of high throughput telescope systems to simultaneously cover the low (0.25 to 10 keV) and high energy (6 to 40 keV) bands. The low-energy Spectroscopy X-ray Telescope (SXT) is optimized to maintain a spectral resolving power of at least 300 across the 0.25 to 10 keV band pass (E/ΔE $\sim$ 3000 at 6 keV) and has a minimum telescope angular resolution of 15″ HPD. The diameter of the field of view is 2.5′ below 10 keV. The high energy system (HXT) with lower spectral resolving power (ΔE $\sim$ 1 keV) overlaps the SXT and primarily is used to measure the relatively line-less continuum from 10 to 40 keV. The diameter of the field of view is 8′ for the HXT. The large collecting area is achieved with a design utilizing several mirror modules, each with its own spectrometer/detector system. The spectral resolving power of the SXT and the effective area of SXT and HXT are shown in Figure 1.

CP470, After the Dark Ages: When Galaxies were Young (the Universe at 2 < z < 5),
edited by Stephen S. Holt and Eric P. Smith

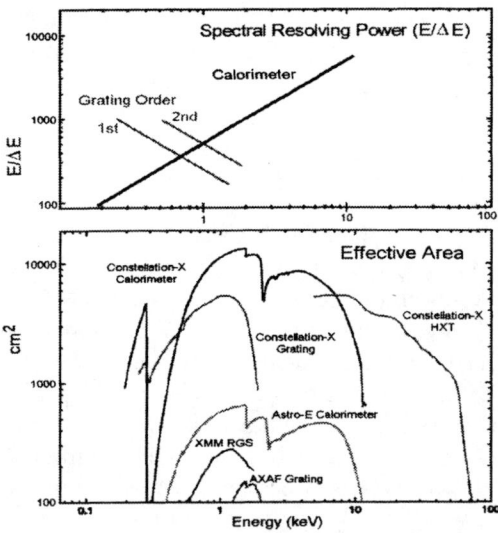

FIGURE 1. Spectral resolving power and the effective area of the instruments onboard Constellation-X.

The SXT uses two spectrometer systems that operate simultaneously to achieve the desired energy resolution: 1) a 2 eV resolution quantum microcalorimeter array, and 2) a set of reflection gratings for energies < 2 keV. The gratings deflect part of the telescope beam away from the calorimeter array in a design similar to XMM except that the direct beam falls on a quantum calorimeter instead of on a CCD. The two spectrometers are complementary, with the gratings optimal for high resolution spectroscopy at low energies and the calorimeter at high energies. The gratings also provide coverage in the 0.3-0.5 keV band where the calorimeter thermal and light-blocking filters cause a loss of response. This low-energy capability is particularly important for high-redshift objects, for which line-rich regions will be moved into this low energy band.

The HXT uses a multilayer coatings on individual mirror shells to provide the first focusing optics system to operate in the 6-40 keV band. Compared to other non-focusing methods such as those used for RXTE, Constellation-X has twice the area, 640 times the energy resolution, 240 times the spatial resolution, and above 10 keV, 100 times the sensitivity. AXAF and XMM, designated as the workhorses of X-ray astronomy in the next decade, will detect photons with energies up to 10 keV.

The technology development program is now underway and is targeting a first launch in 2007-2008, around the time that AXAF will be reaching the end of its projected lifetime. An essential feature of the Constellation-X concept involves minimizing cost and risk by building several identical, modest satellites to achieve a large area. The current baseline is 6 satellites, although other multiple satellite configurations are also under consideration, with the final choice to be made based on a balance of overall cost and risk. The mission will be placed into a high earth or L2 orbit to facilitate high observing efficiency, provide an environment optimal for cryogenic cooling, and simplify the spacecraft design.

SCIENCE GOALS

Constellation-X is a key element in NASA's Structure and Evolution of the Universe (SEU) theme aimed at understanding the extremes of gravity and the evolution of the Universe. We highlight here a few key science areas.

How can we use observations of black holes to test General Relativity? X-ray observations directly probe physical conditions close to the central engine of blackholes where the distortions of time and space predicted by general relativity are most pronounced. Constellation-X will use the spectral features of these objects (e.g. the broad iron $K\alpha$ line discovered by ASCA [1]) to map out the geometry of the inner emission regions and determine the extent to which we can test general relativity.

What is the total energy output of the Universe? Models of cosmic X-ray background predict that the emission at hard X-rays is due to many absorbed AGN [2], with their central engines primarily visible via hard X-rays (and perhaps infrared). If most of the accretion in the Universe is highly obscured, then the emitted power per galaxy based on currently available optical, UV, or soft X-ray quasar luminosity functions may be substantially underestimated. By using hard X-ray spectra to advance our knowledge of the total luminosity of AGN, Constellation-X will bring us closer to knowing the total energy output of the Universe.

What roles do supermassive black holes play in galaxy evolution? Constellation-X measurements of black hole mass and spin for the high z quasar sample will allow understanding of the relative evolution rates of black holes and their host galaxies, and will shed light on when massive black holes formed compared to the galaxy formation epoch.

How does gas flow in accretion disks and how do cosmic jets form? Accretion disks play a fundamental role in many astrophysical settings, ranging from the formation of planetary systems to accretion onto supermassive black holes in AGN. There are, however, many controversies about the nature of viscosity which drives the accretion process, about the stability of the disk at various accretion rates, and about the relevance of advection and mass outflows, and the mechanisms by which jets are formed. Constellation-X will probe the physics of accretion disks to a level not currently possible, by resolving line features from the accretion disk photosphere and by measuring the continuum shape over a broad energy band.

When were clusters of galaxies formed and how do they evolve? To date, cluster abundances have been measured in the X-ray band out to a redshift of about 0.4 but no discernible evolution with z has been seen. Constellation-X spectra of clusters over a range of redshifts will provide crucial information about the presence of primordial gas, including any input from possible pre-galactic generations of stars as well as the contribution from stellar nucleosynthesis as a function of time. The high sensitivity of Constellation-X is essential for extending such studies to the "poorer cousins" of clusters, groups of galaxies. Moreover, by mapping the velocity distribution of hot cluster gas via Doppler shifts in the emission lines, Constellation-X will allow us to examine the effects of collisions and mergers between member galaxies and between separate subclusters and clusters.

Where are the "missing baryons" in the local Universe? Recent observations of the Lyman-α forest show that at large redshifts most of the predicted baryon content of the Universe is in the IGM, while at low redshifts, the baryon content of stars, neutral hydrogen, and X-ray emitting cluster gas is roughly one order of magnitude smaller than that expected from nucleosynthesis arguments. Therefore, a large fraction of baryonic content of the local Universe is considered "missing". Numerical simulations [3] predict that the missing matter may reside in the IGM with a temperature range of $10^5 - 10^7$ K. Such gas in the IGM can be detected with the high sensitivity, high resolution instruments aboard Constellation-X through the absorption lines of metals against the X-ray spectra of background quasars (e.g. OVII and OVIII).

How are matter and energy exchanged between stars and the Interstellar Medium and how is the Intergalactic Medium enriched? The chemical enrichment of the Universe is dominated by star formation and the release of the processed material into the ISM via stellar winds and supernova explosions. Moreover, supernova explosions and enhanced star forming activities can drive hot gas out of the galaxy and enrich the ICM/IGM on megaparsec scales. Detailed, spatially-resolved X-ray spectra reveal the stellar/supernova abundances, the composition of the surrounding ISM, and the interaction of the expanding blast wave with the surrounding material. High throughput instruments such as those aboard Constellation-X are needed to measure the K-lines of less abundant elements such as F, Na, Al, P, Cl, K, Sc, Ti, V, Cr, Mn, Co, Ni, Cu, and Zn. The increased sensitivity of Constellation-X will allow us to extend these studies to external galaxies, beyond the Magellanic Clouds to M1 and M33, for example. This will allow us to further our understanding of the history of star formation and exchange of matter between the ISM and stars.

REFERENCES

1. Tanaka, Y. et al., 1995, Nature, 375, 659.
2. Madau, P., Ghisellini, G., and Fabian, A. C., 1994, MNRAS, 270, L17.
3. Cen, R., and Ostriker, J. P., 1998, astro-ph/9806281.

An Archival Survey of the HDF-South

Kirk D. Borne and E. J. Shaya*
R. A. White and C. Y. Cheung†

*Raytheon ITSS, GSFC Code 631, Greenbelt, MD 20771
†NASA, GSFC Code 631, Greenbelt, MD 20771

Abstract. We present the results of a survey of archival data and catalogued objects in the region around the Southern Hubble Deep Field (HDF-South). The survey encompasses NASA mission logs, astronomical catalogs, and journal tables. The HDF-South (HDF-S) has been the focus of a dedicated HST observing campaign during October 1998. Many astrophysically interesting objects in the vicinity of the HDF-S, including quasars and clusters of galaxies, have been catalogued and observed at a wide range of wavelengths. The byproducts of this and similar user-selected surveys of archival data can be used to study classes of objects that may potentially be represented among the faint objects discovered within the HDF-S. This survey was conducted using a suite of new data search, browse, and visualization tools available at the NASA ADC (Astronomical Data Center: `http://adc.gsfc.nasa.gov/`).

THE ASTRONOMICAL DATA CENTER (ADC)

The Astronomical Data Center (ADC) at the NASA Goddard Space Flight Center (GSFC) is a major archive and distribution center for almost 3000 computer-readable versions of astronomical catalogs, published journal data tables, and observing logs ([1]). The scientific content of these data holdings is very heterogeneous, ranging from data on planets and stars to a significant and comprehensive variety of data on galaxies, quasars, and clusters of galaxies ([2]). The ADC provides astrometric, photometric, morphological, spectroscopic, polarization, kinematic, and multi-wavelength data for stellar and non-stellar objects of all types, plus a variety of "theoretical" data derived from atomic physics, stellar models, and numerical simulations. An example of the wealth of available data on galaxies can be seen by checking out the ADC Quick Reference Page for Galaxies at:

`http://adc.gsfc.nasa.gov/adc/quick_ref/ref_generalgalaxies.htm` 1

Although for brevity and generality we will refer to the ADC's information resources as 'data' in the remainder of this paper, we consider the contents of catalogs and

CP470, After the Dark Ages: When Galaxies were Young (the Universe at 2 < z < 5),
edited by Stephen S. Holt and Eric P. Smith

journal tables to be metadata about astronomical objects, as opposed to the observational data sets from particular telescopes and satellites served by the various astrophysics discipline archival data centers (*i.e.*, HEASARC, IPAC, STScI).

Data Discovery and Visualization Tools at the ADC

Users can access three independent but linked tools for browsing the data at the ADC: the "Viewer", "Catseye", and "IMPReSS" ([3]). The Viewer accesses catalogs and journal tables and allows selection and display of table subsets. CatsEye plots 2-D scatterplots of fields from Viewer tables. IMPReSS plots observation "footprints" of space-based missions on a region of the sky and links directly to the archived data at the discipline-specific astrophysics data centers. Taken together, these tools offer a continuous sequence by which users can browse and visualize a catalog and acquire specific data products of interest from the distributed data archives. Such tools can be tremendously useful both in choosing multi-wavelength data for further analysis and in judging what astronomical objects warrant further observation at particular wavelengths (*i.e.*, for future mission planning).

The Viewer is a Perl-based web interface that allows users to select, subset, sort, save, and download any of the nearly 3000 data sets at the ADC ([4]). After selecting a table to browse, the user selects the table's fields to view and can constrain those fields by numerical range (or regular expression for string fields). The requested table subset is then displayed, and it is automatically saved in a personal work area ("workspace"), for access up to 5 days after last use.

CatsEye is a Perl- and IDL-based web tool that plots selected fields from Viewer-produced tables in the user's workspace, allowing the user to visualize select catalog data. Users select tables and fields from their Viewer-produced tables to plot as X vs. Y scatterplots ([5]). Overplotting of multiple tables is allowed. Users can rescale plots as desired and can click on points within the plot to access the corresponding full ADC catalog record from the chosen table. These records link to both IMPReSS (discussed below) and NED, provided that the record contains celestial coordinates.

IMPReSS is a graphical interface to astronomical observing logs that presents the user with plane-of-the-sky outlines of pointed observations obtained by space-based telescopes ([6]). It searches for observations within a user-selected region of the sky and time period. Once the user selects the missions they are interested in, IMPReSS displays the color-coded aperture footprints (the fields-of-view, or FOV) on the sky. IMPReSS also displays an "observing log" list of all observations which have met the user's search criteria and will link the user directly to the location of observational data offered by the relevant archival data centers. Observation logs for HST-PC, -WFC, & -WFPC-2, EUVE, IUE, and all available missions from HEASARC are available for user searches. The software design allows a search through potentially millions of observations in just seconds. Links to the data are provided through the standard archive access methods offered by each data center.

An Archival Data Search Around the HDF-S Region

The HDF-S region is centered at: RA=22h 32m 56.22s, Dec=-60° 33' 02.7″ (J2000). Dedicated, continuous HST observations of this region of the sky were carried out over a several-day period in October 1998. The HDF-S observing campaign obtained very deep images using STIS, NICMOS, and WFPC2, and it also obtained high-resolution ultraviolet spectra of a QSO at redshift $z = 2.25$. The location of the HDF-S was particularly chosen so as to include this QSO in the STIS field-of-view. Refer to the STScI Web site for further information ([7]).

The results of an ADC IMPReSS search within a wide region of sky around the HDF-S position is displayed in Figure 1. FOV perimeters are plotted for all observations for which archival data are available. The plot shown here is in greyscale, while the actual Web-based search will be in full color, with a color-coded legend to identify each observatory whose data are represented in this plot ([6]).

The observation FOV plot in Figure 1 presents the full collection of archival data that are available for targeted observations in a region of sky around the position of the HDF-S. A perusal of the corresponding data sets indicates that these represent observations of quasars, galaxies, and clusters of galaxies (along with a variety of stellar and other galactic sources). Using the ADC Quick Reference pages ([2]) and Viewer tool ([4]), we can quickly identify various specific objects in this region of sky and obtain numerical data (*e.g.*, positions, fluxes, colors, redshifts) for each of them, including (and certainly not limited to) the primary STIS target: QSO J2233-606. For example, we find quite a few ACO clusters of galaxies around the position of the HDF-S ([8]): ACO 3884, 3890, 3891, 3898, 3906, 3907, 3914, 3919, 3923, and 3927, to name but a few. These clusters may offer candidate absorption-line systems to be searched for within the deep STIS spectrum of the QSO. Such absorption-line systems can in turn serve as probes of large-scale structure, superclusters, and cosmological filaments. Followup redshift measures on the faint galaxies actually detected within the HDF-S deep images may ultimately indicate that some of these objects are in fact (and therefore serve as probes of) large-scale extensions and low-density outer envelopes of the ADC-identified clusters of galaxies. Our IMPReSS archival data-mining search demonstrates that data already exist for some of these structures (*e.g.*, from ASCA, ROSAT, and EINSTEIN). Thus, correlating detailed analyses of these archival data in conjunction with the HST HDF-S data and the ADC catalog data may ultimately yield additional scientific results and insights related to objects within the HDF-S region.

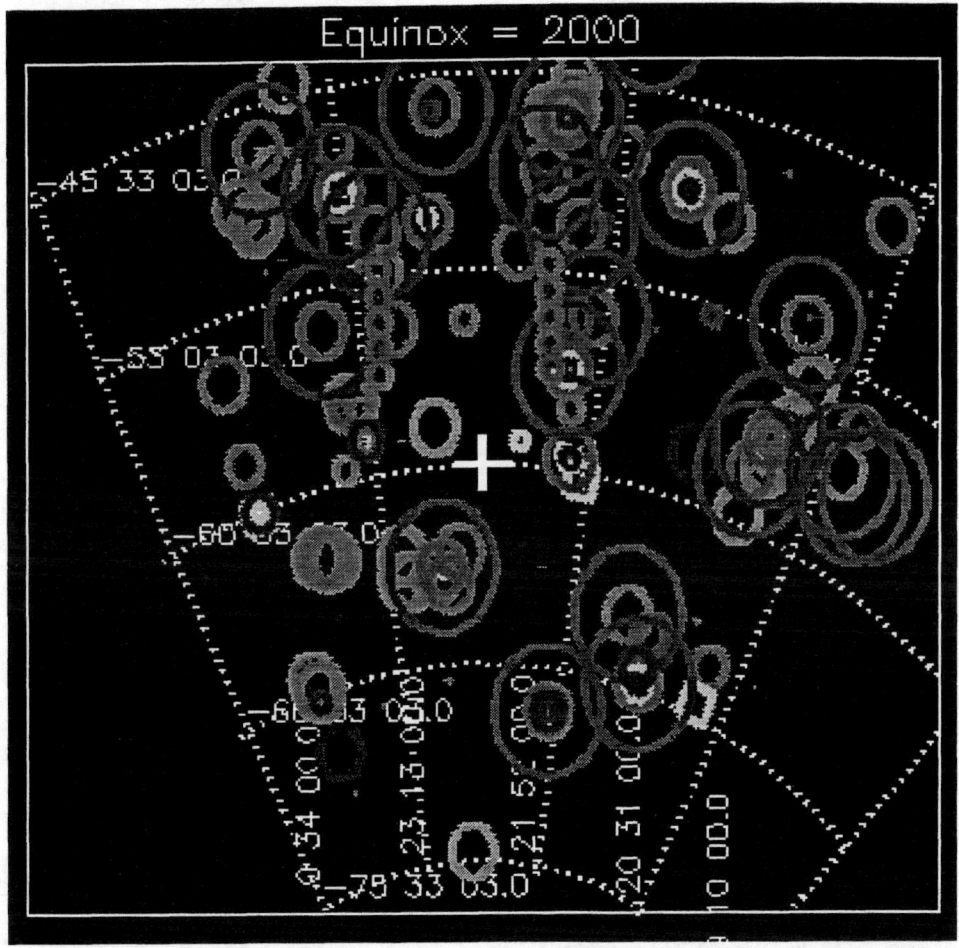

FIGURE 1. Results from an "IMPReSS observation" search within a 15° radius around the position of the HDF-South (identified by the + symbol at the center of the plot). Though shown in greyscale here, the actual results plot that one obtains by using the ADC IMPReSS search tool from a Web browser will be in full color and will be accompanied by a color-coded legend that graphically identifies the full set of 19 space science instruments for which archival data are available (as indicated by the FOV perimeters plotted here). The results screen on the Web also includes a full list of the individual observations shown in the plot, with direct-access links to the archive data centers that hold those particular observational data sets. Missions plotted here include: ASCA, EINSTEIN, EUVE, EXOSAT, GINGA, HST, IUE, ROSAT, RXTE, and SAX. The astronomical targets of these various telescope observations include clusters of galaxies, quasars, individual galaxies, and a variety of stellar (galactic) sources. Interested scientists are requested and encouraged to try IMPReSS for themselves on the WWW to see the power and visualization capabilities of this archival data search tool (http://tarantella.gsfc.nasa.gov/impress/).

REFERENCES

1. http://adc.gsfc.nasa.gov/
2. http://adc.gsfc.nasa.gov/adc/top_refpage.html
3. http://adf.gsfc.nasa.gov/adf/visualization/
4. http://tarantella.gsfc.nasa.gov/viewer/
5. http://tarantella.gsfc.nasa.gov/cgi-bin/catseye/cat_frames.pl
6. http://tarantella.gsfc.nasa.gov/impress/
7. http://www.stsci.edu/ftp/science/hdfsouth/hdfs.html
8. Abell, G. O., Corwin, Jr., H. G., & Olowin, R. P. 1989, ApJS, 70, 1.

11. Conference Summary

Rapporteur's Conference Summary

Daniel W. Weedman

Department of Astronomy and Astrophysics, The Pennsylvania State University, University Park, PA 16802

Abstract. This Conference was highlighted by exceptional observational results for redshifts above two. These results are summarized in context of three major issues that arose for these high redshifts: 1. Why are there no faint quasars?, 2. How does star birth compare to that in the local universe?, 3. How much of what is there is obscured by dust?

INTRODUCTION

Having been pushed to $z = 5$ by the observers, Martin Rees began this Conference by tweaking us about all the things we don't understand yet for epochs beyond z of 5. Although we haven't quite gotten there yet, this was a marvelous meeting to showcase how far the observers have already come. By proving that stars and quasars existed for $2 < z < 5$, they have presented the theorists with the challenge of explaining how these things developed so quickly in the early universe.

Mysterious things are happening during this "end of the dark ages", and we now have the tools to investigate. Observational mysteries ranging from gamma ray bursts to the far-infrared background probably have their explanation from events within $2 < z < 5$. Some answers certainly won't be known until we can push to even higher redshifts. Nevertheless, developing the capability to understand physical processes in the universe for $2 < z < 5$ is an enormous technical achievement, worth celebrating. Using these tools to describe the behavior of our universe during its childhood is an enormous intellectual achievement, worth celebrating even more. Many of the astronomers who deserve credit for these achievements were in attendance.

From the discussions presented, it struck me that there are three particularly intriguing questions which arise, deriving from the discoveries already accomplished. The consequences of each question are several and will be described in more detail. The questions, all applying to $2 < z < 5$, are: 1. Why are there no faint quasars?, 2. How does star birth compare to that in the local universe?, 3. How much of what is there is obscured by dust?

CP470, After the Dark Ages: When Galaxies were Young (the Universe at 2 < z < 5),
edited by Stephen S. Holt and Eric P. Smith

WHY ARE THERE NO FAINT QUASARS?

Kim Weaver presented thorough observational details of AGN, from X-ray to radio, illustrating how quantitative data on the nature of the central engine is being assembled. Such observations continue to build support for the black hole accretion model for AGN and quasars. Roger Blandford presented a clear exposition of why we must care about the number and luminosities of quasars at high redshifts because of the implications regarding black holes. The fundamental questions regard how and when the massive black holes got there, and what were their masses at these early epochs. He summarized the question as, "Which came first - quasar or galaxy?" Did all early galaxies contain black holes at some level, or did only a few contain very massive black holes (explaining the most luminous quasars)? Did galaxies form around black holes? When did the first black holes develop? Attempts to describe quasar "evolution" are the observers' way to get some answers.

Don Schneider was here to remind us that for decades the highest redshift objects known were quasars. He also reminded us of the mystery of quasar evolution, in that the number of bright quasars in the universe seems to peak at z about 2. Charles Steidel was off hunting more high redshift galaxies (with the Lyman break galaxy survey), but Mark Dickinson of that team told us the score for $z > 2$ is now galaxies 597, quasars 18. For quasar hunters, this score seems very strange. Why is it so much easier to find galaxies? At the time of the Conference, quasar fans were also quite chagrined that no quasar yet exceeded the $z = 5$ barrier, while the galaxy hunters gleefully reported reaching 5.74 (found in a Lyman α emission line survey by Esther Hu, reported here by Len Cowie). (Within a couple months after the conference, the quasar hunters began to catch up and found a quasar exceeding $z = 5$ in the early results from the Sloan Digital Sky Survey.) Abraham Loeb reminded us that no confirmed quasar or AGN has been found in the Hubble Deep Field despite the very large number of faint galaxies with $z > 2$.

Do these observational results on quasar deficiencies at high z tell us anything about quasar evolution? For two decades, it has been obvious that there are many more very luminous quasars per unit volume at high redshift compared to low redshift. This can arise from either luminosity evolution or density evolution. The former would mean that all quasars were systematically more luminous in the early universe. The latter would mean that there were many more quasars of all luminosities. The absence of quasars in the HDF demonstrates that the luminosity function for high redshift quasars stops at some luminosity brighter than M = -20. This favors luminosity evolution, because many of today's AGN are much fainter than M = -20. The simplest conclusion which follows is that quasars have undergone primarily luminosity evolution. The most straightforward interpretation would then be that all massive black holes already existed at $z > 2$ and that the systematic luminosity change between then and now is in the efficiency of accretion. Watching quasars "turn-on" is the most intriguing objective of tracing the decline in the numbers of luminous quasars at $z > 2$. Are we observing the era when the black holes themselves were assembling?

Guenter Hasinger presented the most thorough and eloquent discussion of the quasar evolution problem. His data are based primarily on quasars discovered by ROSAT, and he convincingly demonstrated that X-ray quasars keep increasing in number right up to the optical magnitude limit at which redshifts can be obtained. Because 80% of the faint sources are quasars/AGN, he has accumulated excellent statistics on quasar evolution which indicate a combination of luminosity and density evolution. Unfortunately, the ROSAT samples barely reach past $z = 2$ in meaningful numbers so they do not probe the crucial redshift regime containing the drop off for optically discovered quasars. Nevertheless, he has a lot of AGN, and the large fraction which show significant dust reddening lead to his most dramatic extrapolation: that most of the far infrared background arises from shrouded AGN (a topic to which I return subsequently).

HOW DOES STARBIRTH AT HIGH REDSHIFT COMPARE TO LOCAL?

When the Conference was planned, this was the hottest topic because of the exceptional progress of the Lyman dropout group in finding star formation at $z >$ 2. Mark Dickinson, Piero Madau, and James Lowenthal discussed various results from the surveys they have done with Charles Steidel. By comparing to previous measures of star formation in the local universe, there seems consensus that star formation per unit volume at z about one was ten times more intense than it is today. The new result of the Conference - from more extensive Keck surveys - is that it now appears the rate may remain close to the same high rate, even when observed in the rest frame ultraviolet, to redshifts approaching five. Previously, the "Madau diagram" indicated a dropoff in the rate down to local levels by $z = 4$. Though this is still a work in progress, it is important to emphasize that these measurements of star formation to redshifts exceeding four have been an extraordinary observational accomplishment. Refinements will come as the data increase, but the builders and users of the Keck telescope have already achieved a great thing.

The question of how star formation has changed with epoch involves two issues. One is simply the integrated star formation per unit volume of the universe, as measured by the total luminosity from young stars. The second is the question of what is actually changing: is it the number of starburst galaxies, or is it the star formation intensity per galaxy? I like to demonstrate the contrast between local and early epochs by using my favorite prototype luminous local starburst, Markarian 171, and comparing to starbursts seen in the HDF. Reducing an ultraviolet image of Markarian 171 to the size and surface brightness it would have at z about 3 and placing it alongside starbursts at that redshift found in the HDF demonstrates the result. The largest starburst galaxies at high redshift are not significantly larger in extent, but they have higher surface brightness extending over much larger regions. That is, the localized intensity of star formation within a galaxy is greater at high redshift, by a factor of up to ten in surface brightness on kpc scales. Further quantifying such results will lead to progress in understanding how the localized star formation process used to be different.

And for something completely different - what about the gamma ray bursts? Niel Gehrels and Ralph Wijers gave the observational and theoretical summaries. The key proposal is that "starburst" galaxies and "gammaburst" galaxies are the same thing, because collapsing, massive stars trigger the gamma bursts. If proven, by the slow observational accumulation of enough gammaburst galaxies, we may eventually find the observationally hilarious result that the highest redshift objects are found with the highest energy photons.

It is important to note that the discussions of star formation luminosity summarized so far are based only on rest-frame ultraviolet and do not accommodate that fraction of young-star luminosity which is obscured by dust. As reviewed more below, there are now many reasons to suspect that star formation is generally very obscured, perhaps by a typical factor of 10. That means that a correct census of star formation as a function of redshift must also accommodate corrections for dust obscuration.

HOW MUCH IS OBSCURED?

Infrared and submillimeter astronomers have also contributed to dramatic revelations about the nature of the universe at $z > 2$. David Neufeld reviewed some of the fundamental concepts for us. In summary, the universe seems to have been a much dirtier place back then. Far-infrared and submillimeter radiation, from the diffuse background or from individual sources, is only explainable as reradiation from dust particles. The primary radiation source is assumed to be ultraviolet luminosity, either from young stars or from accretion disk activity in AGN. How many such sources are there, where are they, and how are they distributed between starbursts and AGN?

Recent progress has been driven by two independent accomplishments. First was the now-unquestioned measurement by COBE, with DIRBE and FIRAS instruments, of the extragalactic far-infrared diffuse background. This exceeds any reasonable extrapolation of the local infrared galaxy luminosity function, as measured by IRAS. There is an excess source of infrared luminosity at redshifts somewhere beyond unity. COBE could only measure the integrated contribution, but other instruments have found examples of individual sources. The most important have been those detected by SCUBA, " Submillimeter Common-User Bolometer Array". (If this won first prize in the acronym contest, one would hate to see the other entries.) Detections of infrared sources at high redshift have also been made by mid-infrared and far-infrared ISO observations. Ian Smail described the new SCUBA observations, and Ismael Perez-Fournon illustrated some of the latest ISO results. Probably, there are enough discreet sources detectable by SCUBA at $z > 2$ to explain the COBE backgrounds. That leaves us only the puzzle of explaining what these sources are.

We know that starbursts and AGN are dusty, even the samples which are visible in rest-frame ultraviolet. As mentioned previously, Guenter Hasinger argued that most AGN are highly obscured. Tim Heckman and Gerhardt Meurer reviewed the indications that starbursts are obscured in the ultraviolet, by factors of 3 to 10. This means that even those galaxies found with the Lyman break search techniques will also be infrared sources. But the most luminous infrared galaxies will be so obscured as to be invisible in the ultraviolet; Heckman argued that these extremely obscured cases are also the galaxies of highest infrared luminosity and therefore account for most of the far-infrared sources which are observed.

It is reasonable that luminous sources representing the extreme of the infrared galaxy luminosity function are those being found with the far-infrared observations. I also suspect that the bulk of their luminosity arises from AGN components rather than starburst. The template spectrum used to represent a superluminous infrared galaxy has most commonly been that of Arp 220. From all available data, this seems to be purely a starburst. My preferred template is Markarian 231, which shares luminosity about equally between its starburst and its AGN. While the far-infrared fluxes of Mkn 231 and Arp 220 would be similar, there are important differences at other wavelengths. If located at $z = 3$, Markarian 231 would have X-ray flux (0.5 - 2.5 keV) of 1.2×10^{-16} erg cm^{-2} s^{-1}, optical (red) magnitude of 27, mid-infrared (30 μ) flux of 0.21 mJy, and radio (20 cm) flux of 0.046 mJy. (Remember that these are in the observer's frame; rest frame wavelengths are a factor of four shorter.) The observed mid-infrared and X-ray arise primarily from effects of the AGN. Arp 220, though of similar bolometric flux, would be about 10 times fainter in these wavelengths while comparable in optical and radio.

Many of the far-infrared sources are not detectable optically, even to magnitude 26. This means that ground based spectroscopy will not get redshifts. Fortunately, the sources are detectable in the radio with the VLA (at detection limits fainter than 0.1 mJy) so their positions can be accurately determined, making the optical i.d., or lack thereof, unambiguous. If the far-infrared sources have $z < 3$, SIRTF should be able to get redshifts because dusty galaxies contain strong spectral features, between 7 μ and 12 μ, caused by silicate absorption and PAH emission. The SIRTF spectrometer should be able to follow these features to z about 3 for sources with 20 μ flux somewhat fainter than 0.5 mJy (http://ssc.ipac.caltech.edu/sirtf/).

SIRTF can be expected, therefore, to establish a census of dusty galaxies for all $z < 3$. If the number found is adequate to explain the entire DIRBE background, that will mean we have found the sources of luminous but dust-obscured activity in the universe. Comparing with optically-derived surveys in the rest frame ultraviolet, we will be able to determine quite well the fraction of total luminosity arising from obscured galaxies as a function of redshift to $z = 3$. If we are fortunate, we will see a turn down with redshift in counts of luminous infrared sources that will define the epoch at which the peak occurred. Unfortunately, there are not easy spectroscopic discriminants for faint infrared galaxies to choose between AGN and starbursts as the dominate power source. (For high spectral resolution, emission lines of various excitations can do this, but sources must be brighter than several mJy for SIRTF high resolution observations.) SIRTF alone will not be able to chart AGN and starbursts as separate constituents of the dusty early universe, though this sorting will be helped by comparing SIRTF detections with AXAF detections or limits.

CONCLUSION

What can we look for in a realistic future? The answer indicates why this Conference was so cleverly titled: our chances of getting meaningful data on sources with $z > 5$ are very small for many years to come. This means that the epoch you got to know here will be our frontier for a long time. To go further, we need to fulfill more dream concepts, like NGST (Next Generation Space Telescope), MMA (Millimeter Array), and SPECS (Submillimeter Probe of the Evolution of Cosmic Structure). Personally, I am happy to leave something for the future, with confidence that an exciting future will come. When I started doing astronomy, the highest z was 0.5. A factor of 10 is enough for a while!

Appendix A: Conference Programme

After the dark Ages: When Galaxies Were Young
(the Universe at $2 < z < 5$)
College Park, Maryland
12-14 October, 1998

Monday, October 12, 1998

8:15   Welcome – Dan Mote, Jr. – President, University of Maryland

8:30   Session #1: Introduction   Chair: **S. Holt**

       V. Trimble   Beyond the Bright Searchlight of Science:
                           The Quest for the Edge of the World

       M. Rees   Emerging from the Dark Ages

10:00   Coffee & Pastries (provided outside lecture hall in poster area)

10:30   Session #2: The Earliest Structure   Chair: **J. Silk**

       D. Spergel   The CMB and the Origin of Structure
       A. Loeb   The First Stars and Quasars
       N. Gnedin   Reionization of the Universe (15 min.)
       M. Norman   From Cosmological Initial Conditions to Pop III
                           Protostellar Cores: Results of a Direct,
                           Multiscale, 3D Numerical Simulation (15 min.)
       One-Minute Poster Reviews and Discussion

12:30   Lunch (provided at the Conference Center)

2:00   Session #3: First Discrete Structures   Chair: **V. Trimble**

       C. Frenk   Forming Galaxies: Theoretical Expectations
       L. Storrie-Lombardi Quasar Absorption Line Studies of Galaxies
                           and the Intergalactic Medium at $z > 1.5$
       J. Silk   Simulating Galaxy Evolution
       One-Minute Poster Reviews and Discussion

4:00   Tea (provided outside lecture hall in poster area)

4:30   Session #4: Galaxy Formation Renaissance   Chair: **B. Williams**

       R. Thompson   The HST NICMOS Deep Survey
       M. Dickinson   Galaxy Properties at High Redshift
       L. Cowie   Redshift Surveys
       One-Minute Poster Reviews and Discussion

6:30   Poster Session (with refreshments – until 9:30)

Tuesday, October 13, 1998

8:30   Session #5: Largest Structures   Chair: **E. Smith**
        R. Mushotzky Clusters at High Redshift?
        C. Kochanek   Results from the CASTLES Gravitational
                          Lens Survey
        One-Minute Poster Reviews and Discussion

10:00   Coffee & Pastries (provided outside lecture hall in poster area)

10:30   Session #6: Galaxy Formation and Mergers Chair: **G. Hasinger**
        J. Barnes   Mergers and Galaxy Assembly
        R. Windhorst   Clues from Deep HST Images to Galaxy Formation
                            and the Role of Mergers
        R. Blandford   Supermassive Black Holes
        One-Minute Poster Reviews and Discussion

12:30   Lunch (provided at the Conference Center)

2:00   Session #7: QSOs, AGN, and the CXRB   Chair: **R. Blandford**
        D. Schneider   Surveys for High-Redshift Quasars
        K. Weaver   Observations of AGN at Intermediate Redshift
        G. Hasinger   The X-ray Background - Echo of Black Hole
                            Formation?
        One-Minute Poster Reviews and Discussion

4:00   Tea (provided outside lecture hall in poster area)

4:30   Session #8: Star Formation History   Chair: **D. Weedman**
        P. Madau   The Evolution of Luminous Matter in the Universe
        I. Smail   Sub-mm Surveys for Distant Star-Forming Galaxies
        T. Heckman   Starbursts and Cosmogeny
        One-Minute Poster Reviews and Discussion

7:00   Banquet (provided at conference site)

        V. Rubin   Astrophysics from Antarctica:
                            A Visit to the South Pole

Wednesday, October 14, 1998

8:30   Session #9: Gamma Ray Bursts   Chair: **M. Leventhal**
         N. Gehrels   Recent Discoveries in Gamma-Ray Burst Astronomy
         R. Wijers   Gamma-Ray Bursts: How Far, How Useful?
         One-Minute Poster Reviews and Discussion

10:00   Coffee & Pastries (provided outside lecture hall in poster area)

10:30   Session #10: Next Generation Capabilities   Chair: **M. Leventhal**
         E. Smith   The Golden Age for Near IR Astronomy
         D. Neufeld   The Far IR and Sub-mm Frontier
         One-Minute Poster Reviews and Discussion

11:30   Summary
         D. Weedman   Rapporteur

12:30   Lunch (provided at the Conference Center)

         End of conference

Appendix B: List of Attendees

Attendees

Name	Affiliation	Email Address
Alley, Carroll	University of Maryland	coa@kelvin.umd.edu
Arisaka, Katushi	UCLA	arisaka@physics.ucla.edu
Arnaud, Keith	NASA/GSFC	kaa@genji.gsfc.nasa.gov
Audley, Damian	NASA/GSFC	audley@lheamail.gsfc.nasa.gov
Barnes, Joshua	University of Hawaii	barnes@galileo.ifa.hawaii.edu
Barrett, Paul	Universities Space Research Assoc.	barrett@compass.gsfc.nasa.gov
Bazell, David	General Science Corp.	bazell@erols.com
Bennett, Charles	NASA/GSFC	bennett@stars.gsfc.nasa.gov
Bhattacharjee, Pijush	NASA/GSFC	
Blandford, Roger	Caltech	rdb@tapir.caltech.edu
Boldt, Elihu	NASA/GSFC	boldt@lheavx.gsfc.nasa.gov
Borne, Kirk	Raytheon STX/GSFC	borne@xfiles.gsfc.nasa.gov
Bromm, Volker	Yale University	volker@astro.yale.edu
Bullock, James	UC-Santa Cruz	bullock@physics.ucsc.edu
Chambers, Ken	Institute for Astronomy	chambers@ifa.hawaii.edu
Cheung, Cynthia Y.	NASA/GSFC	cynthia.cheung@gsfc.nasa.gov
Chin, Gordon	NASA/GSFC	
Christian, Eric	NASA/GSFC	erc@cosmicra.gsfc.nasa.gov
Cline, Thomas	NASA/GSFC	
Conselice, Chris	U. of Wisconsin-Madison	chris@astro.wisc.edu
Cooray, Asantha	University of Chicago	asante@hyde.uchicago.edu
Coppi, Paolo	Yale University	coppi@astro.yale.edu
Corcoran, Mike	Universities Space Research Assoc.	corcoran@barnegat.gsfc.nasa.gov
Cowen, Ron	Science News	rcowen@sciserv.org
Cowie, Len	University of Hawaii	cowie@uhifa.ifa.hawaii.edu
Crannell, Carol	NASA/GSFC	crannel@gsfc.nasa.gov
De Young, David	NOAO/KPNO	deyoung@noao.edu
Dermer, Chuck	NRL	dermer@burst.nrl.navy.mil
Dickinson, Mark	STScI	
Dinerstein, Harriet	University of Texas	harriet@astro.as.utexas.edu
Drachman, Richard	NASA/GSFC	
Dwek, Eli	NASA/GSFC	eli.dwek@gsfc.nasa.gov
Eisenstein, Daniel	Inst. for Advanced Study	eisenste@sns.ias.edu
Fahey, Dick	NASA/GSFC	fahey@stars.gsfc.nasa.gov
Fan, Xiaohui	Princeton University	fan@astro.princeton.edu
Felten, James E.	NASA/GSFC	felten@stars.gsfc.nasa.gov
Ferruit, Pierre	University of Maryland	pierre@astro.umd.edu
Finkbeiner, Ann	Science	AnnieKF@aol.com

Fiorito, Ralph	Catholic University	rfiorito@rocketmail.com
Fisher, Richard	NASA/GSFC	fisher@c682h.gsfc.nasa.gov
Frenk, Carlos	University of Durham	c.s.frenk@uk.ac.durham
Gehrels, Neil	NASA/GSFC	gehrels@gsfc.nasa.gov
Ghosh, Pranab	NASA/GSFC	pranab@rufus.gsfc.nasa.gov
Gnedin, Nick	University of Colorado	gnedin@casa.colorado.edu
Greyber, Howard D.	Greyber Associates	hgreyber@capaccess.org
Gronwall, Caryl	Wesleyan University	caryl@astro.wesleyan.edu
Gull, Ted	NASA/GSFC	gull@sea.gsfc.nasa.gov
Haarsma, Deborah	Haverford College	dhaarsma@haverford.edu
Haiman, Zoltan	Fermilab	zhaiman@cfa.harvard.edu
Harrington, Patrick	University of Maryland	jph@astro.umd.edu
Hasan, Hashima	NASA Headquarters	hashima.hasan@hq.nasa.gov
Hasinger, Guenther	Astrophysikalisches Institut	ghasinger@aip.de
Hauser, Mike	STScI	hauser@stsci.edu
Heap, Sally	NASA/GSFC	hrsheap@stars.gsfc.nasa.gov
Heaton, Hal	STScI	heaton@stsci.edu
Heckman, Tim	Johns Hopkins University	heckman@pha.jhu.edu
Henry, Dick	Johns Hopkins University	rch@pha.jhu.edu
Hinshaw, Gary	NASA/GSFC	hinshaw@stars.gsfc.nasa.gov
Holt, Steve	NASA/GSFC	steve.holt@gsfc.nasa.gov
Kaplan, George	U.S. Naval Observatory	gkaplan@usno.navy.mil
Kayser, Susan	Nat. Science Foundation	skayser@nsf.gov
Kazanas, Demos	NASA/GSFC	
Kimble, Randy	NASA/GSFC	kimble@stars.gsfc.nasa.gov
Kobulnicky, Chip	U. of California-Santa Cruz	chip@ucolick.org
Kochanek, Chris	Center for Astrophysics	ckochanek@cfa.harvard.edu
Kogut, Alan	NASA/GSFC	kogut@stars.gsfc.nasa.gov
Kondo, Yoji	NASA/GSFC	kondo@iue.gsfc.nasa.gov
Kowitt, Mark	NASA/GSFC	
Kozlovsky, Ben-Zion	USRA/NASA/GSFC	bzk@pair.gsfc.nasa.gov
Kurfess, Jim	NRL	kurfess@osse.nrl.navy.mil
Leisawitz, David	NASA/GSFC	leisawitz@stars.gsfc.nasa.gov
Leiter, Darryl	NASA/GSFC	dleiter@aol.com
Leventhal, Marv	University of Maryland	ml@astro.umd.edu
Loeb, Avi	Harvard University	aloeb@cfa.harvard.edu
Lowenthal, James	Univ. of Massachusetts	james@phast.umass.edu
Loewenstein, Michael	NASA/GSFC	loewenstein@lheavx.gsfc.nasa.gov
Machacek, Marie E.	Northeastern University	mariem@neu.edu
Madau, Piero	STScI	madau@stsci.edu
Maller, Ari	UC-Santa Cruz	maller@physics.ucsc.edu

Maran, Stephen P.	NASA/GSFC	hrsmaran@stars.gsfc.nasa.gov
Marani, Gabriela	NASA/GSFC	marani@milkyway.gsfc.nasa.gov
Marshall, Frank	NASA/GSFC	frank.marshall@gsfc.nasa.gov
Martin, Crystal	STScI	cmartin@stsci.edu
Mather, John	NASA/GSFC	john.mather@gsfc.nasa.gov
McGaugh, Stacy	University of Maryland	ssm@astro.umd.edu
Melott, Adrian	University of Kansas	melott@kusmos.phsx.ukans.edu
Meurer, Gerhardt R.	Johns Hopkins University	meurer@pha.jhu.edu
Miralles, Joan-Marc	Tohoku University	miralles@astr.tohoku.ac.jp
Mitchell, John	NASA/GSFC	
Mundell, Carole G.	University of Maryland	cgm@astro.umd.edu
Mushotzky, Richard	NASA/GSFC	mushotzky@lheavx.gsfc.nasa.gov
Natarajan, Priyamvada	CITA	priya@cita.utoronto.ca
Neff, Susan	NASA/GSFC	neff@stars.gsfc.nasa.gov
Neufeld, David	Johns Hopkins University	neufeld@pha.jhu.edu
Norman, Colin A.	JHU/STScI	norman@stsci.edu
Norman, Mike	University of Illinois	norman@ncsa.uiuc.edu
Norris, Jay	NASA/GSFC	norris@grossc.gsfc.nasa.gov
Offenberg, Joel D.	Raytheon STX	Joel.D.Offenberg.1@gsfc.nasa.gov
Oh, Siang Peng	Princeton University	peng@astro.princeton.edu
Ormes, Jonathan F.	NASA/GSFC	jfo@lheapop.gsfc.nasa.gov
Ostriker, Eve	University of Maryland	ostriker@astro.umd.edu
Ozernoy, Leonid	George Mason University	ozernoy@science.gmu.edu
Parsons, Ann	NASA/GSFC	
Partridge, Bruce	Haverford College	bpartrid@haverford.edu
Perez-Fournon, Ismael	Instituto de Astrofisica de Canarias	ipf@ll.iac.es
Petre, Rob	NASA/GSFC	
Pisarski, Rich	NASA/GSFC	
Polidan, Ronald	NASA/GSFC	polidan@aesop.gsfc.nasa.gov
Rees, Martin	Cambridge University	mjr@ast.cam.ac.uk
Rhodes, Jason	Princeton University	jrhodes@pupgg.princeton.edu
Richards, Eric	NRAO	er4n@virginia.edu
Roman, Nancy Grace		roman@adc.gsfc.nasa.gov
Rose, William K.	University of Maryland	wrose@astro.umd.edu
Rosenbaum, Doris	SMU	drteplitz@aol.com
Rubin, Vera	Carnegie Institution of Washington	rubin@gal.ciw.edu
Safi-Harb, Samar	NASA/GSFC	samar@milkyway.gsfc.nasa.gov
Schneider, Donald	Pennsylvania State U.	dps@astro.psu.edu
Scully, Sean	NASA/GSFC	
Seamans, Joe		jseamans@pgh.net
Serlemitsos, Peter	NASA/GSFC	pjs@astron.gsfc.nasa.gov

Shafer, Richard	NASA/GSFC	
Sharma, Surja	University of Maryland	ssh@astro.umd.edu
Shopbell, Patrick	University of Maryland	pls@astro.umd.edu
Silk, Joe	U. of California-Berkeley	silk@pac2.berkeley.edu
Silverberg, Robert	NASA/GSFC	silverberg@stars.gsfc.nasa.gov
Smail, Ian	University of Durham	Ian.Smail@durham.ac.uk
Smith, Eric	NASA/GSFC	Eric.P.Smith@gsfc.nasa.gov
Spergel, David	Princeton University	dns@astro.princeton.edu
Staguhn, Johannes	University of Maryland	staguhn@astro.umd.edu
Stahle, Caroline	NASA/GSFC	
Stecher, Ted	NASA/GSFC	stecher@uit.gsfc.nasa.gov
Stecker, Floyd W.	NASA/GSFC	stecker@lheapop.gsfc.nasa.gov
Streitmatter, Robert	NASA/GSFC	
Stiller, Bertram		bstiller@capaccess.org
Storrie-Lombardi, Lisa	Carnegie Observatories	lisa@ociw.edu
Struble, M. F.	Lockheed Martin/U. of Pennsylvania	mstruble@lmco.com
Suwall, D. J.	Baltimore Astronomical Society	
Swank, Jean	NASA/GSFC	swank@pcasun1.gsfc.nasa.gov
Szymkowiak, Andrew	NASA/GSFC	andrew.szymkowiak@gsfc.nasa.gov
Taylor, Jason	NASA/GSFC	taylor@milkyway.gsfc.nasa.gov
Teegarden, Bonnard	NASA/GSFC	teegarden@tgrs.gsfc.nasa.gov
Temkin, Aaron	NASA/GSFC	
Teplitz, Harry	NOAO/NASA/GSFC	hit@binary.gsfc.nasa.gov
Teplitz, Vigdor	SMU	teplitz@phyvms.physics.smu.edu
Teuben, Peter	University of Maryland	teuben@astro.umd.edu
Thompson, Rodger	Steward Observatory	rthompson@as.arizona.edu
Titarchuk, Lev	NASA/GSFC	titarchuk@heavax.gsfc.nasa.gov
Trasco, John	University of Maryland	jtrasco@astro.umd.edu
Trimble, Virginia	U. of Maryland and U. of California-Irvine	vtrimble@astro.umd.edu
Valinia, Azita	NASA/GSFC	valinia@rosserv.gsfc.nasa.gov
van der Marel, Roeland	STScI	marel@stsci.edu
Varosi, Frank	Raytheon STX	varosi@gsfc.nasa.gov
Veilleux, Sylvain	University of Maryland	veilleux@astro.umd.edu
Waddington, Ian	Arizona State University	Ian.Waddington@asu.edu
Wadsley, James	University of Washington	wadsley@astro.washington.edu
Weaver, Kim	NASA/GSFC and Johns Hopkins University	kweaver@cleo.gsfc.nasa.gov
Wechsler, Risa	UC-Santa Cruz	risa@physics.ucsc.edu
Weedman, Daniel	Pennsylvania State U.	weedman@astro.psu.edu
Weisheit, Jon	Rice University	jonw@rice.edu
White, Richard A.	NASA/GSFC	richard.a.white.1@gsfc.nasa.gov

White, Nick	NASA/GSFC	white@heagip.gsfc.nasa.gov
Wijers, Ralph A.M.J.	SUNY-Stony Brook	rwijers@astro.sunysb.edu
Williams, Bob	STScI	wms@stsci.edu
Williger, Gerard	NASA/GSFC	williger@fejut.gsfc.nasa.gov
Wilson, Andrew	University of Maryland	wilson@astro.umd.edu
Windhorst, Rogier A.	Arizona State University	Rogier.Windhorst@asu.edu
Windt, David	Bell Labs/Lucent Tech.	windt@bell-labs.com
Wollack, Ed	NASA/GSFC	ed.wollack@gsfc.nasa.gov
Woodgate, Bruce	NASA/GSFC	woodgate@s2.gsfc.nasa.gov
Wootten, Al	NRAO	awootten@nrao.edu
Wu, Chi-Chao	Computer Sciences Corp.	wu@stsci.edu
Yamamoto, Kazuhiro	Hiroshima U.	yamamoto@astro.phys.sci.hiroshima-u.ac.jp
Yaqoob, Tahir	NASA/GSFC	yaqoob@lheavx.gsfc.nasa.gov
Zhang, Will	NASA/GSFC	zhang@xancus10.gsfc.nasa.gov

Author Index

Index

Subject Index

Index

compact stellar remnant, 396
Compton y-parameter, 44
Constellation-X, 254, 331, 434
cosmic infrared background (CIB), 354
cosmic microwave background, 13, 27, 34, 72, 100, 107, 184, 288, 416
CSR, 396

dark age, 13, 27, 34, 299, 445
dark current, 114
dark halos, 88, 191
dark matter, 13, 27, 38, 65, 68, 82, 87, 98, 102, 106, 143, 147, 155, 165, 178, 191, 216, 300
depletion, 84, 418
detector, 10, 114, 133, 160, 235, 251, 288, 335, 359, 371, 384, 403, 414, 428, 434
detectors
    dark current, 114
    Ge:Ga, 412
    read noise, 115
    Si:As, 405
diffraction, 114, 405, 420
Diffuse Infrared Background Experiment, 300, 354, 414, 448
diffuse radio emission, 350
distance scale, 5
dust, 209, 286, 301, 322, 356, 359, 424
dynamics, 13, 68, 72, 92, 102, 107, 267, 326, 382, 390

EGRET, 372
ELG, 335, 341
ellipticals, 5, 55, 89, 120, 143, 157, 163, 191, 202, 228, 271, 304, 320, 322, 392
EUVE, 439
extended radio source, 271
extinction, 119, 135, 163, 182, 218, 256, 301, 324, 336, 345, 356, 359, 364, 389, 415, 424
extra-solar, 130

extragalactic radio background, 346

Fabry-Perot, 422
faint blue galaxies, 202
far-infrared background, 312
fundamental plane, 90, 164, 195

galactic bulges, 48, 332
galactic nuclei, 87, 191, 245, 257, 317, 336, 396
galaxies
    active, 13, 39, 91, 133, 160, 163, 193, 203, 221, 245, 256, 270, 286, 287, 293, 299, 318, 345, 351, 393, 414, 434, 446
        radio, 11, 193, 202, 234, 266, 271, 278, 286, 392
    colors, 38, 113, 122, 165, 180, 206, 225, 235, 282, 301, 316, 324, 351, 355, 360, 440
    dynamics, 13, 68, 72, 92, 102, 107, 267, 326, 382, 390
    elliptical, 5, 55, 89, 120, 143, 157, 163, 191, 202, 228, 271, 304, 320, 322, 392
    emission line, 159, 283, 340
    formation, 11, 27, 56, 63, 84, 87, 102, 135, 149, 165, 191, 203, 239, 256, 299, 335, 359, 436
    high redshift, 266, 286
    kinematics, 84, 104, 143, 266, 320, 326
    luminosity function, 36, 89, 156, 163, 178, 180, 196, 204, 242, 253, 257, 280, 301, 324, 337, 345, 350, 383, 398, 436, 446
    mergers, 22, 52, 69, 88, 103, 119, 139, 191, 202, 221, 225, 267, 271, 317, 389, 437
    protogalaxy, 102, 129, 143, 249, 328, 350, 407
    SED, 290, 318
    starburst, 38, 89, 98, 130, 139, 145, 191, 203, 216, 220, 228,

evolution, 446
high redshift, 193, 234, 282

radio galaxies, 11, 193, 202, 234, 266, 271, 278, 286, 392
radio luminosity function, 280 348
radio-FIR, 346
radio-loud, 10, 173, 238, 251, 292, 306
radio-quiet, 173
read noise, 115
recombination, 13, 28, 34, 65, 74, 99, 275, 307, 330, 350
reddening, 182, 267, 301, 324, 337, 364, 447
reionization epoch, 63
resolving power, 405, 417, 434
rings, 143, 280, 426, 429
ROSAT, 31, 157, 236, 247, 256, 287, 376, 440, 447
Rotation curves, 73
RXTE, 376, 435, 441

Sachs-Wolfe effect, 75
Saturn, 6
SED, see galaxies: SED
sensitivity, 30, 37, 63, 94, 113, 123, 139, 157, 185, 236, 260, 313, 331, 341, 346, 378, 404, 414, 424, 434
Seyfert, 39, 139, 245, 259, 293, 319, 336, 392
Si:As, 405
signal to noise, 83, 114, 184, 293
Sloan Digital Sky Survey, 32, 84, 176, 242, 282, 446
SOFIA, 405, 426
spectral resolution, 37, 236, 405, 421, 430, 434, 449
star formation rate, 65, 79, 91, 135, 168, 192, 206, 263, 287, 308, 318, 335, 340, 345, 350, 362, 364, 366, 380, 384, 426
starburst galaxies, 38, 89, 98, 130, 139, 145, 191, 203, 216, 220, 228, 256, 287, 300, 316, 322, 337, 342, 350, 359, 366, 407, 415, 426, 447
stars
initial mass function, 15, 64, 68, 196, 289, 303, 324, 328, 337
protostar, 58, 418
white dwarf, 283, 365, 392
stellar winds, 17, 325
STIS, 274, 440
Strehl, 404
subgalaxies, 13
submillimeter background, 134
Sunyaev-Zeldovich, 27, 158
supermassive black hole, 245, 305, 396, 434
superwinds, 322
synchrotron radiation, 271, 350, 380

telescope
COBE, 30, 107, 417, 448
FIRST, 242, 412, 421
IRAS, 351
MMA, 421, 424, 450
SIRTF, 346, 406, 420, 449
ASCA, 247, 256, 287, 292–295, 330, 376, 436, 440
BeppoSAX, 256, 371
CGRO, 371, 383, 388
Chandra, 246, 260, 388
Constellation-X, 254, 331, 434
EUVE, 439
Gemini, 404, 426
Hubble Space Telescope, 234, 263, 281, 406, 424
IRIS, 413
ISO, 257, 302, 352, 406, 416, 448
IUE, 323, 360, 439
JCMT, 133, 312, 426
Keck, 12, 13, 79, 123, 143, 158, 204, 216, 257, 280, 287, 299, 317, 341, 364, 377, 404, 426, 434, 447
MDM, 336

MERLIN, 351
Millimeter Array, 319, 421, 424
Next Generation Space Telescope, 34, 406
NGST, 12, 14, 30, 34, 93, 142, 264, 407, 427, 450
performance, 30, 37, 63, 94, 113, 123, 139, 157, 185, 236, 260, 313, 331, 341, 346, 378, 404, 414, 424, 434
ROSAT, 31, 157, 236, 247, 256, 287, 376, 440, 447
RXTE, 376, 435, 441
SOFIA, 405, 426
VLA, 30, 257, 312, 346, 351, 376, 424, 449
VLT, 404
WIRE, 406
WIYN, 336
Tully-Fisher, 90
two-point correlation function, 101, 176

U-dropouts, 326
ULIRGs, 220, 315
ultraluminous galaxies, 92, 136, 222, 300, 312, 426
Ulysses, 374
UV background, 14, 39, 82
UV extinction, 324, 359
UV spectroscopy, 323

Venus, 5
VLA, 30, 257, 312, 346, 351, 376, 424, 449
VLT, 404

WFPC2, 113, 439
white dwarf, 283, 365, 392
WIRE, 406
WIYN, 336

X-ray background, 42, 91, 140, 157, 253, 256, 287, 305, 436
X-ray binaries, 287

X-ray luminosity function, 156, 260
X-ray spectroscopy, 246, 293, 331, 434

zodiacal background, 114

TABLE OF PHYSICAL CONSTANTS

CONSTANT	SYMBOL	MKS	CGS	OTHER
speed of light	c	$3.00 \cdot 10^8$ m/s	$3.00 \cdot 10^{10}$ cm/s	(2.997925)
electron charge	e	$1.60 \cdot 10^{-19}$ coul	$4.80 \cdot 10^{-10}$ esu	
Planck constant	h	$6.63 \cdot 10^{-34}$ J•s	$6.63 \cdot 10^{-27}$ erg•s	
	$\hbar$	$1.05 \cdot 10^{-34}$ J•s	$1.05 \cdot 10^{-27}$ erg•s	
	hc	$1.99 \cdot 10^{-25}$ J•m	$1.99 \cdot 10^{-16}$ erg•cm	
	$\hbar c$	$3.15 \cdot 10^{-26}$ J•m	$3.15 \cdot 10^{-17}$ erg•cm	200 MeV•fm
Boltzmann constant	k	$1.38 \cdot 10^{-23}$ J/K	$1.38 \cdot 10^{-16}$ erg/K	$8.6 \cdot 10^{-5}$ eV/K
	k/h	$2.08 \cdot 10^{10}$ s^{-1}/K	$2.08 \cdot 10^{10}$ s^{-1}/K	
	k/hc	69.5 m^{-1}/K	0.695 cm^{-1}/K	
Gravitational constant	G	$6.67 \cdot 10^{-11}$ $N \cdot m^2/kg^2$	$6.67 \cdot 10^{-8}$ $dy \cdot cm^2/gm^2$	
Gas constant	R	8.314 J/K•mole	$8.31 \cdot 10^7$ erg/K•mole	
Avogadro's number (= R/k)	N	$6.02 \cdot 10^{26}$ amu/kg	$6.02 \cdot 10^{23}$ amu/kg	$6 \cdot 10^{23}$ molecules/mole
electron mass	m_e	$9.11 \cdot 10^{-31}$ kg	$9.11 \cdot 10^{-28}$ gm	0.51 MeV
proton mass	M_p	$1.67 \cdot 10^{-27}$ kg	$1.67 \cdot 10^{-24}$ gm	938 MeV
neutron mass	M_n	$1.67 \cdot 10^{-27}$ kg	$1.67 \cdot 10^{-24}$ gm	939 MeV
pion mass ($=270 \cdot m_e$)	m_π	$2.46 \cdot 10^{-28}$ kg	$2.46 \cdot 10^{-25}$ gm	140 MeV
muon mass ($=207 \cdot m_e$)	m_μ	$1.89 \cdot 10^{-28}$ kg	$1.89 \cdot 10^{-25}$ gm	106 MeV
classical elect radius ($=e^2/mc^2$)	r_c	$2.82 \cdot 10^{-15}$ m	$2.82 \cdot 10^{-13}$ cm	
Compton wavelength ($=h/mc$)	λ_c	$2.43 \cdot 10^{-12}$ m	$2.43 \cdot 10^{-10}$ cm	0.02 Å
Thomson cross-section	σ_T	$6.65 \cdot 10^{-29}$ m^2	$6.65 \cdot 10^{-25}$ cm^2	
Planck length ($=\sqrt{\hbar G/c^3}$)	l_{Pl}	$1.61 \cdot 10^{-35}$ m	$1.61 \cdot 10^{-33}$ cm	
Planck time ($=\sqrt{\hbar G/c^5}$)	t_{Pl}	$5.39 \cdot 10^{-44}$ s	$5.39 \cdot 10^{-44}$ s	
Planck density ($=c^5/\hbar G^2$)	ρ_{Pl}	$5.16 \cdot 10^{96}$ kg/m^3	$5.16 \cdot 10^{93}$ gm/cm^3	
Bohr radius ($=\hbar^2/me^2$)	r_B	$0.53 \cdot 10^{-10}$ m	$0.53 \cdot 10^{-8}$ cm	0.5 Å
Fine structure constant ($=e^2/\hbar c$)	α	$7.30 \cdot 10^{-3}$	$7.30 \cdot 10^{-3}$	1/137
Bohr magneton ($=e\hbar/2m_ec$)	μ_B	$9.27 \cdot 10^{-24}$ J/T	$9.27 \cdot 10^{-21}$ erg/gauss	
Nuclear magneton ($=e\hbar/2M_pc$)	μ_N	$5.05 \cdot 10^{-27}$ J/T	$5.05 \cdot 10^{-24}$ erg/gauss	
Permittivity of vacuum	ε_o	$8.85 \cdot 10^{-12}$ fd/m		$1/4\pi\varepsilon_o = 9.0 \cdot 10^9$